'I measure time and know not what it is that I am measuring.'
St. Augustine. *Confessions*

'We must not introduce into the definition the limitations of our
human understanding.'
C. F. Von Weizsäcker. *The History of Nature*

'I know it seems easy,' said Piglet to himself, 'but it isn't everyone
who could do it.'
Alan Alexander Milne. *The House at Pooh Corner*

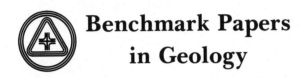

Benchmark Papers
in Geology

Series Editor: Rhodes W. Fairbridge
Columbia University

Published Volumes and Volumes in Preparation

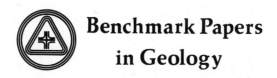

Benchmark Papers
in Geology

—————A *BENCHMARK* TM Books Series—————

GEOCHRONOLOGY:

Radiometric Dating of
Rocks and Minerals

Edited by
C. T. HARPER
The Florida State University

Dowden, Hutchinson
& Ross, Inc.

Stroudsburg, Pennsylvania

Library of Congress Cataloging in Publication Data

Harper, Christopher T   comp.
    Geochronology.

    (Benchmark papers in geology)
    Includes bibliographies.
    1. Geological time--Addresses, essays, lectures.
I. Title.
QE508.H28   551.7'01   73-648
ISBN 0-87933-031-7

Manufactured in the United States of America.

Exclusive distributor outside the United States and Canada:
John Wiley & Sons, Inc.

Acknowledgments
and Permissions

ACKNOWLEDGMENT
Institut Geologic, Bucharest, Romania—*Comptes Rendus (Doklady) de l'Académie des Sciences de l'URSS*
  "Age of the Earth According to Radioactivity Data"

PERMISSIONS
The following papers have been reprinted with the permission of the authors and the copyright owners.

Yale University Press—*Radioactive Transformations*
  "The Production of Helium from Radium and the Transformation of Matter"
Yale University Press—*American Journal of Science*
  "On the Ultimate Disintegration Products of the Radio-active Elements. Part II. The Disintegration
    Products of Uranium"
  "Factors Involved in the Calculation of the Ages of Radioactive Minerals"

The Royal Society, London—*Proceedings of the Royal Society, London*
  "Measurements of the Rate at Which Helium Is Produced in Thorianite and Pitchblende with a Mini-
    mum Estimate of Their Antiquity"
  "The Association of Lead with Uranium in Rock-Minerals and Its Application to the Measurement of
    Geological Time"

Macmillan (Journals) Ltd.—*Nature*
  "Estimates of the Ages of the Whin Sill and the Cleveland Dyke by the Helium Method"
  "The Mass-Spectrum of Uranium Lead and the Atomic Weight of Protactinium"
  "The Mass-Spectrum of Lead from Bröggerite"
  "Origin of Actinium and Age of the Earth"
  "An Estimate of the Age of the Earth"
  "Anomalous 'Common Strontium' in Granite"

Geological Society of America—*Geological Society of America Bulletin*
  "Ages by the Helium Method: II. Post-Keweenawan"
  "Age Measurements by Radioactivity"
  "Measuring Geologic Time by the Strontium Method"
Geological Society of America—*Geological Society of America Special Paper 124*
  "The Cenozoic Time Scale"

The American Physical Society—*Physical Review*
  "The Isotopic Constitution of Lead and the Measurement of Geological Time"
  "Argon 40 in Potassium Minerals"

Microforms International Marketing Corporation—*Geochimica et Cosmochimica Acta*
  "Age of Meteorites and the Earth"
  "Dating Galenas by Means of their Isotopic Constitutions—II"
  "An Interpretation of the Rhodesia and Witwatersrand Age Patterns"

"The Relation of Discordant Rb–Sr Mineral and Whole Rock Ages in an Igneous Rock to Its Time of Crystallization and to the Time of Subsequent Sr^{87}/Sr^{86} Metamorphism"
"A^{40}–K^{40} Dating"
"A^{40}/K^{40} Ratios of Feldspars and Micas from the Same Rock"

American Association for the Advancement of Science—*Science*
"Lead Isotopes and the Age of the Earth"
"Of Time and the Moon"

American Geophysical Union—*Transactions of the American Geophysical Union*
"Discordant Uranium–Lead Ages"
American Geophysical Union—*Journal of Geophysical Research*
"Volume Diffusion as a Mechanism for Discordant Lead Ages"
"Rubidium–Strontium Dating of Shales by the Total-Rock Method"
"The Use of Hornblendes and Pyroxenes for K–Ar Dating"
"Statistical Analysis of the Geomagnetic Reversal Data and the Precision of Potassium–Argon Dating"
"Age of the Steens Mountain Geomagnetic Polarity Transition"
"Potassium–Argon Dating by Activation with Fast Neutrons"

University of Chicago Press—*Journal of Geology*
"Uranium–Lead Isotopic Variations in Zircons: A Case Study"

New York Academy of Science—*Annals of the New York Academy of Science*
"Graphic Interpretation of Discordant Age Measurements on Metamorphic Rocks"

International Atomic Energy Agency, Vienna—*Radioactive Dating*
"New Approaches to Geochronology by Strontium Isotope Variations in Whole Rocks"

John Wiley & Sons Ltd.—*Radiometric Dating for Geologists*
"Potassium–Argon Dating of Igneous and Metamorphic Rocks"
John Wiley & Sons Ltd.—*The Sea*
"Continental Radiometric Ages"

Elsevier Publishing Company—*Earth-Science Reviews*
"Recent Advances in the Application and Interpretation of Radiometric Age Data"

Series Editor's Preface

The philosophy behind the "Benchmark Papers in Geology" series is one of collection, sifting, and rediffusion. Scientific literature today is so vast, so dispersed, and, in the case of old papers, so inaccessible for readers not in the immediate neighborhood of major libraries, that much valuable information has become ignored, by default. It has become just so difficult, or time consuming, to search out the key papers in any basic area of research that one can hardly blame a busy man for skimping on some of his "homework."

This series of volumes has been devised to make a practical contribution to this critical problem. The geologist, perhaps even more than any other type of scientist, often suffers from twin difficulties—isolation from central library resources and an immensely diffused source of material. New colleges and industrial libraries simply cannot afford to purchase complete runs of all their principle reference materials. So it is that we are now making a concentrated effort to gather into single volumes the critical material needed to reconstruct the background to any and every major topic of our discipline.

We are interpreting "Geology" in its broadest sense: the fundamental science of the Planet Earth, its materials, its history, and its dynamics. Because of training and experience in "earthy" materials, we also take in astrogeology, the corresponding aspect of the planetary sciences. Besides the classical core disciplines such as mineralogy, petrology, structure, geomorphology, paleontology, or stratigraphy, we embrace the newer fields of geophysics and geochemistry, applied also to oceanography, geochronology, and paleoecology. We recognize the work of the mining geologists, the petroleum geologists, the hydrologists, the engineering and environmental geologists. Each specialist needs his working library. We are endeavoring to make his task a little easier.

Each volume in the series contains an Introduction prepared by a specialist, the volume editor—and a "state-of-the-art" opening or a summary of the objects and content of the volume. The articles selected, usually some 30–50 reproduced either in their entirety or in significant extracts, attempt to scan the field from the key papers of the last century until fairly recent years. Where the original references may be in foreign languages, we have endeavored to locate or commission translations. Geologists, because of their global subject, are often acutely aware of the oneness of our world. Its literature, therefore, cannot be restricted to any one country and, whenever possible, an attempt has been made to scan the world literature.

To each article, or group of kindred items, some sort of "Highlight Commentary" is usually supplied by the volume editor. This should serve to bring that article into historical perspective and to emphasize its particular role in the growth of the field. References or citations, wherever possible, will be reproduced in their entirety; for by this means the observant reader can assess the background material available to that particular author, or, if he wishes, he too can double check the earlier sources.

A "benchmark," in surveyor's terminology, is an established point on the ground, recorded on our maps. It is usually anything that is a vantage point, from a modest hill to a mountain peak. From the historical viewpoint, these benchmarks are the bricks of our scientific edifice.

Rhodes W. Fairbridge

Preface

Geochronology, involving the measurement of geologic time, provides the essential parameter for all geological investigations. Field geologists, particularly those working with stratified sedimentary rocks and their enclosed fossil remains, were the first to appreciate the immensity and significance of geologic time. Later, physical science and the discovery of the law of radioactive decay provided the most reliable basis for measurement. The results brought a totally new dimension to the study of all natural phenomena. Nature can no longer be viewed as a static system created under catastrophic conditions a few thousand years ago, but is now seen as a temporary product of evolutionary processes operating over time intervals of immense duration.

The forty-three papers I have chosen for this volume were selected specifically to illustrate the historical development of the science of geochronology, and I have further restricted the selection to include only those articles dealing with the measurement of geologic time based on the accumulation of stable radiogenic nuclides. The period covered by these papers extends from 1906 to 1971.

Geochronology involves, of course, much more than all this. Before the establishment of the Rutherford–Soddy law of radioactive decay, problems related to geologic time and its measurement had exercized the minds and imaginations of many distinguished scientists. Questions pertaining to the age of the earth and its origin are as old as man himself, and estimates of time past have been and continue to be based on a wide variety of natural processes. However, the principal methods employed by geochronologists today, and those regarded as being the most reliable, are all based on the inexorable accumulation of stable nuclides, products of the disintegration of naturally occurring long-lived radioactive parents. The rules of the game are supplied by the law of radioactive decay, and the historical development of geochronology based on this law is illustrated by the sequence of papers reprinted in this volume.

The original articles are grouped together into fourteen sections, each section being preceded by a short commentary. One would be presumptuous to usurp more than a small proportion of the total volume, as the original authors have done an excellent job in describing their work. Too much small talk about great contributions seems to me pointless and quite unnecessary. Following each commentary the reader is provided with a list of references to other important articles related to the sectional topic. It is, of course, impossible to reprint every one of the important papers in geochronology which have appeared since 1906, and as my concern has been with presenting the historical development of the subject, I have avoided the temptation to include a discussion of the basic principles of radioactivity and radiometric dating. Several very good texts dealing with these topics and with the modern applications of geochronology have appeared in recent years, and this volume of reprints should supplement, not supersede, such material.

It is my pleasure to acknowledge here the help and many constructive suggestions I have received from colleagues currently active in the field of geochronology, including particularly Emilie Jager, Derek York, and Roy Odom. My gratitude also extends to the various authors and publishers for their permission to reprint the selected materials. I am indebted to Winston Russell and to those graduate students in the Department of Geology at the Florida State University who kept our laboratory facilities running smoothly while I was preparing this volume, and to Gail Sherrer who helped me with the library research.

C. T. Harper
January, 1973

Contents

Contents by Author

Introduction

Geochronology, in its broadest sense, involves both the measurement of time on a scale appropriate for the history of the planet Earth (and other members of the solar system) and interpretation of the dated events in terms of the origin and subsequent history of the materials involved. H. S. Williams (1893) first coined the term, and recommended that a time unit of suitable duration be established, for which he proposed the term ''geochrone.'' The modern science of geochronology is quite specific; it involves the study of naturally occurring, time-dependent isotopic systems, the calculation of radiometric* ages based on the law of radioactive decay, and the interpretation of such ages in terms of geologic events.

The passage of time is recognized by the changes that occur in practically all natural systems, but the measurement of time involves techniques based only on those processes of change which are cyclical in nature or uniformly progressive. Nowadays all our most reliable determinations of the passage of time are based on physical changes associated with the atom. Standard time units, suitable for the activities of man, are based on the cyclic precessional motion of orbital electrons in the magnetic field of a stable atomic nucleus. Longer time intervals, encompassing the origin and evolution of the earth and the solar system, are most accurately measured by the progressive transformations of naturally occurring radioactive nuclides. Such changes are, to say the least, subtle to discern, and this explains why the more obvious (but less precise) changes associated with the whole earth, and not with the atoms of which it is made, have for so long been used as standards of measurement for the passage of time.

---

*The term ''radiometric'' is to be preferred over the term ''absolute'' when referring to geologic time expressed in numerical units. As Arthur Holmes (1962) pointed out, an age does not become absolute by virtue of being expressed in units of time such as years.

1

Cyclic changes associated with the motion of the whole earth in space, although less regular in comparison to the resonance of an atom, have served man for much of his history as a time standard. Unfortunately, uniform changes of a progressive nature of the type that could be used to measure the longer intervals of geologic time are not abundantly obvious in the natural world, and remained for a long time unrecognized.

Time forms an integral part of all geologic studies, and even before suitable clocks became available, the immensity of geologic time had been deduced by geologists following Hutton's (1788, 1795) doctrine that all geologic phenomena could be attributed to the cumulative effects of natural processes similar to those in operation today. All that was needed was time, but time intervals beyond all human cognizance were required if the formation and present distribution of rocks and minerals found at the surface of the earth were to be accounted for by the operation of natural processes consistent with physical laws. The record of geological events and the formation of natural features such as mountains and valleys could be compressed into an intuitively 'sensible' time span only by invoking catastrophic and by implication divine intervention (see Gillispie, 1959; Toulmin and Goodfield, 1965).

Early attempts to actually measure geologic time were based on the deposition of sediments, the accumulation of sodium in seawater, and the rate of cooling of the earth. All suffered from basically the same defect. The rate of change was not uniform and there was the added problem that removal of sedimentary deposits by erosion and of dissolved sodium by precipitation renders these processes in part cyclical and nonprogressive. In contrast to these geologic processes, the progressive transformation of a naturally occurring radioactive nuclide is entirely uniform and predictable, and a constant proportion of unstable nuclei will always decay in a given unit of time irrespective of the environmental conditions. Furthermore, the process cannot under any natural circumstances be reversed.

The modern science of geochronology was conceived by the New-Zealand-born physicist, Ernest (later Lord) Rutherford of Nelson. Working at McGill University in Montreal, he and a young Oxford graduate named Frederick Soddy together established the principles underlying the radioactive transformation of unstable elements (Rutherford and Soddy, 1902a, b, c, d), and during the course of this work suggested that helium might prove to be one of the stable disintegration products of uranium and thorium decay. This suggestion was subsequently confirmed by Ramsey and Soddy (1903). Rutherford quickly realized that the progressive accumulation of stable radiogenic daughter products, specifically helium in uranium-bearing minerals, might be used to measure geologic time, and he presented his ideas in an address* to the International Congress of Arts and Sciences held in St. Louis, Missouri, September 19th –25th, 1904 in conjunction with the centennial celebrations of the Louisiana Purchase. This meeting was attended by many distinguished scientists and Rutherford

---

*This address was subsequently published by Rutherford (1905) under the title "Present problems in radioactivity."

received an invitation to deliver the Silliman lectures at Yale the following year. He accepted, and gave a series of lectures during March and April, 1905, concluding with a discussion of the age of the earth and the sun. The lectures were published by the Yale University Press under the title "Radioactive Transformations" (Rutherford, 1906), and contain the first clear account of the procedures necessary to calculate the age of a uranium-bearing mineral, based in this case on a chemical analysis of the accumulated radiogenic helium.

Helium is not the only stable product of uranium and thorium disintegration, however, and Rutherford considered it probable that the final product of the uranium-series decay was lead. The same conclusion was reached by Bertrand Boltwood, who was working at Yale on the chemical composition of uranium minerals. Boltwood had communicated his ideas to the American Chemical Society in February, 1905, and later published them in the *Philosophical Magazine* (Boltwood, 1905). A method of calculating the age of uranium-bearing minerals from their lead content was suggested to Boltwood by Rutherford, and the first radiometric lead ages were subsequently published by Boltwood in 1907.

Rutherford was characteristically enthusiastic about the new methods for calculating the ages of uranium minerals, and inspired Hon. R. J. Strutt (later Lord Rayleigh) who began a comprehensive study of the helium contents of a wide variety of geologic materials, using methods of gas manipulation developed by Sir William Ramsay and Dr. Travers at the Imperial College of Science in London. The results of Strutt's careful and systematic work were disappointing. Helium was shown to leak from uranium and thorium ore minerals at room temperatures under vacuum in quantities which at first astonished Strutt. Other investigators observed that more helium was liberated by grinding radioactive ore minerals under vacuum. The rate of escape of helium from such minerals under laboratory conditions was always far greater than the rate of production, so that it was quite clear that more favorable conditions must have existed during the lifetime of most minerals; otherwise no helium would have ever accumulated in the crystal lattice. Nevertheless, the helium method of age determination did not, in 1910, look too promising and attention turned toward the lead method.

On April 6th, 1911, Professor Strutt presented a paper to the Royal Society in London on the association of lead with uranium in rock-minerals and its application to the measurement of geological time. This paper was written by one of Strutt's students, a geologist by the name of Arthur Holmes, who between 1911 and 1962 was to publish more than fifty significant contributions on the subject of radiometric age determinations and establish himself as the pioneer of the modern geologic time-scale. Holmes' first paper (1911) reflects this interest, and he began by urging the establishment of a numerical time-scale for the fossiliferous sedimentary successions and the application of radiometric dating to the crystalline basement rocks of the Precambrian.

The earliest radiometric age determinations were based on chemical analyses of rare uranium and thorium minerals* and despite the uncertainties of the new results, it became abundantly clear that Lord Kelvin's (1899) final authoritative estimate of 20–40 million years for the age of the earth, based on the rate of cooling of a nonradioactive earth, was in error by approximately two orders of magnitude. It was also apparent that most of this long and unforeseen history of the earth was imprisoned in Precambrian crystalline rocks which were previously thought to record only the brief turbulent beginnings of the planet. Important though these early results were in establishing the general antiquity of the earth and the duration of Precambrian time, they were of limited general use to geologists for reasons related both to the uncertainties associated with the radioactive decay process itself and to the restricted distribution of suitably radioactive minerals. Furthermore, gain or loss of various proportions of the radioactive parent atoms and/or their daughter products, which causes calculated radiometric ages to become discordant among themselves, or discrepant in relation to the geological evidence, posed a problem for the early geochronologists which is still only partially resolved today. The analyzed minerals or rocks must have remained closed systems with respect to the migration of parent and daughter atoms, if ages calculated on the basis of the observed rates of radioactive disintegration are to yield the true age of the samples.

Unlike Holmes, most geologists remained cautious, if not downright skeptical of the new methods. In the United States, G. F. Becker (1908), using Boltwood's uranium–lead method, calculated ages of several radioactive minerals occurring in pegmatites intruding into the Llano Series in Texas and obtained discordant results, with some calculated ages in excess of 10 billion years. Skeptical would be putting it mildly. Becker poured scorn on the whole idea of determining ages from radioactive decay products. Boltwood cautioned that some of the dated Llano minerals were in an advanced state of alteration, but Becker insisted that weathering effects and hydration should not affect the Pb/U ratios, and suggested instead that the physics of radioactive decay was oversimplified. The Rutherford–Soddy decay law was probably, like Boyle's Law, strictly applicable only in very limited circumstances, and the decay constant (λ), a measure of the proportion of atoms disintegrating in unit time, was probably not a constant at all under geological conditions. Joseph Barrell (1917) rose to the challenge and valiantly tried to defend the method, but the disagreements were enough to cast considerable doubt on the whole subject of radiometric age determination.

The work of Strutt and others had by this time demonstrated that loss of helium from uranium and thorium minerals occurred to such an extent that helium ages could only be regarded as minimum estimates. There was, however, a more fundamental problem. Laboratory tests had indicated that measured rates of radioactive disintegration

---

*A review of much of this early radiometric age work is given in Bulletin No. 80 of the National Research Council, Washington, 1931, entitled "The Age of the Earth." See also Holmes (1913, 1927, 1937).

were unaffected by variations in temperature, pressure, or chemical combination,* but how could one be sure that the observed process could be relied upon to have occurred at the same measured rate in the past, and throughout the whole of geologic time? As is so often the case, the answer was provided by the minerals themselves.

In 1907, John Joly at Trinity College, Dublin, demonstrated that the circular dark spots and concentric rings known as "pleochroic haloes," which are often observed under the polarized-light microscope in certain minerals (notably biotite) surrounding tiny inclusions of zircon, sphene, and other radioactive minerals, were in all probability caused by alpha particles shot out from disintegrating radioactive elements concentrated in the central inclusion. In micas, the radii of normal uranium and thorium haloes varies from 12 to 42 μ, and the haloes were observed to attain just the dimensions expected if the alteration were due to alpha particles traveling a discrete distance from the central inclusion. The sphericity of the resulting halo was seen to be independent of the cleavage or physical structure of the lost mineral.

In 1911, Geiger and Nuttall assembled evidence to show that the distance travelled by an alpha particle in air bears a definite relationship to the decay constant of the disintegrating parent nuclide. Plots of log R (range) versus log λ (decay constant) for each of the uranium and thorium disintegration series give straight-line plots of the type: $\log R = a + b \log \lambda$, where a and b are constants, a being characteristic of the series, and b being essentially the same for all series.

The range of alpha particles in solids is much shorter than in air, depending on density and chemical composition, but for a given mineral the distance travelled by an alpha particle emitted from a disintegrating radioactive parent is proportional to the decay constant of that parent. Thus the radius of the resulting pleochroic halo surrounding any concentration of alpha-emitting radioactive species is, in effect, proportional to the decay constant of that species, and if this radius is found to be constant in minerals of all geologic ages, then the measured decay constant of the alpha emitter must be invariant with time.

During the course of his investigations on pleochroic haloes, which continued through 1923, Joly found that the dimensions of uranium haloes increased in size in biotites of increasing age, suggesting a decrease in the decay constant with time, or possibly the existence of a former, short-lived isotope of uranium now extinct (Joly, 1917, 1923). Later investigations by Henderson and others (1934, 1939), using more precise photometric determinations, showed that Joly had apparently measured the U-238 halo in Precambrian haloes. All recent measurements have indicated that the radii of various haloes are constant whatever the age of the halo, providing evidence that the disintegration rates of known radioactive isotopes have in fact remained constant throughout geologic time. A few known occurrences of anomalous radii,

---

*Chemical combination has since been observed to have a slight effect on the rate of decay involving certain isometric transitions and electron capture.

both dwarf and giant, unrelated to any of the known naturally occurring alpha emitters, may be indicative of extinct radioactivities (see Gentry, 1970, 1971).

The color of a halo depends on the age of the host mineral and the radioactivity of the central inclusion, and attempts were made by Joly and Rutherford in 1913 to estimate the age of natural haloes by comparing their color intensity with that of artificial haloes generated in the same material under artificial conditions. Their results lent support to the few available age determinations based on the accumulation of radiogenic helium and lead, and Joly was forced to doubt his earlier estimate of the age of the earth based on the accumulation of sodium in seawater.

One last note regarding the constancy of radioactive decay involves Dirac's (1938) cosmological principle which can be used to relate the beta-decay constant to the reciprocal square root of the age of the universe. Houtermans and Jordan (1946) suggested that this theory could be tested by comparing the beta-decay and alpha-decay ages of naturally occurring minerals, but such data have recently been shown to be more consistent with a beta-decay rate independent of the age of the universe than with the dependence implied by the Dirac principle (Kanasewich and Savage, 1963).

Twenty-five years after Rutherford had first demonstrated his procedure for determining radiometric ages from the accumulation of stable radiogenic daughter products there were two important advances. Chemical and physical methods of analysis had reached the point where helium ages could be determined for the first time on common rocks and minerals, where uranium and thorium were not among the major elements, but were present only in trace amounts in the host material. Using methods of analysis developed by F. A. Paneth and his co-workers in Berlin, Dubey and Holmes demonstrated in 1929 that helium ages could be measured on whole-rock samples of basalt. Their work initiated a whole series of studies which was to provide the first truly comprehensive geologic time scale. Although largely superseded by the potassium–argon method in the early 1950s, the helium method still remains a useful technique, particularly for very young materials.

The second advance was the application of mass spectrometry to geochronological problems, with the initiation of isotopic analyses of radiogenic lead by Aston. Since its introduction in 1929, the mass spectrometer has become the handmaiden of every practicing geochronologist—a veritable time machine for the exploration of the past. From a knowledge of the isotopic composition of radiogenic lead it is possible to distinguish three stable daughter products, Pb-206, Pb-207, and Pb-208, which are each derived at different rates from separate radioactive uranium and thorium parent isotopes. Thus, for a mineral containing uranium and thorium, it is possible to calculate no less than four radiometric ages from a mass spectrometric determination of each of the time-dependent ratios, Pb-206/U-238, Pb-207/U-235, Pb-208/Th-232, and Pb-206/Pb-207. The calculated ages should be internally consistent if no addition or loss of uranium, thorium or lead has occurred since the time of formation of the analyzed material, except, of course, for the radioactive disintegration of uranium and thorium and the accumulation of stable radiogenic lead. A correction for the presence of nonradiogenic lead can be made on the basis of a measurement of the isotope Pb-204.

Between 1929 and 1939 mass spectrometric techniques developed rapidly, and following the recognition by Nier and co-workers (1938, 1939, 1941) that ordinary lead consisted of a mixture of primeval lead (present at the time of formation of the earth) and radiogenic lead (subsequently derived from the decay of uranium and thorium), isotopic analyses of lead could be used to provide at least a precise age for the earth, and by implication the time of formation of the solar system.

Following 1945, the availability of isotopically enriched materials furnished a new analytical technique which, when wedded to the mass spectrometer, opened the door to a whole new dimension in geochronology. In its simplest form the new method of analysis, known as isotope dilution (Inghram, 1954), involves mixing a known amount of an isotopic standard (the diluent or "spike") with an unknown amount of the isotope in question, and measuring the resulting isotopic composition of the mixture. Provided that the isotopic composition of the diluent and the unknown are sufficiently different, and that both are well mixed, precise quantitative analysis of minute quantities is possible, limited only by uncertainties in the amount of diluent added, its composition, and the precision of the final isotopic analysis.

As a result of these analytical developments, the isotopic uranium–lead method could be applied to a wide variety of accessory minerals, such as zircon, sphene, and apatite, whose uranium and lead concentrations were very much smaller than anything previously analyzed. Moreover, the isotope-dilution method, coupled with the development of new chemical techniques for the separation of minute quantities of the elements involved and progress in gas-handling technology, resulted in the development of two new age methods based on the radioactive decay of the alkali elements, rubidium and potassium. There was an enormous advantage to be gained here, as the alkali elements are widely distributed, occurring in nearly all the common rock-forming minerals and most significantly in the mica and alkali feldspar groups. No longer would the geochronologist be restricted to rare or accessory minerals containing only uranium and thorium.

Mass-spectrometric and isotope-dilution techniques were successfully applied to both the rubidium–strontium and potassium–argon methods during the 1950s, and by the end of the decade radiometric age determinations could be undertaken on a wide variety of minerals and whole-rock aggregates. Since then refinement of the analytical techniques and improved instrumentation continues to extend measurement down to smaller and smaller concentrations using ever decreasing sample sizes, so that today radiometric dating by one method or another can be applied to almost any geologic material.

More and more attention is now being directed toward problems related to the geologic interpretation of radiometric age data. Of particular interest is the widespread occurrence of discordant ages, i.e., ages determined on the same materials by different methods which fail to agree with each other, and ages determined on different cogenetic materials by the same method which do not agree. In many cases, consistent patterns of discordant ages, particularly those found in orogenic belts, are extremely useful in that they yield much information regarding the post-crystallization history of rocks and their constituent minerals. In the case of the isotopic uranium–lead method, discordant Pb-206/U-235 and Pb-207/U-235 ages caused by lead loss during a subsequent (post-crystallization) event can sometimes be used to determine both the original age of crystallization and the time of lead loss. Discordant Rb–Sr mineral ages can occur when redistribution of radiogenic Sr-87 takes place among mineral phases during metamorphism, and not only is it sometimes possible to obtain both the age of metamorphism and the original age, but an independent correction for the presence of common (nonradiogenic) strontium can be obtained by analysis of a suitable number of components. Partial loss of radiogenic Ar-40 from poorly retentive lattice sites within a potassium-bearing mineral can be corrected by step-heating samples and analyzing only that portion of radiogenic Ar-40 released from retentive lattice sites at high temperatures. Potassium from the same sites can be obtained as Ar-39 after fast-neutron irradiation.

In addition to the principal radiometric age methods discussed above, a number of special methods, applicable only to very young materials, are currently enjoying widespread application. Chief among these is the radiocarbon method, based on the decay of C-14, (Libby, 1955) which is applicable to materials ranging in age up to about 40,000 yrs. Beyond this limit, the amount of original C-14 decays to less than 1 per cent of its original amount, and is difficult to detect with any degree of accuracy. Other methods, particularly suitable for dating marine deposits (Koczy, 1963), are based on the observed disequilibrium of the intermediate radioactive daughters of U-238 and U-235 decay. Four of these intermediate daughters have half-lives long enough to be useful for geochronology. They are U-234, Th-230, and Ra-226 in the U-238 series, and Pa-231 in the U-235 series. The remaining daughters in the series have half-lives too short (ranging from a fraction of a second to 22 yr) for geochronological purposes.

Methods based on the tracks produced in solids by the fission products of U-238 (Fleischer et al., 1968) have recently been applied with some success to a variety of geologic materials of various ages. Uranium-238 undergoes spontaneous fission with a decay constant of 6.9×10^{-17} yr^{-1}, and a count of the density of tracks left behind by the fission products together with a measure of the uranium content allows the age of a sample to be calculated. Although the experimental procedures used in fission-track dating are extremely simple, the method is somewhat subjective in that it involves visual identification (after etching) of fossil fission tracks in naturally occurring materials many of which are riddled with lattice defects which can be confused with true fission tracks. For this reason the method is perhaps best applied to young volcanic glasses.

In summary, and as outlined in the selected papers which follow, geochronology, based on Rutherford and Soddy's (1902) law of radioactive transformation, has developed slowly through the insistence of geologists and with contributions from many different physicists and chemists to stand today as one of the most precise and indispensable branches of the earth sciences. The origin of the earth and many of the subsequent events in the history of the earth's crust have been precisely dated. A geologic time scale relating stratigraphically defined time intervals to radiometrically defined ages has been established in detail for Phanerozoic (Harland et al., 1964). Another time scale involving the recorded reversals of the earth's magnetic field is being put together and has already led to determination of the rates of sea-floor spreading. The great antiquity of the continents in contrast to the comparative youth of the ocean basins is now well established.

Research in progress involving detailed studies of the timing of igneous activity, metamorphic recrystallization, and epeirogenic uplift in several orogenic belts is clarifying the sequence of events involved in orogenesis. The frequency of mountain building throughout the history of the earth is becoming more clearly understood, for radiometric dating has shown that orogenesis and metamorphism, once thought to have been confined to brief spasmotic episodes in the history of the earth's crust, occurs at intermittent intervals over long periods of time. Precambrian crystalline shield areas are slowly yielding their hidden record of successive orogenesis and reactivation extending over more than 85 per cent of the total history of the earth. And finally the moon, that bastion of inaccessibility, has been submitted to the scrutiny of the geochronologists and forced to reveal the secrets of her origin and subsequent evolution. There still remains much work to be done.

The reader interested in pursuing the subject of geochronology further is referred to general texts by Faul (1966); Hamilton and Farquhar (1968); Moorbath (1970); and York and Farquhar (1971).

References

Barrell, J. (1917). Rhythms and the measurement of geologic time. *Geol. Soc. Amer. Bull.*, **28**, 745–904.

Becker, G. F. (1908). Relations of radioactivity to cosmogony and geology. *Geol. Soc. Amer. Bull.*, **19**, 113–146.

Boltwood, B. B. (1905). The origin of radium. *Phil. Mag., Ser. 6*, **9**, 599–613.

Boltwood, B. B. (1907). On the ultimate disintegration products of the radioactive elements. Part II. The disintegration products of uranium: *Amer. J. Sci., Ser. 4*, **23**, 77–88.

Dirac, P. A. M. (1938). A new basis for cosmology. *Proc. Roy. Soc., Ser. A*, **165**, 199–208.

Dubey, V. S., and Holmes, A. (1929). Estimates of the ages of the Whin Sill and the Cleveland Dyke by the helium method. *Nature*, **123**, 794–795.

Faul, H. (1966). "Ages of Rocks, Planets, and Stars," 109 pp. McGraw–Hill Book Company, New York.

Fleischer, R. L., Price, B. P., and Walker, R. M. (1968). Charged particle tracks: tools for geochronology and meteorite studies. *In* "Radiometric Dating for Geologists" (E. I. Hamilton and R. M. Farquhar, eds.) p. 417–436, Interscience, London.

Geiger, H., and Nuttall, J. M. (1911). The ranges of the α particles from various radioactive substances and a relation between range and period of transformation. *Phil. Mag., Ser. 6*, **22**, 613–621.

Gentry, R. V. (1970). Giant radioactive halos: Indicator of unknown radioactivity? *Science*, **169**, 670–673.

Gentry, R. V. (1971). Radiohalos: Some unique lead isotope ratios and unknown alpha radioactivity. *Science*, **173**, 727–731.

Gillispie, C. C. (1959). "Genesis and Geology," 306 pp. Harper Brothers, New York.

Hamilton, E. I., and Farquhar, R. M. (eds) (1968). "Radiometric Dating for Geologists," 506 pp. Interscience, London.

Harland, W. B., Smith, A. G., and Wilcock, B. (eds.) (1964). The Phanerozoic time-scale. *Quart. J. Geol. Soc. London* **120s**, 458 pp. See also "The Phanerozoic Time-scale: A Supplement," 356 pp. Geological Society, London, Special Publication No. 5, 1971.

Henderson, G. H., and Bateson, S (1934). A quantitative study of pleochroic haloes, I. *Proc. Roy. Soc. Ser. A,* **145**, 563–581.

Henderson, G. H., and Sparks, F. W. (1939). A quantitative study of pleochroic haloes, IV. *Proc. Roy. Soc. Ser. A,* **173**, 238–264.

Holmes, A. (1911). The association of lead with uranium in rock-minerals, and its application to the measurement of geological time. *Proc. Roy. Soc., Ser. A,* **85**, 248–256.

Holmes, A. (1913). "The Age of the Earth," 196 pp. Harper Brothers, London.

Holmes, A. (1927). "The Age of the Earth, an Introduction to Geological Ideas," 80 pp. Harper Brothers, London.

Holmes, A. (1937). "The Age of the Earth," 263 pp. Nelson and Sons, London.

Holmes, A. (1962). 'Absolute age' a meaningless term. *Nature* **196**, 1238.

Houtermans, F. G., and Jordan, P. (1946). Uber die Annahme der Zetlichen Veränderlichkeit des B-Zerfalls und die Möglichkeiten ihrer experimentallen Prüfung: *Z. Naturforsch.,* **1**, 125–130.

Hutton, J. (1788). Theory of the earth; or an investigation of the laws observable in the composition, dissolution, and restoration of land upon the globe. *Roy. Soc. Edinburgh, Trans.,* **1**, 209–304 (originally read before the Royal Society of Edinburgh in 1785).

Hutton, J. (1795). "Theory of the Earth, with Proofs and Illustrations," Facsim reprint, 2 vols. Hagner, New York, 1959.

Inghram, M. G. (1954). Stable isotope dilution as an analytical tool. *Ann. Rev. Nuc. Sci.,* **4**, 81–92.

Joly, J. (1907). Plechroic haloes. *Phil. Mag., Ser. 6,* **13**, 381–383.

Joly, J. (1917). The genesis of pleochroic haloes. *Phil. Trans. Roy. Soc. Ser. A,* **217**, 51–79.

Joly, J. (1923). Pleochroic haloes of various geological ages. *Proc. Roy. Soc. Ser. A,* **102**, 682–705.

Joly, J., and Rutherford, E. (1913). The age of pleochroic haloes. *Phil. Mag., Ser. 6,* **25**, 644–657.

Kanasewich, E. R., and Savage, J. C. (1963). Dirac's cosmology and radioactive dating. *Can. J. Phys.,* **41**, 1911–1923.

Kelvin, W. Thomson, Lord (1899). The age of the earth as an abode fitted for life. *Phil. Mag., Ser. 5,* **47**, 66–90.

Koczy, F. F. (1963). Age determination in sediments by natural radioactivity, *In* "The Sea," Vol. 3 (M. N. Hill, ed.), pp. 816–831, Interscience, New York.

Libby, W. F. (1955). "Radiocarbon Dating" (2nd ed.). Univ. of Chicago Press, Chicago, Illinois.

Moorbath, S. (1970). "Dating by Radioisotopes," 22 pp. Francis Hodgson, Guernsey.

Nier, A. O. (1938). Variations in the relative abundances of the isotopes of common lead from various sources. *J. Amer. Chem. Soc.* **60**, 1571–1576.

Nier, A. O. (1939). The isotopic constitution of radiogenic leads and the measurement of geological time II. *Phys. Rev.* **55**, 153–163.

Nier, A. O., Thompson, R. W., and Murphey, B. F. (1941). The isotopic constitution of lead and the measurement of geological time III. *Phys. Rev.* **66**, 112–116.

Ramsay, W., and Soddy, F. (1903). Gases occluded by radium bromide. *Nature,* **68**, 246.

Rutherford, E (1905). Present problems in radioactivity. *Pop. Sci. Monthly,* (May), 1–34.

Rutherford, E. (1906). "Radioactive Transformations." Yale Univ. Press, New Haven, Connecticut.

Rutherford, E., and Soddy, F. (1902a). The radioactivity of thorium compounds, I. An investigation of the radioactive emanation. *J. Chem. Soc.,* **81**, 321–250.

Rutherford, E., and Soddy, F. (1902b). The radioactivity of thorium compounds, II. The cause and nature of radioactivity. *J. Chem. Soc.* **81**, 837–860.

Rutherford, E., and Soddy, F. (1902c). The cause and nature of radioactvity, Part I. *Phil. Mag., Ser. 6,* **4**, 370–396.

Rutherford, E., and Soddy, F. (1902d). The cause and nature of radioactivity, Part II. *Phil. Mag., Ser. 6,* **4**, 569–585.

Toulmin, S., and Godfield, J. (1965). "The Discovery of Time," 280 pp. Hutchinson, London.

Williams, H. S. (1893). The elements of the geological time-scale. *J. Geol.,* **1**, 283–295.

York, D., and Farquhar, R. M. (1972). "The Earth's Age and Geochronology," 178 pp. Pergamon, New York.

The Accumulation of
Helium and Lead

I

Editor's Comments on Papers 1, 2 and 3

The first radiometric age determinations were based on chemical analyses of the elements uranium, thorium, helium, and lead, and for more than twenty years chemical methods were to provide the only available means of determining the age of any geologic material. As a consequence of the very small concentrations of accumulated daughter elements present in most rocks and minerals, radiometric dating was restricted to rare minerals containing large concentrations of uranium and thorium.

The idea of dating a radioactive mineral on the basis of its accumulated radiogenic daughter elements was Rutherford's, and the first radiometric age determination, based on the accumulation of helium, was presented by Rutherford himself to the International Congress of Arts and Science held at St. Louis in 1904 (Rutherford, 1905). The New-Zealand-born physicist had just turned thirty-three. He had calculated the age of a rare pegmatite mineral known as fergusonite (a complex oxide of niobium and tantalum, containing rare earths and appreciable amounts of uranium) whose helium and uranium contents had been determined by Ramsay and Travers. Rutherford's estimate for the age of the mineral was 40 million years.

The following year, after his address to the International Congress at St. Louis, Rutherford gave the Silliman Lectures at Yale, which were subsequently published by the Yale University Press under the title "Radioactive Transformations." In these lectures, Rutherford gave a more detailed account of his calculation of the age of the fergusonite, revising the figure to 500 million years, and adding a calculation of the age of a uranium mineral from Glastonbury, Connecticut (Paper 1).

Rutherford was well aware that all the helium generated in situ by radioactive decay might not be quantitatively retained in a mineral, and he commented on the presence of lead in radioactive minerals, referring to the work of B. B. Boltwood at Yale, who had been studying the chemistry of radioactive minerals for some years.* Rutherford pointed out that uranium, with an atomic weight of 238.5, after expulsion of a total of eight alpha-particles (32 mass units) would have an atomic weight of 206.5 very close to the atomic weight of lead. If the production of lead by radioactive decay of uranium could be proved, he said, the accumulation of lead in uranium minerals would provide a far more accurate method of age determination than the helium method, for radiogenic lead would be much less likely to escape from a mineral than helium.

The possibility of a genetic connection between uranium and lead was first suggested by Boltwood in 1905 as a result of a statement by Hillebrand, then the leading authority on the analysis of uranium minerals, who had never in the course of a long experience found uranium unaccompanied by lead (Boltwood, 1905b). In 1907, Boltwood went further, and was able to show from Hillebrand's analyses and his own that the lead-to-uranium ratio in radioactive minerals was nearly constant for minerals having the same stratigraphically assigned age, but increased as the age of the minerals increased. He also published the first radiometric age determinations based on chemically determined Pb/U ratios, crediting Rutherford for first suggesting the method, and assuming, as Rutherford had in the case of helium, a constant rate of production of the radiogenic daughter element (Paper 2).

It is interesting to note here that Boltwood had calculated uranium-lead ages two years prior to his 1907 publication, and sent the results to Rutherford in a letter dated 18th November, 1905.† Luckily, these preliminary calculations were not published, for at that time the accepted equilibrium ratio of radium to uranium was erroneously low, and the original calculated lead ages were all too young. Assuming that the proportion of uranium atoms disintegrating per year (the decay constant) was 10^{-10}, the ages of uranium-bearing minerals were finally calculated according to the equation

$$t = \frac{Pb}{U} \times 10^4 \text{ m.y.}$$

Boltwood found no constant relationship between thorium and lead and concluded "with certainty," that lead was not a disintegration product of thorium. He was also of the opinion that helium was not produced by thorium decay, but had suggested in an earlier publication (Boltwood, 1905b), that argon might be a radiogenic disintegration product. Argon was known to be present in most minerals that contained helium. There was no suggested connection with the radioactivity of

---

*Boltwood was only a year older than Rutherford, and the two struck up a warm friendship while Rutherford was at McGill. Correspondence between the two, edited by Lawrence Badash, has recently been published by the Yale University Press (Badash, 1969). See also Badash (1968) on Rutherford, Boltwood and the age of the earth.
†Badash (ed.), 1969, p. 103. See also Badash, 1968, p. 163.

potassium (see Paper 30), but Boltwood wondered whether some of the low energy alpha particles emitted by radioactive substances might be argon and not helium nuclei.

The same year that Boltwood was calculating U/Pb ages, R. J. Strutt (later Lord Rayleigh) began an extensive series of investigations into the helium method of age determination in England. Unlike Boltwood at Yale, Strutt believed helium to be produced by both uranium and thorium, and he regarded the presence of argon in minerals as being due to atmospheric argon trapped at the time of crystallization. Strutt proceeded with a thorough investigation of the helium contents of a wide variety of mineral substances, including zircon and sphene, which he reported in a succession of nine papers appearing between 1905 and 1910 (e.g., Strutt, 1905, 1908, 1909, 1910). In his last report, reproduced here (Paper 3), Strutt described the results of an experiment designed to determine by direct volume measurement the rate of helium production in thorianite and pitchblende. These experiments confirmed the calculated rate of helium production in uranium minerals, which had previously been based on the observed rate of disintegration of radium existing in equilibrium with uranium, and allowed Strutt to present his helium age calculations with some confidence. A minimum age of 710 million years was obtained for a sphene from Renfrew County, Ontario.

As Rutherford had anticipated, the helium method was plagued by the problem of helium leakage, with the consequence that helium retention ages could only be considered as minimum estimates for the time of formation of the analyzed samples. The problem was investigated by Strutt himself, who was astonished at the quantity of helium diffusing from powdered monazite placed under vacuum at room temperature. As a result of these studies and the investigations of others, the method fell into disrepute for nearly twenty years, and attention focused on the lead method for determining the age of radioactive minerals.

Selected Bibliography

Badash, L. (1968). Rutherford, Boltwood, and the age of the earth: The origin of radioactive dating techniques. *Proc. Amer. Phil. Soc.*, **112**, 157–169.

Badash, L. (ed) (1969). "Rutherford and Boltwood, Letters on Radioactivity," 378 pp. Yale Univ. Press, New Haven, Connecticut.

Boltwood, B. B. (1905a). The origin of radium. *Phil. Mag.*, *Ser. 6*, **9**, 599–613.

Boltwood, B. B. (1905b). On the ultimate disintegration products of the radioactive elements. *Amer. J. Sci., Ser. 4*, **20**, 253–267.

Rutherford, E. (1905). Present problems in radioactivity. *Pop. Sci. Monthly*, (May), 1–34.

Strutt, R. J. (1905). On the radioactive minerals. *Proc. Roy. Soc., Ser. A*, **76**, 88–101.

Strutt, R. J. (1908). On the accumulation of helium in geological time. *Proc. Roy. Soc., Ser. A*, **81**, 272–277.

Strutt, R. J. (1909). The leakage of helium from radioactive minerals. *Proc. Roy. Soc., Ser. A*, **82**, 166–169.

Strutt, R. J. (1910). On the accumulation of helium in geological time, IV. *Proc. Roy. Soc., Ser. A*, **84**, 194–196.

Reprinted from "Radioactive Transformations," E. Rutherford, 187–193 (1906), Yale University Press

1

The Production of Helium from Radium and the Transformation of Matter

ERNEST RUTHERFORD

AGE OF RADIOACTIVE MINERALS

The helium observed in the radioactive minerals is almost certainly due to its production from the radium and other radioactive substances contained therein. If the rate of production of helium from known weights of the different radio-elements were experimentally known, it should thus be possible to determine the interval required for the production of the amount of helium observed in radioactive minerals, or, in other words, to determine the age of the mineral. This deduction is based on the assumption that some of the denser and more compact of the radioactive minerals are able to retain indefinitely a large proportion of the helium imprisoned in their mass. In many cases the minerals are not compact but porous, and under such conditions most of the helium will escape from its mass. Even supposing that some of the helium has been lost from the denser minerals, we should be able to fix with some certainty a minimum limit for the age of the mineral.

In the absence of definite experimental data on the rates of production of helium by the different radioelements, the deductions are of necessity somewhat uncertain, but will nevertheless serve to fix the probable order of the ages of the radioactive minerals.

It has already been pointed out that all the a particles expelled from radium have the same mass. In addition it has been experimentally found that the a particle from thorium B has the same mass as the a particle from radium. This would suggest that the a particles projected from all radioactive substances have the same mass, and thus consist of the same kind of matter. If the a particle is a helium atom, the amount of helium produced per year by a known quantity of radioactive matter can readily be deduced on these assumptions.

The number of products which expel a particles are now well known for radium, thorium, and actinium. Including radium F, radium has five a ray products, thorium five, and actinium four. With regard to uranium itself, there is not the same certainty, for only one product, UrX, which emits only β rays, has so far been chemically isolated from uranium. The a particles apparently are emitted by the element uranium itself; at the same time, there is some indirect evidence in support of the view that uranium contains three a ray products. For the purpose of calculation, we shall, however, assume that in uranium and radium in equilibrium, one a particle is expelled from the uranium for five from the radium.

Let us now consider an old uranium mineral which contains one gram of uranium, and which has not allowed any of the products of its decomposition to escape. The uranium and radium are in radioactive equilibrium and 3.8×10^{-7} grams of radium are present. For one a particle emitted by the uranium, five are emitted by the radium and its products, including radium F. Now we have shown that radium with its four a ray products probably produces .11 c.c. of helium per gram per year. The rate of production of helium by the uranium and radium in the mineral will consequently be $\frac{5}{4} \times .11 \times 3.8 \times 10^{-7} = 5.2 \times 10^{-8}$ c.c. per year per gram of uranium.

Now, as an example of the method of calculation, let us consider the mineral fergusonite which was found by Ramsay and Travers to evolve 1.81 c.c. of helium per gram. The fergusonite contains about 7 per cent of uranium. The amount of helium contained in the mineral per gram of uranium is consequently 26 c.c.

Since the rate of production of helium per gram of uranium and its radium products is 5.2×10^{-8} c.c. per year, the age of the mineral must be at least $26 \div 5.2 \times 10^{-8}$ years or 500 million years. This, as we have pointed out, is a minimum estimate, for some of the helium has probably escaped.

We have assumed in this calculation that the amount of uranium and radium present in the mineral remains sensibly constant over this interval. This is approximately the case, for the parent element uranium probably requires about 1000 million years to be half transformed.

As another example, let us take a uranium mineral obtained from Glastonbury, Connecticut, which was analyzed by Hillebrande. This mineral was very compact and of high density, 9.62. It contained 76 per cent of uranium and 2.41 per cent of nitrogen. This nitrogen was almost certainly helium, and dividing by seven to reduce to helium this gives the percentage of helium as 0.344. This corresponds to 19 c.c. of helium per gram of the mineral, or 25 c.c. per gram of uranium in the mineral. Using the same data as before, the age of the mineral must be certainly not less than 500 million years. Some of the uranium and thorium minerals do not contain much helium. Some are porous, and must allow the helium to escape readily.

A considerable quantity of helium is, however, nearly always found in the compact primary radioactive minerals, which from geologic data are undoubtedly of great antiquity.

Hillebrande made a very extensive analysis of a number of samples of minerals from Norway, North Carolina, and Connecticut, which were mostly compact primary minerals, and noted that a striking relation existed between the proportion of uranium and of nitrogen (helium) that they contained. This relation is referred to in the following words: —

" Throughout the whole list of analyses in which nitrogen (helium) has been estimated, the most striking feature is the apparent relation between it and the UO_2. This is especially marked in the table of Norwegian uraninites, recalculated from which the rule might almost be formulated that, given either nitrogen or UO_2, the other can be found by simple calculation. The same ratio is not found in the Connecticut varieties, but if the determination of nitrogen in the Branchville mineral is to be depended on, the rule still holds that the higher the UO_2 the higher likewise is the nitrogen. The Colorado and North Carolina minerals are exceptions, but it should be borne in mind that the former is amorphous, like the Bohemian, and possesses the further similarity of containing no thoria, although zirconia may take its place, and the North Carolina mineral is so much altered that its original condition is unknown."

Very little helium, however, is found in the secondary radioactive minerals, $i.$ $e.$, minerals which have been formed as a result of the decomposition of the primary minerals. These minerals, as Boltwood has pointed out, are undoubtedly in many cases of far more recent formation than the primary minerals, and consequently it is not to be expected that they should contain as much helium. One of the most interesting deposits of a secondary uraninite is found at Joachimsthal in Bohemia, from which most of our present supply of radium has been obtained. This is rich in uranium, but contains very little helium.

When the data required for these calculations are known with more definiteness, the presence of helium in radioactive minerals will in special cases prove a most valuable method of computing

18

their probable age, and indirectly the probable age of the geological deposits in which the minerals are found. Indeed, it appears probable that it will prove one of the most reliable methods of determining the age of the various geological formations.

SIGNIFICANCE OF THE PRESENCE OF LEAD IN RADIOACTIVE MINERALS

If the a particle is a helium atom, the atomic weights of the successive a ray products of radium must differ by equal steps of four units. Now we have seen that uranium itself probably contains three a ray products. Since the atomic weight of uranium is 238.5, the atomic weight of the residue of the uranium after the expulsion of three a particles would be $238.5 - 12, = 226.5$. This is very close to the atomic weight of radium 225, which we have seen is produced from uranium. Now radium emits five a ray products altogether, and the atomic weight of the end product of radium should be $238.5 - 32, = 206.5$. This is very close to the atomic weight of lead, 206.9. This calculation suggests that lead may prove to be the final product of the decomposition of radium, and this suggestion is strongly supported by the observed fact that lead is always found associated with the radioactive minerals, and especially in those primary minerals which are rich in uranium.

The possible significance of the presence of lead in radioactive minerals was first noted by Boltwood,[1] who has collected a large amount of data bearing on this question.

The following table shows the collected results of an analysis of different primary minerals made by Hillebrande: —

Locality.	Percentage of uranium.	Percentage of lead.	Percentage of nitrogen.
Glastonbury, Connecticut	.70–72	3.07–3.26	2.41
Branchville, Connecticut	74–75	4.35	2.63
North Carolina	77	4.20–4.53	
Norway	56–66	7.62–13.87	1.03–1.28
Canada	65	10.49	0.86

[1] Boltwood: Phil. Mag., April, 1905 ; Amer. Journ. Science, Oct., 1905.

Five samples were taken of the minerals from Glastonbury, three from Branchville, two from North Carolina, seven from Norway, and one from Canada. In minerals obtained from the same locality, there is a comparatively close agreement between the amounts of lead contained in them. If helium and lead are both products of the decomposition of the uranium radium minerals, there should exist a constant ratio between the percentage of lead and helium in the minerals. The percentage of helium is obtained from the above table by dividing the nitrogen percentage by seven. Since probably eight a particles are emitted from the decomposition of uranium and radium for the production of one atom of lead, the weight of helium formed should be $\dfrac{8 \times 4}{206.9} = .155$ of the weight of lead. This is based on the assumption that all the helium formed is imprisoned in the minerals. The ratio actually found is about .11 for the Glastonbury minerals, .09 for the Branchville minerals, and about .016 for the Norway minerals. It will be noted that in all cases the ratio of helium to lead is less than the theoretical ratio, indicating that in some cases a large proportion of the helium formed in the mineral has escaped. In the case of the Glastonbury minerals, the observed ratio is in good agreement with theory.

If the production of lead from radium is well established, the percentage of lead in radioactive minerals should be a far more accurate method of deducing the age of the mineral than the calculation based on the volume of helium, for the lead formed in a compact mineral has no possibility of escape.

While the above considerations are of necessity somewhat conjectural in the present state of our knowledge, they are of value as indicating the possible methods of attacking the question as to the final products of the decomposition of the radioactive minerals. From a study of the data of analyses of radioactive minerals, Boltwood has suggested that argon, hydrogen, bismuth, and some of the rare earths possibly owe their origin to the transformation of the primary radioactive substances.

It does not appear likely that we shall be able for many years to prove or disprove experimentally that lead is the final product of radium. In the first place, it is difficult for the experimenter to obtain sufficient radium for working material, and, in the second place, the presence of the slowly transformed product radium D makes a long interval necessary before lead will appear in appreciable quantity in the radium. A more suitable substance with which to attack the question would be radium F (radiotellurium) or radiolead (radium D).

2

THE

AMERICAN JOURNAL OF SCIENCE

[FOURTH SERIES.]

ART. VII.—*On the Ultimate Disintegration Products of the Radio-active Elements. Part II. The Disintegration Products of Uranium ;* by BERTRAM B. BOLTWOOD.

[Contributions from the Sloane Physical Laboratory of Yale University.]

THE general question of the nature of the ultimate disintegration products of the radio-active elements, as indicated by the occurrence of certain chemical elements in the radio-active minerals, has been discussed in an earlier paper,* and it was there pointed out that lead, bismuth and barium might perhaps be included among the possible disintegration products. As more recent experiments† have indicated, however, that actinium is probably an intermediate product between uranium and radium, the number of possible ultimate products has been correspondingly reduced. In addition to this careful examinations have been made of specially selected samples of typical primary uraninites from Branchville, Conn., and Flat Rock, N. C., and of thorianite from Ceylon, which have led to the conclusion that neither bismuth nor barium can be considered as disintegration products in the main line of descent from either uranium or thorium, at least on the basis of the present disintegration theory.

The conditions essential for the identification of the final disintegration products of uranium from a study of the composition of the natural minerals which contain this element would appear to be the following : In unaltered primary minerals of the same species, and of different species from the same locality, that is, in minerals formed at the same time and therefore of equal ages, a constant proportion must exist between the amount of each disintegration product and the

* This Journal, xx, 253, 1905.   † Ibid., xxii, 537, 1906.

amount of the parent substance with which it is associated. And, in unaltered, primary minerals from different localities, the proportion of each disintegration product with respect to the parent substance must be greater in those minerals which are the older and should correspond with the order of the respective geological ages of the localities in which the minerals have been found.   It also follows that in secondary minerals, namely, in minerals which have been formed by the subsequent alteration of the original, primary minerals, the relative amounts of the disintegration products must be less than in the primary minerals from the same locality, provided, however, that the disintegration products can not be considered as original chemical constituents of the secondary mineral.

It is the purpose of the present paper to show that the above requirements are practically fulfilled by lead and by helium also, in so far as the gaseous nature of the latter element will permit of its retention in the minerals.   The suggestion that lead was one of the final (inactive) disintegration products of uranium was first made by the writer in a paper presented before the New York Section of the American Chemical Society on February 10, 1905, and published later in the Philosophical Magazine.*

The amounts of uranium and lead present in a considerable number of primary uranium minerals have been calculated from the published analyses of these minerals.   The number of such analyses to be found in the literature is not large, and, what is still more unfortunate, with the exception of those made by Hillebrand and a few others, cannot be considered as particularly accurate.   Many of the analyses were made with special objects in view, such as the identification of a given specimen with a species already known or its recognition as a new variety or species.   There is also what is perhaps an unfortunate tendency on the part of many mineralogists to carry out an analysis merely for the purpose of assigning to the mineral some definite chemical formula, which often leads to the overlooking or ignoring of a number of the minor constituents.   And in addition to this there are also the actual analytical difficulties to be taken into account, which may be very considerable in the case of such minerals as samarskite, fergusonite, euxenite and other minerals containing notable proportions of niobium, tantalum and titanium.   Notwithstanding these objections, however, it is necessary to rely very largely on these published analyses, for the simple reason that the greater number of the uranium minerals are extremely rare and the obtaining of suitable samples of the various species and varieties is either extremely difficult or altogether impossible.

* April, 1905.

In the table which follows (Table I) are given the results obtained from the calculation of the ratio of the percentage of lead to that of uranium contained in the different minerals as indicated by the analyses.

No.	Mineral Locality	Per cent U	Per cent Pb	Ratio Pb / U	Analysis by
1.	Uraninite, Glastonbury, Conn.,	70	2·9	0·041	Hillebrand, this Journal, xl, 384, 1890.
2.	Uraninite, Glastonbury, Conn.,	70	3·0	0·043	Hillebrand, *l. c.*
3.	Uraninite, Glastonbury, Conn.,	70	2·8	0·040	Hillebrand, *l. c.*
4.	Uraninite, Glastonbury, Conn.,	72	3·0	0·042	Hillebrand, *l. c.*
5.	Uraninite, Glastonbury, Conn.,	72	2·9	0·040	Hillebrand, *l. c.*
6.	Uraninite, Branchville, Conn.,	74	4·0	0·054	Hillebrand, *l. c.*
7.	Uraninite, Branchville, Conn.,	75	4·0	0·053	Hillebrand, *l. c.*
8.	Uraninite, Branchville, Conn.,	74	4·0	0·054	Hillebrand, *l. c.*
9.	Uraninite, Branchville, Conn.,	66	3·5	0·053	From an analysis by the writer.
10.	Uraninite, Spruce Pine, N. C.,	·77	3·9	0·051	Hillebrand, *l. c.*
11.	Uraninite, Spruce Pine, N. C.,	77	4·2	0·055	Hillebrand, *l. c.*
12.	Uraninite, Spruce Pine, N. C.,	67	3·3	0·049	From an analysis by the writer.
13.	Uraninite, Marietta, S. C.,	71	3·3	0·046	Hillebrand, this Journal, xlii, 390, 1891.
14.	Uraninite, Llano Co., Tex.,	55	9·4	0·17	Hillebrand, *l. c.*
15.	Uraninite, Llano Co., Tex.,	56	9·5	0·17	Hidden and Mackintosh, this Journal, xxxviii, 481, 1889.
16.	Mackintoshite, Llano Co., Tex.,	19	3·4	0·18	Hillebrand, this Journal, xlvi, 98, 1893.
17.	Yttrocrasite, Eurnet Co., Tex.,	2·3	0·44	0·19	Hidden and Warren, this Jour., xxii, 515, 1906.
18.	Samarskite (?) Douglas Co., Colo.,	3·5	0·67	0·19	Hillebrand, Proc. Col. Sc. Soc., iii, 38, 1888.
19.	Samarskite (?) Douglas Co., Colo.,	3·7	0·74	0·20	Hillebrand, *l. c.*
20.	Samarskite (?) Douglas Co., Colo.,	5·1	6·99	0·19	Hillebrand, *l. c.*
21.	Uraninite, Anneröd, Norway,	66	8·4	0·13	Hillebrand, this Journal, xl, 384, 1890.

No.	Mineral	Locality	Per cent U	Per cent Pb	Ratio Pb U	Analysis by
22.	Uraninite,	Anneröd, Nor.,	68	7·8	0·12	Blomstrand, Jour. prakt.
23.	Annerödite,					Chem., xxix, 191, 1884.
		Anneröd, Nor.,	15	2·2	0·14	Blomstrand, Dana's Sys-
24.	Uraninite,					tem of Min., p. 741.
		Elvestad, Nor.,	66	9·3	0·14	Hillebrand, this Jour-
25.	Uraninite,					nal, xl, 384, 1890.
		Elvestad, Nor.,	57	8·0	0·14	Hillebrand, l. c.
26.	Uraninite,					
		Skaartorp, Nor.,	65	8·8	0·13	Hillebrand, l. c.
27.	Uraninite,					
		Huggenäskilen, Nor.,	68	8·8	0·13	Hillebrand, l. c.
28.	Uraninite,					
		Huggenäskilen, Nor.,	76	9·0	0·12	Lorenzen, Nyt. Mag.,
29.	Thorite,					xxviii, 249, 1884.
		Hitterö, Nor.,	8·2	1·2	0·14	Lindström, G. För.
30.	Uraninite,					Förh., v, 500, 1881.
		Arendal, Nor.,	56	9·8	0·17	Hillebrand, l. c.
31.	Uraninite,					
		Arendal, Nor.,	61	10·2	0·17	Hillebrand, l. c.
32.	Uraninite,					
		Arendal, Nor.,	56	9·4	0·17	Lindström, Zeit. f.
33.	Thorite,					Kryst., iii, 201, 1878.
		Arendal, Nor.,	9·0	1·5	0·17	Nordenskiöld, G. För.
34.	Orangite,					Förh., iii, 228, 1876.
		Landbö, Nor.,	7·5	1·2	0·16	Hidden, this Journal,
35.	Xenotime,					xli, 440, 1891.
		Naresto, Nor.,	2·9	0·62	0·21	Blomstrand, G. För.
36.	Hielmite,					Förh., ix, 185, 1887.
		Falun, Sweden,	1·9	0·20	0·10	Weibull, ibid, ix, 371,
37.	Polycrase,					1887.
		Slättakra, Sweden,	7·4	0·85	0·12	Blomstrand, Dana's
38.	Thorianite,					Min., p. 745.
		Sabaragamuwa Province, Ceylon,	9·8	2·1	0·21	Dunstan and Blake,
						Proc. Roy. Soc. Lond.,
39.	Thorianite,					lxxvi (A), 253, 1905.
		Sab. Prov., Cey.,	10·8	2·7	0·25	Dunstan and Blake, l. c.
40.	Thorianite,					
		Sab. Prov., Cey.,	12·8	2·4	0·19	Dunstan and Blake, l. c.
41.	Thorianite,					
		Sab. Prov., Cey.,	11·2	2·7	0·24	Analysis by writer.
42.	Thorianite,					
		Cey.,	11·1	2·3	0·21	Büchner, Nature, lxx,
43.	Thorianite,					169, 1906.
		Galle District, Cey.,	25	2·1	0·086	Dunstan and Jones,
						Proc. Roy. Soc., Lond.,
						lxxvii (A), 546, 1906.

Age of Minerals.

If the quantity of the final product occurring with a known amount of its radio-active parent and the rate of disintegration of the parent substance are known, it becomes possible to calculate the length of time which would be required for the production of the former. Thus, knowing the rate of disintegration of uranium, it would be possible to calculate the time required for the production of the proportions of lead found in the different uranium minerals, or in other words the ages of the minerals.

The rate of disintegration of uranium has not as yet been determined by direct experiment, but the rate of disintegration of radium, its radio-active successor, has been calculated by Rutherford[†] from various data. Rutherford's calculations give 2600 years as the time required for half of a given quantity of radium to be transformed into final products. The fraction of radium undergoing transformation per year is accordingly $2 \cdot 7 \times 10^{-4}$, and preliminary experiments by the writer on the rate of production of radium by actinium[‡] have given a value which is in good agreement with this number. The quantity of radium associated with one gram of uranium in a radio-active mineral has also been determined[§] and was found to be $3 \cdot 8 \times 10^{-7}$ gram. On the basis of the disintegration theory, when radium and uranium are in radio-active equilibrium, an equal number of molecules of each disin-

---

* It should be explained that No. 10 is really a secondary uraninite and is, therefore, not directly comparable with the others.

† Phil. Mag. (6), xii, 367, 1906.   ‡ This Journal, xxii, 537, 1906.

§ Rutherford and Boltwood, this Journal, xxii, 1, 1906.

tegrate per second, and, for our present purposes, we can neglect the difference in atomic weight and simply assume that in any time the weights of radium and uranium which undergo transformation are the same. In one gram of uranium the weight of uranium which would be transformed in one year would therefore be $2 \cdot 7 \cdot 10^{-4} \times 3 \cdot 8 \cdot 10^{-7} = 10^{-10}$ gram, and the fraction of uranium transformed per year would be 10^{-10}

In the table which follows (Table VI) the ages of the minerals included under Table I have been roughly calculated in accordance with the method outlined above. The ages of the minerals in years are obtained by multiplying the average value of the ratio 10^{10}. The general plan of calculating the ages of the minerals in this manner was first suggested to the writer by Prof. Rutherford.

TABLE VI.

Locality.	Age of minerals in million years.
Glastonbury (Portland), Conn.	410
Branchville, Conn.	535
Spruce Pine, N. C.	510
Marietta, S. C.	460
Llano and Burnet Co., Texas	1800
Douglas Co., Colorado	1900
Moss District, Norway	1300
Annerod, Norway	1700
Sabaragamuwa Prov., Ceylon	2200
Galle District, Ceylon	860

The actual values obtained for these ages are, of course, dependent on the value taken for the rate of disintegration of radium. When the latter has been determined with certainty, the ages as calculated in this manner will receive a greater significance, and may perhaps be of considerable value for determining the actual ages of certain geological formations.

Disintegration Products of Thorium.

The available data on the composition of the radio-active minerals serve to throw some light on the nature of the disintegration products of thorium as well as uranium. The relative proportions of uranium and thorium may show large variations in minerals from the same locality without exercising a noticeable effect on the value of the lead-uranium ratio for that locality. It can therefore be concluded with certainty that lead is not a disintegration product of thorium. This fact is particularly emphasized by the composition of the thorite found with the thorianite in the Sabaragamuwa province of Ceylon and in all probability of contemporaneous formation. The constituents of this mineral* are in part as follows: ThO_2, 66·26 per cent; CeO_2, 7·18 per cent; ZrO_2, 2·23 per cent; UO_2, 0·46 per cent. No lead at all is indicated as present, and the amount to be expected from the uranium is only 0·08 per cent, which was probably overlooked in making the analysis. A similar result was obtained in an examination by the writer of a specimen of thorite from Norway, which contained only 0·40 per cent of uranium, 52·0 per cent of ThO_2 and less than 0·10 per cent of lead. No mention is made of the presence of helium in the former of these thorites and in the specimen examined by the writer no indications of the presence of helium in measurable quantities were obtained. Although it has been stated by Ramsay† that the relatively large amount of helium contained in the thorianite from the Sabaragamuwa province is conclusive evidence of the production of helium by thorium, it seems quite probable that the evidence furnished by this mineral is quite the contrary, since it appears to contain only half of the amount of helium which would be produced by the disintegration of the uranium alone.

Summary.

Evidence has been presented to show that in unaltered, primary minerals from the same locality the amount of lead is proportional to the amount of uranium in the mineral, and in unaltered primary minerals from different localities the amount of lead relative to uranium is greatest in minerals from the locality which, on the basis of geological data, is the oldest. This is considered as proof that lead is the final disintegration product of uranium.

It has also been shown that, on the basis of the experimental data at present available, the amounts of helium found in radio-active minerals are of about the order, and are not in excess of the quantities, to be expected from the assumption that helium is produced by the disintegration of uranium and its products only.

The improbability that either lead or helium are disintegration products of thorium has been pointed out.

December 27, 1906.

* Dunstan and Blake, *l. c.*   † Jour. Chem. Phys., iii, 617, 1905.

Reprinted from *Proc. Roy. Soc. (London) Ser. A*, **84**, 379–381, 386–388 (1910)

$$3$$

Measurements of the Rate at which Helium is Produced in Thorianite and Pitchblende, with a Minimum Estimate of their Antiquity.

By the Hon. R. J. STRUTT, F.R.S., Professor of Physics, Imperial College of Science, South Kensington.

(Received July 23, 1910.)

§ 1.—*Introductory.*

The method of deducing a minor limit to the age of minerals from an examination of their radioactive properties has, up to the present time, depended on a measurement of the amount of helium they now contain, and on an indirect calculation of the rate at which it is being produced by the radioactive matter within them.

There is not now much uncertainty about this calculation. Nevertheless, considering the fundamental importance of the question of geological time, it is not superfluous to determine in some favourable case by direct volume-measurement of the gas how much helium is produced per gramme of the mineral per annum, in order to see how long the quantity found in the natural mineral would take to accumulate, and to check the method of calculation to which we must still resort where the much more difficult direct method is impracticable.

A mineral suitable for such experiments must be obtainable by the kilogramme, and very radioactive, so as to give a measurable quantity of helium in a few months. The minerals selected have been thorianite (two varieties) and pitchblende—practically the only ones available.

Some account of preliminary work was given in a former paper.* Much more elaborate and satisfactory experiments have since been carried out. These will now be described.

* ' Roy. Soc. Proc.,' A, vol. 83, p. 98.

The great difficulty of the problem lies in the small quantities of gas which have to be dealt with. In this respect the conditions are far more onerous than in determinations of the helium production by radium ; for a quantity of radium equivalent in activity to a ton of pitchblende or thorianite may be placed in a small vessel, and the helium developed in a given time extracted and measured. In the present investigation, the entire bulk of the original mineral has to be handled, and not merely the radium present in it. It is impracticable to work with more than a few kilogrammes in this way, thus the quantity of helium which can be obtained in, say, six months is small indeed.

Assuming the difficulty of the measurement of small volumes got over, there remains another not less serious. The helium initially present which has accumulated in geological epochs is perhaps 500 million times what the experimenter can grow under his observation. Thus, to make the experiment satisfactory, it is necessary to remove the helium so perfectly that not more than one part in 5,000 millions of the original stock remains.

This could never be done if the mineral were allowed to remain in its original solid condition. For solid minerals only yield their helium slowly and partially by heating.* It is essential, therefore, to get the mineral into solution, and to filter off any slight undissolved residue with the most scrupulous care. Prolonged boiling will then remove helium with the necessary completeness. It is important to use thick filter paper. Fine particles of undecomposed mineral may otherwise get through the pores. The importance of avoiding this will be understood when it is stated that the presence of 1/1000 of a milligramme of undecomposed thorianite was altogether inadmissible in my experiments. Much trouble was incurred through a failure to adopt this precaution in the earlier attempts.

The necessity for dissolving the mineral raises a theoretical question of some importance. Can we be really sure that this does not affect the rate of helium formation ? I shall briefly discuss this question, chiefly for the convenience of those who approach the subject from a standpoint of general scientific interest, rather than as students of radioactivity.

Nearly all the evidence we have at the present time points to the conclusion that the rate of radioactive change is unalterable by anything that man can do. It is true that a few experimenters have thought that they could detect changes in radioactivity at high temperatures, but the experiments of Bronson,[†] which have been pushed further than any others, reveal no such effect up to

---

* Unless, indeed, very high temperatures are used, which would introduce many other difficulties.

† 'Roy. Soc. Proc.,' A, vol. 78, p. 494.

1600° C.   The calorimetric experiments of Curie and Dewar showed no loss of heating effect in radium at liquid hydrogen temperatures.   As the development of heat is quantitatively accounted for by the expulsion of α-particles (helium atoms) with a high velocity, it cannot be doubted that the helium emission is unaltered at these temperatures.   More directly relevant are the experiments of Moore,[*] who found that radium emanation, dissolved in water, decayed at the same rate as when in the gaseous condition.   As the decay is the direct consequence of the emission of α-particles, it is clear that here also helium formation is independent of circumstances, and, in particular, of whether the radioactive body is in solution or not.

Lastly the method is only applicable[†] if we assume that the rate of helium production has been the same throughout the whole geological period which it is sought to measure.   What is there to be said in defence of this assumption?   The critic will naturally object that the radioactive matter present is necessarily diminishing in quantity as it generates helium and other non-radioactive products to which it may give rise.   There must, therefore, have been more of it at the beginning of the geological period considered than at the end, and consequently more rapid production of helium.   The method as here applied is only valid if it can be shown that this diminution is unimportant in the period considered.

A simple argument goes far to establish this for thorianite.   This mineral is a dense substance, consisting almost entirely of the parent radioactive bodies, uranium oxide and thorium oxide.   We cannot suppose that there was ever much more of these bodies in the thorianite crystals than at present, for they do not contain much inactive matter of any kind, which can be assumed to represent the débris of the decayed radioactive bodies.

Apart from this simple argument, we have a good indirect estimate of the rate of decay of uranium,[‡] which shows that it is unimportant in the periods here dealt with.   The decay of thorium is almost certainly slower still.

The other chief line of objection which can be taken is that the rate of formation of helium depends on the physical condition of the radioactive matter.   This objection has already been considered.

*   *   *   *   *   *

---

[*] 'Roy. Soc. Proc.,' A, vol. 80, p. 597.
[†] At least, in the simple form with which alone this paper deals.
[‡] Boltwood, 'Amer. Journ. Sci.,' June, 1908, vol. 25, p. 506.

VOL. LXXXIV.—A.   2 D

§ 3.—*Results of Experiments.*

First Series.—Thorianite from Galle; the variety rich in uranium. Three flasks (numbered 4, 5, 6) set up, each containing 680 grammes in solution.

Experiment.	Volume of helium.
	c.c. × 10^{-6}.
Initial blank test of all three flasks, D_3 invisible	< 1·0
After 205 days' standing—	
Flask No. 4..	15·5
Flask No. 4 + No. 5	34·9
Flask No. 4 + No. 5 + No. 6	45·9
Blank test of all three flasks, D_3 conspicuous	2·9
After standing 129 days more, all three flasks	24·3
Final blank test, D_3 invisible	< 1·0

From these experiments—

Rate of production from all three flasks, per annum $\begin{cases} 8\cdot18 \times 10^{-5} \text{ c.c.} \\ 6\cdot90 \times 10^{-5} \text{ c.c.} \end{cases}$

Mean .... $7\cdot54 \times 10^{-5}$ c.c.

Rate of production per gramme of Galle thorianite, per annum .... $3\cdot7 \times 10^{-8}$ c.c.

Helium initially present, per gramme .... 9·3 c.c.

Time required to produce this .... $2\cdot50 \times 10^8$ years.

Second Series.—Ordinary thorianite containing 13·10 per cent. U_3O_8, and 72·65 per cent. ThO_2. Three flasks (numbered 1, 2, 3) set up, each containing 510 grammes.

Experiment.	Volume of helium.
	c.c. × 10^{-6}.
Initial blank test of all three flasks, D_3 doubtful	< 0·5
After standing 141 days*—	
Flask No. 1..	5·0
Flask No. 1 + No. 2	11·3
Flask No. 1 + No. 2 + No. 3	16·5
Final blank test of all three flasks	< 0·5

* As a matter of fact, the flasks were not all boiled out at quite the same time on this occasion. The numbers given are slightly corrected to compensate for this, so that the results can be studied more easily.

From these experiments—

Rate of production from all three flasks per annum $4·27 \times 10^{-5}$ c.c.

Rate of production per gramme of ordinary thorianite $2·79 \times 10^{-8}$ c.c.

Helium initially present, per gramme .... $7·8$ c.c.

Time required to produce this .... $2·8 \times 10^8$ years.

Pitchblende, from Joachimsthal, 353 grammes. Dissolved (see above, p. 382) in two separate flasks, which were always boiled at the same time, and the gases treated together.

Experiment.	Volume of helium.
	c.c. $\times 10^{-6}$.
Blank test, D_3 invisible	$<0·5$
After 61 days (not a good experiment)..................	$2·0$
After 294 days (good experiment)......................	$9·0$

The last experiment is the only one on which stress can be laid. It gives

Helium per gramme pitchblende, per annum...... $3·16 \times 10^{-8}$.

The pitchblende experiments were not carried so far as those on thorianite, on account of the much greater difficulties of preparing the solution. The best experiment, however, was a very satisfactory one.

The helium measured may be exhibited graphically as a function of the

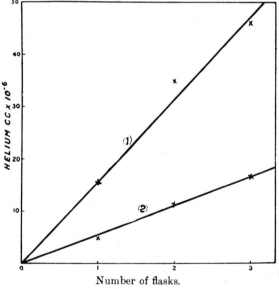

Fig. 3.—(1) Galle thorianite.   (2) Ordinary thorianite.

33

number of flasks emptied, in those cases where the contribution of each was added successively, and shows fair proportionality (fig. 3). This is additional evidence that no appreciable helium was lost in the charcoal, otherwise the first flask would have apparently yielded less than its due contribution.

§ 4.—*Discussion of Results.*

It now remains to compare the rate of formation of helium observed with that calculated theoretically. For the calculation I refer to ' Roy. Soc. Proc.,' A, vol. 81, p. 276, and vol. 83, p. 97.

Mineral.	U_3O_8. Per cent.	ThO_2. Per cent.	Helium production per gramme per annum. C.c. × 10⁻⁸.	
			Observed.	Calculated.
Thorianite. Galle district ...	24 ·50	65 ·44	3 ·70	3 ·46
Ordinary thorianite	13 ·10	72 ·65	2 ·79	2 ·55
Pitchblende	37 ·6	None	3 ·16	3 ·44

The calculation is clearly justified by the direct observations, and can in future be employed with confidence in its substantial correctness.

The present experiments leave, I think, no doubt whatever that some specimens of thorianite are as much as 280 million years old. I will take this opportunity of summarising the data in my previous papers* with regard to the duration of geological time, as deduced by the indirect method. With the verification of this method now presented, I feel justified in stating them numerically without the qualifications before insisted on.

Mineral.	Geological horizon.	Minimum age.
		Years.
Sphærosidenite from Rhine provinces	Oligocene	8,400,000
Hæmatite, Co. Antrim.........................	Eocene	31,000,000
Hæmatite, Forest of Dean	Carboniferous limestone ...	150,000,000
Sphene, Renfrew Co., Ontario	Archæan	710,000,000

These are minimum values, because helium leaks out from the mineral, to what extent it is impossible to say.

Summary.—The rate at which helium has been and is being produced in thorianite has been measured directly with reasonable accuracy, and is found to be in agreement with the rate calculated indirectly. The helium now found in one sample examined would take 280 million years to accumulate. Similar measurements have been made with pitchblende.

* ' Roy. Soc. Proc.,' A, vol. 81, p. 272 ; vol. 83, p. 96 ; vol. 83, p. 298.

The Chemical Lead Method

II

Editor's Comments on Papers 4 and 5

Boltwood's interest in the accumulation of lead in uranium minerals as a means of determining their ages was only a sideline to his main research activities, which were directed toward elucidating the chemistry of the uranium and thorium decay series. The chief proponent of the uranium–lead method was Arthur Holmes, a student of R. J. Strutt's at Imperial College, London. Holmes' first paper, published in 1911 and reproduced here (Paper 4), reflects what was to become a life-long interest in the subject of geochronology. Following Boltwood, he expressed the opinion that radioactive thorium could not be the parent of lead. The association of lead with uranium was so consistent that the possibility of there being a mixture of uranium-derived lead and thorium-derived lead in any radioactive mineral seemed most unlikely. Bismuth was the favored product of thorium decay. Holmes determined the Pb/U ratios of several new minerals and showed how the value of this ratio increased with the stratigraphic age of the samples. He urged the establishment of a radiometric time scale related to the stratigraphic succession, and the application of radiometric dating to the crystalline basement rocks of the Precambrian.

Unfortunately, the accumulation of radiogenic lead was more complicated than either Boltwood or Holmes had first anticipated. Between 1914 and 1915 Holmes and R. W. Lawson reviewed the possible significance of thorium-derived lead and again concluded that lead was probably not the stable end product of thorium decay. Unlike lead–uranium ratios measured in uranium-rich minerals, the ratios of lead to thorium in thorium-rich minerals showed no consistent increase with age. Later, when subsequent work showed that no disintegration products beyond thorium-D (Pb-208) could be recognized and the stability of thorium-derived lead became generally accepted, the observed discrepancies between calculated U/Pb ages and Th/Pb ages were ascribed to preferential leaching of the thorium-derived lead. The lower ages calculated from thorium-rich minerals were, however, preferred by some, notably

John Joly at Trinity College, Dublin, who at the time was concerned with the apparent change in the decay rate of uranium, as evidenced by his study of pleochroic haloes. The younger thorium–lead ages were also more in accord with the older estimates of time based on sedimentation rates and the salinity of the ocean. However, the majority of investigators were of the opinion that thorium–lead ages were of dubious significance and that thorium-rich minerals were to be avoided for dating purposes, because of the apparent loss of radiogenic lead. Lead ages calculated from uranium-rich minerals would yield reliable results, provided a correction was made for the presence of any thorium-derived lead.

Thorium frequently occurs as a minor constituent in uranium-bearing minerals, but the two radioactive parents produce lead at different rates, with uranium producing lead faster than thorium. Assuming a constant rate of lead production with time, it was estimated (from the measured rates of decay) that after 7,400 million years the ratio of uranium-derived lead to uranium would be unity, but the thorium-derived-lead-to-thorium ratio would have a value of only 0.38. So, on the basis of the uranium-derived lead, the age of a radioactive mineral could be calculated from the relationship

$$\text{Age} = \frac{\text{U-Lead}}{\text{U}} \times \frac{7,400 \text{ m.y.}}{1} \tag{1}$$

On the basis of thorium-derived lead, the age of the same mineral could be calculated from the relationship

$$\text{Age} = \frac{\text{Th-Lead}}{\text{Th}} \times \frac{7,400 \text{ m.y.}}{0.38} \tag{2}$$

As there is no way of chemically distinguishing between the two varieties of radiogenic lead, (1) and (2) must be combined, i.e., we have

$$\text{Age} = \frac{\text{U-Lead} + \text{Th-Lead}}{\text{U} + 0.38\,\text{Th}} \times 7,400 \text{ m.y.}$$

$$= \frac{\text{total Pb}}{\text{U} + 0.38\,\text{Th}} \times 7,400 \text{ m.y.}$$

$$= \frac{\text{total Pb}}{\text{U} + k \times \text{Th}} \times C \text{ m.y.} \tag{3}$$

The values adopted for k and C depend on the disintegration rates assigned to the parent uranium and thorium. A correction (the "time-average" correction) could also be made to take into account the fact that the rate of production of the radiogenic daughter depends upon the amount of radioactive parent present. As this decreases with time, the rate of production of the stable daughter decreases exponentially with time.

All the factors involved in calculating radiometric ages from chemical analyses of uranium, thorium and lead in naturally occurring minerals were clearly described by Holmes and R. W. Lawson in 1927 (Paper 5). Precise atomic-weight determinations of the extracted lead were used to correct for the presence of any appreciable quantities of ordinary (nonradiogenic) lead. Ordinary lead has an atomic weight of 207.2, in contrast to uranium-derived lead (206) and thorium-derived lead (208). In 1927 the actinium series was generally regarded as a branch series of the uranium series, about 3 per cent of the decaying uranium parents decaying via protoactinium, and Holmes and Lawson felt justified in ignoring its presence in their age calculations. They conclude their paper with a geologic time-scale based on chemical lead age data, a scale which incidentally bears a strong resemblance to the modern scale, at least for the Paleozoic and younger eras.

Selected Bibliography

Holmes, A., and Lawson, R. W. (1914). Lead and the end product of thorium. (Part I). *Phil. Mag.,* *Ser. 6,* **28**, 823–840.

Holmes, A., and Lawson, R. W. (1915). Lead and the end product of thorium. (Part II). *Phil. Mag.,* *Ser. 6,* **29**, 673–688.

Ellsworth, H. V. (1925). Radioactive minerals as geological age indicators. *Amer. J. Sci., Ser. 5,* **9**, 127–144.

4

The Association of Lead with Uranium in Rock-Minerals, and its Application to the Measurement of Geological Time.

By Arthur Holmes, A.R.C.S., B.Sc., Imperial College of Science and Technology.

(Communicated by Prof. the Hon. R. J. Strutt, F.R.S.   Received March 20,—Read April 6, 1911.)

1. *Introduction.*—The study of radioactive minerals is of great importance from two points of view. Such minerals may be regarded as storehouses for the various series of genetically connected radioactive elements. In them the parent element slowly disintegrates, while the ultimate products of the transformation gradually accumulate. The analysis of these minerals ought, then, in the first place, to disclose the nature of the ultimate product of each series; secondly, a knowledge of the rate of formation of this product, and of the total quantity accumulated, gives the requisite data for a calculation of the age of the mineral.

It has been shown that the disintegration of uranium results in the formation of eight atoms of helium.* In 1907 Boltwood brought forward strong evidence suggesting that lead is the ultimate product of this disintegration.† In this paper it is hoped to produce additional evidence that such is the case, according to the following equation :—

$$U \rightarrow 8He + Pb.$$
$$238{\cdot}5 \quad 32 \quad 2{\cdot}069.$$

---

* See Strutt, 'Roy. Soc. Proc.,' A, 1908, vol. 81, p. 276.
† Boltwood, 'Am. Journ. Sci.,' 1907, p. 77.

On the assumption that helium is produced to this extent, Rutherford has given data* from which it may be calculated that 1 gramme of uranium produces $10·7 \times 10^{-8}$ c.c. of helium per annum. Strutt has verified this theoretical estimate by a direct appeal to experiment.† Actually measuring the annual production of helium, he obtained a corresponding result of $9·9 \times 10^{-8}$ c.c. Accepting the theoretical figure, which is equivalent to $1·88 \times 10^{-11}$ grm., it is easily calculated that the amount of lead which would remain is $1·22 \times 10^{-10}$ grm. per gramme of uranium per annum. If this rate of production were constant, a gramme-molecule of lead would take the place of a gramme-molecule of uranium in 8,200 million years. However, the rate is not constant, but is proportional to the amount of uranium remaining unchanged. If the latter is large compared with the total amount of lead produced, the rate may be taken as nearly constant, and the age of the mineral in which this disintegration has occurred is given by

$$\text{Pb/U} \cdot 8200 \times 10^6 \text{ years,}$$

where Pb and U represent the respective percentages of these elements at the present day. In many cases, however, this constancy cannot be assumed, and it is necessary to substitute for the present-day percentage of uranium its time-average for the period considered. Thus, in the minerals described in this paper, the difference between the uranium now present and that originally present amounts to about 5 per cent., and, in calculating the age, corresponding values are obtained. In this case a sufficiently accurate approximation to the time-average is given by the mean.

For minerals of the same age, the ratio Pb/U should be constant, if all the lead has originated as suggested. Further, for minerals of different ages, the value of Pb/U should be greater or less in direct proportion to those ages.

Collecting all the known analyses of primary uranium-bearing minerals which included a determination of lead, Boltwood‡ showed that the above conditions were generally found to hold. Unfortunately, he omitted to give the geological ages of the several occurrences. In a summary of his analyses, to be given in a later section, these will be indicated as accurately as at present is possible.

2. *Selection of Minerals.*—In order that the suggested relations between lead and uranium should be detectable, and that lead should be confidently used as a reliable age-index, certain assumptions require to be made. The selection of minerals must be such that for them these assumptions are justifiable. They will be considered as follows:—

* Cited by Strutt, 'Roy. Soc. Proc.,' 1908, p. 276.
† Strutt, 'Roy. Soc. Proc.,' A, 1910, vol. 84, p. 388.
‡ 'Am. Journ. Sci.,' 1907, p. 77.

(*a*) That no appreciable amount of lead was present when the mineral was formed.

(*b*) That no lead has originated by any other radioactive process than that suggested.

(*c*) That no lead nor uranium has subsequently been added or removed by external agencies.

(*a*) Previously to the consolidation of a rock magma, the uranium in the latter must, of course, have been generating helium and lead for an unknown period. It is probable that much of the lead then present would, at the time of crystallization, be carried away in hot sulphide solutions to form the hydatogenetic and metasomatic deposits of lead which provide our supplies of that metal. Doubtless, however, a certain amount of lead would be retained in the molecular network of crystals, and consequently analyses of a rock as a whole should give values of Pb/U higher than that corresponding to the period since consolidation. This difficulty may be avoided by considering particular minerals. Thorite, zircon, in some cases apatite and sphene, and other rarer minerals segregate within themselves on crystallization a much larger percentage of uranium than remains to the rest of the magma. Within these minerals lead accumulates to such an extent that the amount originally present becomes negligible.

(*b*) It may be objected that lead may perhaps originate as a product of some element other than uranium. Boltwood shows that it is highly improbable that thorium should give rise to lead, and the results submitted in this paper add further proof to that independence. Wherever lead occurs in primary minerals it is associated with uranium, and there is little doubt that it can be completely accounted for in this way.

(*c*) It may seem unlikely that for periods of hundreds of millions of years a mineral should remain unchanged by external chemical agencies. In the earth's surface materials, making up the belt of weathering, solution is the dominant process. Lower down, in the belt of cementation, re-deposition is more characteristic.* Can we be sure that these processes have not dissolved out lead or uranium at one time, depositing the same elements at another time? In some cases we cannot, but, fortunately for our purpose, many of the uranium-bearing minerals, like zircon, are dense and stable, and capable of withstanding great changes in their environment without undergoing alteration. But an appeal to analysis will rarely fail to dispel this difficulty. If such changes have occurred, it is inconceivable that they would always have affected lead and uranium in the same proportion,

* See Van Hise, "Treatise on Metamorphism," 'Mon. United States Geol. Survey,' 1904, vol. 47.

and hence the results obtained from different minerals should show marked discrepancies. On the other hand, if the analyses give consistent results one can only assume that any alteration has been inappreciable. A microscopical examination of the minerals in question affords a useful guide to the extent of alteration. Unless one can be sure in this way that the mineral is fresh, it is clear that reliable results can only be expected when a series of minerals are examined.

Still another possible objection may be treated here. Under the high temperatures and pressures which rocks have undergone during their geological history, is it safe to assume that radioactive changes proceed at the same rate? All that can be said is that experimental evidence consistently agrees in suggesting that these processes are quite independent of the temperatures and pressures which igneous rocks can have sustained without becoming metamorphosed. Arrhenius has supposed that radioactive processes may be reversed under the conditions prevailing at great depths. This idea has nothing but analogy to support it. There is abundant evidence that molecular changes are reversed at greater depths, *e.g.*, in the upper zones of the earth's crust silicates are replaced by carbonates, while in the lower zones carbonates are decomposed and silicates are formed. But that interatomic changes should reverse, or even proceed more slowly or quickly, there is no evidence.

From these considerations, it is obvious that the only minerals to be chosen are fresh, stable, primary rock-minerals. Secondary and metamorphic minerals could not be relied upon to satisfy the required conditions. There occurs in the Christiania district of Norway,* a geologically depressed area of nearly 4,000 square miles, which is separated on every side by faults from the surrounding Pre-Cambrian gneiss. In this area there is a nearly complete sequence of early palæozoic rocks. Above these strata there are a few beds of red sandstone of Lower Devonian age. Over these beds and intercalated with them are lava flows; and, finally, penetrating the whole mass, representing a later phase of this period of igneous activity, are great intrusions of plutonic rocks. Amongst the earliest of the intrusions is a series of thorite-bearing nepheline-syenites. Brögger believes them to be of Middle or Lower Devonian age, most probably the latter. The minerals occurring in them are, in many instances, notably radioactive, and thus they afford an admirable series in which to investigate the consanguinity of lead and uranium. Several of these minerals were obtained from Brevig, and estimations of these elements in each case were made.

* See Brögger, 'Zeit. für Kryst,' 1890, vol. 16.

3. *Methods of Analysis.*—(a) *Uranium.*—This constituent was estimated by Strutt's method,* in which radium emanation is directly measured, and the constancy of its ratio to uranium used to give the amount of the latter.

From 0·3 grm. to 2·0 grm. of the finely powdered mineral was used for each estimation, according to the relative richness of the mineral in uranium. From preliminary electroscopic tests this could be roughly measured. The powdered mineral was fused with borax in a platinum crucible, and the resultant glass dissolved in dilute hydrochloric acid. After boiling, and standing for several days in a corked flask, the radium emanation was boiled out, collected in a gas-holder, and ultimately transferred to an electroscope. Knowing the normal leak and constant of the electroscope, a measurement of leak sufficed to give the necessary data for the calculation of the equivalent amount of uranium.

Blank experiments were made with the reagents used, and the normal leak was determined at suitable times throughout the investigations. In no case was any appreciable difference observed. Two solutions of each mineral were made, and two estimations of each. Without exception, the results obtained agreed closely.

(b) *Lead.*—Several methods of estimating lead were attempted, but the most constant and reliable results were found to be attained by weighing it as sulphate, and in cases when the quantity of lead present was too small for the gravimetric method, colorimetric estimations were made.

Gravimetric method.—Quantities varying up to nearly 100 grm. of the finely powdered mineral were intimately mixed with four or five times as much fusion mixture, and fused in a platinum basin. On allowing the melt to cool completely, the cake could usually be easily separated by treating with boiling water. A second heating and cooling always resulted in a successful separation. The cake was broken up by boiling with water in a beaker. Dilute hydrochloric acid was gently heated in the platinum basin to remove any still adherent portions of the cake. The contents of the basin were then washed into the beaker, and more hydrochloric acid added. The solution thus formed (with a colloidal mass of silica) was evaporated to dryness, and, dilute hydrochloric acid having been added, this was repeated a second time. On again adding dilute acid and heating, the silica was easily filtered off, leaving a clear solution. From the latter lead was precipitated as sulphide, by heating and adding ammonium sulphide. The precipitate was collected on as small a filter paper as possible, dried and ignited. The residue was treated with a little nitric acid, and boiled to convert any reduced lead to nitrate. Sulphuric acid was finally added, and the whole heated until all nitric

* 'Roy. Soc. Proc.,' A, 1906, p. 473 *et seq.*

acid fumes had ceased. A tiny white precipitate then remained. This was collected on a very small filter, of which the weight of the ash was accurately known, washed with alcohol, dried, ignited, and weighed with the greatest possible accuracy.

*Colorimetric method.**—Standard solutions containing known quantities of lead, as nitrate, were prepared by dissolving the lead compound in a slightly acidified solution containing ammonium acetate and grape sugar, known as the " diluting solution.

The lead to be estimated having been concentrated in a nitric acid solution as already described, the latter was evaporated to dryness, or nearly so. The residue was then taken up with a little of the diluting solution. This was treated with a known quantity of well-diluted ammonium sulphide, the liquids being contained in a graduated glass vessel. A brown coloration was produced. One of the standard solutions was similarly treated in an exactly similar vessel. The diluting solution was then added to one vessel or the other until the colours produced in both were indistinguishable, care being taken that the amount of ammonium sulphide in each was proportionate to the respective volume. After a little practice with solutions of known strengths, this could be done with confidence, and concordant results were obtained. Tested by the sulphate methods slightly higher results were given in general.

The colorimetric method obviously assumes the absence of copper and bismuth.

From Vogt's estimates of the average amount of these metals in 100 grm. of rock :—†

	Grm.
Lead	$0.000x$
Copper	$0.0000x$
Bismuth	$0.00000x$

it might be anticipated that the latter two would not have much influence. Tests were, however, applied to detect any very small quantities which might be present. By testing for copper‡ with hydrobromic acid, about 0.00002 grm. was probably the greatest amount indicated. Schneider's test applied for bismuth§ failed to detect that element. To the fourth decimal the amount of lead was therefore unaffected by its non-separation from copper or bismuth.

The smallest amount of lead estimated, viz., 0.0003 grm. in 100 grm. of

* V. Harcourt, 'Journ. Chem. Soc.,' 1910, vol. 97, p. 841.
† 'Zeit. prakt. Geol.,' 1898.
‡ 'Select Methods of Chemical Analysis,' Crookes, 1905, p. 295.
§ *Ibid.*, 1905, p. 355.

felspar, approaches the limit to which the colorimetric method can be applied quantitatively, although smaller quantities than this can easily be detected.

4. *Experimental Results.*—The results obtained are tabulated below :—

Mineral.	Uranium. Grm. per 100 grm. mineral.	Lead. Grm. per 100 grm. mineral.	Pb/U.
Thorite (1)	10 ·1040	0 ·4279	0 ·042
Orangite (1)	1 ·2437	0 ·0570	0 ·046
Orangite (2)	1 ·1825	0 ·0542	0 ·046
Thorite (2)	0 ·4072	0 ·0196	0 ·048
Homelite	0 ·2442	0 ·0121	0 ·049
Zircon	0 ·1941	0 ·0085	0 ·044
Pyrochlore (1)	0 ·1923	0 ·0120	0 ·062
Pyrochlore (2)	0 ·1855	0 ·0093	0 ·050
Biotite	0 ·1602	0 ·0069	0 ·043
Tritomite	0 ·0631	0 ·0026	0 ·041
Freyalite	0 ·0526	0 ·0028	0 ·053
Mosandrite	0 ·0432	0 ·0024	0 ·056
Eagerine	0 ·0253	0 ·0015	0 ·060
Astrophyllite	0 ·0140	0 ·0007	0 ·050
Catapleite...........................	0 ·0132	0 ·0009	0 ·068
Nepheline...........................	0 ·0010	0 ·0004	0 ·400
Felspar	0 ·0006	0 ·0003	0 ·500

With the exception of pyrochlore, specimen (1), and astrophyllite, the number of lead estimations varied from two to five. Of the minerals named, only one determination was made, owing to lack of material. It will be noticed that with few exceptions the value of Pb/U increases as the uranium content decreases. This may be due to the possibility of the lead originally present in the magma having a gradually increasing relative importance as the lead generated from uranium decreases in amount. Thus it would seem in the case of nepheline and felspar that the lead so generated is of no importance whatever when compared with that originally present. Such minerals are, of course, valueless in age-estimations, and of the results given here only eight of the first nine will be used for determining the age. Omitting that of pyrochlore (1), since the single rather anomalous determination of lead could not be verified by a second estimation, these results give 0·046 as their mean, and if the uranium percentage be replaced by its approximate time-average the mean becomes 0·045. This gives an age of 370 million years, and is probably the most reliable estimate that can be deduced from the evidence.

5. *Summary of Analyses collected by Boltwood.*[*]—(*a*) The analysis of five specimens of uraninite from Glastonbury (Conn.) gives a ratio of Pb/U = 0·041. The minerals occur in a pegmatite associated with a

---

[*] Boltwood, 'Am. Journ. Sci.,' 1907, p. 77.

granite intruding Lower Carboniferous strata, and probably itself of Carboniferous age.

(*b*) Uraninite from Branchville (Conn.) gives four closely agreeing ratios, 0·053. The geological evidence here is similar to that at Glastonbury, with the exception that the intruded strata are of Silurian or Ordovician age.

(*c*) Material from dykes of pre-Carboniferous age in Carolina gives less consistent results, of which the mean ratio is 0·05.

(*d*) In Llano Co. (Texas), there occurs a group of metamorphosed sedimentary rocks of early Algonkian age. Into these the Burnet granites are intrusive, and are therefore somewhat younger than the schists and quartzites. The ratio of minerals from these igneous rocks is 0·160.

(*e*) Another group of minerals from Burnet Co. (Texas) and Douglas Co. (Colorado) gives a ratio of 0·175.

Geological evidence is similar to that of Llano Co., and it is impossible to say whether or not the rocks are older.

(*f*) The pre-Cambrian rocks of Sweden are divided by Högbom into three main divisions, Jotnian, Jatulian, and Archæan, in order of increasing age. Above the Archæan, but younger than the Jatulian, is a series of igneous massives known as the Sen-archæan granites, and with these are associated the famous uranium-bearing pegmatites of Scandinavia.

In a series of 17 minerals from these pegmatites, taken from all parts of Norway and Sweden, there appear to be two clearly marked groups. One gives a ratio of 0·125 and the other of 0·155. Amongst these rocks geological correlation is very speculative, but it is agreed that there is nothing by which any difference in age could be detected, and provisionally the two groups are regarded geologically as one.

(*g*) The greatest ratio is given by thorianite from Ceylon, for which $Pb/U = 0.20$. Here the only evidence for the pre-Cambrian age of the minerals is derived from the similarity of the rocks to those of the fundamental complex of India. These latter underlie a vast series of sedimentary strata considered to be of pre-Cambrian age.

It should be observed that in calculating the above ratios U represents the time-average, and not the amount actually present. The difference is, however, not great.

6. *Conclusion.*—Evidence has been given to prove that the ratio Pb/U is nearly constant for minerals of the same age, the slight variability being what theoretically one would anticipate.

For minerals of increasing geological age the value of Pb/U also increases, as the following table clearly shows :—

Geological period.	Pb/U.	Millions of years.
Carboniferous	0 ·041	340
Devonian	0 ·045	370
Pre-carboniferous	0 ·050	410
Silurian or Ordovician	0 ·053	430
Pre-Cambrian—		
a. Sweden .... {	0 ·125	1025
	0 ·155	1270
b. United States .... {	0 ·160	1310
	0 ·175	1435
c. Ceylon	0 ·20	1640

Wherever the geological evidence is clear, it is in agreement with that derived from lead as an index of age. Where it is obscure, as, for example, in connection with the pre-Cambrian rocks, to correlate which is an almost hopeless task, the evidence does not, at least, contradict the ages put forward. Indeed, it may confidently be hoped that this very method may in turn be applied to help the geologist in his most difficult task, that of unravelling the mystery of the oldest rocks of the earth's crust; and, further, it is to be hoped that by the careful study of igneous complexes, data will be collected from which it will be possible to graduate the geological column with an ever-increasingly accurate time scale.

In conclusion, I wish to express my thanks to those gentlemen who in any way have helped to make this investigation possible. I am indebted to Profs. Sir T. H. Holland, Brögger, and Högbom, and to the Director of the United States Geological Survey, for information regarding the geological position of many of the occurrences cited in §5 ; and to Dr. Prior for his permission to make preliminary electroscopic tests of several minerals in the collection of the Natural History Museum. Finally, I owe my best thanks to Prof. Strutt, at whose suggestion the work was attempted, for his ever-ready help and criticism, and for his kindness in obtaining for me the suite of minerals and allowing me the use of apparatus in their investigation.

Copyright © 1927 by Yale University
Reprinted from *Am. J. Sci., Ser. 5*, **13**, 327–344 (1927)

5

FACTORS INVOLVED IN THE CALCULATION OF THE AGES OF RADIOACTIVE MINERALS.

ARTHUR HOLMES AND ROBERT W. LAWSON.

I. HISTORICAL REVIEW.

The end-product of the uranium-radium disintegration series is now known to be a stable isotope of lead of atomic weight 206.0; and in the case of the thorium series the evidence indubitably points to another stable isotope of lead as the end-product of disintegration, the atomic weight being about 208. The accumulation of these end-products in primary unaltered radioactive minerals makes it possible to determine the age of the latter, provided that the rates of production of lead from uranium and thorium are known. Given the fundamental measures of rate, it only remains to analyze each suitable mineral for uranium, thorium and lead, and to ascertain from every possible source of evidence that the lead is, in fact, of radioactive origin. Atomic weight determinations are of greatest value in settling this point, while the paragenesis of the minerals themselves generally indicates quite definitely whether or not there is any probability of the original presence of appreciable quantities of ordinary (207.2) lead. When ordinary lead is known to be present, as for example in associated galena, its amount can generally be evaluated if the atomic weight of the isotope mixture is known. The assumption of uniformity of rate of disintegration throughout geological time has been called in question by Joly, but the cumulative evidence of pleochroic halos, atomic weights, and analyses of minerals strongly supports the view that the assumption adversely criticized by Joly is valid, and that at the worst it cannot lead to errors of more than a few per cent.[1] The above topics have been widely discussed in recent years, with the notable exception of the rates of production of lead, and it is this aspect of the subject with which we propose to deal in the present paper.

[1] NOTE: The references are given at the end of the paper.

The general principle of the radioactive method of age determination was originally suggested by Boltwood,[2] and developed by Holmes in 1911 and later years.[3] In the early stages of the exploration of the method thorium was not taken into account, and the first discussion of the possible significance of thorium-lead appeared in 1914.[4] The stability of thorium-lead was then seriously doubted, but later work has demonstrated the non-accumulation of any possible further disintegration product, and it is now generally accepted that thorium-lead is stable.

If U, Th, and Pb respectively denote the percentage contents of a suitable mineral in uranium, thorium and lead of radioactive origin, then the age of the mineral is given to a first approximation in millions of years by the formula

$$\text{Age} = \frac{\text{Pb}}{\text{U} + k.\text{Th}} \cdot C,$$

where k and C are constant factors, k being the amount of uranium which is equivalent in lead-producing capacity to 1 gm. of thorium, and $1/C$ being the amount of lead produced by 1 gm. of uranium in one million years.

This formula fails to take into account the obvious fact of the wearing out of part of the original uranium and thorium during the lifetime of the mineral. The correction involved is small, rising to about ten per cent for the oldest known minerals, and will be considered in a later section of this paper.

The values to be adopted for k and C must obviously depend upon the disintegration constants of the parent elements uranium and thorium. Unfortunately there is an appreciable divergence in the values used by different authors, as a consequence of which the calculated ages are not always directly comparable. It is greatly to be desired that uniformity should be attained in this respect, and we feel that the time is now ripe for the adoption of agreed values of k and C by the various workers in this field of research, particularly in view of the increasing attention now being devoted to its cultivation.

Before attempting to arrive at the best attainable values of the factors on current data and their possible limits, we propose to review critically the observational evidence on which the factors have been based from time to time since the method was first proposed. In all cases the estimation of the rate of production of lead depends essentially on the quantitative relations between the ultimate products of disintegration. For the uranium and thorium series respectively these may be expressed by the following atomic transformation schemes:

$$\underset{238}{U} \longrightarrow \underset{32}{8\ He} + \underset{206}{Pb}\ (RaG)\ ;\ \underset{232}{Th} \longrightarrow \underset{24}{6\ He} + \underset{208}{Pb}\ (ThD).$$

Thus, if the rate of production of helium is known, that of lead can readily be calculated.

(a) In his early work on the association of lead and uranium in rock-minerals, Holmes[5] used Strutt's direct determinations of the rate of production of helium from uranium and thorium in minerals.[6] The value of C so obtained was 8,200. As the part played by thorium as a lead-producer was then unknown, no value of k was considered.

(b) In their 1914 paper Holmes and Lawson[7] used slightly different data, and arrived at the values $C = 8,230$, and $k = 0.4$. The data for uranium were taken from *Le Radium* (11, 4, 1914), whereas those for thorium were calculated from the results of Rutherford and Geiger's work on the counting of the α-particles emitted by thorium.

(c) Applying Rutherford and Geiger's counts of the α-particles emitted by uranium as well as by thorium,[8] Lawson in 1917 derived the revised values[9] $C = 7,900$ and $k = 0.384$.

. (d) In 1920 Holmes gave a summary of the subject, but he dealt only with dominantly uranium minerals.[10] In this paper, the value of C was taken as 7,500, this result having been derived from the work of Hess and Lawson referred to in Section III, and by the use of the value $Ra/U = 3.33 \times 10^{-7}$ for the ratio of radium to uranium in old uranium minerals.

(e) Harold Jeffreys deals with the age problem in his book *The Earth* (Cambridge, 1924). He refers to a table of results in Holmes' paper mentioned in the preceding paragraph, and states: "In his (i.e. Holmes') table the value of $1/\kappa$ has been taken as 7.5×10^9 years, whereas the revised value of 6.6×10^9 years obtained by Lawson and Hess has been adopted here." (Jeffreys uses κ for the disintegration constant of uranium.) In Holmes' paper the value 7.5×10^9 is in reality $C \times 10^6$ and not $1/\kappa$, whereas the figure 6.6×10^9 adopted by Jeffreys is $1/\kappa$ and not $C \times 10^6$. Thus what Jeffreys uses as though it were C is merely the period of average life of uranium (divided by a million), calculated by him from the rate of emission of α-particles from radium, as measured by Lawson and Hess. There is, moreover, undue approximation in Jeffreys' value, for the period of average life of uranium on the data adopted should be 6,370 million years.   Thus, instead of $C = 6,600$, the value should have been $C = 6,370 \times 238/206 = 7,370$, which allows for the fact that 238 parts of uranium can produce only 206 parts of lead (RaG).

(f)   H. V. Ellsworth, in a recent paper,[11] makes use of a more exact formula for calculating the age of a mineral, but the values of C and k which he adopts are taken from Lawson's 1917 paper mentioned above under (c).   He also gives an alternative value of $C = 8,300$, using different data.

(g)   These same values were also adopted by C. W. Davis[12] in a recent paper.

(h)   In a German pamphlet on radioactivity and geology[13] Hahn also adopts Lawson's value of k, but derives C from the value $\lambda = 1.4 \times 10^{-10}$ per year for the disintegration constant of uranium.   This value of λ we believe to be about ten per cent too small.   Hahn's value $C = 8,200$ is thought for the same reasons to be too high.

(i)   In a paper read at the 24th Annual Meeting of the Lake Superior Mining Institute, August, 1925, Professor A. C. Lane stated that a recent re-determination of k gave the value 0.357 instead of 0.38.   No reference, however, is given.

(j)   Holmes has recently returned to the subject in a critical paper on "Estimates of Geological Time with Special Reference to Thorium Minerals and Uranium Haloes."[14]   He adopted Jeffreys' value of C, not realizing that it was as we have seen, erroneous.   For k he used the value 0.37 derived from approximate values of the disintegration constants of uranium and thorium, and he stated that the possible range of k was from 0.35 to 0.38.

(k)   Finally, L. A. Cotton, dealing with certain Australian minerals,[15] used Lawson's value of $k = 0.384$, and the approximated round-figure value $C = 8,000$.

Whereas the different values that have been used for k range only between 0.36 and 0.4, the values adopted for C have fluctuated between 6,600 and 8,300.   Admitting that the 6,600 should have been 7,370, the range is still wide, and the necessity of a reconsideration in the light of the most reliable data is obvious.

II.   LIMITING VALUES OF THE FACTORS
C AND k.

The above survey shows that three different, but not altogether independent, methods of arriving at the values of C and k have been used.

(a)   We may utilize the direct observations that have been made (as, for example, in the pioneer work of Strutt) of the growth of helium from uranium and thorium.   This method is likely to lead to rather larger values for C, owing to the difficulty—if not impossibility—of extracting from large quantities of material all the helium generated therein.   Moreover, unless the volume of helium extracted from a uranium preparation is of the same order of magnitude as that extracted from a thorium preparation of similar bulk, the respective results are unlikely to be in error by the same amount.   An accurate determination of k by this method is thus improbable, and consideration of Strutt's experiments suggests that the derived value of k is likely to be too small.   The values actually calculated from Strutt's results are

$$C = 8{,}200 \qquad \text{and} \qquad k = 0.30,$$

C being probably too high, and k too low.

(b)   Less directly, the disintegration constants of uranium and thorium may be used to give C and k.   The value for uranium (the half-value period T_U) is known with reasonable accuracy to lie within the limits 4,420 and 4,900 million years. For thorium, unfortunately, the half-value period, T_U, is still very uncertain, the current estimates ranging from 13,000 to 22,000 million years.

The above extreme values for uranium give $C = 7{,}370$ and 8,160, which can be regarded as the limits for this factor.   For reasons to be given later, we believe the smaller value to be the more reliable.

The limiting values for k work out as follows, for the possible combinations of data:

For $T_U = 4{,}900 \times 10^6$, and $\begin{cases} T_{Th} = 1.3 \times 10^{10}; & k = 0.39. \\ T_{Th} = 2.2 \times 10^{10}; & k = 0.23. \end{cases}$

For $T_U = 4{,}420 \times 10^6$, and $\begin{cases} T_{Th} = 1.3 \times 10^{10}; & k = 0.35. \\ T_{Th} = 2.2 \times 10^{10}; & k = 0.21. \end{cases}$

The extreme values for k are thus 0.21 and 0.39, and of these the larger, as we shall see, is likely to be the more accurate.

AM. JOUR. SCI.—FIFTH SERIES, VOL. XIII, No. 76.—APRIL, 1927.
23

(c)   The third method of evaluating the factors is based on experiments involving the counting of the α-particles emitted by uranium and thorium, or by known quantities of their daughter elements. The results obtained differ inappreciably from those of method (b), for the results of counting experiments are used in deriving the disintegration constants of uranium and thorium.

III.   EVALUATION OF THE FACTORS
C AND k.

(A)   *Evaluation of* C.   Rutherford and Geiger have determined[8] by the scintillation method the number of α-particles emitted per second from 1 gm. of uranium and thorium respectively.   Their counts correspond to the annual production from 1 gm. of uranium (in equilibrium with its disintegration products) of 11.0×10^{-8} cc. of helium; and from 1 gm. of thorium (similarly in equilibrium) of 3.14×10^{-8} cc. of helium.   These values imply an annual production of 1.26×10^{-10} gm. of uranium-lead (RaG) and of 0.485×10^{-10} gm. of thorium-lead (ThD).   From the first of these two values we find $C = 7,930$, and this was how the value used by Lawson in 1917, referred to above in I (c), was obtained.

Rutherford and Geiger took account of the inefficiency of their scintillation screens in the usual way, by counting the number of scintillations observed when a known number of α-particles from RaC were incident on the screen.   Nevertheless in the light of later experience it is probable that a still heavier correction should have been applied to the relatively feeble scintillations from uranium and, if this criticism be justified, the true value of C should be less than 7,930.   Moreover, in calibrating their screens, these authors made use of the value 3.4×10^{10} for the number of α-particles emitted per second by 1 gm. of radium or the equivalent amount of RaC, and we shall see in what follows that this would involve an error in the same sense.

In their determination of the number of α-particles emitted per second by radium, Rutherford and Geiger[16] used the electrical method of counting.   For 1 gm. of radium they obtained the above value 3.4×10^{10}, which was later[17] corrected to 3.57×10^{10} to take account of an error in the value of their radium standard preparation.   Since 1 gm. of uranium corresponds to an equilibrium amount of 3.40×10^{-7}

gm. of radium, we can combine the two results and find $C = 7{,}690$.

At a later date Hess and Lawson made a careful re-determination of the rate of emission of α-particles by radium, using the electrical method of counting.[18] They obtained the revised value 3.72×10^{10} per gm. of radium per second, and this result, combined with the ratio $Ra/U = 3.40 \times 10^{-7}$, yields the value of $C = 7{,}370$.

In 1924 Geiger and Werner[19] obtained the smaller value of 3.40×10^{10} α-particles per second per 1 gm. radium. They used the scintillation method of counting, and their result has been criticized by Hess and Lawson,[20] who gave strong reasons indicating that it should be several per cent higher. Geiger and Werner's experiments yield for C the value 8,080, which, according to Hess and Lawson, should be reduced by several per cent.

From the above discussion it appears most probable that the true value for C is not far from that calculated from the results obtained by Hess and Lawson. Moreover, the evidence of the heat production by radium alone, and by radium together with its short-lived products, lends strong support to these results.[21] Using the result of Hess and Lawson (3.72×10^{10}), the calculated heat production due to the α- and γ-rays from 1 gm. of radium without disintegration products amounts to 25.47 cals./hour, whereas the experimental values are 25.1 (Rutherford and Robinson) and 25.2 (Hess). In the latter determinations the γ-rays were not entirely absorbed, so that these results should be somewhat larger. Equally good agreement is obtained when we consider the heat production of radium together with its short-lived products. The calculated value for 1 gm. of radium works out to be 137.7 cals./hour, assuming Hess and Lawson's result and the recent estimates of Ellis and Wooster and of R. W. Gurney for the respective contributions of the γ- and the β-rays. By Rutherford and Robinson this heat production was found experimentally to be 135 cals./hour, and by St. Meyer and Hess to be 137 cals./hour, both results being a shade too low owing to the fact that in all probability an underestimate was made of the effect of the γ-rays, which were not entirely absorbed in the experiments. This striking concordance of evidence provides an independent reason for adopting the results of Hess and Lawson for our present purpose in preference to those of Rutherford and Geiger or of Geiger and Werner.

At the present time the most reliable value of the factor C is therefore 7,370; but as the third figure is somewhat uncertain, we advocate the round-figure value of

$$C = 7,400.$$

(B) *Evaluation of k.* Since the lower estimate of the half-value period of uranium given in Section II is that obtained from Hess and Lawson's counting experiments, it is (for the reasons just given) regarded as the most reliable at present obtainable. The lower figure for the half-value period of thorium ($T_{Th} = 1.3 \times 10^{10}$ years) is now gaining wide acceptance,[22] and if we adopted this value, it would seem that the value $k = 0.35$ cannot be far removed from the truth.

Probably the most reliable value of k at present attainable is still that which was deduced by Lawson in 1917. As already stated, Rutherford and Geiger's scintillation experiments on uranium and thorium indicate that per annum

1 gm. uranium gives 1.26×10^{-10} gm. uranium-lead, and

1 gm. thorium gives 0.485×10^{-10} gm. thorium-lead;

whence it follows that 1 gm. of thorium produces the same amount of lead as 0.384 gm. of uranium; i.e. $k = 0.38$, since the third figure is uncertain. Now although the individual results for uranium and thorium are probably both slightly low, both counts were made by the same method, and the results are therefore strictly comparable. A slight correction to each of the results makes no appreciable difference to the ratio between them, as in each case it is in the same direction and of the same relative order.

It appears then, that the most trustworthy value of k obtainable with the data now available is still

$$k = 0.38.$$

(C) *The Most Probable Formulae.* We therefore conclude that the most probable expression for the age of a mineral (omitting the time-correction) is, on present data

$$\text{Approximate Age} = \frac{Pb}{U + 0.38\ Th} \times 7,400 \text{ million years.}$$

For the purpose of completeness we also give the most probable expression for the age of a mineral by the helium method, on the same data. As before, U and Th are the percentage contents of the mineral in uranium and thorium, but He is the volume in cc. of helium at N. P. T. in 100 gms. of the mineral.

$$\text{Approximate Age} = \frac{\text{He}}{\text{U} + 0.29\,\text{Th}} \times 8.5 \text{ million years.}$$

To transform the volume of helium in this expression into its percentage by weight (or *vice versa*) is easily accomplished from the relation

1 cc. of helium has a mass of 0.0001788 gm., or,

1 gm. of helium has a volume of 5,590 cc.

It remains to add that it is never possible to measure He, but only part of He, since helium escapes from minerals as soon as they are exposed to the air, and particularly during the grinding preparatory to analysis. This method can therefore never give more than a minimum result, usually, unfortunately, a very low minimum.

(D) *Effect of the Actinium Series.* In the preceding discussion we have purposely refrained from a consideration of the actinium series. There have been various suggestions of a hypothetical isotope of uranium as the parent of the actinium series, which, on this view, would be distinct from the main uranium-radium series. Recent observations of pleochroic halos would seem rather to favor this view,[23] but as they appear to involve inconstancy of the Ac/U ratio in minerals, as well as atomic weights of the end-product (lead) in conflict with observation, it seems certain that confirmatory evidence of an independent nature will have to be forthcoming before such an interpretation of the so-called X- or Z-halos will gain general acceptance. Apart from this isolated case, however, no experimental evidence has been obtained which definitely supports the theoretical suggestion of an independent Ac-series, and the view most favored by radioactivists is that the actinium series is a branch series (about 3%) of the uranium series, the branching probably occurring at U_{II}. Moreover, the end-product of the actinium series is probably a stable isotope of lead, and the evidence of the atomic weights of lead derived from uranium minerals indicates that actinium-lead probably has the same atomic weight as uranium-lead, and almost certainly is not more than 207. Results obtained from the formulae already given cannot, then, be affected by neglect of the possible complication due to actinium by more than quite a small percentage, and we are therefore justified at present in ignoring actinium in age determinations by the radioactive methods.

IV.   THE TIME-AVERAGE CORRECTION.

So far it has not been taken into account that the amounts of uranium and thorium now found in a mineral are necessarily less than those which were present when the mineral originally crystallized.   In the above formula for the age of a mineral, U and Th should ideally represent the time-average of the percentages present during the life of the mineral.   By definition, the values of C and k in the formula do not vary with time; but the fact that uranium and thorium do not disintegrate at the same rate can be conveniently expressed by assuming a virtual variation of k with time, to take account of the different relative quantities of uranium and thorium originally present in the mineral.   Since uranium disintegrates about three times as rapidly as thorium, the effect is as though k were smaller in the early history of the mineral.   It can be easily shown that for a mineral of age 2,000 million years, the average virtual value of k during the life of the mineral would be 0.35, instead of the value 0.38 found in Section III.   For a mineral of this age, and containing 60% uranium and 20% thorium, however, the effect of this virtual difference in the value of k is only one per cent, so that we are clearly justified in neglecting it in the case of minerals suitable for age determinations.

The first attempt to obtain a formula embodying the time-average correction for minerals containing both uranium and thorium was made in 1914 by Holmes and Lawson[1] (l.c. p. 826), but in its application it is not so convenient as those about to be given.

In their book "Radioaktivität" (p. 447) St. Meyer and Schweidler give a formula involving the time-average correction for uranium minerals.   As applied to minerals containing both thorium and uranium, this formula can be written

(A) $$\text{Age} = \frac{\text{Pb}}{\text{U} + 0.38\,\text{Th}} \times 7{,}400 \times \{1 - \tfrac{1}{2}\,\lambda\,t\} \text{ million years.}$$

Here λ is the disintegration constant of uranium $= 1.57 \times 10^{-10}$ year^{-1}, and t is the age of the mineral in years.   For t it is sufficient for a first approximation to substitute the age as obtained from the approximate formula given in Section III.   The corrected age obtained in this way is slightly low, but a second application of the formula using this second value for t gives a result imperceptibly different from the true age.

Ellsworth (l. c.) has used a different formula to find the correct age. The simple formula for the disintegration of uranium, viz. $U_t = U_0 . e^{-\lambda t}$, can be expressed in the form

$$t = \frac{\log. U_0 - \log. U_t}{0.434\,\lambda}; \text{ where } \lambda = 1.57 . 10^{-10}.$$

This formula gives the age of a mineral containing both thorium and uranium when applied in the following way. U_0 is the original amount of uranium in the mineral, and when thorium is present we have

$$U_0 = \text{present U} + \text{the U-equivalent of the Th} + 1.155$$
$$\text{times the Pb}$$
$$= (U + 0.38\,\text{Th} + 1.155\,\text{Pb});$$
$$\text{and } U_t = (U + 0.38\,\text{Th}).$$

The expression for the age of the mineral thus becomes

(B) $$\text{Age} = \frac{\log. (U + 0.38\,\text{Th} + 1.155\,\text{Pb}) - \log. (U + 0.38\,\text{Th})}{6.82 \times 10^{-5}} \text{ million years.}$$

The factor $1.155 = (206 + 32)/206$ embodies both the helium and lead formed during the life of the mineral, for we know the relative masses of these elements formed.

By a slight readjustment of Ellsworth's formula we can derive a third expression for the age. Clearly

$$t = \frac{1}{\lambda} \log_e \frac{U_0}{U_t} = \frac{1}{\lambda} \log_e \left\{ \frac{U + 0.38\,\text{Th} + 1.155\,\text{Pb}}{U + 0.38\,\text{Th}} \right\};$$

or,

$$\text{Age (in years)} = 6.37 \times 10^9 \times \log_e (1 + x)$$

$$= 6.37 \times 10^9 \left(x - \frac{x^2}{2} + \frac{x^3}{3} \dots \right).$$

In this expression we have $x = 1.155 \dfrac{\text{Pb}}{U + 0.38\,\text{Th}}$, and it

is necessary to take only the first three terms inside the bracket. Putting x outside the bracket, and substituting the value of x just given, we obtain our final expression for the age (in millions of years) :

(C) $$\text{Age} = \left(\frac{\text{Pb}}{U + 0.38\,\text{Th}} . 7,400 \right) \times \left(1 - \frac{x}{2} + \frac{x^2}{3} \right),$$

$$= \text{Approximate Age } \left(1 - \frac{x}{2} + \frac{x^2}{3} \right).$$

Of these three expressions for the age, all of which embody the time-average correction, we shall in what follows adopt the expression (C), because it is simpler in its application than are the others. As an illustration of the application of these formulae, however, we shall now apply them in turn to the Villeneuve (Quebec) uraninite analyzed by Hillebrand (U. S. Geol. Survey Bull. 90, p. 23, 1892). For this mineral, we have Pb = 10.46; U = 64.74; Th = 5.63;

$$\text{Lead Ratio} = \frac{\text{Pb}}{\text{U} + 0.38 \text{ Th}} = 0.156.$$

Whence,

Approximate Age = 1,154 millions of years.

The values of the corrected age as obtained by the three formulae derived above work out as follows:

Formula (A):

Corrected Age = 1,154 x 0.909 = 1,050 million years.

If, as described above, we make a second application of formula (A) using this corrected value of *t,* we obtain the value 1,060 million years.

Formula (B):

$$\text{Corrected Age} = 14,700 \times \log. \frac{78.96}{66.88} = 0.0721 \times 14,700$$
$$= 1,059 \text{ million years.}$$

Formula (C):

Corrected Age = 1,154 x 0.921 = 1,063 million years.

As was to be expected, the values of the corrected age obtained by use of the three formulae given are practically identical. Moreover, from what has been previously stated they can be applied without error even when an appreciable amount of thorium is present in the mineral, for the virtual variation of *k* is of little importance unless we are dealing with very old minerals containing a large percentage of thorium. In general, however, as Holmes has pointed out, it is not advisable to use essentially thorium minerals for age determinations, owing to the uncertainty caused by the differential leaching of the lead isotopes derived from uranium and thorium respectively. We conclude, then, that the time-average correction can be made most conveniently by use of the formula:

$$\text{Corrected Age} = \text{Approximate Age} \left(1 - \frac{x}{2} + \frac{x^2}{3}\right),$$

$$\text{where } x = 1.155 \frac{Pb}{U + 0.38 \, Th}$$

V. THE GEOLOGICAL TIME SCALE.

Although at present the geological ages of many radioactive minerals are far from satisfactorily known, the lead-ratios of suitable minerals are everywhere in accordance with such evidence as is available. To indicate the calculated ages that result from the use of the formulae and factors adopted in this paper the following table has been compiled from analyses to which references are given. The ages are in units of millions of years, and the geological and atomic weight data are given in notes.

TABLE I.

Reference Nos.	Minerals and Localities	Lead Ratios	Approximate Age	Corrected Age
24.	Brannerite, Idaho.	0.005	37 m.y.	36.9 m.y.

The mineral occurs in a placer deposit derived from the granite at the head of Kelly Gulch. The intrusion is probably part of the great central batholith of Idaho, the age of which is post-Cretaceous and pre-Miocene and probably late Oligocene.

25.	Pitchblende, Gilpin Co., Colorado.	0.008	59 m.y.	58.7 m.y.

The ore-deposits in which the pitchblende occurs are probably of late Cretaceous or early Eocene age.

26.	Pitchblende, Joachimstal.	0.028	207 m.y.	204 m.y.

The geological age is Permo-Carboniferous. On account of the difficulty of correlating Permian rocks it is not practicable at present to attempt a closer identification of the epoch of ore-deposition.

The uncorrected ratio is 0.042. Hönigschmid has determined the atomic weight of the lead isotopes from selected samples of the pitchblende as 206.4. As only ordinary and uranium-lead are present (thorium being negligible), this implies that only two thirds of the lead is uranium-lead, giving a corrected ratio of 0.028.

27.	Uraninites, North Carolina.	0.033	244 m.y.	239 m.y.

The geological age is late Palaeozoic, probably late Carboniferous. Here again ordinary lead is present. The atomic weight of the mixed isotopes is 206.4 (Richards and Lembert). The uncorrected ratio of 0.05 is thus reduced to 0.033, thorium-lead being negligible.

28.   Uraninites,   0.039   289 m.y.   283 m.y.
         Glastonbury, Conn.

The geological age cannot be stated more closely than Devonian to Permian.

29.   Various minerals,   0.04   296 m.y.   290 m.y.
         Brevik, S. Norway. (approx.)

These minerals constitute a very unsatisfactory series, for most of them are rich in thorium, and thorium-lead appears to have been leached out in variable proportions. The geological age is usually considered to be Middle Devonian, but a Lower Carboniferous age is not ruled out.

30.   Uraninites,   0.052   385 m.y.   374 m.y.
         Branchville, Conn.

The geological age is Ordovician to Permian and probably Devonian.

31.   Coracite,   0.062   459 m.y.   443 m.y.
         Lake Superior.

The geological age and even the exact locality are unknown at present. The environment, however, suggests Keweenawan or Ordovician as the only likely possibilities. This ratio will be of interest when some definitely Keweenawan minerals have been analyzed.

32.   Pitchblendes,   0.081   599 m.y.   573 m.y.
         Katanga, Central Africa.

The granites with which the ore-deposits are associated are intrusive into rocks that are older than the oldest fossiliferous series; that is, they are pre-Devonian. From the evidence of a doubtful fossil in Angola they have been considered Ordovician, but some geologists, including J. W. Gregory, think they are late pre-Cambrian like the Torridonian of Scotland. The Katanga ores are free from thorium, and the lead gives an atomic weight of 206.048 (Hönigschmid & Birckenbach). Richards and Putzeys, however, found 206.20, which would reduce the ratio to 0.068. From the point of view of the age question the former value is probably the more important, for Richards and Putzeys state that their lead was derived from a mixture of curite, kasolite, dewindtite, and stasite, all alteration products.

33.   Thorianites,   0.083   614 m.y.   587 m.y.
         Ceylon.

From a comparison of the Ceylon sequence with that of India and that of East Africa, a late pre-Cambrian age is indicated, though the direct geological evidence does not preclude an early Palaeozoic age.

The lead present is a mixture of uranium- and thorium-lead, and several atomic weight determinations have been made by Hönigschmid. The ratio given is based on these as indicated in the paper referred to (33).

61

34.   Uraninites,   0.091   673 m.y.   640 m.y.
Morogoro, Tanganyika.

The geological age is probably older than that of the Katanga pitchblende by about the length of a geological system, and should thus fall somewhere in the Upper pre-Cambrian.

The lead is nearly pure uranium-lead, its atomic weight being 206.046 (Hönigschmid & St. Horovitz).

35.   Bröggerites,   0.13   962 m.y.   898 m.y.
Moss, Norway.

This ratio is the average of a long series ranging from 0.123 to 0.136, and it is not impossible that two successive periods of granitic or pegmatitic intrusion may be implied. All the pegmatites, however, belong to near the close of the Middle pre-Cambrian, being definitely post-Kalevian.

The atomic weight of the lead is 206.06 (Hönigschmid & St. Horovitz).

36.   Cleveites, etc.,   0.155   1,147 m.y.   1,056 m.y.
Arendal, Norway.

This ratio is also the average of a long series, here ranging from 0.145 to 0.165. Geologically the pegmatites are known to be Middle pre-Cambrian; there is no evidence (apart from the ratios) to show whether they belong to the post-Kalevian or the post-Bottnian intrusions. The ratios suggest the latter.

37.   Radioactive Ore,   0.144   1,065 m.y.   987 m.y.
Olary, S. Australia. (approx.)

The ore is of early Proterozoic, that is, of Middle pre-Cambrian age. The atomic weight of lead from associated carnotite was determined by Richards as 206.34. This indicates that only part of the lead is uranium-lead (thorium being negligible), and the uncorrected ratio 0.195 is thus reduced to the more probable value 0.144, which may, however, be too low.

38.   Uraninites,   0.15   1,100 m.y.   1,024 m.y.
Ontario, Canada.

The age is in the Middle pre-Cambrian, and the results have been fully discussed in the important paper by Ellsworth referred to (38).

39.   Uraninites,   0.16   1,184 m.y.   1,087 m.y.
Llano Co., Texas.

The geological age is in the Middle pre-Cambrian. For an extended discussion, removing former misconceptions as to these minerals, see the paper by Barrell referred to (39).

40.   Samarskites,   0.16   1,184 m.y.   1,087 m.y.
Douglas Co., Col.

The geological age is the same as that of the Llano Co. uraninites. As the minerals are as rich in thorium as in uranium, the highest of three ratios (0.14, 0.15 and 0.16) is given, on the grounds that thorium minerals have commonly lost part of their lead, and so tend to give minimum results.

41. Mackintoshite, etc.   0.187   1,384 m.y.   1,257 m.y.
     Wodgina, W. Australia.

The pegmatites in which these minerals occur are thought by Sir Edgeworth David to belong to the Warrawoona Group, probably the oldest of the three main divisions of the Australian pre-Cambrian rocks. They may thus represent the close of the Lower pre-Cambrian. As the minerals are rich in thorium, the ratio is likely to be too low rather than too high, if the usual rule be followed. An atomic weight determination would be of special interest in this case.

42.

Zircons from the oldest gneissose granites of Mozambique have given a still higher ratio, but in view of the small quantities of lead involved the result cannot at present be considered to do more than indicate the pre-Cambrian age of the associated rocks.

Summing up the above results, it is clear that we can now tentatively construct a geological time-scale of the following kind, based on the lead-ratios Pb/(U + 0.38 Th).

TABLE II.

Geological Eras	Lead-Ratios.	Ages in millions of years	
		Approximate	Corrected
Cainozoic {	0.000	..	..
	0.005	37	36.9
	0.008	59	58.7
Mesozoic {	0.01	74	73.5
	0.02	148	146
	0.03	222	218
	0.04	296	289
	0.05	370	360
Palaeozoic {	0.06	444	430
	0.07	518	498
	0.08	592	567
Upper	0.09	666	635
Pre-Cambrian {	0.10	740	700
	0.11	814	767
	0.12	888	831
	0.13	962	897
Middle	0.14	1036	961
Pre-Cambrian {	0.15	1110	1026
	0.16	1184	1089
	0.17	1258	1150
	0.18	1332	1212
Lower	0.19	1406	1273
Pre-Cambrian	0.20	1480	1336

It should be noticed, as Ellsworth has pointed out, that, apart from the question of geological time and the age of the earth, the chief value of the results for geological purposes lies in the ratios themselves. These clearly express comparative ages on which correlations can be based. One of the greatest developments of the method in the future will be the correlation of the Pre-Cambrian rocks in all parts of the world where suitable minerals can be found for the purpose.

The absolute ages given in the above list are based on the factor $C = 7,400$. The last column gives the ages corrected for the time-average values of equivalent uranium and serves to show that it is only for the older minerals that the correction becomes notably greater than possible errors due to other causes.

VI.   CONCLUSIONS.

(1)   The age of a suitable radioactive mineral is given to a first approximation in millions of years by the formula

$$\text{Approximate Age} = \frac{\text{Pb}}{\text{U} + k.\text{Th}} \cdot C.$$

(2)   A review of all the evidence shows that on present data the best values for the factors are

$$k = 0.38; \text{ and } C = 7,400;$$

and it is suggested that until unequivocably better data are available, all workers in this field should agree to use these values, in order that the ages of minerals as calculated by different authors may be directly comparable.

(3)   The correction for the wearing out of uranium and thorium during the life-time of a mineral is considered, and a simplified formula is derived which gives satisfactorily the corrected age in terms of the approximate age (see (1)). Thus

$$\text{Corrected Age} = \text{Approximate Age} \left(1 - \frac{x}{2} + \frac{x^2}{3} \right);$$

$$\text{where } x = 1.155 \cdot \frac{\text{Pb}}{\text{U} + 0.38 \text{ Th}}.$$

(4)   The ages of various minerals and suites of minerals, ranging from the Tertiary to the Pre-Cambrian, are given as calculated from the factors here advocated, both by the approximate formula and by the time-corrected formula. The results are concordant among themselves and justify the construction of a tentative geological time-scale as a basis for future discussion and correlative purposes.

DURHAM, ENGLAND,
AND
SHEFFIELD, ENGLAND.

REFERENCES.

1. Holmes, A., Phil. Mag., **1**, 1055, 1926.
2. Boltwood, B. B., this Journal, **20**, 253, 1905; **23**, 77, 1907.
3. Holmes, A., Proc. Geol. Assoc., **26**, 289, 1915; Bull. Geol. Soc. Am., **28**, 845, 1917.
4. Holmes, A., and Lawson, R. W., Phil. Mag., **28**, 823, 1914; **29**, 673, 1915.
5. Holmes, A., Proc. Roy. Soc., (A), **85**, 248, 1911; "The Age of the Earth," (Harper's Library), 139, 1913.
6. Strutt, R. J., Proc. Roy. Soc., (A), **84**, 379, 1910.
7. Holmes, A., and Lawson, R. W., Phil. Mag., **28**, 823, 1914.
8. Rutherford, E., and Geiger, H., Phil. Mag., **20**, 691, 1910.
9. Lawson, R. W., Die Naturwissenschaften, **5**, 429, 452, 610, 705, 1917; Wien. Ber., **126**, 721, 1917.
10. Holmes, A., Discovery, **1**, 118, 1920.
11. Ellsworth, H. V., this Journal, **9**, 127, 1925.
12. Davis, C. W., this Journal, **11**, 201, 1926.
13. Hahn, O., Was lehrt uns die Radioaktivität über die Geschichte der Erde? (Springer, Berlin), 1926.
14. Holmes, A., Phil. Mag., **1**, 1055, 1926.
15. Cotton, L. A., this Journal, **12**, 41, 1926.
16. Rutherford, E., and Geiger, H., Proc. Roy. Soc., (A), **81**, 141, 162, 1908.
17. Rutherford, E., Phil. Mag., **28**, 320, 1914.
18. Hess, V. F., and Lawson, R. W., Wien Ber., **127**, 405, 1918.
19. Geiger, H., and Werner, A., Zs. f. Phys., **21**, 187, 1924; Verh. d. phys. Ges. Berlin, **5**, 12, 1924.
20. Hess, V. F., and Lawson, R. W., Phil. Mag., **48**, 200, 1924.
21. Lawson, R. W., Nature, **116**, 897, 1925.
22. See Kovarik, A. F., and McKeehan, L. W., Radioactivity, **22**, 1925, for a critical summary of the evidence.
23. Joly, J., J. Chem. Soc., **125**, 897, 1924; Wilkins, T. R., Nature, **117**, 719, 1926; Iimori, S., and Yoshimura, J., Sci. Papers of the Inst. of Phys. & Chem. Research, Tokyo, No. 66, **5**, 11, 1926.
24. Wells, R. C., J. Frankl. Inst., **189**, 230, 1920; see also, Hess, F. L., and Wells, R. C., 16th Ann. Rep. Mine Insp. Idaho, Boise, 1915, p. 29.
25. Econ. Geol., **11**, 266, 1916.
26. Kirsch, G., Wien. Ber., **131**, 551, 1922.
27. Hillebrande, U. S. G. S. Bull., **78**, 65, 1891; Boltwood, B. B., this Journal, **23**, 77, 1907.
28. Hillebrande, loc. cit.
29. Holmes, A., and Lawson, R. W., Phil. Mag., **28**, 832, 1914.
30. Hillebrande, loc. cit.
31. Walker, T. L., Univ. Toronto Studies, Geol. Series, No. 17, 1924.
32. Davis, C. W., this Journal, **11**, 209, 1926; see also Schoep, A., Bull. Soc. Geol. Belgique, **33**, 172, 1924.
33. Holmes, A., Phil. Mag., **1**, 1061, 1926.
34. Hönigschmid, O., and St. Horovitz, Wien. Ber., **123**, 2407, 1914.
35. Hillebrande, loc. cit.; Gleditsch, E., Norsk. Videns. Akad., Oslo, I Mat. natur. Kl., No. 3, 1925; Hönigschmid, O., and St. Horovitz, loc. cit.
36. Holmes, A., Phil. Mag., **1**, 1064, 1926; Gleditsch, E., op. cit., p. 34.
37. Cotton, L. A., this Journal, **12**, 42, 1926.
38. Ellsworth, H. V., this Journal, **9**, 143, 1925.
39. Barrell, J., Bull. Geol. Soc. Am., **28**, 745, 1917.
40. Dana's Mineralogy, 7th Edn., 740, 1903.
41. Cotton, loc. cit.
42. Holmes, A., Q. J. G. S., **74**, 86, 1918.

The Helium Method

III

Editor's Comments on Papers 6, 7, and 8

The limitations posed by the necessity for chemical separation and gravimetric determination of radiogenic lead restricted the application of the uranium–lead method to rare radioactive minerals containing appreciable amounts of uranium and thorium. The common rocks and their constituent minerals contain so little uranium and thorium that chemical determination of the lead produced in situ by radioactive decay was impossible. The problem was compounded by the presence of both uranium- and thorium-derived lead, which could not be separated chemically, and also by the existence of appreciable amounts of nonradiogenic lead mixed with the radiogenic. Where relatively pure uranium-bearing minerals could be found (i.e., with very low thorium concentrations), the presence of thorium-derived lead could be ignored, and a correction could be applied for nonradiogenic lead based on an accurate atomic weight determination (see p. 48). But all this only served to compound the already complex analytical procedures and further restricted application of the chemical-lead method.

Applied to highly radioactive minerals in the early years of radiometric geochronology, the helium method was found to be unreliable because of a tendency of radiogenic helium to escape from such minerals, either during the natural lifetime of the material or during the preparation of the sample for analysis. The problem was that literally too much helium was being generated in minerals containing high concentrations of the radioactive parents, uranium and thorium. For every atom of radiogenic lead generated by the parent uranium or thorium, between six to eight atoms of helium could be produced, representing the accumulation of all the alpha particles emitted during the radioactive series decay. Considerable concentrations of helium would be generated over geologically long periods of time and it was small wonder that not all of it could be retained within the crystal lattice of the radioactive parent mineral.

In common rocks and minerals, on the other hand, the concentration of uranium and thorium was so low that accumulations of radiogenic helium were much more likely to be detected than accumulations of radiogenic lead, and as the volumes of helium produced would be small (10^{-4} to 10^{-6} cm^3 g^{-1}), there was a good chance of their being quantitatively retained. Another consideration favoring helium was the observation that it would be extremely unlikely for any significant amount of this gas to be retained in any magma, so there would be no need to worry about nonradiogenic contamination such as existed in the case of uranium–lead age determinations. The atmosphere itself contains only a negligible proportion of helium (5.2 ppm by volume) and by 1928 gas-handling and vacuum techniques were quite adequate for the successful separation and volumetric determination of small quantities of radiogenic helium. Uranium and thorium were determined indirectly from their emanation products, radon (Rn-222) and thoron (Rn-220), found in equilibrium with chemical separations of radium (Ra-226) and thorium-X (Ra-224). Knowing the concentrations of radium and thorium-X, concentrations of the parent uranium and thorium could be calculated assuming radioactive equilibrium had been maintained in the analyzed rock or mineral sample (Urry, 1936 a, b).

Progress in the field of helium analysis was largely due to Professor Fritz H. Paneth and his co-workers in Berlin,* who were successful in detecting quantities of helium as small as 10^{-10} cm^3, and devised methods for quantitatively measuring helium in quantities as small as 10^{-8} cm^3 (see Urry, 1933; Holmes and Paneth, 1936, for a review of these methods). Much of Paneth's work in Berlin was directed towards determining the helium ages of iron meteorites (Paneth et al., 1930), but his methods could also be used to determine the ages of terrestrial rocks.

Seizing the initiative, Arthur Holmes arranged for V. S. Dubey to undertake helium age determinations on material from two dolerite specimens collected from the Whin Sill and the Cleveland Dyke in North England, which had been analyzed chemically in the course of earlier petrologic investigations. Dubey was to analyze the samples for radium and thorium by the emanation method in Professor H. Mache's laboratory at the Radium Institute in Vienna, while the helium determinations were to be done by Dr. Günther in Paneth's laboratory in Berlin.

The resulting ages, reported on May 25, 1929 (Paper 6), were in excellent agreement with the stratigraphic evidence for the time of intrusion of the respective rock units, and Dubey and Holmes expressed their excitement about the possibilities of being able to date any fresh igneous rock with the helium method.† Holmes' dream of constructing a geologic time-scale based on radiometric ages determined on common rocks of known stratigraphic position seemed for the first time to be within the bounds of possibility. Because of their wide distribution is space and time, a time scale

---

*See *Nature* **123**, 879–881 (1929) for an early editorial report on Paneth's work.

†Dubey (1930) subsequently completed helium age determinations on some basic rocks from India.

based on igneous rocks dated by the helium method would be much more detailed than any provided by the chemical lead method, which was of necessity restricted to rare radioactive minerals.

William D. Urry, who had worked with Paneth in Berlin, arrived at MIT in 1932 and began a comprehensive series of helium age determinations designed to make the dream come true. The results of Urry's research were published in two papers by the Geological Society of America between 1935 and 1936. The second of these, published on August 31, 1936, reports the results obtained on thirty-nine samples of mostly basaltic rocks, carefully selected from all over the U.S., Canada and Europe on the basis of their known stratigraphic positions, which ranged from Cambrian to Upper Tertiary (Paper 7). A time scale based on the helium ages of these samples was found to be in excellent agreement with the available lead age determinations.

Holmes was delighted, and published a combined time scale including both the helium and reliable lead age determinations the following year in the revised version of his book on the age of the earth (Holmes, 1937). Commenting on the remarkably good agreement between the helium and lead age determinations relative to the stratigraphic position of the analyzed samples, Holmes noted, "this test of internal consistency . . . must be regarded as the final proof that the ages calculated from lead- and helium-ratios are at least of the right order and that no serious error is anywhere involved" (1937, p. 179).

Unfortunately, even as Holmes was writing these words, a cooperative investigation was under way at MIT which was to reveal a faulty calibration of the ionization chamber used by Urry for the measurement of radon in conjunction with his helium age determinations, which caused a drastic reevaluation of the helium data. Urry's calculated helium ages were in fact all too high, the initial agreement with the lead determinations was purely fortuitous, and the optimistic outlook for the helium method was soon to be replaced by new doubts as to the quantitative retention of radiogenic helium in all rocks and minerals.

The problem was first noticed by Robley Evans, who had arrived at MIT as an assistant professor of Physics in 1934, and began there a study of the radioactivity of terrestrial materials and methods of age determination in association with Clark Goodman, Norman Keevil (a Canadian), and Professor A. C. Lane, who for many years had been chairman of the National Research Council's Committee on the Measurement of Geological Time by Atomic Disintegration.* Evans and his collaborators developed a new physical method for determining helium ages, which became known as the alpha–helium method. This involved a determination of the rate of helium production (the "activity ratio") from the observed rate of alpha-particle emission (measured electrically from a finely powdered deposit of the rock sample) and a measurement of the total volume of accumulated helium extracted from a separate

---

*The reports of this committee under Lane's chairmanship were published annually between 1925 and 1946. John P. Marble succeeded Lane as chairman in 1947.

portion of the sample by direct fusion at 2000°C in a special, evacuated, graphite resistance furnace (the graphite resistor doubling as a crucible) which had been developed by Evans at the California Institute of Technology before he moved to MIT.

Both Paneth and Urry had used a carbonate-flux fusion technique for the liberation of helium,* and determined the concentration of uranium and thorium in the residue by measuring the emanation products in equilibrium with their parents. This was the standard radon–thoron–helium method, which had the advantage that helium and its radioactive parents were all determined on the same aliquot of sample. The radon–thoron–helium method was superior where low activities (less than 0.4 alpha emissions h^{-1} mg^{-1}) were encountered, but there were no chemical manipulations involved with the alpha–helium method, which thus avoided both the problem of loss and/or contamination of the active elements and the necessity of calibration with standard radium and thorium solutions. Furthermore, with the alpha–helium method the decay constants for uranium (U-238), thorium (Th-232) and actino-uranium (U-235) were needed only to make a small correction for very old samples.†

When ages obtained by the new alpha–helium method were compared with Urry's published ages obtained on the same samples by the radon–thoron–helium method, the alpha–helium ages, which were consistent among themselves, were found to be much lower. At first the problem was thought to be due to contamination of the thin film of rock powder used for alpha-counting, but numerous blank runs discounted this and the problem was finally traced to the calibration of the radium apparatus used by Urry in the radon–thoron–helium method.

In the determination of radium by the emanation method, an electrometer was used for the ionization measurements, and it was necessary to calibrate this instrument with a solution of known radium content in order to reduce the observations to an absolute basis. Urry (1933; 1936a, b) had used a calibration constant of 9.45 mV per 10^{-13} g radium, but Keevil, working with Urry's electrometer, in 1937 determined a much lower calibration constant of 5.3 mV per 10^{-13} g and later, using more reliable standards, a value of 4.44 ± 0.50 mV per 10^{-13} g was accepted. The ratio between the old and new calibrations was 2.13!

The results of this intercalibration work and a comparison of the two helium age methods was reported in a joint paper involving all the participants (Evans, Goodman, Keevil, Lane, and Urry, 1939) and two years later Goodman and Evans published a general review of all the age determination methods based on radioactive decay. This review included some comments on the possibilities inherent in new methods based on the radioactive decay of rubidium and potassium, but the helium method was discussed in considerable detail (Paper 8). Using the lower calibration constants

---

*Paneth did experiment with a vacuum induction furnace used with and without various fluxes; see Holmes and Paneth (1936).

†A related method is the alpha–lead method developed after World War II by E. S. Larsen, Jr., of the U.S. Geological Survey with Keevil and H. C. Harrison; see p. 163.

in their calculations, Evans and his colleagues found good agreement between redetermined radon–thoron–helium ages and alpha–helium ages determined on the same sample, but Urry's original (1933–36) radon–thoron–helium ages were still discrepantly high even when recomputed using the new factors. The unknown source of error may have been leakage of helium from the hand specimens in the 3–5 years intervening between measurements, but whatever the cause, Goodman and Evans felt it necessary to discard all Urry's helium age determinations and presented a new, revised time scale based on the remaining helium and lead age data (Paper 8, Fig. 10).

The uncertainties inherent in the helium method were further investigated by Patrick Hurley, who had joined the MIT group working on helium age determinations as a research associate in 1938. Hurley began a comprehensive study of the distribution and migration of helium in rocks and minerals, which culminated in a report published in 1954. In this paper he presented a historical review of the helium age method and went on to discuss possible reasons for the low observed helium ratios, including radioactive contamination and helium loss due to radiation damage.

As a method of age determination applicable to a wide range of common rocks and minerals, the helium method has now been largely superseded by the potassium–argon method (see p. 300). However, it continues to receive the attention of geochronologists, and was applied with some success by Hurley and others to the accessory minerals, magnetite, zircon, and sphene. More recently, Damon and Kulp (1957) adopted a stable isotope dilution technique (using He-3) to determine the radiogenic helium content of zircon with a high degree of accuracy, and a uranium–helium method has been applied with some success to young marine carbonate shells (Fanale and Schaeffer, 1965) and fossil bones (Turekian et al., 1970).

Selected Bibliography

Damon, P. E., and Kulp, J. L. (1957). Determination of radiogenic helium in zircon by stable isotope dilution technique. *Trans. Amer. Geophys. Union,* **38,** 945–953.

Dubey, V. S. (1930). Helium ratios of the basic rocks of the Gwalior Series. *Nature,* **126,** 807.

Evans, R. D., and Goodman, C. (1944). Alpha–helium method for determining geological ages. *Phys. Rev.* **65,** 216–227.

Evans, R. D., Goodman, C., Keevil, N. B., Lane, A. C., and Urry, W. D. (1939). Intercalibration and comparison in two laboratories of measurements incident to the determination of the geological ages of rocks. *Phys. Rev.* **55,** 931–946.

Fanale, F. P., and Schaeffer, O. A. (1965). Helium–uranium ratios for Pleistocene and Tertiary fossil aragonites. *Science,* **149,** 312–317.

Holmes, A., and Paneth, F. A. (1936). Helium-ratios of rocks and minerals from the Diamond Pipes of South Africa. *Proc. Roy. Soc., Ser. A,* **154,** 385–413.

Holmes, A. (1937). "The Age of the Earth," 263 pp. Nelson and Sons, London.

Hurley, P. M. (1954). The helium age method and the distribution and migration of helium in rocks. "Nuclear Geology" (H. Faul, ed.), pp. 301–329. John Wiley & Sons, New York.

Lane, A. C., and Urry, W. D. (1935). Ages by the helium method: I. Keweenawan. *Geol. Soc. America Bull.,* **46,** 1101–1120.

Paneth, F. A., and Urry, W. D. (1931). Heliumuntersuchungen VIII. Uber eine Methode zur quantitativen Bestimmung kleinster Heliummengen. *Z. Phys. Chem.*, **152**, 110–149.

Paneth, F. A., Urry, W. D., and Koeck, W. (1930). The age of iron meteorites. *Nature*, **125**, 490–491.

Turekian, K. K., Kharkar, D. P., Funkhouser, J., and Schaeffer, O. A. (1970). An evaluation of the uranium–helium method of dating fossil bones. *Earth Planet. Sci. Lett.*, **7**, 420–424.

Urry, W. D. (1933). Helium and the problem of geological time. Chem. Rev., **13**, 305–343.

Urry, W. D. (1936a). Determination of the thorium content of rocks. *J. Chem. Phys.*, **4**, 34–40.

Urry, W. D. (1936b). Determination of the radium content of rocks. *J. Chem. Phys.*, **4**, 40–48.

6

and therefore quantities of the order found in rocks and ordinary rock-forming minerals are readily determinable.

With these considerations in mind the helium method has been successfully applied to two north of England rocks (the Whin Sill and the Cleveland Dyke) that have recently been under detailed petrological investigation (A. Holmes and H. F. Harwood : *Min. Mag.*, **21**, pp. 493-542 ; 1928 ; and **22**, pp. 1-52 ; 1929). The determinations of radium and thorium were carried out (by V. S. D.) in the laboratories of Prof. H. Mache, at the Radium Institute, Vienna, while those of helium were done in Prof. Paneth's laboratories in Berlin (also by V. S. D.). The materials used for these determinations were in each case parts of the specimens already analysed chemically and mineralogically in the course of the investigation to which reference has been made. The following are the results obtained :

Rocks Investigated.	$Ra \times 10^{12}$ gm./gm.	$U \times 10^{6}$ gm./gm.	$Th \times 10^{6}$ gm./gm.	$He \times 10^{6}$ c.c./gm.
Whin Sill, Scordale Beck, Westmorland (No. 551) .	0·27	0·81	3·0	36·0
Cleveland Dyke, Bolam, Co. Durham (No. 402) .	0·61	1·83	6·1	11·0

The approximate age (omitting a negligible time-correction for the wearing out of uranium and thorium during the life-time of the rock) is given by the formula

$$\frac{He}{U + 0.29\ Th} \times 8.5 \text{ million years,}$$

where U and Th are the percentage contents of the rock in uranium and thorium, and He is the volume in c.c. of helium at N.T.P. in 100 gm. of the mineral (A. Holmes and R. W. Lawson : " Factors involved in the Calculation of the Ages of Radioactive Minerals ", *Amer. Jour. Sci.*, April 1927, pp. 334-5).

From the data of the above table the ages are found to be :

Whin Sill . . . 182 million years.
Cleveland Dyke . . 26 million years.

The Whin Sill was injected into the Carboniferous rocks of the north of England in very late Carboniferous times. The Cleveland Dyke was injected in post-Liassic time, and the recognition of its definite status as an outlying member of the Mull dyke-swarm points more closely to an early or middle Tertiary age. The numerical ages are thus seen to be in excellent agreement with the geological evidence. They also conform quite satisfactorily with the scanty results based on lead-ratios. The latter give 192 million years for the late Carboniferous (Joachimstal pitchblende provisionally corrected by atomic weight evidence for primary lead) ; 36 million years for the (?) late Miocene (brannerite from Idaho, uncorrected) ; and 66 and 52 million years for the late Cretaceous (pitchblendes from Colorado and Wyoming respectively, also uncorrected).

While it is probable that the helium results may be slightly low, it must be remembered that there is no real proof of this, for the lead-ratios cited are themselves not yet so securely founded as one could wish. There may be traces of primary lead in the Tertiary and late Cretaceous pitchblendes of North America, and, if so, the figures given would be too high. The Joachimstal evidence, while generally consistent, suffers from the fact that the specimens analysed were

---

Estimates of the Ages of the Whin Sill and the Cleveland Dyke by the Helium Method.

THE helium method of measuring geological time, originally devised by Lord Rayleigh, has hitherto been applied only to minerals or other materials that were found to be relatively rich in the radioactive elements, uranium and/or thorium. It is already well known that the results obtained are to be regarded as minima, on account of the special tendency of accumulated helium to escape from such specimens during their exposure to the atmosphere and during their preparation for analysis. This tendency is necessarily most marked in old and richly radioactive minerals, like uraninite and thorianite, in which large quantities of helium have been generated (for example, 10 c.c. per gm.). In very feebly radioactive materials, like ordinary igneous rocks, the amount of helium is correspondingly minute (for example, about 10^{-4} to 10^{-5} c.c. per gm.), and its proportion to that of other gases (about 1 c.c. per gm.) is very low. The ordinary gases of an igneous rock are not appreciably extractable by a pump, even when specimens are ground *in vacuo* ; nor, as a rule, do they begin to escape in appreciable quantity as a result of heating, until the temperature exceeds 300° C. It is therefore to be anticipated that the loss of helium from specimens of close-grained igneous rocks awaiting analysis will be much less serious than that from radioactive minerals.

The technique introduced by the late Sir William Ramsay, and developed by Prof. Collie and by Lord Rayleigh, for the determination of minute traces of helium has recently been still further improved by Prof. F. Paneth and, independently, by Dr. R. W. Lawson. It is now possible to measure with a reasonable degree of accuracy the helium accumulated in ordinary igneous rocks, even if their geological ages date from epochs no more remote than those of the Tertiary. In the case of plateau basalts, possibly some 40 million years old, the average radium and thorium content is such that the accumulated helium should be of the order 10^{-5} c.c. per gm. With modern methods amounts down to 10^{-9} c.c. can be estimated,

not the specimens from which lead was separated for atomic weight determination.

Clearly there is a vast field of geological research now open to investigation by the long neglected helium method. If our initial hopes are realised— and these preliminary results provide ample encouragement—a method is now available for dating all fresh igneous rocks which have not been heated up or metamorphosed since they came into place. There should not be the slightest difficulty, for example, in distinguishing Carboniferous dykes and sills from those of Tertiary age. It should be equally easy to settle with certainty the controversy as to whether the Carrock Fell complex belongs to the Ordovician or to some later epoch of igneous activity. There are many such problems awaiting solution in every country where igneous rocks occur. Moreover, since igneous rocks suitable for the helium method are far more abundant and far better distributed in time than are radioactive minerals suitable for the lead method, there is now available a practical means of effecting long-distance correlations and of building up a geological time-scale which, checked by a few reliable lead-ratios here and there, should become far more detailed than could ever be realised by means of the lead method alone.

Further work is in progress on the north of England rocks, and it is our intention as soon as possible to begin the systematic prosecution of this extremely promising line of research. Dr. R. W. Lawson has consented to collaborate in the work by making the helium determinations and by carrying out a quantitative investigation on the possibilities of escape of helium in various circumstances.

V. S. Dubey.
Department of Geology and Mining,
    Gwalior State, India.

Arthur Holmes.
The University, Durham, May 6.

BULLETIN OF THE GEOLOGICAL SOCIETY OF AMERICA

VOL. 47, PP. 1217-1234.   AUGUST 31, 1936

7

AGES BY THE HELIUM METHOD:
II. POST-KEWEENAWAN

BY WILLIAM D. URRY

CONTENTS

INTRODUCTION

In Part I of this series,[1] the validity of the application of the helium method to the determination of geologic time expressed in years was discussed from the analyses of a suite of Keweenawan basalts. The year was defined as the present period of the earth's revolution around the sun, although, of course, variation of its orbital velocity may have been appreciable over geologic intervals.[2] The geologic age was defined as the number of years since the beginning of accumulation of the stable end-products of the radioactive series. The parent elements of these series are present to the extent of at least a few atoms per billion in all rocks, and a determination of the amount of either of the stable end-products, lead or helium, and of the parent elements present, provides all the data necessary for computing the length of time of accumulation. It was shown, however, that the lead method is applicable only to the radio-active minerals like uraninite, whereas the helium method can be applied to a wide range of medium- to fine-grained basic igneous rocks, and possibly to nearly all types of igneous material. The elapsed time determined by the helium method dates from the original cooling of the rock, except as it may have been modified by metamorphism, leaching, or

---

[1] A. C. Lane and W. D. Urry: *Ages by the helium method: I. Keweenawan*, Geol. Soc. Am., Bull., vol. 46 (1935) p. 1101-1120.

[2] A. C Lane: *Rating the geologic clock*, 16th Intern. Geol. Cong., Washington, 1933, Rept., vol. 1 (1936) p. 145-167.

(1217)

TABLE 1.—*Analyses and computed ages*

Horizon	Laboratory designation	Description	Helium $\times 10^5$ cc/gram	± 0.02	Radium $\times 10^{13}$ gram/gram	± 0.08	Thorium $\times 10^6$ gram/gram	± 0.12	Age in millions of years	±
Pliocene	P	Basalt, Steens Mt., Oregon	0.19	0.02	2.10	0.08	2.32	0.12	13	1
Miocene	M1	Basalt, Douglas Creek Canyon, Wash.	0.23	0.03	2.17	0.06	2.02	0.20	17	1.5
	M2	Picture Gorge, Oregon	0.21	0.02	2.16	0.04	2.10	0.23	15.5	1
	M3	"	0.27	0.03	2.20	0.07	2.52	0.23	18	1.5
Oligocene	Ja1	Basalts, Lower Silesia, Germany.	0.48	0.04	1.57	0.06	2.84	0.12	34	2
	Ja2	"	0.70	0.06	2.98	0.08	3.09	0.12	36	2
	Ja3	"	0.68	0.06	3.36	0.08	4.00	0.12	29	2
Early Eocene	T	Tinguaite, Montreal.	6.44	0.26	20.40	0.17	14.5	0.29	57	1.5
Early	LH93	Little Hatchet Mts., N. Mex.	0.92	0.07	3.55	0.07	4.48	0.13	36	2
	LH67	Dikes cutting the Nevadan granodiorite.	0.93	0.08	1.20	0.06	2.20	0.12	85	3
Cretaceous	J57	"	1.17	0.09	1.68	0.08	1.58	0.16	110	7
	J29	"	1.24	0.08	3.10	0.07	1.06	0.11	90	4
	L80	"	2.87	0.14	4.10	0.07	5.00	0.10	98	3
Late Jurassic	J28a	Autolith in Nev. granodiorite	2.16	0.12	2.90	0.08	3.10	0.12	110	5
	J28g	Nevadan granodiorite	1.97	0.12	2.60	0.10	2.94	0.15	110	5
Jurassic ?	J86 (I)	Diabase host for the Nevadan granodiorite	1.76	0.11	1.65	0.08	1.98	0.16	150	8
	J86 (II)	"	1.89	0.12	1.67	0.08	1.90	0.15	165	8
	J21	"	2.83	0.14	2.41	0.06	2.67	0.12	170	6
Triassic	B1	Palisade intrusive, N. J.,	1.60	0.10	1.44	0.08	1.79	0.12	155	8
	B2	"	1.66	0.10	1.30	0.08	1.77	0.14	165	9
	B3	"	1.68	0.10	1.38	0.08	1.80	0.12	165	8
	K1 (I)	West Rock sill, New Haven, Conn.	1.82	0.13	1.96	0.15	1.30	0.38	170	12
	K1 (II)	Buttress dike, New Haven, Conn.	1.88	0.13	2.13	0.16	1.21	0.35	170	12
	K2	"	1.52	0.09	1.55	0.07	1.06	0.11	175	9
	NS1	Cape Spencer flow, N. S., Can.	1.33	0.08	1.05	0.07	1.53	0.11	160	8
	H1	1st Watchung Mt., N. J.	1.13	0.08	1.00	0.06	0.90	0.10	180	11
Permian	1440	Museum specimen, Germany	1.34	0.09	5.66	0.09	5.84	0.10	36	1.5
	SaP	Melaphyre, Saar, Germany	0.98	0.07	4.53	0.08	7.90	0.12	25	1
	BP	Brighton, Mass.	1.28	0.09	0.59	0.07	1.12	0.11	230	15
Devonian	Del	Newbury volcanic, Mass.	4.03	0.17	1.98	0.07	2.13	0.12	300	10
Ordovician	L5	Diabase in the Martinsburg shale, Pa.	2.61	0.13	1.29	0.07	0.91	0.10	355	15
	L69	"	3.88	0.17	1.30	0.06	1.83	0.12	375	15
	L6	Basalt at base of Martinsburg shale, Pa.	1.90	0.11	2.25	0.09	1.98	0.18	140	7
	La19	"	1.15	0.08	1.16	0.06	1.27	0.11	145	8
	C1	Metabasalt, Quebec, Can.	0.62	0.06	0.54	0.07	0.49	0.10	>185	15
	CP	Chapel Pond, Adirondacks, N. Y.	0.78		0.56		<0.10		>345	<400
	SK	Stark's Knob, N. Y.	0.485		0.34		<0.08		>340	<410
Cambrian	U1 (I)	Basalt, Unicoi formation, Tenn.	1.62	0.11	0.87	0.08	0.15	0.03	465	30
	U1 (II)	"	1.51	0.10	0.85	0.08	0.15	0.03	440	30

excessive heating. The Keweenawan suite of rocks represented a wide range of such conditions, and from their study there was defined a set of standard conditions to apply in selecting material for the present purpose. For complete details of the helium method and questions regarding the validity and range of application, the reader is referred to Part I of this study.[3]

For the physico-chemical determination of Th and Ra in rocks, see a recent paper by Urry.[4]

TIME SCALE

PURPOSE

The abundance of material suitable for study by the helium method, compared to the scarcity of the radioactive minerals which alone are suited to the lead method, should result in a far more detailed time scale. The geologic correlation of a particular flow horizon is attainable with more precision than is usually the case with a deposit of radioactive minerals. In all but a few cases the secondary nature of radioactive minerals causes doubt as to their particular period of formation.

Accordingly, with the generous help of many geologists in North America and Europe, it has been possible to apply the helium method to small groups of basic igneous rocks, conforming to the standards already mentioned, and representative of most of the geologic horizons above the Keweenawan. The object in view is two-fold; first, to compare the resulting time scale with other scales, quantitatively in the case of the best age determinations derived by the lead method, and at least in order of magnitude with other scales such as that based on the rate of accumulation of sediments; and second, as a prerequisite to the important application of the method in long distance time correlations, doubtful or impossible from geologic relations. A few pre-Cambrian rocks have also been examined but the small number does not warrant the extension of the time scale beyond the Keweenawan horizon. They show, however, that the method is applicable to the older rocks, and a solution of many pre-Cambrian problems, difficult at present owing to the lack of fossils, may be anticipated in the near future.

MEASUREMENT AND DESCRIPTION OF MATERIAL

Table 1 records for all specimens the measured quantities of helium and its parent elements, as well as their computed ages, in millions of years, which have been plotted to scale in Table 2. The computation of

---

[3] A. C. Lane and W. D. Urry: *op. cit.*

[4] W. D. Urry: *Determination of the thorium content of rocks,* Jour. Chem. Phys., vol. 4 (1936) Th, p. 34-40; Ra, p. 40-48.

the age and the probable error are discussed in Part I of this study.[5]
Geological and mineralogical notes on the specimens were kindly prepared by Professor A. C. Lane and follow Table 1. All the material
used for establishing the time scale has been carefully selected with a
view to the accuracy with which such material can be placed in a definite
horizon on the basis of geologic evidence.

* * * * * * *

---

[5] A. C. Lane and W. D. Urry: *op. cit.*

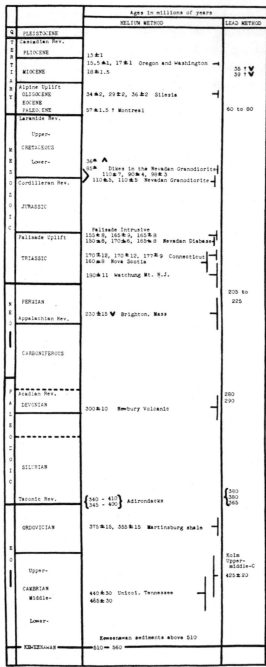

		Ages in millions of years		LEAD METHOD
		HELIUM METHOD		
Q	PLEISTOCENE			
T E R T I A R Y	Cascadian Rev.			
	PLIOCENE	13±1		
		15.5±1, 17±1 Oregon and Washington		
	MIOCENE	18±1.5		35 ?, 39 ?
	Alpine Uplift			
	OLIGOCENE	34±2, 29±2, 36±2 Silesia		
	EOCENE			
	PALEOCENE	57±1.5 ? Montreal		60 to 80
M E S O Z O I C	Laramide Rev.			
	Upper-			
	CRETACEOUS			
	Lower-	36▲		
		85▲ Dikes in the Nevadan Granodiorite		
		110±7, 90±4, 98±3		
		110±5, 110±5 Nevadan Granodiorite		
	Cordilleran Rev.			
	JURASSIC			
	Palisade Uplift	Palisade Intrusive 155±8, 165±9, 165±8		
		150±8, 170±6, 165±8 Nevadan Diabase		
	TRIASSIC	170±12, 170±12, 177±9 Connecticut		
		160±8 Nova Scotia		
		180±11 Watchung Mt. N.J.		
				205 to 225
N E O	PERMIAN	230±15 Brighton. Mass		
	Appalachian Rev.			
	CARBONIFEROUS			
P A L E O Z O I C	Acadian Rev.			
	DEVONIAN	300±10 Newbury Volcanic		280, 290
	SILURIAN			
	Taconic Rev.	{340 – 410, 345 – 400} Adirondacks		{380, 380, 365}
	ORDOVICIAN	375±15, 355±15 Martinsburg shale		
	CAMBRIAN Upper-			Kolm Upper-middle-C 425±20
	Middle-	440±30 Unicoi. Tennessee		
		465±30		
	Lower-			
		Keweenawan sediments above 510		
	KEWEENAWAN	510 — 560		

TABLE 2.—*Post-Keweenawan time scale*

DISCUSSION

GEOLOGIC SEQUENCE

Of the thirty-nine results reported here, only five are incompatible with the correct geologic sequence as determined from the field relations. All five specimens should be rejected as unsuitable on strict adherence to the standard conditions set for selecting material. (*See* Appendix.) The Permian specimen 1440 was collected from the surface in 1890. A gradation from a brownish exterior ring, about half the diameter of the specimen, to a dark-gray, fresher central nucleus suggests some alteration during museum handling. Specimen SaP, though from a Saar coal mine, is considerably altered, rich in limonite, and so friable as to soil the hands. Specimens L6 and La19, from the Ordovician basaltic flow at the base of the Martinsburg shale, exhibited considerable limonite throughout the entire one-pound specimens. Not even the 5 grams necessary for analysis could be selected from fresh material. A lack of more suitable mtaerial in these horizons at the time prompted the work on the four specimens above. The determination of the apparent age of the Mt. Orford metabasalt was carried out to open up the question of the interpretation of the result for a metamorphosed rock. A. F. Buddington believes he can furnish dike material both metamorphosed and unmetamorphosed but of the same age. Such material should throw light on the interpretation of results obtained from such rocks as the Mt. Orford metabasalt.

The post-Keweenawan time scale in Table 2 has been drawn up from the remaining thirty-four results. Geological periods to the left are boxed approximately to scale, the divisions being based on the evidence presented by these helium method results on basic rocks. The Triassic box is enlarged by about one-fifth to give room for the many results, though it is not absolutely certain that the Palisade is Triassic. A given result is placed as nearly as possible opposite the position in a given period prescribed by the geologic evidence. A question mark denotes doubt as to the position of a particular specimen from the geologic evidence. V denotes a possibility of the formation being older than represented by the position given in the chart owing to uncertain field relations, and ∧ vice versa. The Newbury volcanic represents a single determination and hence the Devonian period is not well defined. The close of the Carboniferous and the division between the Cambrian and the Ordovician are determined from Schuchert's data on the sediments. No attempt is made here to close the pre-Cambrian other than to suggest that in all likelihood the Keweenawan bridges this division. In accordance with recent authors,[22] the latest Keweenawan sediments are placed in the lower Cambrian. The term Keweenawan, therefore, should not be applied as a separate chronologic unit. The preliminary helium results reported by Holmes [23] have not been included in Table 2, but even though different pieces of a hand specimen were used for the various necessary determinations, his results fit well into the scheme. In the right-hand column of Table 2 are the results of lead method determinations on the radioactive minerals which can be placed with some degree of certainty in a given geological period. They can be identified by reference to Table 3. Probable errors given after each age are repeated dia-

---

[22] C. K. Leith, R. J. Lund, and Andrew Leith: *Pre-Cambrian rocks of the Lake Superior region*, U. S. Geol Surv., Prof. Pap. 184 (1935).

[23] E. W. Brown, A. Holmes, A. Knopf, A. F. Kovarik, and C. Schuchert: *Physics of the earth. IV. The age of the earth*, Nat. Res. Council, Bull. 80 (1931) p. 413. A. Holmes (*personal communication*) writes ". . . the felsite tested by Dubey (Bull. 80) may not be as 'low' as it appeared to be at the time. More recent work has shown that the occurrence is not a plug, but an acid flow of very much younger age than the Deccan traps on which it lies."

grammatically on the right side of the helium column. Unfortunately, authors have seldom quoted probable errors for lead ratio determinations, nor are the ranges of the geological uncertainties indicated. With the examination of a suite of Carboniferous and Devonian rocks now made possible through the generous cooperation of Bernard Smith, Director, Geological Survey of Great Britain, with the establishment of the date of the Laramide Revolution and with more determinations in the Cambrian, the second purpose will have been achieved in a broader sense, and long distance correlations will be possible.

COMPATIBILITY OF TIME SCALES

Geologists will be interested primarily in any device for objective measurement of the time coordinate insofar as such a device provides a tool for correlation purposes. In this respect any method which supplies a relative, as opposed to an absolute, scale will suffice, as for example, the study of fossils in sedimentary stratigraphy. Table 1 illustrates the fact that the helium method does provide such a relative scale for the igneous rocks, but if at any time a comparison is to be made with time coordinates measured by other means, for example, in estimating the compatibility of astronomical and geological history, it at once becomes imperative to know the correspondence factor between the methods involved; or better, to establish the method on an absolute scale. Methods of estimating time from geologic evidence will necessarily be based on the transference of matter in large aggregates of atoms, i.e., sand grains and colloidal suspensions, by such processes as erosion and sedimentation, salt concentration changes in waters, and biological evolution. Physical methods may be based upon considerations of the behavior of single atoms, such as radioactivity, or of the energy of aggregate systems, such as heat and gravitation. Erosion and sedimentation may alternate in cycles but radioactive disintegration is a unidirectional process. This accounts for the fact that time measurements based on this phenomenon show the greatest consistency. Nevertheless, a comparison of the radioactive time measurements with the evidence presented by the sediments, apparently the only geologic evidence offering any hope, is not inapt, especially because the work on varves by De Geer and Antevs, Bradley, and Korn,[24] demonstrates the possibility of determining rates of accumu-

[24] Ernst Antevs: [*The last glaciation*, Am. Geog. Soc., Research Ser., no. 17 (1928)] gives a bibliography from the work of De Geer presented to the 11th International Geological Congress. Both he and De Geer have papers in the compte rendu of the 16th International Geological Congress.
   See also: W. H. Bradley: *Non-glacial marine varves*, Am. Jour. Sci., 5th ser., vol. 22 (1931) p. 318-329.
   W. W. Rubey: *Lithologic studies of fine-grained Upper Cretaceous sedimentary rocks of the Black Hills region*, U. S. Geol. Surv., Prof. Pap. 165 (1930) p. 1-54.
   H. Korn: *Schichtung und Absolute Zeit*, Geol. Rundschau, vol. 26 (1935) p. 137-139.

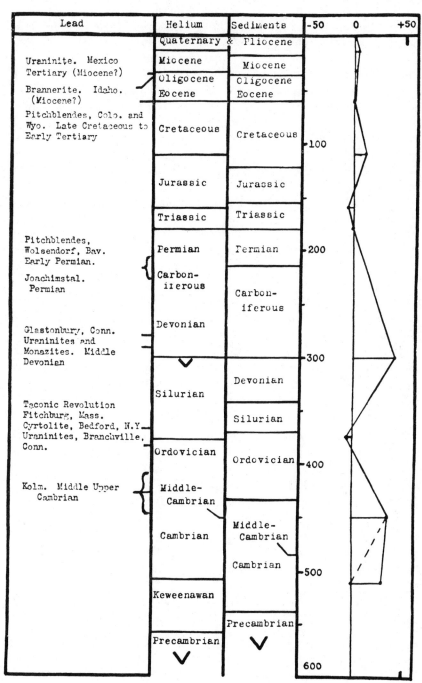

Lead	Helium	Sediments	-50	0	+50
	Quaternary & Pliocene				
Uraninite. Mexico Tertiary (Miocene?)	Miocene	Miocene			
Brannerite. Idaho. (Miocene?)	Oligocene	Oligocene			
	Eocene	Eocene			
Pitchblendes, Colo. and Wyo. Late Cretaceous to Early Tertiary	Cretaceous	Cretaceous	100		
	Jurassic	Jurassic			
	Triassic	Triassic			
Pitchblendes, Wolsendorf, Bav. Early Permian.	Permian	Permian	200		
Joachimstal. Permian	Carbon-iferous	Carbon-iferous			
Glastonbury, Conn. Uraninites and Monazites. Middle Devonian	Devonian		300		
	Silurian	Devonian			
Taconic Revolution Fitchburg, Mass. Cyrtolite, Bedford, N.Y. Uraninites, Branchville, Conn.		Silurian			
	Ordovician	Ordovician	400		
Kolm. Middle Upper Cambrian	Middle-Cambrian				
	Cambrian	Middle-Cambrian			
		Cambrian	500		
	Keweenawan				
		Precambrian			
	Precambrian		600		

TABLE 3.—*Determination of time scales by three methods*

lation of sediments for the particular geologic periods under consideration. A general reconciliation of the geologic evidence with the radioactivity results available at the time was summarized by Schuchert.[25]
There is considerable danger in tabulating the helium and lead scales
against the sediment scale because the sediment column in Table 3 was
made by Schuchert to correspond in total length with the radioactive
evidence (from lead ratios, helium values not being available at that
time). Bearing this in mind, however, a comparison of the individual
periods is legitimate. Although the helium and lead methods are based
fundamentally on the same physical phenomenon—radioactivity—the
only ultimate difference being in the choice of the end-product used as
the "clock-hand," a comparison is invaluable in showing that there is
no longer any large discrepancy between these time scales, when suitable
material is chosen for establishing them. For reasons given in Part I
of this study,[26] it is impossible at present to use both methods on the
same specimen, nor has a radioactive mineral suited to the lead method
been found, whose age, relative to that of a nearby basic rock suited
to the helium method, was known from the field relations. In Table 3
the sediment and helium columns are plotted on the vertical scale in
millions of years. In the differential diagram in the right-hand column,
a positive difference denotes an earlier date for a given division between
periods on the basis of the sediments than on the basis of the helium
method; a negative difference, a later date. Positions of the "lead ages"
are shown in the left-hand column.[27] The exact positions of these radioactive minerals in the geologic column can not always be determined
from the field relations. For reasons given in Part I of this study, it
might be expected in general that the lead method would give ages higher
than the true, and conversely for the helium method. However, the
general agreement of the two scales does not bear this out.

The sediment scale shows the close of the Devonian at the same date
as that given by the helium method for the opening of this period. The
helium age is based on a single result for the Newbury Volcanic which
may be somewhat low. On the other hand the Middle Devonian (?)
Glastonbury lead ratios are also above Schuchert's close of the Devonian.

[25] Charles Schuchert: *Geochronology or the age of the earth on the basis of sediments and life*, in *Physics of the earth. IV. The age of the earth*, Nat. Res. Council, Bull. 80 (1931) p. 10-63.

[26] A. C. Lane and W. D. Urry: *Ages by the helium method: I. Keweenawan*, Geol. Soc. Am., Bull., vol. 46 (1935) p. 1101-1120.

[27] E. W. Brown, A. Holmes, A. Knopf, A. F. Kovarik, and C. Schuchert: *Physics of the earth. IV. The age of the earth*, Nat. Res. Council, Bull. 80 (1931) p. 305-310, 341, 347, 348.

A. C. Lane: *Reports of the National Research Council Committee on measurement of geologic time* (1931-1935).

O. B. Muench: *Analysis of cyrtolite for lead and uranium*, Am. Jour. Sci., 5th ser., vol. 21 (1931) p. 350-357.

The Middle Cambrian also appears to be later as measured by the helium method. The Kolm, however—represented by the best age determination by the lead method and the only material suited to that method that has been precisely dated by geologic evidence—of the middle Upper Cambrian corresponds closely to Schuchert's close of the Cambrian. A discrepancy occurs at the opening of the Cambrian, but this is entirely due to uncertainty as to how much of the Keweenawan should be termed Cambrian.

Thus the helium method, applicable in great detail to such abundant material in the earth's crust as the basic rocks, which are capable of a higher degree of precision in chronological interpolation from the field relations than is usually the case with radioactive minerals, adds greatly to our data corroborating the absolute value of the radioactive time scale.

EXTENSION OF APPLICABILITY

Problems of immediate concern center around the possible extension of the method to more acidic rocks in one direction and to coarser material in another. This will be tested by two suites of rocks, all of one age, in which one variable at a time is progressively changed. It is desirable to investigate the possibilities of determining rates of sedimentation during each period by systematic studies of the age of flows (not sills) separated by a known thickness of sediments. The necessary increase in accuracy for such an investigation will involve several independent analyses from each flow. If such plans are feasible, especially in terrains where cross-checking by varves is possible, the data should be of especial interest. An extension of the studies to the pre-Cambrian rocks has already begun. An investigation of the suitability of chemically precipitated sediments for these determinations is also projected.

SUMMARY

The ages of numerous specimens of fresh igneous rocks obtained by the helium method as given in Table 1 are consistent with those obtained for radioactive minerals by the lead ratio. (*See* Tables 1 and 2.) The chronological sequence of the rocks investigated is in general agreement with the same sequence established from purely geologic data. From the helium data it is possible to block out roughly the geologic periods. The evidence of the sediments confirms the relative duration of the periods, but it must be emphasized that, owing to our present incomplete knowledge of changing rates of sedimentation in the past, the sediment scale *per se* should not be regarded as an absolute measure of time.

From the helium time scale which, it is believed, closely approximates an absolute scale, Tertiary rocks are less than 60 million years old; Mesozoic, between 60 and 200 million years; and Paleozoic (including Keweenawan), between 200 and 560 million years old. With unknown specimens taken in conformity with the conditions of sampling described in the Appendix, the application of the helium method can differentiate certainly between the eras and in most cases between periods. The position in a given period is not as well defined.

ACKNOWLEDGMENTS

It cannot be overemphasized that in a border-line research of this nature the close cooperation of geologist, physicist, and chemist is of vital importance. It is therefore fitting that all the contributors of specimens be regarded as active collaborators, and to them I wish to express my indebtedness. To Dr. A. C. Lane I am especially indebted for continued help and guidance.

It is a pleasure once more to express my thanks to the Geological Society of America for its continued interest and active financial support made possible through the Penrose Bequest. Analyses of doubtful specimens reported herein, not specifically covered by the terms of the Penrose grant, have been financed by a fund made available by the late Everett Morss.

APPENDIX

DESCRIPTION OF MATERIAL SUITABLE FOR AGE DETERMINATION

The selected rock should be of a basic type containing less than 65 per cent SiO_2, and as fresh as possible. Limonite stains throughout the hand specimen are an important indication of unsuitable material. A rock should be fine-grained but not to the extent of having a predominantly glassy base. Surface exposures are in most cases unsuitable, leading to an uncertainty in the interpretation of the result. The most suitable sources of material are outlined below in order of preference.

Mines.—The best material is from new drift facings as deep as possible in the mine and free from circulating waters. Older drifts will often yield suitable material. Recent drill cores, but not churn drillings, have been successfully employed.

Quarries.—If possible, the quarry should be active and specimens should be taken from freshly broken surfaces. Material should not be collected if the quarry has been inactive for more than thirty years, unless it is possible to blast out a new surface 1 to 2 feet deep. Specimens should not be taken within 3 feet of the present topographic surface, and a somewhat deeper limit, 8 to 10 feet, is desirable in unglaciated regions.   If possible, specimens should be taken from unjointed material.

Road and railroad cuts.—Road cuts are sources of suitable material, especially in glaciated regions, if not over four years old.   The conditions applying to quarry material should be followed in the selection of road and railroad cut material.

MASSACHUSETTS INSTITUTE OF TECHNOLOGY, CAMBRIDGE, MASS.
MANUSCRIPT RECEIVED BY THE SECRETARY OF THE SOCIETY. MAY 26, 1936.
READ BEFORE THE GEOLOGICAL SOCIETY, DECEMBER 26, 1935.
CONTRIBUTION FROM THE RESEARCH LABORATORY OF PHYSICAL CHEMISTRY, MASSACHUSETTS INSTITUTE OF TECHNOLOGY, No. 369.

Reprinted from *Geol. Soc. Am. Bull.*, **52**, 521–529 (1941)

8

Age Measurements by Radioactivity

D. GOODMAN AND R. D. EVANS

AGE MEASUREMENTS

General statement.—We have seen that the reliability of the helium method can be evaluated to some extent on the basis of independent physical and geological knowledge, but in the final analysis this method, or any other geological tool, must be judged by its success in actual application. The road of research is seldom smooth, and this is particularly true in the development of radioactive methods of age determination. Considerable work has been done during the past 3 decades, but much of it has been discarded as newer developments disclosed earlier misconceptions and discrepancies. Our present position can best be appreciated by a review of the past progress.

Early investigations.—Following the suggestion by Rutherford in 1905 that the accumulation of helium in radioactive minerals might provide a means of measuring geological time, Strutt made an extensive survey of the helium, uranium, and thorium content of a number of the more radioactive terrestrial materials. He obtained helium age ratios for phosphatic deposits and hematite (1908, p. 272), other iron ores (1909, p. 96), zircons (1909, p. 298), and sphenes and other minerals (1910, p. 194). From these results a broad correlation between the geological age of a mineral and the helium age ratio was established. However, Strutt (1909, p. 166) was the first to point out that, owing to leakage of helium from all these minerals, the age ratios observed could only represent minimal estimates of the true age. These helium age ratios were found to be only 30 to 70 per cent of the values based on the accumulation of lead in corresponding radioactive minerals. For this reason, it has been necessary to discard most of the excellent helium age researches of Strutt.

Waters (1909, p. 677; 1910, p. 903) made preliminary helium age studies of some common rock minerals. Unfortunately, the electrical detection methods used were too insensitive to measure other than the more radioactive mineral inclusions. Because these materials showed evidence of helium leakage, the work was discontinued.

Following the development of methods of measuring minute quantities of helium, Paneth and his coworkers determined helium age ratios for a large number of meteorites (Paneth and Urry, 1930, p. 127). The magnitude of the age ratios obtained, together with some laboratory tests on the materials, indicates that iron meteorites have a remarkably high retentivity for helium, even at elevated temperatures. Similar measurements were attempted on native metals, but their extremely low radioactivities prevented application of the method. Subsequently, Paneth has collaborated with Holmes (1936, p. 385) on an investigation of the rocks composing the Kimberley diamond pipes. These measurements are mainly of value in emphasizing the type of material *not* to be used for helium age studies.

The first real stride in the application of the helium method to rocks can be said to have begun with the work of Dubey and his collaborators. These results, summarized in Table 3, illustrate a number of interesting points. The felsite (1) shows a much lower age ratio than the Cleveland Dike (2), although the two rocks are of approximately the same geologic age. The age ratios of (2 - 5) are in the proper sequence, but the actual values observed are somewhat lower than the corresponding lead ages. Specimens (5 - 7) were selected to test the effect of texture on the retention of helium as well as to estimate the geologic age of the Gwalior series of India. In the original publication Dubey (1930, p. 807) listed (5) as a fine-grained basalt and both (6) and (7) as medium-grained dolerite from a Morar-group sill. Holmes, (National Research Council, 1931, Table LXXVI, p. 415), presumably on the basis of a more exact

89

description, subsequently indicates (6) as medium-grained and (7) as "of coarser grain than 6." With this revision in the texture, it is seen that the age ratios for (6 - 8) decrease with increasing grain size. In the light of these observations, it seemed desirable at that time to concentrate on close-grained, crystalline basaltic or other basic rocks.[21]

TABLE 3.—*Age measurements of igneous rocks*

Rock specimen	Geologic age	Helium content in 10^{-5} cc/g	Activity index Rn-Th method	Age ratio in 10^6 years
1. Acid felsite.........	Miocene...............	*0.11	*0.68	5
2. Cleveland Dyke......	Oligocene or Miocene...	0.11	0.12	28
3. Deccan basalt.......	Early Eocene..........	0.12	0.10	37
4. Whin Sill..........	Late Carboniferous.....	0.36	0.057	196
5. Gwalior basalt......	Late pre-Cambrian.....	0.55	0.034	500
6. Morar dolerite......	Late pre-Cambrian.....	0.42	0.033	389
7. Morar dolerite......	Late pre-Cambrian.....	0.31	0.047	203

1-7 Geologic notes, references and further discussion, National Research Council, 1931, p. 411-415.
* Mean of two determinations.

Following this work, researches were initiated in 1931 by Dr. W. D. Urry in collaboration with Professor A. C. Lane on an extensive suite of Keweenawan rocks and minerals,

"with a view to co-ordinating age results with certain geological and mineralogical aspects pertaining to the specimens, in addition to the important question of the age and duration of the Keweenawan" (Lane and Urry, 1935, p. 1119).

The magnitude and scope of this work were indeed admirable. As subsequent events have proved, however, it was premature. Age ratios were obtained for more than 30 rock specimens and minerals. With only a few exceptions, excellent agreement with geologic conditions was observed. The magnitude of the age ratios also appeared to be in accord with equivalent lead ages. On the strength of this encouragement age ratios were determined on 39 selected rock specimens ranging in geologic age from Cambrian to Pliocene. Urry (1936, p. 1226) reports that of these results

"only five are incompatible with the correct geologic sequence as determined from field relations. All five specimens should be rejected as unsuitable on strict adherence to the standard conditions set for selecting material."

The helium time scale represented by the summation of these measurements was also found to agree quite well with the somewhat less extensive

[21] Holmes (National Research Council, 1931, p. 414) states as follows: "A systematic programme of research has already been started by Dr. Dubey in India and by the writer in collaboration with Prof. Dr. F. Paneth, and Dr. R. W. Lawson who have arranged to make helium determinations and to investigate quantitatively the escape of helium under various conditions." Unfortunately these investigators have found it possible to complete only partially this admirable program.

TABLE 4.—*Helium age ratios of basic igneous rocks*

Sample No.	Rock specimen	Location	Geologic age	Age ratio in 10^6 years
1B	Porphyritic basalt	Yellowstone Park	Tertiary	5
37B	Tower Falls basalt	Yellowstone Park	Recent	5.5 ± 1.5
3B	Crescent Hill basalt	Yellowstone Park	Tertiary	9 ± 1
I	Overhanging Cliff	Yellowstone Park	Miocene	11 ± 4
5I	Cleveland Dike	Durham, England	Oligocene or Miocene	17 ± 3
4B	Giant's Causeway	England	Lower Tertiary	22 ± 4
2	Deccan basalt	Mt. Pawagarh, India	Early Eocene	37
7B	Geode Creek basalt	Yellowstone Park	Tertiary	36 ± 4
1	Kimberlite	South Africa	Late Cretaceous	58
6B	Oldwick basalt	Oldwick, N. J.	Triassic	57
J	Dike 3,200' level	Eustis Mine, Quebec	Paleozoic (?)	54 ± 3
113	Earlier dike	Eustis Mine, Quebec	Paleozoic	81 ± 8
27B	Monchiquite dike	Kilchattan, Scotland	Post-Devonian	83 ± 10
26B	Monchiquite dike	Riasg Buidhe, Scotland	Post-Devonian	105 ± 8
3	Whin Sill	Westmoreland, England	Late Carboniferous	196 ± 6
32B	Dalhousie andesite	Nova Scotia	Devonian	160 ± 25
43B	Gogebic dike	Michigan	Late pre-Cambrian	190 ± 25
24B₁	Long Lake dike	Montana	Late pre-Cambrian	290 ± 35
14B	Noranda diabase	Quebec	Post-Timiskaming	250 ± 20
4	Gwalior basalt	India	Late pre-Cambrian	500
R	Basalt	Ropruchei, Karelia	Pre-Cambrian	600 ± 35
25B	Beartooth trap	Montana	Pre-Cambrian	1050 ±100
22B	Stillwater norite	Quad Creek, Montana	Early pre-Cambrian	1830 ±250

Analysts: Except for the following samples, the measurements were made by the authors: Dubey, *et al.* (Nat. Res. Council, 1931, p. 413–415) 2, 3 and 4; Keevil (1938, p. 147): I, J, 113, R; Goodman and Keevil (Evans, *et al.*, 1939, p. 942): 6B, 14B, 22B, 24B, 26B, 27B and 43B; Holmes and Paneth (1936, p. 400): 1.

Notes: I. Another specimen (10B) of this rock mass analyzed by the authors showed evidence of a high proportion of neon present. The age ratio, including neon and helium, was 375 ± 35 m.y. This neon may originate in air absorbed by the rock at the time of solidification.

51. An earlier determination by Dubey and Holmes (1929, p. 794) on a separate sample of this rock gave an age ratio of 28 m.y.

3. Recent measurements by the authors on a separate sample (7I) of this rock gave a substantially higher age ratio than that observed by Dubey and Holmes. However, there is evidence of extreme inhomogeneity in the distribution of radioactivity in this rock, and the sample at hand was too small to allow check measurements to be made.

5I and 3. These two rocks are intermediate in composition between acidic and basic but have been included in the present table because the age ratios observed agree more closely with basic rocks than with acidics.

lead time scale. This mutual compatibility of the two radioactive methods of age determination appeared as the final conclusive proof of the reliability of both methods when properly applied. While there were a number of gaps left in the structure and much work remained to be done in the pre-Cambrian, in general, the geologic time scale seemed a definitely established framework on which future research would build.

Concurrent with these developments (1929-1936), Evans was engaged in the development and application of sensitive detection instruments for the measurement of terrestrial radioactivities. (*See* Evans, 1935, p. 99-112, for key references.) The logical extension of this work was in the direction of age measurements. The physical methods used offered a number of analytical advantages that should make analyses more accurate and more routine and, hence, more generally available for geologic problems.

Intercalibration measurements.—In February 1936 collaboration between the authors began on helium age studies. The first step taken was to test the reliability of the alpha-counting method of determining radioactivity and the direct-fusion method of measuring helium, using portions of hand samples of rock previously analyzed by Urry. These preliminary determinations, on samples 19B, 34B, and 3I, gave age ratios substantially lower than those obtained by Urry. The differences appeared to be primarily due to a lack of agreement on the radium contents. However, it also appeared possible that the discrepancies in radium results might be due to inhomogeneities in the samples. Accordingly, a set of Keweenawan rocks, 39B to 43B inclusive, were measured by the alpha-helium method. These age ratios were in even more drastic disagreement with Lane and Urry's (1935, p. 1108) extensive list of ages for the Keweenawan. Thus it seemed advisable to make a number of direct comparisons between the alpha-helium and the radon-thoron helium methods. Portions of the same granulated rock samples and of the standard solutions used for calibration purposes were measured independently in the two laboratories.[22] The results of this work already have been summarized in the preceding general discussion. For the purposes of the present review, we need only repeat that, because of apparently unsystematic errors in the radioactivity measurements, all of Urry's helium age ratios had to be discarded.

Summary of age data.—The helium age data remaining after this drastic, but necessary, re-evaluation are summarized in Tables 4 and 5.

---

[22] A detailed description of this intercalibration and comparison has been given by the participants, Evans, Goodman, Keevil, Lane, and Urry (1939, p. 931-946).

TABLE 5.—*Helium age ratios of acidic igneous rocks*

Sample No.	Rock specimen	Location	Geologic age	Age ratio in 10^6 years
5	Mt. Pawagarh felsite	Kathiawar, India	Lower Eocene	5
28A	Obsidian Cliff	Yellowstone Park	Tertiary	8 ± 1
23A	Rhyolite	Yellowstone Park	Tertiary	22 ± 3
15A	Fitchburg granite	Fitchburg, Mass.	Carboniferous	45 ± 5
Q	Quincy granite (series)	Quincy, Mass.	Mississippian	66 ± 2
4A$_1$	Chelmsford granite	Chelmsford, Mass.	Mississippian	72 ± 6
11A	Granite	Franklin, Maine	Middle Paleozoic	60 ± 6
7A	Rockport granite (fine)	Cape Ann, Mass.	Middle Paleozoic	88 ± 5
8A	Rockport granite	Cape Ann, Mass.	Middle Paleozoic	100 ± 8
NB5	Hampstead granite	Hampstead, N. B.	Devonian	109 ± 5
Ga	Granite	Gananoque, Ont.	Late pre-Cambrian	150 ± 20
17A	Kingston granite	Kingston, Ont.	Late pre-Cambrian	280 ± 30
4I	Granodiorite	E. La Motte Tp., Que.	Late pre-Cambrian	195 ± 15
T2A	Granite	Taschereau, Que.	Late pre-Cambrian	250 ± 20
6I	Granodiorite	Varsan, LaCorne Tps., Que.	Pre-Cambrian	60 ±100

Analysts: Dubey (Nat. Res. Council, 1931, p. 413): 5; Keevil (1938, p. 146): Q, NB5, Ga and T2A; Goodman and Keevil (Evans, et al., 1939, p. 938–942): remaining samples.

Notes: Q—Age measurements in good agreement with the above value have also been made by the authors on a separate sample (12A) of this well-known rock.

4I and 6I—Intermediate in composition between acidic and basic.

The values given include the earlier results of Dubey and Holmes from Table 3 and one measurement by Holmes and Paneth. The remaining age ratios include those obtained in connection with the interchecking work mentioned above, other concurrent measurements by Keevil (working in Urry's laboratory), and subsequent determinations by the authors. The helium age ratios fall into two groups, one for basic rocks (Table 4) and one for acidic rocks (Table 5). These have been placed in their appropriate positions on the left hand side of the time scale shown in Figure 10. The field evidence, on which the geological age is based, is in some cases only very approximate, and in these instances the position assigned is the midpoint of the range within which the age can be defined. The lead ages shown on the right side are also divided into two groups, one for lead ratios and one for lead isotope ratios. These two methods yield results that are in general agreement, although there are notable exceptions, for example the Permian pitchblende from Bohemia and the Cambrian Swedish Kolm. These lead age methods and the results obtained by their application are discussed in some detail in the next section. For our present purpose, the important generalizations to be drawn from Figure 10 are: (1) With only a few exceptions the helium age ratios lie in the proper geologic sequence when the results for acidic rocks and for basic rocks are considered separately, (2) the age ratios for acidic rocks are consistently lower than the age ratios for corresponding basic rocks, and (3) the helium age ratios, even for the basic rocks, are substantially lower than corresponding lead ages on radioactive minerals. These three general inconsistencies in helium age ratios indicated quite clearly that there was some fundamental failure in the helium method. A suggestion of the possible difficulty had come from Keevil's (1938, p. 406-416) investigation of the Quincy granite. Keevil observed no lateral or vertical variation of age ratio in this rock body, but he did find marked differences in the age ratios for the three main mineral components of this rock. A theoretical treatment of the probability of helium leakage from minerals (Keevil, 1940, p. 311) with particular consideration of feldspar, which was the Quincy rock mineral that gave the lowest age ratio, led Keevil [23] to conclude that the helium retentivity of practically all igneous rock minerals is sufficient to ensure the retention of essentially all the helium produced in them during geologic time. On the strength of this conviction, Keevil (1938, p. 406) attributed the observed differences in age ratios to leakage of helium resulting from alteration of the feldspar (chiefly microperthite). He states (p. 412): "if the feldspar crystals were perfect, no loss of helium could occur, and even with a normal amount of

---

[23] This investigation was made in 1937 although its publication was delayed until 1940.

94

HELIUM METHOD — Igneous rock age ratios in millions of years			LEAD METHODS — Radioactive mineral age ratios in millions of years	
Acidic rocks	Basic rocks		Lead ratios	Isotope ratios
Mt. Pawagarh felsite........ 5	Porphyritic basalt........ 5			
Obsidian Cliff......... 8	Tower Falls basalt........ 5			
Yellowstone rhyolite...... 22	Crescent Hill basalt....... 9			
	Overhanging Cliff........ 11	TERTIARY	34 Uraninite, Mexico	
	Cleveland Dike........ 17		Brannerite, Idaho	
	Giant's Causeway........ 22			
	Deccan basalt........ 37			
	Geode Creek basalt....... 36			
	Kimberlite......... 58	CRETACEOUS	70 Pitchblende, Colorado	
		JURASSIC	123 Ishikawaite, Japan	140
	Oldwick basalt........ 57	TRIASSIC		
		PERMIAN	220 Thorite, Norway	
	Whin Sill......... 196		Jachymov, Bohemia	
Fitchburg granite....... 45		CARBONIFEROUS	232 Uraninite, N. Carolina	
Quincy granite........ 66	Monchiquite dike........ 83			
Chelmsford granite...... 72	Monchiquite dike........ 105			
	Eustis Dike......... 54	DEVONIAN	269 Pitchblende, Silesia	
Franklin granite........ 60	Eustis Earlier dike........ 81		278 Various Minerals, Conn.	
Rockport granite (fine).... 88	Dalhousie andesite....... 160	SILURIAN		
Rockport granite....... 100			349 Uraninite, Mass.	
Hampstead granite...... 109		ORDOVICIAN	Cyrtolite, New York....	300
			366	375
			371 Uraninite, Conn.	
Gananoque granite....... 150	Gogebic dike........ 190	CAMBRIAN	400 Kolm, Sweden.......	770
Kingston granite........ 280	Long Lake dike........ 290		580 to 995 Pitchblende, Katanga	610 to 665
La Motte granodiorite..... 195	Noranda diabase........ 250		531 Ceylon Thorianite....	485
Taschereau granite...... 250	Gwalior basalt........ 500		765 Besner uraninite......	825
	Ropruchel basalt........ 600		803 Morogoro uraninite.....	595
			882 Pied des Monts clevette..	905
			900 Broggerite, Norway.....	
Varsan granodiorite...... 860	Beartooth trap........ 1050		1085 Norwegian clevette......	1090
			1077 Wilberforce uraninite....	1035
	Stillwater norite........ 1830		1251 Great Bear Lake pitchblende......	1420
			1500 Uraninite, S. Dakota	
			1570 Huron Claim uraninite....	
			1850 Uraninite, Russia	2200

FIGURE 10.—*Radioactive time scales*

95

crystal imperfection, the permeability would be low." However, Keevil did suggest that if further work on fresh minerals substantiated these observations (p. 416), "absolute helium age determinations will be obtained only from certain minerals and not from the rock as a whole."

Recent developments.—In view of these numerous uncertainties in the applicability of the helium method, it appeared essential that a really fundamental investigation be made of the basic requirements underlying this method. The preceding review of our present knowledge concerning the first three requirements constitutes a part of this work. The progress made to date in collaboration with Dr. P. M. Hurley on the evaluation of the fourth requirement is summarized in the paper by Hurley and Goodman. Definite differences in helium retentivity have been found for various common rock minerals. These differences offer a very logical explanation for the variations in age ratios observed for basic and acidic rocks, owing to the higher feldspar content of the latter. It can also be stated with considerable surety that all helium age ratios on rocks containing feldspar as an essential mineral are too low owing to the leakage of helium from this mineral. Since all the acidic rocks in Table 5 and most of the basic rocks in Table 4 contain appreciable proportions of feldspar, the large majority of these age ratios must be considered only as minimal values. Hence, these results are of little use in the establishment of a helium time scale or for correlation purposes. However, there is a possibility that the helium retentivity of even the more permeable rock minerals can be established as a fairly specific quantity. If this can be done, a knowledge of the mineral composition of some of the rocks previously studied may enable estimates to be made of the true helium age. Until this investigation has been made, it appears necessary to include in the helium time scale only those measurements made on selected rock minerals. This restriction limits the acceptable age ratios to those reported by Hurley and Goodman. Fortunately, the suitable minerals appear to be rather widely distributed in igneous rocks. Magnetite, in particular, offers considerable promise. Age-ratio measurements made on magnetites from widely varying horizons appear to be in a sequence consistent with geologic knowledge and in quite good agreement with corresponding lead ages. By the continuance of this work, it is anticipated that a new helium time scale can be established on a sound scientific basis. Although complicated by the necessity for greater selection in the choice of samples, the new helium method should provide a valuable aid to geologists in the solution of the numerous important problems awaiting a reliable dating method for rocks.

Lead Isotopes

IV

Editor's Comments on Papers 9, 10, and 11

In the same year (1929) that Dubey and Holmes published the first successful analyses of the radiogenic helium contents of common igneous rocks (Paper 6), F. W. Aston, working at the Cavendish Laboratory in Cambridge, published the first *isotopic* analyses by mass of a radiogenic daughter element, in this case lead (Paper 9). These results lead immediately to a new estimate for the age of the earth calculated by Rutherford (Paper 11), and three months later C. N. Fenner and C. S. Piggot reported the first isotopic mineral ages based on Aston's data (Paper 10). All three papers represent the product of a fruitful collaboration between the Geophysical Laboratory in Washington, D.C., and the Cavendish Laboratory in Cambridge, and mark one of the most important advances in the field of geochronology.

The concept of isotopes, a term proposed by Soddy in 1913, was first applied to radioactive substances which had identical chemical, but different physical properties. For instance, ionium, discovered by Boltwood in 1906 and recognized as a member of the uranium series and parent of radium, was chemically identical to thorium but had a different atomic weight and decayed more rapidly. Both thorium and ionium would occupy the same place in the periodic table of elements and were appropriately termed isotopes (*iso*—equal, *topos*—place). By 1919, F. W. Aston had improved an early positive-ray apparatus of J. J. Thomson's at the Cavendish Laboratories, Cambridge, and showed conclusively that not only the radioactive elements, but stable elements also, consisted of isotopes whose masses were almost exact integers. Aston worked first with the inert gases, neon, argon and krypton, and with chlorine gas, and by 1925 the existence of stable isotopes of many of the elements had been established. Would Aston's mass spectrograph reveal the existence of isotopes of uranium, thorium, and lead?

Few doubted that it would. The accepted atomic weight of ordinary lead (207.90), was intermediate between the atomic weight of lead extracted from pure uranium minerals (206.05) and that extracted from pure thorium minerals (207.90), and two radiogenic lead isotopes of mass 206 and 208 were anticipated from the decay of uranium and thorium respectively. For the purposes of age determination, the lead present in uranium-rich minerals, where complications arising from the radiogenic additions of thorium-derived lead could be safely ignored, was commonly corrected for the presence of admixtures of "ordinary" (nonradiogenic) lead on the basis of atomic weight determinations. For example, if the atomic weight of lead extracted from a uranium-rich mineral was determined as 206.4, it would be considered a mixture of one-third ordinary lead and two-thirds uranium-derived lead. The gross chemical lead : uranium ratio could then be reduced by one-third to give the required proportion of radiogenic lead for age calculation.

Lead, however, proved difficult to analyze with the mass-spectrographic equipment available in the 1920s, and the first attempts by Aston in 1925, using finely divided metallic lead, produced a totally inadequate mass spectrum. The following year, C. S. Piggot of the Geophysical Laboratory in Washington, D.C., wrote to Aston in Cambridge suggesting a cooperative venture involving the mass analysis of a volatile organic lead compound which might be prepared in the U.S. and sent to Cambridge for analysis. After some discussion it was finally decided to try lead tetramethyl, which at that time was being manufactured by the US Chemical Warfare Service. In 1927, Piggot obtained a sample of pure lead tetramethyl from the Chemical Warfare Service and sent it to Aston. Mass-spectrographic analysis of the lead tetramethyl vapor revealed the presence of three isotopes of lead, Pb-206, 207, and 208, in the ratio 4 : 3 : 7, giving good agreement with the accepted atomic weight of lead, 207.2.

Having demonstrated the existence of three lead isotopes in an ordinary lead sample, the next step was to repeat the experiment using the lead extracted from a radioactive mineral. Piggot obtained "from a trustworthy dealer" a mineral sample bearing the label "Uraninite, var. Bröggerite, Karlshus, Raade, Smaalenene, east of Kristianiafiord, Norway." This sample was analyzed at the Geophysical Laboratory in Washington by C. N. Fenner, and contained 61.158 per cent uranium, 4.377 per cent thorium, and 8.818 per cent lead. Fifteen grams of lead was then extracted as the chloride, and a portion (5 g) converted to tetramethyl by S. C. Witherspoon of the Chemical Warfare Service. A sealed tube containing the precious liquid was sent to Aston early in 1928.

On arrival in Cambridge this first tube was found to have been broken in transit, and the lead tetramethyl unfortunately lost. Luckily, however, 10 g of the original chloride of radiogenic lead extracted from the Bröggerite sample still remained in Washington, and a second sample of tetramethyl lead was synthesized and reached Aston safely in the summer of 1928. The mass-spectrographic results were reported on March 2, 1929 (Paper 9). Radiogenic lead from the Bröggerite was highly enriched in lead-206. The mass-analysis revealed 86.8 per cent Pb-206 and only 3.9 per cent

Pb-207, and Fenner and Piggot were able to report the first age calculations to be based on the new isotopic analyses on May 25, 1929 (Paper 10).

The calculated uranium–lead isotope age (combining lead-206 and lead-207) was 908.6 million years, which compared favorably with the chemical lead–uranium age (919 million years), but the thorium–lead isotope age (1313 million years) was not in agreement with the uranium age, and because of this Fenner and Piggot were somewhat disappointed with the results of the experiment. They suggested that more precise isotopic analyses would be required before the techniques of mass spectrography could be usefully employed in geochronology. They were quite right of course, but there was one more significant deduction to be made from the mass-spectrographic data. After he had obtained his photographic record of the Bröggerite lead spectrum, Aston showed his results to Rutherford and pointed out that the presence of lead-207 in the spectrum could not be due to contamination by ordinary lead, as there was no significant lead-208. It followed that lead-207 must represent the end product of a separate decay series—the actinium series. Knowing the mass number of the stable end product, all the mass numbers of the intermediate radioactive daughters could be calculated from the known disintegration characteristics. Protactinium must, on this basis, have a mass number of 231.

Rutherford went even further (Paper 11). It had already been suggested by several investigators that the parent of the actinium series was not protactinium, but an isotope of uranium. On the basis of Aston's results, Rutherford suggested that the mass of this uranium isotope must be 235, and estimated that the present abundance ratio U-238/U-235 was 357 (the modern accepted figure is 137.7). He found that the decay constant of U-235 could be estimated in proportion to the decay constant for U-238 from the relative number of radioactive daughter atoms in the uranium and actinium decay series, assuming that the Bröggerite sample was 10^9 years old. Knowing the decay constants of both uranium isotopes, and assuming that these isotopes were originally present in equal numbers at the time of formation of the earth, he then calculated the elapsed time required for the U-238/U-235 ratio to increase from a value of unity to the present-day value. The elapsed time would be in effect the age of the earth. Rutherford's calculations resulted in a figure of 3.4×10^9 years, and in 1929 this figure, the first to be based on isotopic analyses, was about twice the age of the oldest known radioactive minerals.

Selected Bibliography

Aston, F. W. (1919). A positive ray spectrograph. *Phil. Mag., Ser. 6*, **38**, 707–714.

Aston, F. W. (1923). "Isotopes". 152 pp. Arnold and Co., London.

Aston, F. W. (1927). The constitution of ordinary lead. *Nature*, **120**, 224.

Aston, F. W. (1933). The isotopic constitution and atomic weight of lead from different sources. *Proc. Roy. Soc., Ser. A*, **140**, 535–543.

Boltwood, B. B. (1907). The origin of radium. *Nature*, **76**, 544–545, 589.

Dubey, V. S., and Holmes, A. (1929). Estimates of the ages of the Whin Sill and the Cleveland Dyke by the helium method. *Nature*, **123**, 794–795.

Piggot, C. S. (1928). Radium and geology. *J. Amer. Chem. Soc.*, **50**, 2910–2916.

Piggot, C. S. (1928). Lead isotopes and the problem of geologic time. *J. Wash. Acad. Sci.*, **18**, 269–273.

Piggot, C. S. (1933). Isotopes of uranium, thorium, and lead, and their geophysical significance. *Phys. Rev.* **43**, 51–59.

Soddy, F. (1913). Intra-atomic charge. *Nature,* **92**, 399–400.

Von Grosse, A. (1932). On the origin of the actinium series of radioactive elements. *Phys. Rev.* **42**, 565–570.

Von Grosse, A. (1934). On the origin of the actinium series of radioactive elements II. *J. Phys. Chem.* **38**, 487–494.

Reprinted from *Nature*, **123**, 313 (1929)

9

Letters to the Editor.

[*The Editor does not hold himself responsible for opinions expressed by his correspondents. Neither can he undertake to return, nor to correspond with the writers of, rejected manuscripts intended for this or any other part of* NATURE. *No notice is taken of anonymous communications.*]

The Mass-Spectrum of Uranium Lead and the Atomic Weight of Protactinium.

IT will be recalled (NATURE, Aug. 13, 1927) that the identification of the isotopes of ordinary lead was made by means of a sample of its tetramethide kindly supplied to me by Mr. C. S. Piggot, of the Geophysical Laboratory, Washington. He has since succeeded in the much more troublesome task of preparing the similar compound of a rare uranium lead from Norwegian bröggerite. His reasons for this work have already been published (C. S. Piggot, " Lead Isotopes and the Problem of Geologic Time," *Jour. Wash. Acad. Sci.*, May 19, 1928). The first tube of uranium lead methide despatched to me a year ago was unfortunately broken in transit, but the second reached Cambridge safely last summer. At that time I was endeavouring to work out a photometric method of measuring the relative abundance of isotopes. This work is by no means complete, but has recently reached a stage which justified an attempt on the mass-spectrum of this very precious material. The procedure was the same as with ordinary lead methide, but the general conditions of the discharge tube, etc., were not so favourable, so that the spectra obtained are weaker.

The mass-spectrum consists of a strong line at 206, a faint one at 207, and a still fainter one at 208. The last is barely visible to the eye, but easily distinguish- able on the photometer curves. The impossibility of eliminating mercury limits the search for lighter isotopes, but there is not the least indication of 203 or 205. Unfortunately, the experimental conditions all conspire to make the determination of the true relative intensities of the lines from the curve of photometer wedge readings too complex to be really trustworthy. Calling the intensity of the strong line 100, the mean of the best plates gives $10\cdot7 \pm 3$ and $4\cdot5 \pm 2$ for 207 and 208 respectively. As the only curve available for transforming wedge readings into intensities is one derived from krypton, these figures are probably both too high. They correspond to percentages $86\cdot8 : 9\cdot3 : 3\cdot9$, and as the packing fraction is indistinguishable from that of mercury ($0\cdot8 \times 10^{-4}$), the mean atomic weight deduced is $206\cdot19$, rather higher than that determined chemically for other uranium leads. These figures have been communi- cated to Mr. Piggot, and when combined with the analyses of the mineral should enable its age to be fixed with considerable certainty.

There is, however, another point of view from which these results are of fundamental interest in connexion with the radioactive elements. The line 207 is of peculiar significance. It cannot be due to the presence of lead as an impurity, for in ordinary lead 208 is about twice as strong as 207, neither can it be the product of radium or thorium. It is difficult to resist the natural conclusion that it is the end product of the only other known disintegration, namely, that of actinium. If this is so it settles the mass numbers of all the members of this series, that of protactinium being 231. Extrapolation of the packing fraction curve suggests an atomic weight on the oxygen scale of $231\cdot08$. F. W. ASTON.

Cavendish Laboratory,
Cambridge, Feb. 16.

No. 3096, VOL. 123]

10

Letters to the Editor.

[The Editor does not hold himself responsible for opinions expressed by his correspondents. Neither can he undertake to return, nor to correspond with the writers of, rejected manuscripts intended for this or any other part of Nature. *No notice is taken of anonymous communications.]*

The Mass-Spectrum of Lead from Bröggerite.

In the issue of Nature for Mar. 2, 1929, Dr. Aston gives the results of his determination of the mass-spectrum of a sample of lead in the form of its tetramethyl compound, of which the lead had been extracted by us from a sample of Norwegian bröggerite. We obtained the lead in the form of chloride, and took particular care to have it free from impurities. The conversion of the chloride into tetramethyl was kindly carried out for us by Mr. S. C. Witherspoon, and care was taken to test all chemicals and reagents used to see that they were free from lead.

Dr. Aston discusses his results and reaches interesting conclusions, and a further discussion is given by Sir Ernest Rutherford. It may be of interest to consider the matter further, in the light of our analysis of the mineral.

The specimen was obtained from a trustworthy dealer and bore the label " Uraninite, var. Bröggerite, Karlshus, Raade, Smaalenene, east of Kristianiafiord, Norway ". It appeared to be homogeneous except for a little pink feldspar, mica, and quartz, and was of an iron-grey colour and of the general appearance of massive magnetite, but with some crystal faces. Close examination showed no evidence of its having been acted upon by weathering processes. Our analysis is as follows :

$$U_3O_8 = 72 \cdot 12 \text{ per cent} \left. \begin{array}{l} \text{equi-} \\ \text{valent} \\ \text{to} \end{array} \right\} \begin{array}{l} U = 61 \cdot 158 \text{ per cent.} \\ Th = 4 \cdot 377 \quad ,, \quad ,, \\ Pb = 8 \cdot 018 \quad ,, \quad ,, \end{array}$$

$ThO_2 = 4 \cdot 98$,, ,, $PbO = 8 \cdot 64$,, ,,

We have confidence in the essential accuracy of these figures.

For calculating the age we used the formula given by the International Critical Tables of the National Research Council :

$$\text{Age} = \frac{\log (U + 0 \cdot 38 \text{ Th} + 1 \cdot 156 \text{ Pb}) - \log (U + 0 \cdot 38 \text{ Th})}{6 \cdot 5} \times 10^{11} \text{ years.} \quad \text{(I)}$$

This gives an age of $919 \cdot 5 \times 10^6$ years for this mineral. Changes which might be made because of some variation in the values of the disintegration constants involved in the factor $6 \cdot 5$ of the formula are not likely to be of large amount. The calculated age is in good agreement with previous determinations by others on uranium minerals from the same general locality. We may now compare this value with results obtained by making use of Dr. Aston's figures in connexion with our analytical results.

Dr. Aston gives the figures 86·8, 9·3, and 3·9 as the percentage values obtained for Pb^{206}, Pb^{207}, and Pb^{208} present in the lead tetramethyl. Of these

isotopes of lead, the first and second have presumably been derived from uranium and its isotope actino-uranium, and the third from thorium. In analysis, uranium^{238} and actino-uranium are necessarily determined together as 'uranium', and their disintegration has resulted in Pb^{206} and Pb^{207} respectively. Calculations of age from these elements, disregarding thorium and Pb^{208}, should give practically the same result as the original calculation, and these results in turn should agree with the result that thorium and Pb^{208} give. For uranium plus actino-uranium we express the formula as

$$\text{Age} = \frac{\log (U + 1 \cdot 156 \ Pb^{206+207}) - \log U}{6 \cdot 5} \times 10^{11} \text{ years (II)}$$

and get

$$\text{Age} = 908 \cdot 4 \times 10^6 \text{ years.}$$

This may be considered a satisfactory agreement with the $919 \cdot 5 \times 10^6$ years previously obtained.

For thorium and its lead we have

$$\text{Age} = \frac{\log (0 \cdot 38 \text{ Th} + 1 \cdot 156 \ Pb^{208}) - \log (0 \cdot 38 \text{ Th})}{6 \cdot 5} \times 10^{11} \text{ years.} \qquad \text{(III)}$$

From this calculation, however, we get the result

$$\text{Age} = 1313 \times 10^6 \text{ years,}$$

which is widely different from the previous figures.

It is pertinent to inquire as to the probable cause of the discrepancy.

In Dr. Aston's account he expresses some uncertainty as to relative intensities of the lead lines, and gives a margin of possible error of ± 2 for Pb^{208}. In view of the small total quantity of Pb^{208}, this means a large percentage error, the possible variation running from 5·9 to 1·9 per cent, and corresponding ages (calculated by formula III) running from 1900×10^6 to 671×10^6 years. The limits of error, therefore, include the value $919 \cdot 5 \times 10^6$ deduced from the original calculations, but if this is accepted as correct, Dr. Aston's figure for Pb^{208} apparently requires correction to bring it into harmony. The limits of error he himself sets likewise point to the desirability of greater refinement of photometric measurement in order to make the results serve for age calculations. Instead of 3·9 per cent of Pb^{208} given by him, our figures indicate 2·64 per cent, which is obtained by substituting in formula III the age $919 \cdot 5 \times 10^6$ years and the analytical value of thorium, and solving for Pb^{208}.

There is, however, another aspect of this matter which should be considered. Formula III involves the factor 0·38, accepted as expressing the disintegration equivalence of thorium in terms of uranium. It may be thought that it is this factor which should be revised, as there has been some variation in determinations of the value of this quantity among different experimenters. As a basis for judgment in this matter we may make a new calculation of the conversion factor from the data supplied by Dr. Aston. For this purpose we combine formulæ II and

III in the form

$$\frac{U + 1 \cdot 156 \ Pb^{206+207}}{U} = \frac{xTh + 1 \cdot 156 \ Pb^{208}}{xTh} \qquad (IV)$$

and solve for x.

Such a calculation does not involve the correctness of the constants in the uranium series, but only the value of the conversion factor required to get identical results for the uranium series and the thorium series.

Proceeding in this manner, we get the result 0·57. Possibly it may be regarded as an open question whether the accepted value 0·38 obtained by direct measurement by physicists does not require correction to bring it into closer accord with the figure 0·57 derived from Dr. Aston's work, but in reading Dr. Aston's letter we are left with the impression that Dr. Aston himself does not wish to be held too strictly to the numerical values that he gives.

Furthermore, previous work by one of us (*Amer. Jour. Sci.*, November 1928) has given support to the substantial correctness of the figure 0·38. Two minerals from a certain deposit in Brazil were analysed, after taking means to remove weathered products. One was a uranium mineral carrying little thorium, and the other was a thorium mineral carrying almost no uranium. From the results the ages were calcu-lated, using for thorium the equivalence ratio 0·38. The ages found for the two were in close agreement.

The investigation, of which the results have been reported by Dr. Aston, was suggested (by C. S. P.) in the hope of obtaining a direct determination of uranium lead (Pb^{206}) and thereby improving the accuracy of the existing formula for calculating ages. It was also hoped that the uranium-thorium equivalence factor (0·38) could be independently determined and perhaps improved, in order that the determination of geological ages might be rendered more certain. From a consideration of the matter in the light of the analysis, it seems probable that a higher degree of precision in the measurement of the intensity of lead lines will be necessary in order to attain these ends. We hope that future work by Dr. Aston will bring this about. In any event, we are happy to know that our sample has been useful to Dr. Aston in finding fairly conclusive evidence of the existence of actino-uranium.

C. N. FENNER.
C. S. PIGGOT.

Geophysical Laboratory,
Washington, D.C.,
Mar. 25.

11

Origin of Actinium and Age of the Earth.

By the kindness of Dr. Aston, I have had the opportunity of inspecting his photographs showing the isotopes of lead obtained from the radioactive mineral bröggerite. As he concludes, it seems highly probable that the isotope of mass 207 is mainly due to actinium lead, and that the actinium series has its origin in an isotope of uranium—a suggestion independently put forward by several investigators on other evidence. Since six α particles are emitted in the successive changes from protactinium to the end product actinium lead, the atomic weight of protactinium should be 231. The direct determination of the atomic weight of this element number 91 now in progress in the laboratory of Prof. Hahn in Berlin should afford a crucial test of the accuracy of this deduction.

In the light of this new knowledge and of the measurements made by Dr. Aston of the relative intensities of the lead isotopes in the mineral, it may be of interest to consider its bearing on the origin of actinium and other problems. We shall first discuss the probable mass of this new isotope, which for convenience will be called actino-uranium. It seems simplest to suppose that its mass is 235, and that it undergoes first an α and then a β ray transformation into protactinium. The β ray body is probably to be identified with uranium Y, discovered by Antonoff, which has generally been regarded as the immediate parent of protactinium. On this view, the successive transformations follow the order αβαβ, where the α and β changes alternate, and differ in this respect from the main uranium series which follow the order αββα. It is of course possible to assume that actino-uranium has a mass 239 and number 92, and is converted into a mass 235 of number 92 in consequence of an α ray change followed by two β ray transformations, but no evidence has been obtained of the existence of such β ray bodies, although a careful search has been made for them by Hahn and others.

An estimate of the period of transformation of the new isotope of uranium can be deduced on certain probable assumptions. The ratio K' of the number of atoms of actinium lead to those of uranium lead can be deduced approximately from Aston's measurements, and we also know the ratio K—about 3/100—of the number of atoms delivered in a mineral into the actinium series compared with the number passing into the radium series. If λ_1, λ_2 are the constants of transformation of actino-uranium and the main uranium isotope respectively, it can easily be deduced

that $K'/K = \dfrac{\lambda_2}{\lambda_1} \cdot \dfrac{e^{\lambda_1 t} - 1}{e^{\lambda_2 t} - 1}$, where t is the age of the mineral from which the lead is derived. We shall suppose for the purpose of calculation that t is 10^9 years—an average estimate of the age of old primary uranium minerals. Taking as a low estimate that $K' = 7/100$, it can be deduced from the equation that $\lambda_1/\lambda_2 = 10.6$. Since the half-value period of transformation of uranium is 4.5×10^9 years, it follows that the period of actino-uranium is 4.2×10^8 years. A larger value of K' lowers the period, while a higher value for the age of the mineral raises it.

Taking the period as 4.2×10^8 years, it is seen that the amount of actino-uranium is only about 0.28 per cent of the main uranium isotope—an amount too small to influence appreciably the atomic weight of uranium as ordinarily measured. The amount of actino-uranium at the time of its formation taken as 10^9 years age comes out to be 1.44 per cent.

There is another interesting deduction that can be made from these estimates. It is natural to suppose that the uranium in our earth has its origin in the sun, and has been decaying since the separation of the earth from the sun. From the work of Aston, it is known that with two exceptions the most abundant isotope in an even numbered element is of even atomic weight. If it be supposed that uranium, like other heavy elements, is formed from stellar matter, it is likely that actino-uranium of odd atomic weight would be formed in smaller quantity than the main isotope of even atomic weight. Even, however, if we suppose they were formed in equal quantity, it can be shown that it would require only 3.4×10^9 years to bring down the amount to the 0.28 per cent observed to-day.

If we suppose that the production of uranium in the earth ceased as soon as the earth separated from the sun, it follows that the earth cannot be older than 3.4×10^9 years—about twice the age of the oldest known radioactive minerals. In addition, if the age of the sun is of the order of magnitude estimated by Jeans, namely, 7×10^{12} years, it is clear that the uranium isotopes which we observe in the earth must have been forming in the sun at a late period of its history, namely, about 4×10^9 years ago. If the uranium could only be formed under special conditions in the early history of our sun, the actino-uranium on account of its shorter average life would have practically disappeared long ago. We may thus conclude, I think with some confidence, that the processes of production of elements like uranium were certainly taking place in the sun 4×10^9 years ago and probably still continue to-day.

E. RUTHERFORD.

Primeval Lead

V

Editor's Comments on Paper 12

Following Aston's (1927, 1933) pioneer work on the isotopic composition of lead isotopes, Rose and Stranathan (1936) pointed out that the ratio of the two radiogenic lead isotopes, Pb-207 and Pb-206, generated by decay of the two parent uranium isotopes, U-235 and U-238, must increase systematically with time due to the different decay rates of the parents. The age of any uranium-bearing mineral could thus be determined simply by extracting the radiogenic lead and measuring the Pb-207/Pb-206 ratio. Rose and Stranathan determined Pb-207/Pb-206 ratios from densitometer measurements of the hyperfine structure of the lead emission spectra (using the λ5372 line) for a number of radiogenic leads, and calculated Pb-207/Pb-206 ages in good agreement with chemical uranium–lead ages.

An even more important step from a point of view of geochronology was provided by further developments in mass spectrometry and the lead isotopic analyses of Alfred Nier (1938; 1939a, b). Working in the Department of Physics at the University of Minnesota as an assistant, and later associate, professor between the years 1938 and 1941, Nier discovered that there existed systematic variations in the relative abundances of the stable isotopes of common lead which were quite unsuspected. The atomic weight of common ore lead is measureably constant, and it had always been assumed that this implied a constant isotopic composition. Mass-spectrometric analysis showed that common lead samples with higher Pb-206 per cent abundances tend to have lower Pb-208 abundances (see Russell and Farquhar, 1960, p. 14) but increases in Pb-206 relative to Pb-204 are often accompanied by increases in the Pb-208/Pb-204 ratio, so that the isotopic variations fortuitously result in reasonably constant mean mass numbers and atomic weights.

Isotopic fractionation effects could not account for the magnitude of the variations observed in such a heavy element as lead (atomic no. 82), and Nier suggested that the isotopic composition of common lead extracted from lead ore minerals such as galena consisted of a mixture of *primeval* lead of fixed isotopic composition, which existed at the time of formation of the earth, and radiogenic lead, which was subsequently generated by decay of uranium and thorium. Both components were dispersed in the earth prior to their concentration and emplacement as a separate lead ore deposit. Nier's isotopic data were of great importance in that they provided numerical evidence for the evolution of lead isotopes on which the first really precise determinations of the age of the earth could be based. He also determined the relative abundance of the two parent uranium isotopes, U-235 and U-238, discovered the geochronologically important radioactive isotope of potassium, K-40, and following World War II demonstrated, with L. T. Aldrich, the accumulation of radiogenic Ar-40 in potassium-bearing minerals (Paper 31).

In the case of lead, Nier began by analyzing the isotopic variations in common lead extracted from lead ore minerals (mostly galena) from various sources, and immediately followed this by a mass-spectrometric determination of the present-day abundances of the uranium isotopes, from which he was able to provide an accurate value for their decay constants. He then determined the isotopic composition of a large number of radiogenic lead samples extracted from various uranium and thorium minerals, and recommended that in the event of chemical alteration, age determinations based on the radiogenic Pb-207/Pb-206 ratio, as shown by Rose and Stranathan (1936), would be more reliable than either Pb-206/U-238 or Pb-208/Th-232 ages. In the last report of the series (Paper 12), Nier, together with R. W. Thompson and B. F. Murphey, reports the results of isotopic analyses of lead separated from both radioactive (U + Th) minerals and lead ore minerals. One of the thorium minerals (No. 28) a monazite from the Huron district, gave a calculated Pb-207/Pb-206 age of 2570 ± 70 million years. A uraninite sample (No. 14) from the same locality gave a similar Pb-207/Pb-206 age of 2200 million years (Nier, 1939b, p. 159). These determinations represent the first substantial isotopic evidence in support of the great antiquity of the Precambrian.

Selected Bibliography

Aston, F. W. (1927). The constitution of ordinary lead. *Nature, 120*, 224.

Aston, F. W. (1933). The isotopic constitution and atomic weight of lead from different sources. *Proc. Roy. Soc., Ser. A, 140*, 535–543.

Nier, A. O. (1938). Variations in the relative abundances of the isotopes of common lead from various sources. *J. Amer. Chem. Soc., 60*, 1571–1576.

Nier, A. O. (1939a): The isotopic constitution of uranium and the half-lives of the uranium isotopes. I. *Phys. Rev., 55*, 150–153.

Nier, A. O. (1939b). The isotopic constitution of radiogenic leads and the measurement of geological time. II. *Phys. Rev., 55*, 153–163.

Rose, J. L., and Stranathan, R. K. (1936). Geologic time and isotopic constitution of radiogenic lead. *Phys. Rev., 50*, 792–796.

Reprinted from THE PHYSICAL REVIEW, Vol. 60, No. 2, 112–116, July 15, 1941
Printed in U. S. A.

The Isotopic Constitution of Lead and the Measurement of Geological Time. III

ALFRED O. NIER, ROBERT W. THOMPSON AND BYRON F. MURPHEY
Department of Physics, University of Minnesota, Minneapolis, Minnesota
(Received June 5, 1941)

12

A mass spectrographic measurement of the relative abundances of the isotopes in eight samples of radiogenic lead and thirteen samples of common lead has been made. As five of the radiogenic lead samples originated from minerals containing both uranium and thorium, three independent determinations of the age could be made. One of the samples was the oldest so far studied and appears to have an age close to two billion years. The common lead samples were found to have large variations in the relative abundances of the same sort as were reported in a previous investigation of twelve other samples.

THE well-established fact that uranium and thorium disintegrate naturally to form, ultimately, isotopes of lead makes possible the measurement of the geological age of uranium and thorium minerals. As the decay constants of uranium and thorium are accurately known, a knowledge of amounts of lead, uranium, and thorium contained in a given mineral enables one to compute its age provided the atomic weight, or alternately, the isotopic constitution of the lead, is known. As common lead is often present as an impurity it is important to know its isotopic constitution also. In this paper will be discussed the results obtained from the isotopic analyses of a number of radiogenic and common lead samples.

APPARATUS AND PROCEDURE

The mass spectrometer used in this work was essentially identical with one described earlier.[1] As a source of lead ions, lead iodide vapor is bombarded with low energy electrons. The mass analyses of the ions is accomplished by a 180° magnetic analyzer. As the analyzed ion currents are measured with an electrometer tube amplifier, rapid and accurate measurements of the relative abundances are possible. The general procedure used in analyzing lead has already been described.[2,3]

[1] A. O. Nier, Phys. Rev. **52**, 933 (1937).
[2] A. O. Nier, J. Am. Chem. Soc. **60**, 1571 (1938).
[3] A. O. Nier, Phys. Rev. **55**, 153 (1939). Several errors have been found in this paper. Equation (5) should have the coefficient 1/139 instead of 139. In Table I, for sample 9, column 3 should have 3.09 instead of 0.309; for sample 20, columns 17 and 18 should have 0.0106 and 235, respectively, instead of 0.016 and 355. In Table II the headings 206 and 208 should be interchanged. In footnotes 23 and 29 the initials should be A. D. rather than E. W. Bliss.

ANALYSIS OF RADIOGENIC LEAD

Because the intermediate members of the radioactive families have short half-lives in comparison with the ages of the minerals with which we are concerned one can write the following equations relating the numbers of atoms of radiogenic lead, thorium and uranium isotopes:

$$N(\mathrm{Pb}^{206}) = N(\mathrm{RaG})$$
$$= N(\mathrm{U\ I})(\exp[\lambda(\mathrm{U\ I})t]-1), \quad (1)$$

$$N(\mathrm{Pb}^{207}) = N(\mathrm{AcD})$$
$$= N(\mathrm{AcU})(\exp[\lambda(\mathrm{AcU})t]-1), \quad (2)$$

$$N(\mathrm{Pb}^{208}) = N(\mathrm{ThD})$$
$$= N(\mathrm{Th})(\exp[\lambda(\mathrm{Th})t]-1). \quad (3)$$

A chemical analysis of the mineral gives the relative amounts of U, Th and Pb. An isotopic analysis of the lead, after being corrected for common lead contamination, yields the relative amounts of RaG, AcD and ThD. Thus, as the decay constants are known, three independent determinations of a mineral's age are possible provided it contains both uranium and thorium in sufficient amounts to permit an accurate chemical analysis.

It is customary to apply Eqs. (1) and (3) directly whereas (2) is applied indirectly by dividing (2) by (1) to give

$$\frac{\mathrm{AcD}}{\mathrm{RaG}} = \frac{N(\mathrm{AcU})}{N(\mathrm{U\ I})} \frac{\exp[\lambda(\mathrm{AcU})t]-1}{\exp[\lambda(\mathrm{U\ I})t]-1}. \quad (4)$$

As $\lambda(\mathrm{AcU})$ is not known directly it may be evaluated by means of the expression

$$R = \lambda(\mathrm{AcU})N(\mathrm{AcU})/\lambda(\mathrm{U\ I})N(\mathrm{U\ I}), \quad (5)$$

112

TABLE Ia. *Isotopic abundances in radiogenic lead.*

1	2	3	4	5	6	7 ThD	8 AcD	9 RaG
No.	MINERAL AND SOURCE*	208	ISOTOPIC ABUNDANCES 207	206	204	RELATIVE TO TOTAL Pb206 =100†		
8a. Curite[1] Katanga, Africa		0.178 ±0.005	6.18	100	—			
22. Uraninite[2] Parry Sound, Canada		1.52	7.40	100	0.0062 ±0.0008	1.28	7.30	99.9
23. Samarskite[3] Glastonbury, Conn.		21.3	7.60	100	0.167 ±0.01	14.9	5.03	96.9
24. Pitchblende[4] Woods Mine, Colo.		38.0	19.4	100	1.01 ±0.05	−0.6	3.8	81.3
25. Pitchblende[4] Gilpin County, Colo.		44.5	22.0	100	1.17 ±0.03	−0.2	4.0	78.4
26. Thucholite[5] Parry Sound, Canada		6.90	5.88	100	0.024 ±0.005	5.98	5.51	99.5
27. Monazite[6] Mt. Isa Mine, Australia		100	1.03 ±0.05	5.82	0.041 ±0.003	98.4	0.40	5.05
28. Monazite[4] Huron Claim, Manitoba, Canada		100	2.11 ±0.05	11.6	0.011 ±0.002	99.6	1.94	11.4
29. Monazite[4] Las Vegas, New Mexico		100	1.25 ±0.05	10.1	0.028 ±0.002	98.9	0.82	9.6

TABLE Ib. *Analysis and ages of minerals.*

		10	11	12	13	14	15	16	17	18
No.	MINERAL AND SOURCE	MINERAL ANALYSIS % U	% Th	% Pb	RATIO OF NOS. OF ATOMS RaG/U I	ThD/Th	AcD/RaG	AGES IN 106 YR. FROM RaG/U I	ThD/Th	AcD/RaG[7]
22. Parry Sound Uraninite		69.0	2.95	10.8	0.166	0.0483	0.0731 ±1%	1003	945	1030 ±15
23. Conn. Samarskite		6.91	3.05	0.314	0.0396	0.0134	0.0518 ±3%	253	266	280 ±60
24. Woods Mine Pitchblende		72.3	0.11	1.063	0.0088	—	0.0467 ±25%	57	—	—
25. Gilpin Co. Pitchblende		38.28	0.053	0.643	0.0091	—	0.051 ±25%	59	—	—
26. Parry Sound Thucholite		4.63	0.903	0.187	0.0414	0.0123	0.0553 ±2%	265	245	430 ±40
27. Mt. Isa Monazite		0.0	5.73	0.285	—	0.051	0.0792 ±25%	—	1000	1190 ±400
28. Huron Claim Monazite		0.281	15.63	1.524	0.628	0.0955	0.170 ±4%	3180	1830	2570 ±70
29. Las Vegas Monazite		0.122	9.39	0.372	0.303	0.0392	0.0855 ±10%	1730	770	1340 ±200

* References for this column refer to analysis of minerals appearing in columns 10, 11, and 12.
† Except for samples 27, 28 and 29 in which the amounts are relative to total Pb208 =100.
[1] This sample was one referred to in reference 3 of text and was analyzed here only as a check on the present apparatus.
[2] H. V. Ellsworth, *Canadian Geologic Survey*, Economic Geology Bull. 11, 268 and 174 (1932).
[3] R. C. Wells, *Report of the Committee on Determination of Geologic Time* (National Research Council, April, 1935, and September, 1936). The atomic weight of the lead from this sample was found by Baxter, Faull and Tuemmler, J. Am. Chem. Soc. 59, 702 (1937) to be 206.34. The value computed from the isotopic abundances is 206.36 if the P. F. of lead is taken as +1.55, reference 2 of text.
[4] O. B. Muench, *Report of the Committee on the Measurement of Geologic Time*, 1939–1940, p. 88.
[5] O. B. Muench, J. Am. Chem. Soc. 59, 2269 (1937).
[6] Private Communication to Professor A. C. Lane, from R. Blanchard and F. E. Connah.
[7] The estimates of error are based upon error in AcD/RaG ratio which in turn depends upon the errors given in columns 4 and 6.

where R is the present day ratio of the activities of the actinium and uranium series. Since $N(\text{U I})/N(\text{AcU})$ has been shown to be 139.0[4] Eq. (4) may be written as

$$\frac{\text{AcD}}{\text{RaG}} = \frac{1}{139} \frac{\exp[139R\lambda(\text{U I})t]-1}{\exp[\lambda(\text{U I})t]-1}. \quad (6)$$

[4] A. O. Nier, Phys. Rev. 55, 150 (1939).

In Tables Ia and Ib are listed the minerals investigated together with isotope analyses and other pertinent data needed for age computations. Unless otherwise designated the numbers in columns 3, 4, 5, and 6 are believed to be correct within one percent.

Since Pb^{204} may, as far as is known, be attributed entirely to common lead, it is used as an index of the common lead impurity in the samples. Accordingly the data in columns 7, 8, and 9 are found by correcting the measured abundances for common lead. In each case common lead was assumed to have a constitution of 204 : 206 : 207 : 208 :: 1 : 18.5 : 15.4 : 38.2. This set of values corresponds to a type of common lead which would be an average between samples 13 and 14 of Table II, the two samples exhibiting the widest differences in isotopic abundances. Actually, except for samples 24 and 25 so little common lead is present that it makes little difference which type is assumed in making the corrections. In samples 24 and 25 so much common lead is present that the AcD/RaG ratio cannot be computed with sufficient accuracy to have significance even if a better assumption could be made as to the type of common lead present.

In making the age computations the decay constants of Th and U I were chosen as 4.99×10^{-11} yr.$^{-1}$,[5] and 1.535×10^{-10} yr.$^{-1}$,[6] respectively.

In employing Eq. (6) for computing the ages from the AcD/RaG ratio R was assumed to be 0.046 as in the earlier work.[3] It is interesting to compare the ages as computed by the three different methods. It is to be recalled that a close agreement between the three methods may be considered as a good indication that the mineral has not suffered alterations. Conversely, discrepancies in the numbers in columns 16, 17, and 18 are indicative of some alteration, unless the quantities in columns 10 or 11 are so small that the limit of chemical accuracy becomes important, as for the uranium in a monazite and thorium in some uraninites and pitchblendes.

The close agreements in samples 22 and 23 indicate no serious alterations have taken place. Unfortunately, samples 24 and 25 contain so little thorium and so much common lead that age computations are possible only from the RaG/U I ratio. The presence of the large amounts of common lead in these is especially unfortunate as an accurate computation of AcD/RaG cannot

be made. It had been hoped that as AcD/RaG is not critically dependent upon t for low ages in Eq. (6), it would have been possible to use Eq. (6) in reverse manner to obtain an accurate check upon the assumed value of R.

It is not surprising that the AcD/RaG age for sample 26 does not agree with the other two ages. An earlier investigation (reference 5, Table I) of the material indicated that alterations probably had taken place.

As sample 27 contained very little uranium a determination of the amount had not been made. Thus the age from RaG/U I could not be computed. Unfortunately the mass spectrometer does not have sufficient resolution to permit a more accurate measurement of Pb^{207} in the presence of so large a Pb^{208} ion current. Thus the AcD/RaG ratio cannot be measured with the accuracy that might be desired. This same difficulty was encountered in sample 29.

Sample 28 is especially interesting as it comes from a region known to be very old geologically. The analysis of a sample of uraninite from the same locality gave[3] for the age in millions of years as computed from RaG/U I, ThD/Th and AcD/RaG the values 1570, 1252 and 2200, respectively. It is to be recalled that it has been suggested[3] that in the case of the alteration of a mineral AcD/RaG is least affected and hence should serve as the most reliable index of the age. From the present results it appears that the high RaG/U I age may be accounted for by a leaching of uranium. It must be mentioned that such an alteration would give an AcD/RaG age which was also somewhat high. Thus one has some justification in assuming the mineral to be close to two billion years old. A comparison of the three age determinations for the two samples shows that only the AcD/RaG ages are in reasonable agreement.

ANALYSIS OF COMMON LEAD

In Table II are shown the isotopic abundances obtained for a group of common lead samples having different geographical origins. The abundance of Pb^{204} has been arbitrarily chosen as a reference, as it is not produced, as far as is known, by natural radioactive decay. An examination of the table reveals that in spite of a

---

[5] A. F. Kovarik and N. I. Adams, Phys. Rev. **53**, 928 (1938).
[6] This value is based on the new determination of the rate of emission of alpha-particles from uranium given by Kovarik and Adams, J. App. Phys. **12**, 296 (1941). For method of calculation see reference 4.

TABLE II. *Isotopic abundances in common lead.*

No.	Source of Lead[1] Locality	Isotopic Abundances Referred to Pb²⁰⁴ =1.000			Excess over Ivigtut Galena			Excess 207 / 206	208 / 206	Mean[2] Mass Number
		206	207	208	206	207	208			
13. Galena[3] Ivigtut, Greenland		14.54 14.75	14.60 14.70	34.45 34.5						207.261
14. Galena III[4] Joplin, Missouri		22.37 22.28	16.10 16.20	41.8 41.8	7.70	1.50	7.36	0.19	0.96	207.203
15. Galena[5] Tetreault Mine, Canada		16.27	15.16	35.6	1.62	0.50	1.08	0.31	0.67	207.239
16. Bournonite[6] Casapalca, Peru		18.65 18.69	15.49 15.40	38.15 38.15	4.02	0.80	3.68	0.20	0.91	207.225
17. Galena Casapalca, Peru		18.86 18.83	15.71 15.61	38.75 38.5	4.20	1.01	4.16	0.24	0.99	207.226
18. Galena[7] Clausthal, Harz Mts., Germany		18.46	15.66	38.6	3.81	1.01	4.13	0.27	1.08	207.233
19. Galena[8] Przibram, Bohemia		17.95	15.57	37.9	3.31	0.92	3.43	0.28	1.04	207.235
20. Galena[9] Franklin, New Jersey		17.15	15.45	36.53	2.50	0.80	2.06	0.32	0.82	207.233
21. Galena[10] Mexico		18.71	15.70	38.5	4.06	1.05	4.08	0.26	1.00	207.227
22. Native Lead[11] Langban, Sweden		15.83	15.45	35.6	1.18	0.80	1.16	0.68	0.99	207.249
23. Galena[12] Arizona		19.22	16.17	39.15	4.57	1.53	4.68	0.33	1.03	207.226
24. Galena[13] Freiberg, Saxony		18.07	15.40	38.0	3.42	0.75	3.48	0.22	1.02	207.234
25. Galena[14] Alpine Trias, Austria		17.75	16.21	38.05	3.10	1.56	3.58	0.50	1.15	207.237
					Average			0.32	0.97	
8a. Galena II 8b. Joplin, Missouri		21.72 21.62	15.74 15.74	40.3 40.4						
14a. Galena III Joplin, Mo.		22.48	16.15	41.3						

[1] As the bulk of the samples listed below were obtained from the Harvard Mineralogical Museum through the courtesy of Dr. H. Berman, the number given in the footnotes describing the individual samples is the Harvard catalog number.
[2] The chemical atomic weight may be computed from the mean mass number by subtracting 0.025 (see reference 2 of text).
[3] 80014. Occurred in pegmatite, associated with cryolite and siderite; Lindgren's *Mineral Deposits*, p. 765.
[4] Galena, Charles Palache collector; 8a and 8b are from the father of G. P. Baxter.
[5] Galena, Tetreault Mine near Montaubon les Mines, Port Neuf Company, Quebec, from J. N. Herring, Superintendent (letter April 15, 1938). Recommended by J. A. Dresser as an old vein, not pre-Grenville.
[6] From Casapalca Mine, M vein, 600 foot level, Peru; Tertiary or later? From L. C. Graton.
[7] 88919 late Paleozoic, page 574 Lindgren, with siderite.
[8] 81731 Mesothermal deposits from Przibram, Bohemia.
[9] Galena. See Palache reports on Franklin Furnace.
[10] 91074 Galena and anglesite, Durango, Mexico. General type discussed by Lindgren, pages 598–599.
[11] Native lead with iron ores. Langban, Sweden; Precambrian (Wahl). See Palache and Flink, *American Mineralogist*, 1926, p. 195.
[12] 90486 Galena, Sonora Mine, Castle Dome, Arizona, with fluorite; H. H. Chen, collector.
[13] 81725 Mesothermal deposit, like 91074 probably of younger formation, Freiberg, Saxony; Lindgren, page 577.
[14] 84456, Galena in sedimentary rocks,—dolomites of the Austrian Triassic.

nearly constant atomic weight (see column marked "Mean Mass Number") the isotopic abundances vary between rather wide limits. That these variations are not the result of errors in analyses may be seen from the fact that in the five cases, samples 13, 14, 16, 17, and 8, where check runs were made at a later time, the results are almost identical with the original ones. In the repeat analyses the samples were given to the operator of the mass spectrometer as un-

knowns so there was no chance for prejudice. It may be seen that, in general, results are reproducible within one-half percent. As all samples are studied under conditions as nearly identical as possible, systematic discriminations in the mass spectrometer should not play a part in comparing samples. The absolute values of the abundances should be correct within two percent.

The analysis of Joplin galena III, sample 14,

was intended as an intercheck between the present and earlier results. However, it was not until the mass spectrometer tube had been dismantled and rebuilt that it was learned that Joplin III was not identical with either of the previous[2] Joplin galenas studied. Hence, the analysis of Joplin III, 14a, was repeated in the rebuilt tube. This was followed by a re-examination of Joplin II, (from earlier work[2]), 8a. No. 8b gives the analysis found previously for Joplin II. From the figures it may be inferred that there was no appreciable discrimination or systematic differences between the performances of the several mass spectrometer tubes used and that the new results may be compared directly with the older ones.

It is quite apparent that the present samples exhibit large variations similar to those found in the earlier work. Those samples containing relatively more Pb^{206} also contain relatively more Pb^{207} and Pb^{208}. As an explanation for this it was proposed[2] that one could consider any common lead sample as made up of "uncontaminated" or "primal" lead to which had been added a certain quantity of radiogenic lead composed of approximately equal amounts of uranium and thorium lead. The added radiogenic lead could not be due to U and Th impurities, as a simple calculation reveals that the samples did not contain enough to account for the results in any of the cases. As the Great Bear Lake galena[2] contained the least amounts of Pb^{206}, Pb^{207} and Pb^{208}, it was chosen, tentatively, as "uncontaminated." Its isotopic constitution is approximately the same as sample 22 of the present work. In the present investigation sample 13, the Ivigtut galena was found to contain even less Pb^{206}, Pb^{207} and Pb^{208} than the Great Bear Lake galena. Thus it has been chosen as "uncontaminated" for the new comparisons. The justification for the assumption that the "excess" is made up of equal quantities of ThD and RaG is seen in the close agreement of the numbers in the excess 208/206 column with unity. If one assumes that these excess isotopes were generated in the period one to two billion years ago then the average ratio 0.97 corresponds to a Th/U ratio in the source as would be measured today of 3.4. It is interesting to note that recent average values for the Th/U ratio for granitic and basaltic rocks range from 2.6 to 4.3.[7, 8] The results of the isotopic study are thus not in disagreement with the theory that ore lead has its origin in the igneous rocks or magmas of these rocks.

The average excess Pb^{207}/Pb^{206} ratio of 0.32 is difficult to account for in view of the fact that the relative rates of production of Pb^{207} and Pb^{206} have varied only between 0.046 (at present) and 0.24[3] two billion years ago, the time usually taken as the earth's beginning. The fact that the Ivigtut galena has an isotopic constitution so far different from any other sample so far studied may mean that it is anomalous in some way. If the Great Bear Lake galena were used as "uncontaminated" as it was in the earlier work,[2] one would find very little difference in the excess Pb^{208}/Pb^{206} ratio whereas the excess Pb^{207}/Pb^{206} would be near to 0.1, a value somewhat lower than one might expect for old uranium lead.

It is a pleasure to acknowledge the interest taken by Professor Alfred C. Lane, Chairman of the Committee on the Measurement of Geologic Time, through whose efforts many of the specimens were obtained. The radiogenic lead iodides and some of the common lead iodides were very kindly prepared by Professor G. P. Baxter of the Harvard Chemistry Department. The bulk of the common lead samples was obtained from Professor H. Berman of the Harvard Mineralogical Museum and were converted to PbI_2 by Dr. Charles Sage of the University of Minnesota. The senior author expresses his appreciation to the graduate school for financial support.

---

[7] N. B. Keevil, Economic Geology **33**, 685 (1938).
[8] R. D. Evans, J. App. Phys. **12**, 297 (1941).

The Age of the Earth

VI

Editor's Comments on Papers 13–17

The problem of determining the age of the earth involves establishing both the initial state and the final state of any evolving system which came into existence at the same time as the earth. Provided the rate at which that system changes is known precisely, the age of the earth can be determined.

By 1941, the oldest reliably dated minerals were those analyzed by Nier and his associates, which gave isotopic Pb-207/Pb-206 ages of approximately 2500 million years, but the age of the earth as a whole was rightly believed to be greater than the age of any of the individual mineral constituents of the earth's crust. There was no a priori reason to suppose that the oldest existing rocks and their constituent minerals were formed at the same time as the earth. In fact it appeared most unlikely that the original mineral constituents of the earth's crust would have survived or avoided subsequent geological recycling.*

It was Nier's isotopic analysis of common lead that contained evidence pertaining to the time of formation of the earth, for the isotopic composition of the primeval lead component in common lead ore was fixed at the time of formation of the earth and the rate of change of that isotopic composition depended only upon the rate of addition of radiogenic lead, i.e., on the concentration of uranium and thorium disseminated in the source region of the lead ore. If this were known, and the primeval isotopic composition could be determined, it should be possible to calculate the age of the earth.

The idea was not altogether new. In 1921, Henry Norris Russell, Professor of Astronomy at Princeton, calculated an age of 8000 million years for the earth

---

*The oldest radiometrically dated rocks known today come from West Greenland, and give Rb–Sr ages of 3980 ± 170 million years (Black et al., 1971), about 500 million years younger than the accepted age of the earth.

assuming that all the lead in the earth's crust was derived from the radioactive decay of uranium and thorium. This estimate was later revised to 3200 million years by Holmes (1926), using a lower estimate for the average concentration of lead in igneous rocks. There was also Rutherford's (1929; Paper 11) figure of 3400 million years based on the different decay rates of U-238 and U-235 and the resulting systematic increase in the ratio U-238/U-235 with time. Nier's common lead isotope abundance measurements provided a new and important source of information, for the lead ore minerals from which samples of common lead had been extracted were all essentially free of the radioactive parent elements uranium and thorium, and all additions of radiogenic lead to the primeval component of these common leads must have occurred *before* that lead was separated from its radioactive parents, concentrated by the ore-forming processes into a lead mineral deposit, and effectively, at least as far as the isotopic composition was concerned, fossilized. There was one invariable isotopic constituent of common lead, the nonradiogenic isotope Pb-204, and changes in the isotopic composition of common lead caused by the additions of radiogenic lead to the primeval constituent could be conveniently monitored relative to the invariant Pb-204 isotope.

The first reported calculation of the age of the earth based on Nier's common lead isotope analyses was Gerling's in 1942, which is reproduced here (Paper 13). The procedure he adopted is essentially an extension of the radiogenic Pb-207/Pb-206 method for age determination. Taking the very low isotopic ratios (relative to Pb-204) obtained from the galena deposits from Ivigtut, Greenland (Paper 12; Nier et al., 1941, sample 13) as being the best representative of primeval lead, Gerling argued that the difference between the Ivigtut ratios and those of the other lead ores quoted represented the excess ''admixture'' of radiogenic lead which must have been generated through radioactive decay of uranium and thorium during the time interval between the time of formation of the primeval Ivigtut ore and the time of formation of the other ore deposits. The average Pb-207/Pb-206 ratio for this radiogenic component for seven ore deposits having an average age of 130 million years was 0.253, and the ratio U-238/U-235 at this time was 126. Assuming that all the lead now found in the seven different ore deposits originally (i.e., at the time of formation of the earth) had the isotopic composition of the Ivigtut deposits and that it was disseminated throughout a source region in association with radiogenic uranium and thorium, Gerling calculated that 3100 million years would be required to generate the Pb-207/Pb-206 ratios now observed in these deposits. Adding the average age of the deposits, 130 million years, he obtained a minimum age of 3230 million years for the age of the earth. Using the isotopic composition of a Precambrian (1250 million years old) galena from the great Bear Lake (Nier, 1938, sample 1), Gerling found that 2700 million years were required to produce the excess radiogenic lead in this sample relative to Ivigtut, giving a total of 3940 million years for the age of the earth.

The next step was taken by Holmes (Paper 14) and independently by Houtermans (Paper 15) in 1946, and both subsequently developed their ideas in a series of papers (Holmes, 1947, 1949, 1950; Houtermans, 1947, 1953). They pointed out that the equation used by Gerling to calculate the age of the earth is a linear equation of the type $dy/dx = R$, i.e.,

$$\frac{y - b_0}{x - a_0} = \frac{1}{\alpha} \cdot \frac{e^{\lambda' t_0} - e^{\lambda' t_m}}{e^{\lambda t_0} - e^{\lambda t_m}} = \frac{\text{number of atoms of Pb-207 generated between } t_0 \text{ and } t_m}{\text{number of atoms of Pb-206 generated between } t_0 \text{ and } t_m}$$

where
x, y = measured ratios Pb-206/Pb-204, Pb-207/Pb-204
a_0, b_0 = primeval ratios Pb-206/Pb-204, Pb-207/Pb-204
α = present-day ratio U-238/U-235
λ and λ' = decay constants for U-238 and U-235
t_m = time of formation of ore deposit
t_0 = time of formation of the earth.

On an $x - y$ plot of Pb-207/Pb-204 against Pb-206/Pb-204 for lead extracted from ore deposits of the same age (t_m), a linear array of data points passing through the primeval ratios would result if the common lead samples contained variable proportions of radiogenic lead mixed with the primeval constituent. The age of the earth could be calculated from the slope of this line if the time of mineralization (t_m) and the primeval abundances relative to Pb-204 were known. Source regions with relatively high concentrations of the radioactive parent uranium will produce lead with high values of Pb-206/Pb-204 and Pb-207/P-204, while source regions with lower concentrations of uranium will produce lead with lower values of Pb-206/Pb-204 and Pb-207/Pb-204, but for all ores having the same age, the radiogenic component of the common lead would have a constant Pb-207/Pb-206 ratio.

Putting it another way, for common lead ores of the same age the radiogenic Pb-207/Pb-206 ratio will be constant, but the concentration of radiogenic Pb-206 and Pb-207 generated in the source region will depend upon the concentration of uranium disseminated there, all concentrations being measured relative to Pb-204. Thus, at any one time, i.e., for all ore deposits of the same age, the total (radiogenic + primeval) measured Pb-207/Pb-204 and Pb-206/Pb-204 ratios will plot on a linear array, the slope of which will be related to the time of mineralization (t_m) and the age of the earth (t_0). Houtermans (1947, p. 323) called this line an isochrone* and pointed out that the isotopic composition of primeval lead could be determined independently from the intersection of two or more isochrones.

The age of the earth (t_0) could thus be calculated from the slope of any isochron passing through two or more common lead samples of the same geologic age provided that both the geologic age (time of mineralization) and the primeval isotopic abundances were known. Holmes (1947, p. 128) found values for t_0 ranged from 2000 million years to more than 4000 million years but clustered about a mode of 3350 million years and concluded that this value for the age of the earth "is unlikely to be seriously wrong" (Holmes, 1950, p. 239).

Bullard and Stanley (1949) modified Holmes' computation using a least-squares

---

*Now referred to as an isochron, see p. 246.

118

approach and arrived at a figure of 3290 million years for the age of the earth, while Alpher and Herman (1951) determined a maximum age for the earth's crust of 5300 million years based on Nier's data. Additional lead isotopic analyses on ore minerals provided by Collins, et. al. (1953) confirmed the earlier results and provided a new figure of 3500 million years for the age of the earth.

Both Holmes' and Houtermans' original computations were based on Nier's isotopic analyses of terrestrial lead ore deposits and depended on finding two or more common lead ore deposits of the same age but with different proportions of radiogenic lead that would define an isochron passing through the primeval abundances. By 1956 new isotopic analyses determined by Claire Patterson and his associates at the California Institute of Technology on the minute amounts of lead found in meteorites and deep-sea sediments provided better defined information and a more precise estimate for the age of the earth (Paper 16; see also Patterson et al., 1955).

Patterson found that the very low Pb-207/Pb-204 and Pb-206/Pb-204 ratios of 9.46 and 10.34 obtained from lead extracted from the Canyon Diablo iron meteorites provided more realistic data for the elusive isotopic composition of primeval lead than Nier's terrestrial Ivigtut samples, which had much higher Pb-207/Pb-204 and Pb-206/Pb-204 ratios of 14.54 and 14.60. The U-238/Pb-204 ratio in the Canyon Diablo material was extremely low (0.025) indicating that no observable change in the initial lead isotopic ratios could have occurred since the meteorite formed. Using the results from three stone and two iron meteorites, including the primeval Canyon Diablo data, Patterson showed that the Pb-207/Pb-204 versus Pb-206-Pb-204 isochron passing through all five points corresponded to an age of 4550 ± 70 million years. The position of any lead sample on this isochron is determined only by the U-238/Pb-204 ratio in the system from which the lead evolved. To clinch his argument, Patterson showed that modern terrestrial lead extracted from recent marine deposits gave isotopic ratios which fell right on the 4550 million year isochron, confirming this age as the age both of meteorites and the earth.

Tilton and Steiger (Paper 17) in 1965 added a final note of caution, pointing out that modern terrestrial leads extracted from young volcanic rocks were showing quite variable isotopic compositions indicative of chemical inhomogeneities in the source regions. They preferred, therefore, to calculate the age of the earth from the slope of an isochron passing through meteoritic lead and lead extracted from 2700 million year old deposits in the Canadian shield. Their calculations resulted in a figure of 4750 ± 50 million years for the age of the earth, some 200 million years older than Patterson's estimate and 220 million years older than the figure of 4530 ± 30 million years determined on the basis of isotopic analyses by Ostic, et al. (1963) of conformable lead ore deposits. It would seem probably that the time of formation of the earth based on lead isotopic data cannot be more closely defined.*

---

*A new approach suggested by Ulrych (1967) utilizing the Wetherill concordia plot (see Papers 19 and 20) gives a figure of 4530 ± 40 million years for the age of the earth based on isotopic analyses of lead extracted from oceanic basalts. The method is independent of the age of the samples analyzed but requires a knowledge of their uranium and lead concentrations.

Selected Bibliography

Alpher, R. A., and Herman, R. C. (1951). The primeval lead isotopic abundances and the age of the earth's crust. *Phys. Rev.*, **84**, 1111–1114.

Black, L. P., Gale, N. H., Moorbath, S., Pankhurst, R. J., and McGregor, V. R. (1971). Isotopic dating of very early Precambrian amphibolite facies gneisses from the Godthaab district, West Greenland. *Earth Planet. Sci. Lett.*, **12**, 245–259.

Bullard, E. C., and Stanley, J. P. (1949). "The Age of the Earth," p. 33. Suomen Geodeettisen Laitoksen Julkaisuja, No. 36. Finnish Geodetic Institute, Helsinki.

Collins, C. B., Russell, R. D., and Farquhar, R. M. (1953). The maximum age of the elements and the age of the earth's crust. *Can. J. Phys.*, **31**, 420–428.

Holmes, A. (1926). Rock-lead, ore-lead, and the age of the earth. *Nature*, **117**, 482.

Holmes, A. (1947). A revised estimate of the age of the earth. *Nature*, **159**, 127–128.

Holmes, A. (1949). Lead isotopes and the age of the earth. *Nature*, **163**, 453–456.

Holmes, A. (1950). The age of the earth. *Rept. Smithsonian Institution for 1948*, pp. 227–239.

Houtermans, F. G. (1947). Das Alter des Urans. *Z. Naturforsch.*, **29**, 322–328.

Houtermans, F. G. (1953). Determination of the age of the earth from the isotopic composition of meteoritic lead. *Nuovo Cim., Ser. 9*, **10**, (12), 1623–1633.

McCrady, E. (1952). The use of lead isotope ratios in estimating the age of the earth. *Trans. Amer. Geophys. Union*, **33**, 156–170.

Murthy, V. R., and Patterson, C. C. (1962). Primary isochron of zero age for meteorites and the earth. *J. Geophys. Res.*, **67**, 1161–1167.

Nier, A. O., Thompson, R. W., and Murphey, B. F. (1941). The isotopic constitution of lead and the measurement of geologic time. III. *Phys. Rev.*, **60**, 112–116.

Ostic, R. G., Russell, R. D., and Reynolds, P. H. (1963). A new calculation for the age of the earth from the abundances of lead isotopes. *Nature*, **199**, 1150–1152.

Patterson, C. (1964). Characteristics of lead isotope evolution on a continental scale in the earth. *In* "Isotopic and Cosmic Chemistry," pp. 244–268. North-Holland, Amsterdam.

Patterson, C., Tilton, G., and Inghram, M. (1955). Age of the earth. *Science*, **121**, 69–75.

Russell, H. N. (1921). A superior limit to the age of the earth. *Proc. Roy. Soc., Ser. A*, **99**, 84–86.

Ulrych. T. J. (1967). Oceanic basalt leads: a new interpretation and an independent age for the earth. *Science*, **158**, 252–256. See also comments by Oversby, V. M., and Gast, P. W. (1968). Oceanic basalt leads and the age of the earth. *Science*, **162**, 925–927, and Ulrych's reply, p. 928.

Comptes Rendus (Doklady) de l'Académie des Sciences de l'URSS
1942. Volume XXXIV, № 9

13

AGE OF THE EARTH ACCORDING TO RADIOACTIVITY DATA

By E. K. GERLING

(Communicated by V. G. Chlopin, Member of the Academy, 20.II.1942)

The first attempt to estimate the age of the earth from the content of radioactive elements (uranium, thorium and lead) in its crust was made by Russel [1] in 1921. Later Holmes [2], Hevesy [3] and Starik [4] became interested in the subject. The computations of these authors were based on the assumption that all lead was of radioactive origin. Besides, Starik derived the age of the earth from the ratio AcD/RaG in the common lead. In this way he found a figure of 3×10^9 years which was in agreement with what he had obtained from the average content of U, Th and Pb in the earth crust, while taking the decay of AcU into consideration (3×10^9—4×10^9). None of these computations, however, can claim sufficient accuracy or firm foundation. So Russel and Holmes have relied upon data that are now out of date, not to mention the fact that they failed to take due account of the decay of AcU. Nor has Hevesy paid any attention to the latter circumstance.

More reliable data for the content of U, Th and Pb have been used by Starik who has also included in his calculations the decay of AcU along with that of U and Th. Yet, the values he obtained for the age of the earth are liable to objection on the ground that the radioactive origin of the whole of the lead in the earth crust, which is his basic assumption, has never been proved and is even conflicting with observation. If, indeed, all lead contained in the earth crust were of radioactive origin, one should expect that computation and experiment would not yield diverging data for the isotopic composition of this element. As a matter of fact, they do. This is brought out by Table 1, in which the calculated figures for the isotopic composition of lead are facing those determined experimentally by Aston. The divergence is seen to be great and is probably due to the presence of not-radioactive lead in the earth crust.

Table 1

Isotopic Composition of Lead

Calculated	Found by Aston
Pb_{206}—19%	Pb_{204}— 1.5 %
Pb_{207}—68%	Pb_{206}—27.75%
Pb_{208}—13%	Pb_{207}—20.20%
	Pb_{208}—49.55%

259

A striking case in point is that of the lead with atomic weight 208. To account for the difference it has been supposed that the content of Th in the earth crust was determined inaccurately and is actually higher. Computations show that it should be increased 4 times. But then the ratio Th/U, which at present is generally put at 3, must be raised correspondingly and attain a figure of 10 to 12. Such a ratio of Th/U in the rocks is, however, highly improbable and will not accord with the experiment. Moreover, it has recently been shown by Nier [5] and Keevel [6] that when lead ores are in process of formation, RaG and ThD are captured from granitic and basaltic magmas in a proportion 1 : 1, which again gives nearly 3 : 1 for the relative content of Th and U in rocks. So we have no ground for the time being to put another value in place of 3 for the ratio of Th to U.

Of late there has been put in doubt the correctness of the decay constant for AcU as determined by Grosse [7] and used by Starik in his calculations. Nier [3] has found that adopting this constant leads to a considerable difference between the age value calculated with reference to RaG/U and that calculated with reference to AcD/RaG. This difference he ascribes to inaccurate determination of the decay constant for AcU. He has obtained agreeing results in those two computations when putting the decay constant of AcU at $9.72 \times 10^{-10} g^{-1}$. This may be taken to mean that the decay of AcU proceeds much more slowly than was believed earlier. If so, the calculations made by Starik are obviously inaccurate. But on the other hand, adopting the new constant of decay one obtains too long an age of the earth. It will therefore be apparent that recent investigations have furnished no corroborative evidence of the assumption that all the lead of the earth crust has radioactive origin.

In 1941 there appeared the work of Nier [5] on isotopic composition of leads of different age and origin. Some of his data have made the basis for calculations presented in this paper. The data obtained by Nier are given in Table 2.

It may be seen from this table that sample 1 from Ivigtut contains lead of atomic weight 206, 207 and 208 in unusually small quantities. It therefore was supposed by Nier to include very little lead of radioactive origin, if any at all. The isotopic composition of the Ivigtut lead served to compute the admixture of radioactive leads in other samples. The results of these computations are referred to in columns 5, 6 and 7. In column 8 is given the admixture ratio 207/206 which for seven samples gave an average of 0.253. This ratio can be placed at the basis of age calculations, for it does not remain constant throughout the geological periods, but rather varies continually. The following values have been used in the calculations:

$$\lambda U - 1.535 \times 10^{-10} g^{-1}, \quad \lambda AcU - 9.72 \times 10^{-10} g^{-1}, \quad \frac{Ac D}{RaG} - 0.253.$$

Average age of lead samples, 130×10^6 years. 130×10^6 years ago the ratio U_{238}/U_{235} was equal to 126; the activity ratio $R = 0.0505$.

By substituting these values in the formula

$$\frac{Ac D}{RaG} = \frac{e^{\frac{U_{238}}{U_{235}} \cdot R \lambda_U T} - 1}{\frac{U_{238}}{U_{235}} (e^{\lambda_U T} - 1)} \tag{1}$$

we get $T = 3.1 \times 10^9$ for the age value.

Taking the age of lead to equal 130×10^6 years on the average, we obtain for the age of the earth at this day a value of 3.23×10^9 years, which we should regard as minimum. Another way of computing the earth age was by

260

Table 2

Nier's Data on Isotopic Composition of Lead

Name and age	Isotopic composition $Pb_{204}=1.0$			Admixtures			Admixture ratio
	206	207	208	206	207	208	207/206
Galenite, Ivigtut, Greenland . . .	14.65	14.65	34.47				
Vanadinite, Tucson Mts., Ariz. Miocene	18.4	15.53	38.1	3.75	0.88	3.63	0.232
Cerussite, Wallan Ida, late Cretaceous	15.98	15.08	35.07	1.33	0.43	0.60	0.323
Lead-silver ore, Coeur d' Alene Ida	16.1	15.13	35.45	1.45	0.48	0.98	0.331
Galenite, Metalline falls, Washington, late Cretaceous . .	19.30	15,73	39.50	4.65	1.08	5.08	0.232
Galenite, Joplin, late Carboniferous	21.63	15.79	40.5	6.98	1.14	6.03	0.163
Cerussite, Eifel, Germany, Carboniferous . .	18.20	15.46	37.7	3.55	0.81	3.23	0.228
Galenite, Nassau, Germany, Carboniferous . .	18.10	15.57	37.85	3.45	0.92	3.38	0.263
			Mean	3.56	0.82	3.27	0.253
Galenite, Great Bear lake, pre-Cambrian . . .	15.93	15.30	35.3	1.28	0.65	3.53	0.50

using the galenite sample from the Great Bear lake, which has lived no less than 1.25×10^9 years. The values applied in this calculation are

$$\frac{Ac\ D}{RaG} = 0.50; \quad \frac{U_{238}}{U_{235}} = 52.1; \quad R = 0.123.$$

The substitution into (1) gives $T = 2.7 \times 10^9$ years. Adding up the time elapsed since the formation of the galenite, which is 1.25×10^9 years, we obtain a 3.95×10^9 year old earth. From these computations the age of the earth is not under $3 \times 10^9 - 4 \times 10^9$ years. This is certainly not too much, since the age of certain minerals, calculated with reference to AcD/RaG, was put at 2.2×10^9 years ([3]) and even 2.5×10^9 years ([5]).

Radium Institute.   Received
Academy of Sciences of the USSR.   20.II.1942.

REFERENCES

[1] A. S. R u s s e l, Proc. Roy. Soc., A, **99**, 84 (1921).   [2] A. H o l m e s, Bull. Nat. Research Council, Washington, **80** (1931).   [3] G. H e v e s y, Fortschr. d. Miner., Kristallogr. und Petrogr., **16**, 147 (1932).   [4] И. Е. С т а р и к, Радиоактивные методы опредeл. геологич. времени (1938).   [5] A. N i e r, Phys. Rev., **60**, 2 (1941). [6] N. V. K e e v e l, Economic Geology, XXXIII, 7.(1938).   [7] A. V. G r o s s e, Phys. Rev., **42**, 565 (1932); J. Phys. Chem., **38**, 487 (1934).   [8] A. N i e r, Phys. Rev., **55**, 2 (1939).

261

14

AN ESTIMATE OF THE AGE OF THE EARTH

By Prof. ARTHUR HOLMES, F.R.S.

Grant Institute of Geology, University of Edinburgh

EVER since the publication by Nier and his co-workers[1] of the relative abundances of the isotopes in twenty-five samples of lead from common lead minerals of various geological ages (Table 1), I have entertained the hope that from these precise data it might be possible to fathom the depths of geological time. The calculations involved are, however, somewhat formidable, and a systematic investigation became possible only recently, with the acquisition of a calculating machine, for which grateful acknowledgment is made to the Moray Endowment Research Fund of the University of Edinburgh. The results have fully justified expectation and indicate that the age of the earth, reckoned from the time when radiogenic lead first began to accumulate in earth-materials, is of the order 3,000 million years.

In his first paper, Nier pointed out that those samples of lead "which contain relatively more Pb^{206} also contain relatively more Pb^{207} and Pb^{208}". The abundances of these isotopes, listed in Table 1, are all relative to $Pb^{204} = 1$. Since Pb^{204} is not generated by any naturally radioactive element, it can be taken as an invariable constituent of the *primeval lead* occurring in the earth at the time of the earth's origin. Calling the lead with the lowest relative abundances (No. 19) "the least contaminated lead", Nier suggested that all the other samples could be regarded as made up of this "least contaminated lead" plus additions of Pb^{206} and Pb^{207} generated from uranium, and of Pb^{208} generated from thorium. The minerals from which the samples of lead were extracted are all essentially free from radioactive elements, and hence the excess isotopes must have been generated before the minerals were formed. This point is, of course, of fundamental importance. It can easily be calculated that if the excess isotopes in the Joplin leads (Nos. 9–11) had been formed in the ore itself, the necessary amounts of the radioactive elements would have been 6·2 gm. U I, 24 gm. AcU and 20 gm. Th per gm. of galena—impossible amounts, hundreds of millions of times greater than any actual traces that may locally be present. Post-deposition contamination of lead being thus completely ruled out, it follows that before its concentration in ores, the lead must have been

TABLE 1.   ISOTOPIC ABUNDANCES ($Pb^{204} = 1$) OF LEAD (NIER *et al* FROM MINERALS OF VARIOUS AGES.

No.	Source of lead, locality and geological age of ore deposit				
	Pb^{206}	Pb^{207}	Pb^{208}	Total (incl. $Pb^{204} = 1$)	Pb^{207}/Pb^{206}
1	Galena (2)*, Casapalca Mine, Peru. Late Tertiary, 25 m.y.				
	18·85	15·66	38·63	74·14	0·831
2	Bournonite (2), Casapalca Mine, Peru. Late Tertiary, 25 m.y.				
	18·67	15·45	38·15	73·27	0·828
3	Wulfenite and Vanadinite, Tucson Mts., Arizona. Miocene, 25 m.y.				
	18·40	15·53	38·1	73·03	0·832
4	Galena, Sonora, Castle Dome, Arizona. Tertiary, 25 m.y.				
	19·22	16·17	39·15	75·54	0·841
5	Galena, Freiberg, Saxony. Tertiary, 25 m.y.				
	18·07	15·40	38·0	72·47	0·833
6	Galena and Anglesite, Durango, Mexico. Tertiary, 25 m.y.				
	18·71	15·70	38·5	73·91	0·839
7	Galena, Metaline Falls, Washington. Laramide, 60 m.y.				
	19·30	15·73	39·5	75·53	0·815
8	Cerussite (2), Wallace, Idaho. Laramide, 60 m.y.				
	16·04	15·11	35·26	67·41	0·942
9	Galena I, Joplin, Missouri. Late Mid. Cretaceous, 190 m.y.				
	21·65	15·88	40·8	79·33	0·733
10	Galena II (4), Joplin. Late Mid. Cretaceous, 100 m.y.				
	21·65	15·74	40·36	78·75	0·727
11	Galena III (4), Joplin. Late Mid. Cretaceous, 100 m.y.				
	22·38	16·15	41·63	81·16	0·722
12	Galena in Dolomite, Austria. Triassic, 175 m.y.				
	17·75	16·21	38·05	73·01	0·913
13	Galena, Nassau. Late Carboniferous/Early Permian, 220 m.y.				
	18·10	15·57	37·85	72·52	0·860
14	Galena, Eifel. Carb./Permian, 220 m.y.				
	18·20	15·46	37·7	72·36	0·850
15	Galena, Saxony. Carb./Permian, 220 m.y.				
	17·36	15·46	37·38	71·20	0·890
16	Galena, Clausthal, Harz Mts. Carb./Permian, 220 m.y.				
	18·46	15·66	38·6	73·72	0·848
17	Galena, Przibram, Bohemia. Carb./Permian, 220 m.y.				
	17·95	15·57	37·9	72·42	0·868
18	Galena, Yancey Co., N. Carolina. Late Carboniferous, 220 m.y.				
	18·43	15·61	38·2	73·24	0·847
19	Galena (2), Ivigtut, Greenland. Late Pre-Cambrian, 600 m.y.				
	14·65	14·65	34·48	64·78	1·000
20	Galena, Franklin, New Jersey. Pre-Cambrian (—)				
	17·15	15·45	36·53	70·13	0·901
21	Galena, Tetreault Mine, Quebec. Pre-Cambrian, 800 m.y.				
	16·27	15·16	35·60	68·03	0·932
22	Galena, Broken Hill, N.S. Wales. Pre-Cambrian, 1,200 m.y.				
	16·07	15·40	35·5	67·97	0·958
23	Cerussite, Broken Hill, N.S. Wales. Pre-Cambrian, 1,200 m.y.				
	15·93	15·29	35·25	67·47	0·960
24	Native Lead, Långban, Sweden. Pre-Cambrian (—)				
	15·83	15·45	35·60	67·88	0·976
25	Galena, Great Bear Lake, Canada. Pre-Cambrian, 1,330 m.y.				
	15·93	15·30	35·3	67·47	0·960

* Numbers in brackets indicate the number of determinations made. In all such cases the results closely agreed and the average figures are those here given.

dispersed through radioactive source-materials in which the primeval lead was slowly modified by additions of radiogenic lead. Calculation of the requisite amounts of uranium and thorium (relative to the lead in common rocks) indicates that the source materials could have been any of the common crustal rocks, granitic types and their derivatives being the most probable in most cases.

Several years ago[2], I showed, using the data then available on the uranium, thorium and lead contents of rocks, that the atomic weights of ore-lead derived from granitic or basaltic sources should vary with the age of the ore in accordance with the varied proportions of Pb^{206}, Pb^{207} and Pb^{208} generated in the source-rocks up to the time when the lead ore was formed. Since the recorded atomic weights of lead from ores of widely different ages failed to show the expected variation with time, it was inferred that ore-lead could not have come from granitic or basaltic sources. This inference must now be withdrawn, not only because Nier's work shows that the isotopic constitutions and the resultant atomic weights do, in fact, vary with time, but also because the data for uranium and thorium in rocks on which I relied have since been shown to be misleading[3]. Nier and his collaborators[1] have already pointed out that their results "are not in disagreement with the theory that ore-lead has its origin in the igneous rocks or magmas of these rocks". Allowing for certain subsidiary possibilities, such as the upward migration of primeval or 'old' lead from deeper sources, and the probability that some sedimentary leads (for example, No. 12) may be of marine origin, I am in complete accord with Nier's conclusion, which is, indeed, fundamental to the present study of some of the further implications of his work.

Fig. 1, in which the relationships between the data of Table 1 are graphically displayed, clearly shows that, with a few minor irregularities, Pb^{206}, Pb^{207} and Pb^{208} vary systematically among themselves and with the totals of the isotopic abundances. I, I' and I'' represent the results for the "least contaminated lead" (Ivigtut, No. 19), and J, J' and J'' those for the lead with the highest abundances (Joplin III, No. 11). Assuming provisionally that lead of the J type has evolved from lead of the I type by radiogenic additions, it is possible to construct curves I–J, I'–J' and I''–J'' showing the gradual changes in the abundances of the respective isotopes which would be brought about by the presence of appropriate amounts of uranium and thorium in the source-rocks. The derivation of these curves is illustrated by Fig. 2. The curve RIJ represents the cumulative growth of Pb^{206} during the last few thousands of millions of years within, say, 1 gm. of a source now containing 1×10^{-6} gm. uranium. $RI'J'$ represents the corresponding increase of Pb^{207}. The relative positions of the two curves can be fixed by the following considerations :

(a) The age, t_m, of the Joplin ores is about 100 m.y.[4].
(b) $Pb^{206} = Pb^{207}$ in lead of Ivigtut constitution, whence it follows that the curves must cross at a point $I = I'$.
(c) The position of I must be such that the ratio IJ'/IJ = excess Pb^{207}/excess Pb^{206} in Joplin III lead, which ratio is found to be 0.194 from the following data :

Lead	Pb^{204}	Pb^{206}	Pb^{207}	Pb^{208}
Joplin III (11)	1	22.38	16.15	41.63
Ivigtut (19)	1	14.65	14.65	34.48
Excess isotopes in Joplin III	–	7.73	1.50	7.15

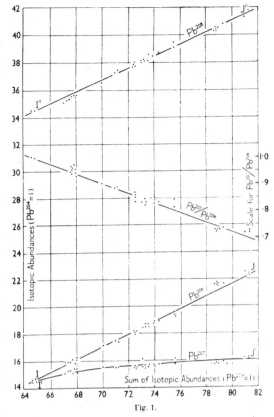

Fig. 1.

The time t corresponding to I can be found by trial ; for it must be such that, in lead generated from a given amount of uranium,

$$\frac{Pb^{207} \text{ generated from } t \text{ to } t_m}{Pb^{206} \text{ generated from } t \text{ to } t_m} = 0.194.$$

The value of t that gives this ratio when $t_m = 100$ m.y. is 2,760 m.y. The positions of I and the two curves being now determined against an arbitrary vertical scale, the latter can be readily transformed into one of isotopic abundances, since I', J' and J are also fixed, and we know that the point $U = 14.65$, $J' = 16.15$ and $J = 22.38$. Completing the vertical scale of Fig. 2, the corresponding points for Pb^{208} can be inserted, since $T = 34.48$ and $J'' = 41.63$. I'' is necessarily vertically above I. The resulting curve $I''J''$ coincides exactly with that representing the cumulative growth of Pb^{208} in 1 gm. of a source now containing 3.17×10^{-6} gm. thorium ; that is, the present value of the ratio of thorium to uranium in the source material of Joplin III lead would be 3.17. The latest estimates of this ratio from direct determinations are about 3.4 for granitic rocks and 3.3 for basaltic rocks[3].

From the data employed in constructing Fig. 2, the very similar curves IJ, $I'J'$ and $I''J''$ are added to Fig. 1, with isotopic abundances as abscissæ instead of time. It will be seen that they are very nearly the best curves that could be drawn through the points plotted. This coincidence demonstrates that the excess isotopes are essentially of radioactive origin ; and that in the source-materials the average values of the ratio lead/uranium have been every-

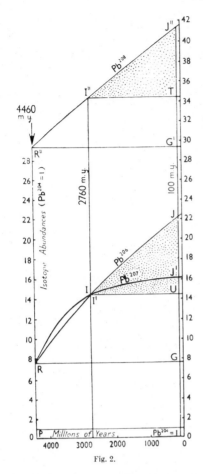

4460
m.y.

Fig. 2.

should pass through y. This consideration suggests a method for determining x and y and the corresponding value of t_0, the age of the earth. Using the following symbols, with t_m for the age of the lead ore,

	Pb²⁰⁴	Pb²⁰⁶	Pb²⁰⁷	Pb²⁰⁸
Ore-lead (= source-lead)	1	a	b	c
Primeval lead	1	x	y	z
Radiogenic lead (from t_0 to t_m)	—	$a-x$	$b-y$	$c-z$

we have :

$$\frac{b-y}{a-x} = r = \frac{\text{Pb}^{207} \text{ generated by 1 gm. U from } t_0 \text{ to } t_m}{\text{Pb}^{206} \text{ generated by 1 gm. U from } t_0 \text{ to } t_m}$$

Preliminary tests having shown that t_0 is within the range 2,760–3,100 m.y., r can be systematically computed for various values of t_0 such as 2,760, 2,900, 3,000 and 3,100 m.y. For each assigned value of t_0 we have :

$$ar - xr = b - y \text{ for one sample of lead, and}$$
$$a'r' - xr' = b' - y \text{ for another sample ;}$$

whence, $x = \dfrac{b - b' + a'r' - ar}{r' - r}$; $y = b + rx - ar$

By plotting x and y against time, two converging lines are obtained. Taking another pair and constructing the corresponding lines, it is then possible to find graphically the value for t_0 at which the respective values for x and y coincide or most nearly coincide. The method is very sensitive to small variations in the data, and it is therefore one more remarkable that in a fair proportion of combinations the value for t_0 where the x-lines cross is almost exactly the same as that given by the intersection of the y-lines. When the two values for t_0 differ appreciably, their average is taken. In some combinations the x-lines (and/or the y-lines) are coincident or parallel and no result is possible. The following examples illustrate the method and the range of results obtained :

	PAIR A		PAIR B	
	Broken Hill (23)—Joplin (11)		Broken Hill (23)—Tucson (3)	
$t_0 = 3,100$ m.y.				
r	0·3106	0·2385	0·3106	0·2349
ar	4·948	5·338	4·948	4·332
x		11·37		11·44
y		13·87		13·89
$t_0 = 3,000$ m.y.				
r	0·2939	0·2244	0·2939	0·2209
ar	4·682	5·023	4·682	4·065
x		12·50		11·74
y		14·32		14·06

The solutions for the above combination are listed below, together with the results from a few other combinations of pair A with other pairs.

Combination of Pair A with	t_0 (m.y.)		x	y
Pair B, Broken Hill (23)—Tucson (3)	3090	A	11·48	13·9
		B	11·47	13·9
Pair C, (23)—Peru (2)	3040	A	12·03	14·1
		C	11·98	14·1
Pair D, (23)—Nassau (13)	3000	A	12·50	14·3
		D	12·48	14·2
Pair E, (23)—Clausthal (16)	2995	A	12·56	14·3
		E	12·61	14·3
Pair F, (23)—Mexico (6)	2915	A	13·54	14·6
		F	13·52	14·6

The average result of eleven such combinations with Broken Hill (23) in common is $t_0 = 2,990$ m.y.

As already mentioned, some of the combinations fail to give results. This brings out the important fact that not all the samples of lead can be regarded as 'normal' in the sense of being products of continuous contamination in a single source. For example, a lead ore might be due to concentrations of rock-lead plus lead from older ore deposits which became involved in the cycle of magmatic or geochemical activity responsible for the lead ore in question. In such a case the total of the isotopic abundances would be

where nearly the same at any given time, as also have been the average values of thorium/uranium. It is obvious, moreover, that in the course of its evolution the Joplin III lead must have passed through a stage at which Pb²⁰⁶ was equal to Pb²⁰⁷ and that the point representing this stage cannot be far from I and I'.

The continuation of the curves for Pb²⁰⁶ and Pb²⁰⁷ beyond I and I' in Fig. 2 brings them to a point R, at $t = 4,460$ m.y., where they again coincide. This corresponds to the time at which, on the assumptions made concerning Joplin lead, U I and AcU first began to disintegrate. The age of the earth must therefore lie between the approximate limits of 2,760 m.y. and 4,460 m.y. Similarly, the constitution of the primeval lead must fall between the approximate limits determined by the ordinates of R and I, and R'' and I''.

Primeval lead	Pb²⁰⁴	Pb²⁰⁶	Pb²⁰⁷	Pb²⁰⁸	Pb²⁰⁷/Pb²⁰⁶
Upper limit, at $t = 2,760$ m.y.	1	14·65	14·65	34·48	1
Real values (to be determined)	1	x	y	z	>1
Lower limit, at $t = 4,460$ m.y.	1	7·8	7·8	29·5	1

Inspection of Figs. 1 and 2 shows that it would be possible to construct similar curves through the points representing any given pair of lead samples with a suitable difference in age and constitution. The curves for different pairs would in general have different vertical scales, but ideally the various sets of curves for Pb²⁰⁶ should all pass through the point x on the corresponding scales, and those for Pb²⁰⁷

ess than the normal figure for the age of the ore, the ratio total Pb^{207}/total Pb^{206} would be greatly increased, and the ratio radiogenic Pb^{207}/radiogenic Pb^{206} would also be increased. The lead samples of Idaho (8) and Saxony (15) show these peculiarities. Again, a lead ore might come from a source to which additions of the radioactive elements had been made at some time after the origin of the source but before the date of ore formation ; or from a source from which radiogenic lead became concentrated in a higher proportion than primeval lead. In both these cases the total abundances would be increased, while the ratios mentioned above would be decreased. The Joplin leads (9–11) reveal these peculiarities very strikingly, though Joplin III is more nearly normal than the other two. Ivigtut (19) and, to a less degree, Tetreault (21) show similar abnormalities, but with modifications that are probably to be referred to a low uranium content in the source. The Austrian sedimentary lead (12) has so high an abundance of Pb^{207} that it must have had an entirely different kind of origin from all the other samples investigated. A marine origin, with selective concentration of Pb^{207} by organic intervention, may be provisionally suggested. Rejecting all the 'suspect' samples (and those for which a numerical age cannot at present be estimated) leaves us with those listed in Table 3.

From the Broken Hill (23) series of results already exemplified, and also from two similar series with Broken Hill (22) and Great Bear Lake (25) respectively in common, all the suspect samples except Joplin III were omitted. Finally, a series was worked out from all the possible combinations of the normal samples (that is, excluding Joplin III). The average solutions, together with the average solutions for a series from which the suspect samples were not excluded, are listed in Table 2.

TABLE 2.

Series of combinations	Number of good combinations	x	y	t_0	Range of t_0
General (including suspect samples)	27 (incomplete)	12·00	14·20	2950	2700–3150
Broken Hill (22) in common (including Joplin III)	14	11·45	14·06	2960	2760–3140
Broken Hill (23) in common (including Joplin III)	11	12·41	14·29	2990	2900–3090
Great Bear Lake (25) in common (including Joplin III)	7	12·58	14·28	3025	2900–3125
Normal samples only (excluding Joplin III)	13	12·52	14·29	3015	2725–3150

It is of interest to notice that the values for t_0 vary less widely than those for x and y, and that the value for t_0 derived from the first series differs but slightly from the last result, which is probably to be regarded as the best. It may be concluded with a high degree of probability that the age of the earth is not far from 3,000 m.y. Adopting this estimate, the corresponding values for x and y are 12·50 and 14·28.

From these values the abundance of Pb^{208} in primeval lead (z) and the present-day value of the ratio thorium/uranium in the source materials can now be approximately determined. Writing p for the ratio thorium/uranium, we have, for any sample of lead :

$$\frac{c - z}{a + b - x - y} =$$

$\frac{Pb^{208}\text{ generated by 1 gm. Th (now) from } t_0 \text{ to } t_m}{Pb^{206} + Pb^{207}\text{ gen. by 1 gm. U (now) from } t_0 \text{ to } t_m} \times \frac{Th}{U}$
$= Rp$ (R being calculable for the interval concerned).

Hence, $c - z = Rp(a + b - x - y)$
for one sample of lead, and
$$c' - z = R'p(a' + b' - x' - y')$$
for another sample.

Solving for p and z in 24 successive pairs, the average values for p and z are found to be $z = 31\cdot82$ and $p = Th/U = 3\cdot83$ (with a range from 2·5 to 5·24). In granitic rocks recently analysed for both uranium and thorium[3], Th/U is found to average about 3·4.

An approximate average for the uranium content of the source materials of the various samples of lead can be arrived at by transforming the relative abundances into actual amounts of lead. The latest work on the lead content of granitic rocks[6,7] indicates that the older average of Hevesy and Hobbie[8] ($Pb = 30 \times 10^{-6}$ gm./gm.) is too high, and that 20×10^{-6} is a more probable value. Goldschmidt and Hörmann[9] find the same figure to be characteristic of sandstones and shales. Taking the total of the abundances for late Tertiary lead (Table 1) as 74, we can write

$$74n = 20 \times 10^{-6} ; \text{ whence } n = 0\cdot27 \times 10^{-6}.$$

The amount of Pb^{206} generated by the uranium I in 1 gm. of source-rock from t_0 to t_m is then given by $(a - x) \times 0\cdot27 \times 10^{-6}$. The present amount of uranium I in the source (assuming it to be granitic) is found by dividing this amount of Pb^{206} by the amount generated by 1 gm. uranium I (now) from t_0 to t_m. The results are tabulated in Table 3. The average content of uranium I (= 0·993 uranium) is $3\cdot23 \times 10^{-6}$, which corresponds well with the average uranium data for actual granites. Recent analyses give $2\cdot77 \times 10^{-6}$ (Keevil[10], 1938), $3\cdot82 \times 10^{-6}$ (Evans and Goodman[3], 1941) and $3\cdot35$ (Keevil[3], 1944). Almost equally consistent results are found by assuming a basaltic source with $Pb = 5 \times 10^{-6}$ gm./gm.[6,8] ; uranium then is equal to $0\cdot81 \times 10^{-6}$, against an actual average of $0\cdot88 \times 10^{-6}$ (Evans and Goodman[3] and Keevil[3]). However, neither field associations nor the lead contents of basaltic rocks favour a basaltic source for lead ores. The results are internally consistent with the inference that the granitic rocks and their derivatives are the main source of lead ores. There are reasons for suspecting that the Joplin ores may have had an ultrabasic source, but this possible exception need not be discussed here.

TABLE 3.

Lead samples	Excess Pb^{20} $(a-x)$	$n(a-x)$ $\times 10^{-6}$	Pb^{206} from 1 gm. UI	$\dfrac{n(a-x)}{Pb^{20}\text{ from 1 gm. UI}}$
Peru (1)	6·35	1·71	0·49905	$3\cdot43 \times 10^{-6}$
Tucson (3)	5·90	1·59	,,	3·19
Mexico (6)	6·21	1·68	,,	3·37
Metaline Falls (7)	6·80	1·84	0·49438	3·72
Nassau (13)	5·60	1·51	0·47278	3·19
Clausthal (16)	5·96	1·61	,,	3·41
Bohemia (17)	5·45	1·47	,,	3·11
N. Carolina (18)	5·93	1·60	,,	3·39
Broken Hill (22)	3·57	0·96	0·32778	2·93
Broken Hill (23)	3·43	0·93	,,	2·84
Great Bear Lake (25)	3·43	0·93	0·30755	3·03

* The last column gives the amount of UI per gm. of source-rock, assuming that the latter contains 20×10^{-6} gm./gm. of lead ; and that $n = 0\cdot27 \times 10^{-6}$ and $x = 12\cdot5$.

The results of the analysis of Nier's isotopic abundances summarized in this preliminary announcement depend, of course, on the accuracy of the data in Table 1 and of the disintegration constants adopted in the calculations :
$\lambda_{UI} = 1\cdot52 \times 10^{-10}$/year ; $\lambda_{AcU} = 9\cdot72 \times 10^{-10}$/year ; and $\lambda_{Th} = 4\cdot99 \times 10^{-11}$/year. Adopting these figures, and assuming that the 'normal' samples of

ore-lead are concentrations of rock-lead, it is shown that the age of the earth is not far frcm 3,000 m.y. ; that corresponding to an age of exactly 3,000 m.y. the constitution of the earth's primeval lead is about

Pb^{204}	Pb^{205}	Pb^{207}	Pb^{208}
1	12·50	14·28	31·82

and that the most probable sources of ore-lead (with the provisional exception of the Joplin ores) are granitic rocks now containing, on average, about $3\cdot3 \times 10^{-6}$ gm./gm. uranium, with the ratio thorium/uranium about $3\cdot8$. It is hoped to publish a detailed account of the investigation elsewhere.

[1] Nier, A. O., J. Amer. Chem. Soc., 60, 1571 (1938); Nier, A. O., Thompson, R. W., and Murphey, B. F., Phys. Rev., 60, 112 (1941).

[2] Holmes, A., Econ. Geol., 32, 764 (1937) ; and 33, 829 (1938).

[3] Evans, R. D., and Goodman, C., Bull. Geol. Soc. Amer., 52, 459 (1941); Keevil, N. B., Amer. J. Sci., 242, 309 (1944).

[4] Bastin, E. S., et al., Geol. Soc. Amer., Spec. Pap., 24, 131 (1939).

[5] Wegmann, C. E., Medd. om Gronland, 113 (No. 2), 135 (1938).

[6] Sandell, E. B., and Goldich, S. S., J. Geol., 51, 99 and 167 (1943).

[7] Rosenqvist, I. Th., Amer. J. Sci., 240, 356 (1942).

[8] Hevesy, G., and Hobbie, R., Nature, 123, 1038 (1931).

[9] Goldschmidt, V. M., Skr. Norsk. Videns. Akad., Oslo, I Mat.-Nat. Kl., 1937, No. 4, 94 (1938).

[10] Keevil, N. B., Econ. Geol., 33, 685 (1938).

15

The Isotopic Abundances in Natural Lead and the Age of Uranium†

F. G. Houtermans

The isotope analyses reported by Nier[2] for "ordinary" lead allow extrapolation to the composition of "natural" lead of nonradiogenic origin, on the condition that all lead, as was assumed by Nier, can be regarded as natural lead contaminated by radiogenic lead of primary rocks. Primary rocks in this sense are rocks that retained both lead and uranium in appreciable quantities before the lead minerals crystallized out. This working hypothesis is supported by the dependence of the ratios of the numbers of atoms $\beta = (^{207}Pb)/(^{204}Pb)$ and $\gamma = (^{208}Pb)/(^{204}Pb)$ on $\alpha = (^{206}Pb)/(^{204}Pb)$ for the samples investigated by Nier, and by the fact that the data obtained here give a value of 2.7 for the present Th/U weight ratio in the parent rocks of the lead ores, in good agreement with Keevil's average value[3] of 2.8 for granitic rocks. On the above assumption, if the age of the uranium, p the age of the lead sample, and λ and λ' the decay constants of UI and AcU, then

$$\frac{\beta - \beta_0}{\alpha - \alpha_0} = \frac{Ac\ D}{Ra\ G} = \frac{1}{139} \cdot \frac{e^{\lambda'w} - e^{\lambda'p}}{e^{\lambda w} - e^{\lambda p}} \tag{1}$$

This applies even if the view advocated by A. Holmes[4] that the "ore lead" is largely derived from the natural lead of a deeper nonradioactive layer, or at least of one that contains much less uranium, is correct. Equation (1) means that points in an α versus β plot corresponding to minerals of the same age must lie on straight lines (isochrones), which must intersect at the point α_0, β_0, corresponding to natural lead. To determine the three unknowns α_0, β_0, and w, therefore, we need at least three dated lead minerals, two of which should if possible have different ages. The values $w = (2.9 \pm 0.3) \times 10^9$ years, $\alpha_0 = 11.52 \pm 0.60$, and $\beta_0 = 14.03 \pm 0.20$ are found

†Translated from Die Isotopenhäufigkeiten im Natürlichen Bloi und das Alter des Urans, *Naturwissenschaften*, **33**, 185–186, 219 (1946).

graphically, and $\gamma_0 = 31.6 \pm 0.60$ is found from the α versus γ plot;[5] this corresponds to the following isotopic abundances of true "natural" lead:

Pb	204	206	207	208
%	1.72 ± 0.05	19.81 ± 1.00	24.13 ± 0.43	54.34 ± 1.10

These figures must be regarded as provisional values; a more accurate calculation will require a greater volume of statistical material. The above working hypothesis does not appear to be entirely valid for a galena from the Austrian Triassic and possibly for the Carboniferous galena from Joplin, Missouri; it may also fail for the galena from Ivigtut, Greenland.

Conversely, (1) offers a possibility of age determinations on nonradioactive lead minerals if the values of α_0, β_0, and w are assumed to be known. An age of $(1700 \pm 100) \times 10^9$ years is found in this way for a native lead from Langban, Sweden. This type of age determination to some extent supplements the age determination on a simultaneously crystallized sample of a uranium mineral. In the former the lead formed between the formation of the uranium and the crystallization of the mineral is used for the age determination, while in the second case the lead formed since the separation of the uranium mineral is used.

The "age of the uranium" found agrees with the usual time scale obtained from astrophysical and radioactive data,[6] but disagrees with the age found by Paneth et al.[7] in determinations on meteorites by the helium method, which gave ages of up to 7.8 $\times 10^9$ years. The w determined here gives the age of the solar system, i.e., the time of formation of the terrestrial uranium, if it is assumed that no relative enrichment of uranium in relation to lead has occurred with the formation of the earth's crust, or that between the formation of the uranium and the formation of the solid crust of the earth, there has been no period long in relation to 10^8 years in which uranium and lead were intimately mixed, e.g., in the gaseous and liquid phases. However, if such an enrichment of uranium has occurred (and this is suggested by the abundance of lead in relation to uranium in iron meteorites[8] and by the well known argument, connected with the geothermal gradient, that uranium and thorium are considerably enriched in the outermost parts of the crust in relation to the overall average for the earth), then by definition, as is shown by a simple calculation, even if this separation was incomplete, w gives the time at which this uranium separation occurred, i.e., the age of the solid crust of the earth.

References and Notes

1. Communication to the Conference of the German Physical Society in Göttingen on October 4, 1946; a fuller report will appear shortly in Z. Naturforsch.
2. A. O. Nier, J. Amer. Chem. Soc., **60**, 1571 (1938) and A. O. Nier, R. W. Thompson, and B. F. Murphey, Phys. Rev., **60**, 112 (1940).
3. N. B. Keevil, Econ. Geol., **33**, 685 (1938).
4. A. Holmes, Econ. Geol., **32**, 764 (1937), and **33**, 829 (1938).

5. The values used by Nier for the decay constants of UI, AcU, and Th, i.e., $\lambda = 1.535 \times 10^{-10} a^{-1}$; $\lambda = 6.394.\lambda$ [sic]; and $\lambda_{Th} = 4.99 \times 10^{-11} a^{-1}$, were used for the calculation.
6. Z. B. H. Kienle, *Naturwissenschaft.*, **31**, 149 (1943); S. Meyer, *Mittg. Inst. Radiumforsch. Wien.*, No. 393, No. 407; F. F. Koszy, *Nature*, **151**, 24 (1943); Wefelmeier, unpublished.
7. W. J. Arrol, R. B. Jacobi, and F. A. Paneth, *Nature*, **140**, 235 (1942).
8. V. M. Goldschmidt, *Geochem. Verteilungsgesetze IX*, Oslo (1938).

Addendum

After the above report had been set in type . . . I learned of a recent publication by A. Holmes [*Nature*, **157**, 680 (1946)] in which the quantitative conclusions of the present work were again reached on the basis of Nier's measurements by the use of the relation contained in Eq. (1). Holmes finds, partly with the use of fuller age data for the various lead samples, $\alpha_0 = 12.50$, $\beta_0 = 14.28$, $\gamma_0 = 31.82$, and $w = 3.0 \times 10^9$ years, though the evaluation method differs somewhat from that mentioned here. It is interesting to note that Holmes in this report expressly advocates the view that all the ore lead analyzed so far can be explained as resulting from a differentiation process from parent rocks or their magmas, and he considers the arguments that previously seemed to point to an admixture of "natural" lead as being refuted. The dispersion of the points of an α versus β plot for samples of the same age along an isochrone must thus be regarded as due merely to dispersion of the primary ratio of uranium to natural lead in the parent rocks of the ores, since according to (1) the present ratio of the numbers of atoms in the parent rock must be $\mu_U = (UI)/(^{204}Pb) = (\alpha - \alpha_0)/(e^{\lambda w} - e^{\lambda p})$, a state of affairs that can be checked experimentally by comparison with the corresponding data for natural eruptive rocks.

Geochimica et Cosmochimica Acta, 1956, Vol. 10, pp. 230 to 237. Pergamon Press Ltd., London

16

Age of meteorites and the earth

Claire Patterson

Division of Geological Sciences
California Institute of Technology, Pasadena, California

(*Received 23 January* 1956)

Abstract—Within experimental error, meteorites have one age as determined by three independent radiometric methods. The most accurate method (Pb^{207}/Pb^{206}) gives an age of $4 \cdot 55 \pm 0 \cdot 07 \times 10^9$ yr. Using certain assumptions which are apparently justified, one can define the isotopic evolution of lead for any meteoritic body. It is found that earth lead meets the requirements of this definition. It is therefore believed that the age for the earth is the same as for meteorites. This is the time since the earth attained its present mass.

It seems we now should admit that the age of the earth is known as accurately and with about as much confidence as the concentration of aluminium is known in the Westerly, Rhode Island granite. Good estimates of the earth's age have been known for some time. After the decay-constant of U^{235} and the isotopic compositions of common earth-leads were determined by Nier, initial calculations, such as Gerling's, roughly defined the situation. Approximately correct calculations were made by Holmes and by Houtermans on the basis of bold assumptions concerning the genesis of lead ores. Subsequent criticism of these calculations created an air of doubt about anything concerning common leads and obscured the indispensable contributions which these investigators made in establishing the new science of the geochemistry of lead isotopes. When the isotopic composition of lead from an iron meteorite was determined, we were able to show that a much more accurate calculation of the earth's age could be made, but it still was impossible to defend the computation. Now, we know the isotopic compositions of leads from some stone meteorites and we can make an explicit and logical argument for the computation which is valid and persuasive.

The most accurate age of meteorites is determined by first assuming that meteorites represent an array of uranium-lead systems with certain properties, and by then computing the age of this array from the observed lead pattern. The most accurate age of the earth is obtained by demonstrating that the earth s uranium-lead system belongs to the array of meteoritic uranium-lead systems.*

The following assumptions are made concerning meteorites: they were formed at the same time; they existed as isolated and closed systems; they originally contained lead of the same isotopic composition; they contain uranium which has

---

* C. Patterson: N.R.C. Conference on nuclear processes in geologic settings, 1955 September meeting, Pennsylvania State University. Except for minor disagreements, this paper is probably a concrete expression of the attitudes of most investigators in this field, both here and in Europe. The author is grateful to his colleagues, Clayton, Ingram, Tilton, Wasserburg, and Wetherill, for their criticisms which helped clarify this paper.

230

the same isotopic composition as that in the earth. On the basis of these assumptions various leads might be expected to evolve as a result of different original U/Pb ratios in separate meteorites, and an expression* for any pair of leads derived from such an array is:

$$\frac{R_{1a} - R_{1b}}{R_{2a} - R_{2b}} = \frac{(e^{\lambda_1 T} - 1)}{k(e^{\lambda_2 T} - 1)} \tag{1}$$

where $R_1 = Pb^{207}/Pb^{204}$ and $R_2 = Pb^{206}/Pb^{204}$ for leads from different meteorites a and b, $k = U^{238}/U^{235}$ today (137·8), $\lambda_1 = U^{235}$ decay-constant (9·72 $\times$ 10^{-10} yr^{-1}), $\lambda_2 = U^{238}$ decay-constant (1·537 $\times$ 10^{-10} yr^{-1}), and T = age of the array.

The isotopic compositions of leads isolated from three stone and two iron meteorites are listed in Table 1 (PATTERSON, 1955). Because the radiogenic and nonradiogenic leads may occur in different mineral environments in a stone meteorite and the sample dissolution procedures may be chemically selective, the lead ratios for the first three meteorites in Table 1 have estimated errors from the absolute of about 2%. The lead ratios for the last two meteorites in Table 1 have estimated errors from the absolute of about 1%.

Table 1. The isotopic compositions of lead in meteorites

Meteorite	Pb Composition		
	206/204	207/204	208/204
Nuevo Laredo, Mexico	50·28	34·86	67·97
Forest City, Iowa	19·27	15·95	39·05
Modoc, Kansas	19·48	15·76	38·21
Henbury, Australia	9·55	10·38	29·54
Canyon Diablo, Arizona	9·46	10·34	29·44

These leads cover an extreme range in isotopic composition and satisfy expression (1), yielding, within experimental error, a unique value of T. This is illustrated in Fig. 1, where it is shown that the Pb^{206}/Pb^{204} and Pb^{207}/Pb^{204} ratios from meteorite leads lie on a straight line whose slope corresponds to an age of 4·55 $\times$ 10^9 yr. The dotted lines indicate how stone meteorite leads have evolved. It is clear that the assumptions of the age method are justified by the data. Errors in the lead data and in the decay-constants contribute about equally to the overall error in the calculated age, which amounts to about 1½%. The age for the meteorite array is calculated to be 4·55 $\pm$ 0·07 $\times$ 10^9 yr.

The assumptions have not been shown to be unique. The data can be explained

---

* A similar form of this expression was first used by A. NIER in 1939. F. HOUTERMANS has termed the expression an "isochron." References for the constants are: (k) M. INGHRAM; Vol. 14 Manhattan Project Tech. Sev., Div. 2, Gaseous Diffusion Project, Chap. V, p. 35 (1946); (λ_1, λ_2) E. FLEMING, A. GHIORSO, and B. CUNNINGHAM: *Phys. Rev.* **88**, 642 (1952).

231

by other qualifying or even contradictory assumptions. Most of these can be excluded as improbable. One common criticism should be mentioned: the time of a process of division or agglomeration of meteoritic material (without differentiation) cannot be distinguished by this age method. It seems probable that any such process of division or agglomeration would be accompanied by chemical differentiation. Any meteorite which had a differentiation history after its initial formation would fall off the isochron. The five meteorites in Table 1 represent a most extreme range of differentiation which occurred during the initial process of

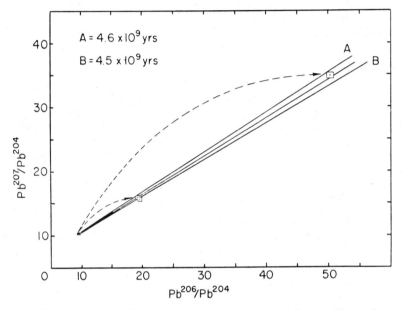

Fig. 1. The lead isochron for meteorites and its estimated limits. The outline around each point indicates measurement error.

formation. This criticism is not serious as far as meteorites are concerned, since if it were valid the lead-lead isochron would date the occurrence of differentiation processes; however, it is important with respect to the age of the earth and will be mentioned later.

At the present time, the next most accurate meteorite age is determined by the A^{40}/K^{40} method. The argon ages of six stone meteorites, three of them determined by WASSERBURG and HAYDEN (1955), and three of them determined by THOMSON and MAYNE (1955), are listed in Table 2. The age of *Forest City* has been redetermined without change by REYNOLDS and LIPSON (1955). Two sets of ages are calculated on the basis of the two reasonable limits of the e^-/β^- branching ratio.

The 0·085 branching ratio is the value obtained by studies of old potassium minerals dated by uranium-lead techniques. The 0·125 branching ratio is the value obtained by counting techniques and by direct measurements of the amounts of decay products. The difference between the two values can be accounted for by systematic loss of radiogenic argon in the old potassium minerals. If one assumes

232

134

that a fixed amount of about 20% of radiogenic argon is lost from all stone meteorites, i.e. using a branching ratio of about 0·10, then there is agreement of lead and argon ages for the same stone and an indication that the stones have existed as cold and solid bodies since they were formed. Argon meteorite ages different from the ones mentioned here have been reported by GERLING (1951), and PAVLOVA GERLING and RIK (1954). Since errors in the data presented by GERLING and PAVLOVA cannot be evaluated with any certainty, we cannot be concerned by differences between ages calculated by them and ages calculated from other data. Because of logarithmic behaviour, values for calculated ages of these old samples are insensitive to changes in the e^-/β^- branching ratio. For this reason only disagreements of about 15% between A^{40}/K^{40} and Pb^{207}/Pb^{206} meteoritic ages can be accounted for by a twofold change in the branching ratio. Large age differences must therefore be reconciled on the basis of other experimental errors. Measurements of the amounts of nonradiogenic argon in radiogenic and nonradiogenic argon mixtures are subject to large uncertainties, and for the first four meteorites in

Table 2. A^{40}/K^{40} ages of meteorites

Meteorite	Age $\times 10^{-9}$		Investigators
	$(e^-/\beta^- = 0·085)$	$(e^-/\beta^- = 0·125)$	
Beardsley, Kansas	4·8	4·2	WASSERBURG and HAYDEN
Holbrook, Arizona	4·8	4·2	WASSERBURG and HAYDEN
Forest City, Iowa	4·7	4·1	{WASSERBURG and HAYDEN {REYNOLDS and LIPSON
Akabu, Transjordan	4·4	3·8	THOMSON and MAYNE
Brenham Township, Kansas	4	3	THOMPSON and MAYNE
Monze, Northern Rhodesia	2	2	THOMSON and MAYNE

Table 2, nonradiogenic argon corrections were small. For the last two meteorites in Table 2, nonradiogenic argon corrections were extremely large and the errors in calculated age are excessive. The isotope dilution determination of potassium, used by WASSERBURG and HAYDEN, is nearly an absolute method, while the flame-photometric determination of potassium, used by THOMSON and MAYNE, requires a natural absolute standard which they did not use.

The age of meteorites has been determined by the Sr^{87}/Rb^{87} method. The concentrations of rubidium and strontium and the isotopic compositions of strontium have been determined in two stone meteorites by SCHUMACHER (1955). The Rb/Sr ratio in one stone was so low that any change in isotopic composition of strontium due to radioactivity would be within experimental error. The Rb/Sr ratio in the other stone (Forest City, Iowa) was considerably higher and sufficient

233

to cause a 10% difference in the relative abundance of Sr^{87} when the isotopic compositions of strontium from both stones were compared.

The value for the decay-constant of Rb^{87} is in question at the present time. Reported values range from 4·3 to 6·7 $\times$ 10^{10} yr for the half-life. Part of the difficulty in the counting techniques of measuring the half-life arises from the fact that the frequency of β^-s at the low end of the energy spectrum increases rapidly with no appearance of a maximum. Measurements of decay products in terrestrial rubidium minerals dated by uranium-lead technique involve errors of open chemical systems. SCHUMACHER's experiment probably constitutes an ideal case of the geological measurement of the half-life of Rb^{87}, since the ages have been determined by lead methods and the possibility of open chemical systems are remote. His methods of measurement are at least as accurate as the radiometric methods. One would therefore use his data to calculate the half-life of Rb^{87}, using the Pb-Pb isochron age of meteorites. The half-life of Rb^{87}, as determined by these data, is 5·1 $\times$ 10^{10} yr, and is probably the most reliable value at present. The half-life determined by the geological method on terrestrial minerals (5·0 $\times$ 10^{10} yr) agrees well with this.*

Because of the overwhelming abundance of nonradiogenic helium in iron meteorites and the large errors associated with the determination of the concentrations of uranium and thorium in iron and stone meteorites, the age of meteorites by the helium method is not accurate to much better than an order of magnitude (PANETH et al., 1953; DALTON et al., 1953). It has been reported that iron meteorites and the metal phases of stone meteorites were outgassed of helium as of about 5 $\times$ 10^8 yr ago, while the silicate phases of stone meteorites were not (REASBECK and MAYNE, 1955). Such an event would be highly significant and would require detailed evolutionary theory for meteorites. Recent neutron activation (REED and TURKEVICH) and nuclear emulsion (PICCIOTTO) analyses of iron meteorites show that the concentrations of uranium in these bodies are very low, and that the uranium concentrations used for helium age calculations of iron meteorites may be erroneously high. The question is unresolved at present, but it seems reasonable to believe that investigations of meteoritic helium will become vitally important to cosmic-ray studies and may be decisive in meteorite evolution theory, but cannot be used for accurate meteorite-age calculations at the present time.

The Canyon Diablo lead listed in Table 1 was isolated from troilite where the U^{238}/Pb^{204} ratio was shown by direct analysis to be 0·025 (PATTERSON et al., 1953). This ratio is accurate to at least an order of magnitude, and it is so small that no observable change in the isotopic composition of lead could have resulted from radioactive decay after the meteorite was formed. Since stone meteorites were cold and solid during their lifetime, it is unlikely that lead transport could have occurred between iron and stone meteoritic phases if they existed in one body. This iron-meteorite lead is therefore primordial and represents the isotopic composition of primordial lead at the time meteorites were formed. Using the isotopic composition of primordial lead and the age of meteorites, expressions can be

---

* A value recommended by the work of the geochronology laboratory at the Dept. of Terrestrial Magnetism, Carnegie Institute of Washington.

234

written for a representative lead which is derived today from any system belonging to the meteoritic array:

$$Pb^{206}/Pb^{204} = 9.50 + 1.014 \; U^{238}/Pb^{204} \qquad (2)$$

$$Pb^{207}/Pb^{204} = 10.36 + 0.601 \; U^{238}/Pb^{204} \qquad (3)$$

If any two of the three ratios above can be independently measured in the earth's uranium-lead system, and they satisfy expressions (2) and (3), then this system belongs to the meteoritic array and must have its age. Two of the ratios can be measured in a sample of earth lead, but the problem of choosing such a sample is complex because the ratio of uranium to lead varies widely in different rocks and minerals whose ages are short compared to the age of the earth.

One approach is to partition the earth's crust into separate chemical systems of uranium and lead and consider their interactions. Such systems may range from minerals to geochemical cycles. Nearly all of the lead-isotope data concerns either minerals in which the uranium-to-lead ratio is very high (uraninites, etc.) or minerals in which this ratio is essentially zero (galenas). The approximate times of formation of some galenas have been determined, and of these, two dozen or so lately formed galenas may be used as a measure of earth lead (*Nuclear Geology* (1954), W. Faul, Ed.). The isotopic compositions of lead in some recent oceanic sediments have also been determined (PATTERSON, GOLDBERG, and INGHRAM 1953), and these may be used as a measure of earth lead.

Any of these samples will be improper or biased if they are derived from a system of uranium and lead which is only partially closed and is subject to slow but appreciable transport from other systems with different U^{238}/Pb^{204} ratios. In this respect, the sample which may represent the system of largest mass is probably the more reliable. One sample of oceanic sediment lead probably represents more material than a dozen galenas. The isotopic composition of this sediment lead is $Pb^{206}/Pb^{204} = 19.0$ and $Pb^{207}/Pb^{204} = 15.8$, which satisfies expressions (2) and (3) surprisingly well. It is doubtful if these figures are grossly biased, since a few measurements of uranium and the isotopes of lead in rocks with widely different U^{238}/Pb^{204} ratios indicate rather good mixing to be the first-order effect on the isotopic composition of lead in the earth's crust (PATTERSON, TILTON, and INGHRAM, 1955).

Independent of the absolute abundances of lead isotopes, a rough measure of the rates of change of the lead-isotope abundances in the earth's crust may be obtained from the isotopic composition of galenas of different ages. These rates of change are defined by the ratios of uranium and thorium to lead in the material from which the galenas are derived. From the observed rate of change of Pb^{206} the U^{238}/Pb^{204} ratio in the earth's crust is found to be 10 (COLLINS, RUSSELL, and FARQUHAR, 1953). This value satisfies expression (2) and (3) for sedimentary lead with unexpectedly good agreement.

In Fig. 2 it is shown that oceanic sediment lead (open circle) falls on the meteoritic lead isochron. Most of the lately formed galenas fall within the dotted outline, although a few are widely aberrant. The position of a lead along the isochron is determined by the U^{238}/Pb^{204} ratio in the system from which the lead

235

137

evolves. The arrow indicates the position on the isochron which sediment lead should occupy as predicted by the isotopic evolution of dated ore leads. Independently measured values for all three ratios adequately satisfy expressions (2) and (3), and therefore the time since the earth attained its present mass is $4\cdot55 \pm 0\cdot07 \times 10^9$ yr.

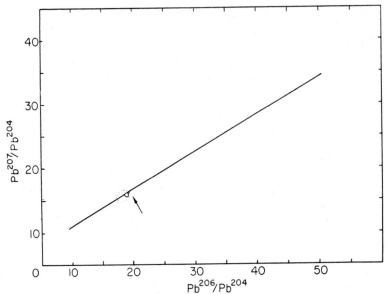

Fig. 2. The relationship between common earth leads and the meteoritic lead isochron.

If the earth is a late agglomeration without differentiation of meteoritic material then it can have any age less than meteoritic material. Rather than arguing that such a process would be accompanied by chemical differentiation (and a change of the U/Pb ratio), it seems reasonable to believe instead that such a late agglomeration process would be less probable than one where both meteorites and the earth were formed at the same time. It is a fact that extreme chemical differentiation occurred during the process which led to the mechanical isolation of the mass of material of which the earth is made, and since changes in this mass were accompanied by chemical differentiation, the Pb/Pb meteorite isochron age properly refers to the time since the earth attained its present mass.

REFERENCES

COLLINS C., RUSSELL, R., and FARQUHAR R. (1953) Canad. J. Phys. 31, 402.
DALTON J., PANETH F., REASBECK P., THOMSON, S., and MAYNE K. (1953). Nature 172, 1168.
GERLING E. and PAVLOVA T. (1951) Doklady Akad. Nauk, U.S.S.R. 77, 85.
GERLING E. and RIK K. (1954) Meteoritika 11, 117.
PANETH F., CHACKETT, K., REASBECK P., WILSON E., DALTON J., GOLDEN J., MARTIN G., MERCER E., and THOMSON S. (1953) Geochim. et Cosmochim. Acta 3, 257.
PATTERSON C., BROWN H., TILTON G., and INGHRAM M. (1953) Phys. Rev. 92, 1234.
PATTERSON C., GOLDBERG E., and INGHRAM M. (1953) Bull. Geol. Soc. Amer. 64, 1387.
PATTERSON C. (1955) Geochim. et Cosmochim. Acta 7, 151.

236

Patterson C., Tilton, G., and Inghram M. (1955) *Science* **121**, 69.

Picciotto E. Nuclear Physics Centre, University of Brussells, manuscript.

Reynolds J. and Lipson J. Epipoleological Society, spring 1955 meeting, U.C.L.A.

Reasbeck P. and Mayne K. (1955) *Nature* **176**, 186.

Reed G. and Turkevitch A. Inst. Nuclear Studies, University of Chicago, manuscript.

Schumacher E. N.R.C. conference on nuclear processes in Geologic Settings (communicated by H. Urey and M. Inghram), 1955 Sept. meeting, Penn. State University. (Copies of his manuscript are available).

Thompson S. and Mayne K. (1955) *Geochim et Cosmochim. Acta* **7**, 169

Wasserburg G. and Hayden R. (1955) *Phys. Rev.* **97**, 86.

Wasserburg G. N.R.C. Conference on nuclear processes in Geologic Settings, 1955 Sept. meeting, Penn. State University.

Nuclear Geology (1954) H. Faul, Ed., J. Wiley, N.Y.

Reports

Lead Isotopes and the Age of the Earth

Abstract. *Calculations based on comparison of the isotopic composition of lead from iron meteorites with that of various modern terrestrial leads have placed the age of the earth at around 4550 million years. However, recent data from young volcanic rocks reveal that modern terrestrial lead can have a wide range in isotopic composition. The variations in its composition mean that one or more of the assumptions used in the age calculation have been violated. We modified the usual approach by comparing meteorite lead with lead from rocks 2700 million years old, from the Canadian Shield. Using this method and the same constants and assumptions utilized in the earlier calculations we calculated an age of 4750 ± 50 million years for the earth. The earth may be approximately 200 million years older than previously thought; alternatively, primordial terrestrial lead may not have had the same isotopic composition as lead in iron meteorites does.*

Calculations of the age of the earth are based on an attempt to determine when terrestrial lead had the same isotopic composition as does lead observed today in the troilite phase of certain iron meteorites. The ratio of uranium to lead in the troilite phase is so low that the decay of the uranium in the mineral would not have changed isotopic composition of lead measurably over many billions of years. On the other hand readily measurable increments of radiogenic lead have been produced over the past several billion years in the outer portions of the earth that we can sample and study. With this knowledge, an age for the earth can be calculated under the following assumptions: (i) The isotopic composition of terrestrial lead at the time of formation of the earth was identical to that in meteoritic troilite. (ii) The uranium-lead ratios in the outer portion of the earth were established in a short time, relative to the age of the earth. (iii) These systems have been chemically closed with respect to uranium and lead since they were established—that is, changes in the concentrations of these elements have been due solely to radioactive decay. The difference between the isotopic composition of troilite lead and that of terrestrial lead determines the increments of radiogenic Pb^{206} and radiogenic Pb^{207} in the terrestrial lead, and these increments determine an age for the earth according to the equation

$$\frac{b - b_0}{a - a_0} = \frac{R (e^{\lambda' t} - 1)}{(e^{\lambda t} - 1)}$$

where a and b are the Pb^{206}/Pb^{204} and Pb^{207}/Pb^{204} ratios for present-day terrestrial lead; a_0 and b_0 are the same ratios for troilite lead; R is the present ratio of U^{235}/U^{238}; λ' is the decay constant for U^{235}; λ is the decay constant for U^{238}; and t is the age of the earth. The method is analogous to that used to determine the Pb^{207}-Pb^{206} age of a mineral, when the correction for primary lead is usually inferred from measurement of the isotopic composition of lead from presumed cogenetic potassium feldspar or galena.

Several authors have used this method with isotopic values for meteorite and modern terrestrial lead from various sources and have found ages of around 4550 million years for the earth (1). Lead from young volcanic rocks, particularly basalts, and lead ores have been used to determine the isotopic composition of modern terrestrial lead. Since the isotopic composition of lead in surface rocks is in general quite variable, basic volcanic rocks, which are believed to originate, in the outer mantle or lower crust of the earth, seem the most likely of any rocks that can be obtained to contain lead fulfilling the closed-system requirement. Volcanic rocks from the ocean basins are especially promising as the

earth's crust is thin in such areas, and contamination with crustal lead is much less likely to occur here than on continents.

Recent papers (2, 3) on the isotopic composition of lead in young volcanic rocks from the ocean basins show that the lead from these rocks is of variable isotopic composition. Values for the age of the earth which vary from 4420 to 4650 million years can be calculated from these data. Obviously some, if not all, of the rocks contain lead whose history does not fulfill the three assumptions stated above. Close study of the data shows that the closed-system requirement has been violated (2). Variations in the isotopic composition of lead in the basalts show that the uranium-lead ratios in the source materials have differed greatly during the past billion years or so. Possibly other assumptions have been violated as well.

Since the closed-system assumption does not hold for lead in young rocks, study of the isotopic composition of lead in the oldest rocks that can be obtained seems desirable. The error introduced into the age calculation by the failure of the closed-system assumption should be less when older rocks are used, providing their age and the isotopic composition of lead in the rocks at the time of crystallization can be accurately determined. In making such a study it is necessary to utilize lead from granitic rocks and galena ore bodies since basalts in early Precambrian assemblages always display more or less alteration. Such a substitution seems justified since the isotopic composition of lead in modern galena (4) and that in young potassium feldspars (5) lie within the range of values covered by lead in young oceanic basalts.

In studying old rocks, the isotopic composition of lead from both granitic rocks and galena ore bodies should be determined for several reasons. The age of granites and pegmatites can be determined by the conventional methods of geochronology, whereas the age of ore bodies must usually be inferred. On the other hand the concentration of lead in minerals of granitic rocks is in the parts-per-million range, creating two difficulties not present with ore samples. First it is necessary to make corrections for radiogenic lead produced by the uranium and thorium in a mineral, even in potassium feldspars, which have the lowest uranium-lead and thorium-lead

Table 1. Mineral ages from Algoman granites and pegmatites at Manitouwadge, Ontario. MG-17-1 and MG-17-d are light- and dark-colored fractions separated from a single sample of granite. The light fraction contains 257 parts of uranium per million; the dark fraction, 701 ppm of uranium. Observed Sr^{87} was 92 percent radiogenic for MG-20; 47 percent radiogenic for MG-45.

Rock	Mineral	Age (million years) obtained by various methods				
		Pb^{206}/U^{238}	Pb^{207}/U^{235}	Pb^{207}/Pb^{206}	Pb^{208}/Th^{232}	Sr^{87}/Rb^{87}
MG-17-1	Zircon	2500	2610	2700	2710	
MG-17-d	Zircon	2420	2560	2670	2300	
MG-17	Biotite					2630
MG-20	Microcline					2590
MG-45	Microcline					2560

ratios of any of the minerals available in granitic rocks. Second, the low lead concentrations increase susceptibility to contamination from external sources. Contamination of lead in potassium feldspars by relatively mild metamorphic episodes (ones that did not seriously alter rubidium-strontium ages of the mineral) has been demonstrated in two localities (6, 7). Galena contains negligible amounts of uranium and and thorium, making corrections for internally produced radiogenic lead unnecessary. Since lead is a major constituent of galena, it is difficult to contaminate ore by lead from external sources.

Data in our studies are from granitic rocks and lead ores from the Superior Province of North America, where mineral ages have generally been found to be 2500 to 2700 million years (8). Doe, Tilton, and Hopson (7) separated lead from five potassium feldspars taken from granites and pegmatites in the vicinity of Rainy Lake in northern Minnesota. The feldspars were taken from Algoman rocks, which are the younger, post kinematic rocks of the area. Rubidium-strontium ages of 2550 million years have been reported for these granites (9), while Anderson

and Gast (10) determined from lead isotopes an age of 2650 million years for zircon from an Algoman granite in northern Minnesota. The Minnesota feldspar samples used for the lead work were separated from the rocks used in the age determinations by Goldich et al. (11). Biotite from the rocks gives potassium-argon age values of 2400 to 2600 million years, indicating that the area has not experienced severe thermal metamorphism since that time. In interpreting the Minnesota data on lead, it was necessary to make substantial corrections for radiogenic lead produced internally over the past 2600 million years by the uranium in the feldspars. Since galena does not occur in the Rainy Lake area, the isotopic composition of feldspar lead was compared with that of lead from galena at Manitouwadge, Ontario, about 300 km to the northeast (4). The Pb^{206}/Pb^{204} and Pb^{207}/Pb^{204} ratios of the galena and corrected ratios in feldspar lead agreed rather closely. The Pb^{208}/Pb^{204} ratios of galena and of feldspars agreed within error limits without corrections. These results suggested that the isotopic composition of galena lead was a good approximation of the isotopic compo-

sition of lead in the granite when it crystallized. It was also apparent that if the Manitouwadge galena were assigned an age of 2600 million years, and if this value were used to calculate the age of the earth, an age of more than 4550 million years would be obtained.

We now report additional studies at Manitouwadge that enable data from granite and from ore to be compared with greater confidence. We have again determined the isotopic composition of lead and the mineral ages for samples taken from the Algoman granites of the area according to Pye's classification (12). The mineral ages are given in Table 1. The ages for zircon, as determined by various methods, are nearly concordant and indicate an age of approximately 2700 million years. If the age discordances are interpreted according to a continuous diffusion mechanism for loss of lead (13), an age of 2750 million years results. Since biotite age values are very sensitive to metamorphism (14), the age found for this mineral shows that no severe metamorphism has affected the rocks in this area for the past 2600 million years. We are also determining rubidium-strontium age values for the granites using total rock specimens. Of four granites studied thus far, three fit a 2690-million-year isochron within the limits of experimental error, and yield a value of 0.701 for the initial ratio Sr^{87}/Sr^{86} when the Sr^{86}/Sr^{88} ratio is normalized to 0.1194. The fourth sample plots off the isochron in the direction of a younger age. The combined data point to an age of 2600 to 2750 million years for the Algoman granites. The ore bodies were presumably emplaced at about the same time, although after the granites (12).

The isotopic composition of lead from four feldspar and two galena samples are given in Table 2. The isotopic compositions are for the samples as prepared, without acid-washing unless indicated. The isotopic ratios are shown graphically in Fig. 1, along with the Rainy Lake samples of Doe, Tilton, and Hopson (7). The line through the points representing ratios in feldspar is a "secondary isochron," giving the pattern expected if the isotopic ratios of lead in the feldspars and galenas were identical 2700 million years ago, but have changed due to the decay of the small amounts of uranium present in the feldspars. Within error limits, the feldspars conform to this hypothesis. Even more convinc-

Table 2. Isotopic composition of lead at Manitouwadge, Ontario.

Source of lead	Pb^{206}/Pb^{204}	Pb^{206}/Pb^{207}	Pb^{206}/Pb^{208}	Pb^{207}/Pb^{204}	Pb^{208}/Pb^{204}
Feldspar					
MG-19	13.69	0.9341	0.4076	14.66	33.59
MG-19 (acid-washed)	13.57	.9284	.4057	14.62	33.45
MG-45	13.61	.9245	.4059	14.72	33.53
MG-20	13.92	.9481	.4148	14.68	33.56
MG-20 (acid-washed)	13.71	.9366	.4089	14.64	33.53
MG-41	14.16	.9616	.4211	14.73	33.63
Galena					
Geco Mine	13.30	.9166	.3960	14.51	33.59
Willroy Mine	13.30	.9149	.3961	14.54	33.58
Average from three mines*	13.40	.9202	.3997	14.57	33.54
Standard					
Caltech lead standard, average of five determinations	16.63	1.0726	.4574	15.51	36.35

* Data from Ostic (16).

Table 3. Age of the earth calculated for various assumed ages of Manitouwadge galena. Constants used in calculations are: for primordial lead, $Pb^{206}/Pb^{204} = 9.54$ and $Pb^{207}/Pb^{204} = 10.27$; for Manitouwadge galena, $Pb^{206}/Pb^{204} = 13.30$ and $Pb^{207}/Pb^{204} = 14.52$; and for decay constants, $U^{238} = 1.54 \times 10^{-10}$ yr^{-1} and $U^{235} = 9.72 \times 10^{-10}$ yr^{-1}.

Age of galena (million years)	Age of earth (million years)
2600	4800
2700	4750
2800	4700

ing evidence linking the feldspar and galena lead is the close agreement of all the Pb^{208}/Pb^{204} ratios in Table 2. From these observations we conclude that the galena data may be taken to characterize the isotopic composition of lead associated with the igneous activity that produced the Algoman granitic rocks at Manitouwadge 2600 to 2700 million years ago. The granitic rocks in the Rainy Lake district were probably formed with lead of very similar isotopic composition.

The age of the earth can be calculated from the Manitouwadge lead data using the same assumptions outlined at the beginning of this report, along with the additional requirement of assigning an age to the rocks. The

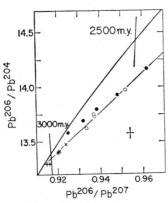

Fig. 1. Pb^{206}/Pb^{204} ratios plotted against Pb^{206}/Pb^{207} ratios for leads from Superior Province, North America. $\bigcirc$, granites and pegmatites from Rainy Lake and northern Minnesota; $\bullet$, pegmatites near Manitouwadge, Ontario; $\times$, Manitouwadge galena described by Ostic (16); $+$, Manitouwadge galena, this report. Cross at lower right represents approximate analytical uncertainty. Lines labeled in millions of years are isochrons based on an age of 4550 million years for the earth and values of 9.54 and 10.27 for the initial Pb^{206}/Pb^{204} and Pb^{207}/Pb^{204} ratios, respectively. See text for explanation of line through points.

increments of radiogenic Pb^{207} and Pb^{206} in the granitic or galena lead relative to meteoritic lead give a measure of the time elapsed from the formation of the earth to the formation of the granite. This interval is calculated by the equation given in the introductory paragraph of this report, substituting for the term R the value appropriate for the isotopic composition of uranium at the assumed time of crystallization of the granite. When this value is added to the age of the granite, an age is obtained for the earth. Calculated values for the age of the earth resulting from three different assignments of age for the granites are given in Table 3. All values for the age of the earth given in Table 3 are greater than those obtained from the calculations we have mentioned which have used modern lead. In this report we use an age of 2700 ± 100 million years for the granites, which yields a value of 4750 ± 50 million years for the age of the earth. We emphasize that the uncertainty is a formal one, based on the assumed correctness of the constants and mathematical model. The extent of actual uncertainty is difficult to estimate, but is larger than the formal one.

Since the Manitouwadge–Rainy Lake data represent an extremely small part of the areal extent of the Superior Province, additional data are required to ascertain whether the results are of more than local significance. Isotopic data for lead from galena ores given in the compilation by Russell and Farquhar (15) permit a preliminary evaluation of this problem. Their data show that galena deposits along a 1100-km traverse across the Superior Province from Noranda, Quebec, to Kenora, Ontario, generally contain lead quite similar in isotopic composition to that at Manitouwadge. We think, therefore, that the data presented here characterize lead from a substantial portion of the Superior Province.

Ostic, Russell, and Reynolds (4) have reported a somewhat different method of calculating the age of the earth that is based on the curvature of the line describing lead evolution as defined by a number of galena samples with ages varying from 0 to approximately 3000 million years. They obtained a value of 4540 million years for the age of the earth, one in good agreement with earlier estimates. Their calculation is strongly influenced by the choice of Manitouwadge lead for the sample in range of 2500 to 3000 mil-

lion years. Using the compilation by Russell and Farquhar (15) we compared data for the isotopic composition of lead from galenas of the Southern Rhodesian Shield of Africa (where mineral ages of 2600 to 2700 million years have also been found) with data for galenas from the Superior Province in the Canadian Shield of North America. Such a comparison shows that the African leads appear to have evolved in sources with higher uranium-lead ratios than did the North American galenas. Use of the African galenas in place of Superior Province galena would increase the age for the earth obtained by the method of Ostic, Russell, and Reynolds. The evidence from young volcanic rocks that many leads have not had a closed system for isotopic evolution and the fact that the African and North American leads are different introduce uncertainty into their results.

There are two simple ways of interpreting the new experiments. The earth may be approximately 200 million years older than previously supposed, having an age of about 4750 million years. If the difference between our results and the earlier ones is due to failure of the closed-system assumption over the past 2700 million years, the failure is in the direction of increase of uranium with respect to lead in the source material for the lead in granites and galenas. If the increase also occurred in the source material prior to 2700 million years ago, calculations show that an age of over 4800 million years is possible. However, the age calculation also contains the assumption that the isotopic composition of lead in the earth was initially the same as that in the troilite phase of iron meteorites. This assumption may be incorrect. We emphasize that the calculation leading to the value of 4750 million years is as good as, if not better than, any of the previous calculations of the age of the earth, particularly those based on modern terrestrial lead, where errors due to failure of the closed-system assumption may be quite large. In any event, interesting effects have been found that indicate the desirability of obtaining new data for the isotopic composition of lead from very old rocks and ores on all continents.

G. R. TILTON*
R. H. STEIGER†
Geophysical Laboratory,
Carnegie Institution of Washington,
Washington, D.C. 20008

References and Notes

1. C. Patterson, G. Tilton, M. Inghram, *Science*
   121, 69 (1955); C. Patterson, *Geochim. Cosmochim. Acta* **10**, 230 (1956); V. Rama Murthy
   and C. C. Patterson, *J. Geophys. Res.* **67**,
   1161 (1962).
2. P. W. Gast, G. R. Tilton, C. Hedge, *Science*
   145, 1181 (1964).
3. M. Tatsumoto, *Trans. Amer. Geophys. Union*
   46, 165 (1965); C. C. Patterson, in *Recent Researches in the Fields of the Hydrosphere, Atmosphere and Nuclear Geochemistry*, Y.
   Miyake and T. Koyama, Eds. (Maruzen,
   Tokyo, 1964), pp. 257–261.
4. R. G. Ostic, R. D. Russell, P. H. Reynolds,
   Nature **199**, 1150 (1963).
5. C. Patterson and M. Tatsumoto, *Geochim.
   Cosmochim. Acta* **28**, 1 (1964).
6. B. R. Doe and S. R. Hart, *J. Geophys. Res.*
   68, 3521 (1963).
7. B. R. Doe, G. R. Tilton, C. A. Hopson, *ibid.*
   70, 1947 (1965).
8. R. K. Wanless, R. D. Stevens, G. R. Lachance, R. Y. H. Rimsaite, C. H. Stockwell, H.
   Williams, *Can. Dep. Mines Tech. Surv., Geol.
   Surv. Can. Paper 64-17*, 1964; A. E. J. Engel,
   Science **140**, 143 (1963); G. R. Tilton and S.
   R. Hart, *ibid.*, p. 357.
9. C. E. Hedge and F. G. Walthall, *Science* **140**,
   1214 (1963).
10. D. H. Anderson and P. W. Gast, *Geol. Soc.
    Amer. Special Paper*, in press.
11. S. S. Goldich, A. O. Nier, H. Baadsgaard, J.
    H. Hoffman, H. W. Krueger, *The Precam-
    brian Geology and Geochronology of Minne-
    sota* (Univ. of Minnesota Press, Minneapolis,
    1961).
12. E. G. Pye, *Rep. Ontario Dep. Mines* **66**, part
    8 (1957).
13. G. R. Tilton, *J. Geophys. Res.* **65**, 2933 (1960).
14. S. R. Hart, *J. Geol.* **72**, 493 (1964).
15. R. D. Russell and R. M. Farquhar, *Lead Iso-
    topes in Geology* (Interscience, New York,
    1960).
16. R. G. Ostic, "Isotopic investigation of con-
    formable lead deposits" thesis, University of
    British Columbia, 1963.
17. We thank D. Timms, chief geologist at the
    Willroy Mine, Manitouwadge, Ontario, for
    assistance in the collection of samples at
    Manitouwadge. The assistance of our col-
    leagues, L. T. Aldrich, G. L. Davis and J.
    B Doak is greatly appreciated.
* Present address: Department of Geology,
    University of California, Santa Barbara, Cal-
    ifornia.
† Present address: Division of Earth Sciences,
    California Institute of Technology, Pasadena,
    California.

20 October 1965

Anomalous Leads and the Age of Lead Mineralization

VII

Editor's Comments on Paper 18

The model for the evolution of common lead isotopes formulated by Holmes and Houtermans (Papers 14 and 15) lead to the recognition of "anomalous" lead ores, i.e., deposits that did not obey the restriction that they evolved from primeval lead through the addition of radiogenic lead produced in a single source region containing the radioactive parent elements uranium and thorium. Leads that did obey such a requirement were termed "ordinary" leads. Anomalous leads were of two types, containing either too little or too much radiogenic lead, so that their "model" ages calculated on the basis of a single-stage evolution in a source region containing a fixed concentration of uranium (i.e., a closed system) would be either too old or too young—in some cases even negative.

The problem of anomalous leads is a complex one, but in certain cases, particularly where the cause of the anomaly can be clearly demonstrated, information concerning the time of mineralization can be extracted from the isotopic data. For instance, where mixing of leads derived from two sources has occurred a linear isotopic relationship will exist between Pb-207/Pb-204 and Pb-206/Pb-204 ratios, forming a "secondary isochron." In the following paper (No. 18), Russell and Farquhar summarize many of the difficulties inherent in the phenomenon of anomalous leads and point the way to further resolution of the problem. They suggest that all deposits of lead occurring in veins may to a lesser or greater extent be anomalous because, by migrating from the source regions through veins in crustal rocks, common lead could easily have assimilated additional radiogenic lead derived not from the source region, but from the country-rocks themselves. Alternatively, the vein lead may be remobilized and reemplaced in much younger rocks. The isochron concept of Holmes and Houtermans could thus be misleading, as variable Pb-206/Pb-204 and Pb-207/Pb-204 ratios could be caused by the ore deposition process itself and may not necessarily be related to the evolution of lead in separate source regions having different uranium concentrations.

Only one type of lead ore deposit fitted a single growth curve indicative of uncontaminated lead samples extracted from a single source region (the mantle) with uniform uranium concentration. These were the conformable lead ore deposits described by the Australian economic geologist R. L. Stanton. Galena samples from conformable ore deposits yielded lead isotopic compositions which, when plotted on a Pb-207/Pb-204 versus Pb-206/Pb-204 diagram, defined a precise growth curve representing the evolution of lead isotopic composition in the earth's mantle. One end of the growth curve is fixed by the primeval isotopic abundances determined on lead extracted from iron meteorites; the other end of the curve lies on the "zero isochron" used by Patterson to define the age of meteorites and the earth (Paper 16). Galenas of the conformable type can then be dated by reference to the growth curve simply by measuring the ratio of any two lead isotopes. The age obtained represents the time of extraction from the mantle source region which, for conformable ore deposits, will be close to the time of deposition of the sedimentary beds in which the conformable leads are found.

Further applications and extensions of the ideas expressed by Russell and Farquhar have been subsequently reported by Moorbath (1962), Kanasewich (1968), and Russell (1972). Reviews of the whole field of lead isotopes applied to geologic and geochronologic problems have been written by Russell and Farquhar (1960) and by Doe (1970).

Selected Bibliography

Doe, B. R. (1970). "Lead Isotopes," 137 pp. Springer-Verlag, New York.

Kanasewich, E. R. (1968). The interpretation of lead isotopes and their geological significance. *In* "Radiometric Dating for Geologists" (E. I. Hamilton and R. M. Farquhar, eds.), pp. 147–223. Interscience, London.

Moorbath, S. (1962). Lead isotope abundance studies on mineral occurrences in the British Isles and their geological significance. *Phil. Trans. Roy. Soc., Ser. A,* **254**, 295–360.

Ostic, R. G., Russell, R. D., and Stanton, R. L. (1967). Additional measurements of the isotopic composition of lead from stratiform deposits. *Can. J. Earth Sci.,* **4**, 245–269.

Russell, R. D. (1972). Evolutionary model for lead isotopes in conformable ores and in ocean volcanics. *Rev. Geophys. Space Phys.,* **10**, 529–549.

Russell, R. D., and Farquhar, R. M. (1960). "Lead Isotopes in Geology," 243 pp. Interscience, New York.

Russell, R. D., Farquhar, R. M., Cumming, G. L., and Wilson, J. T. (1954). Dating galenas by means of their isotopic constitutions. *Trans. Amer. Geophys. Union,* **35**, 301–309.

Geochimica et Cosmochimica Acta, 1960, Vol. 19, pp. 41 to 52. Pergamon Press Ltd. Printed in Northern Ireland

18

Dating galenas by means of their isotopic constitutions—II

R. D. Russell and R. M. Farquhar

Geophysics Laboratory, Department of Physics, University of British Columbia, Canada
and Geophysics Laboratory, Department of Physics, University of Toronto, Canada

(*Received 5 April* 1959; *revised August* 1959)

Abstract—Recently a rather large number of analyses of common lead samples has become available, and these in turn have led to a better understanding of the precise nature of the physical and geological processes resulting in the observed variations of lead isotope ratios. Specifically, there now appear to be three classes of leads, each of which has lead isotope ratios that vary in a remarkably simple pattern. These are meteoritic leads, anomalous leads, and leads from certain conformable lead ore deposits. The meteoritic leads and conformable leads both appear to have developed in surroundings where thorium and uranium had very similar properties, and in both cases the thorium to uranium ratio corresponds to a present value of $3 \cdot 73 \pm 0 \cdot 03$. This is taken to be evidence that both classes developed under predominantly reducing conditions. In contrast, the presence of thorium to uranium ratio present during the production of anomalous leads is extremely variable, suggesting an oxidizing environment. It has been suggested that all vein leads may be anomalous to some degree and therefore should be interpreted with this possibility in mind.

The methods for dating galenas proposed previously by the present writers are re-examined in view of these developments and it is shown that they may lead to more explicit information on the age and history of lead ores than has usually been obtained.

INTRODUCTION

A FEW years ago we became authors of a paper dealing with a method for estimating the ages of galenas for which the lead isotope ratios were known (RUSSELL *et al.*, 1954). At the time that paper was written there appeared to be sufficient isotopic data available to justify the supposition that the isotopic behaviour of leads was rather regular and predictable and that the isotope ratios could be used to estimate the ages of the minerals. It also appeared that the relationships between the isotope ratios and ages of lead minerals were very imperfect and very large uncertainties were assigned to the age values obtained. These ranged as high as ± 800 million years.

Since the publication of the above paper there have been a number of developments that bear on this subject and it is worthwhile to reassess this problem in the light of our present knowledge. These developments have resulted to a considerable degree from the very extensive effort applied to dating galenas by F. G. HOUTERMANS and his associates at the University of Bern, Switzerland, and by ourselves and others at the University of Toronto, Canada, as well as new discoveries in closely allied fields made by other scientists. They include the great advances in the understanding of the nature and behaviour of *anomalous* leads (CAHEN *et al.*, 1958; RUSSELL *et al.*, 1957; FARQUHAR and RUSSELL, 1957), the recent determination of the ratios of the lead isotopes in meteorites (PATTERSON, 1956) and the recognition of a certain class of lead ores that may have had a simpler history than many vein leads and therefore may be easier to interpret (STANTON and RUSSELL, 1959). It is the primary purpose of this paper to discuss the dating of galenas in the light of these developments.

There have been two procedures suggested for dating common leads, one by the present writers in the paper cited (RUSSELL *et al.*, 1954) and that proposed by

41

HOUTERMANS (EBERHARDT *et al.*, 1955). The first was based on a mathematical treatment developed by ALPHER and HERMAN (1951) and assumed that the lead isotopes had been produced in a single region uniform in lead, uranium and thorium, and which was common to all leads. Since this was clearly an oversimplification, the deviations of measured isotope ratios from those predicted by the simple theory had to be ascribed to the effects of more or less complex geological processes causing small, but more or less random deviations from the theory. The magnitude of these deviations was estimated and used to determine uncertainties for the ages calculated.

The method of HOUTERMANS is based on the model used by GERLING (1942), HOLMES (1946) and HOUTERMANS (1946) for calculating the age of the earth in which it is assumed that the lead in each ore developed in particular source rocks —each characterized by particular thorium/lead-204 and uranium/lead-204 ratios. This model is much more flexible than the previous one, because the predicted isotope ratios are a function not only of the age of the sample, but of the relative abundances of uranium and thorium in the source rock for that particular ore. Ages calculated from these assumptions have often been called "model ages". This model assumes that the particular source rocks associated with each ore remained a closed system from a time approximately 4500 million years ago until the formation of the ore, and thus requires that the outer part of the earth be thought of as divided into a very large number of compartments among which there is negligible transfer of uranium, thorium and lead for periods of the order of thousands of millions of years. This is perhaps possible if the origin of the lead is well below the surface rocks (perhaps sub-crustal) but is a very difficult assumption in the case of surface rocks which are exposed to various geological processes. A much higher precision has been claimed for this method than for the method of RUSSELL *et al.* and in many cases the results seem to have justified this.

Both groups recognize the importance of anomalous leads and it is implicitly assumed that such leads can be identified and will not be interpreted on the basis of the simple models described above. It is important to remember that anomalous leads themselves do not present a source of embarrassment in interpreting the isotope ratios of common leads, for when recognized, these leads can themselves be interpreted by other procedures and often provide more valuable information on the history of an ore deposit than could be obtained without them. It is their recognition that is sometimes difficult.

It will be seen in the remaining part of this paper that the present writers now assign a greater and more fundamental role to anomalous leads than has been ascribed to them before.

THE ORIGIN OF LEAD ORES

Now that the isotope ratios of a large number of lead ores have been studied (the order of 1000), it is possible to pick out three classes of leads for which the isotope ratios are apparently related in a remarkably regular fashion, although the nature of the regularity is different in each case. These classes are:

(1) Meteoritic leads
(2) Conformable leads.
(3) Anomalous leads.

42

The extreme importance of the isotope ratio determinations of meteoritic leads by PATTERSON (1956) has been widely acknowledged. Five analyses are available; two for lead in troilites separated from iron meteorites, and three for lead in stone meteorites. PATTERSON has given convincing reasons for supposing

Table 1. Isotopic compositions of lead in meteorites (after Patterson)

Meteorite	Lead composition		
	206/204	207/204	208/204
Nuevo Laredo, Mexico	50·28	34·86	67·97
Forest City, Iowa	19·27	15·95	39·05
Modoc, Kansas	19·48	15·76	38·21
Henbury, Australia (Troilite)	9·55	10·38	29·54
Canon Diablo, Arizona (Troilite)	9·46	10·34	29·44

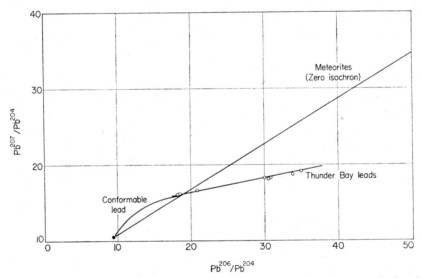

Fig. 1. The relationship between meteoritic lead isotope abundances and abundances of a typical suite of anomalous leads. The points shown represent anomalous leads from the Thunder Bay re ion, Ontario.

that these analyses represent leads grown since a common starting time in closed systems having quite different uranium/lead-204 and thorium/lead-204 ratios, and that the meteorites share with the earth a common origin time t_0 and common primeval lead abundance values. The meteoritic lead isotope ratios, which are given in Table 1. show a remarkable linear relationship. In Fig. 1 the ratios Pb^{207}/Pb^{204} and Pb^{206}/Pb^{204} are plotted and in Fig. 2 the ratios Pb^{208}/Pb^{201} and

43

Pb^{206}/Pb^{204} are plotted. The first straight line has a slope that is a function only of t_0 (with the assumptions stated) and gives a t_0 value of 4500 million years as previously reported by PATTERSON. The slope of the meteorite-lead line in Fig. 2 is insensitive to t_0, but depends critically on the relative thorium–uranium abundance. The fact that this plot is such a good straight line is rather remarkable

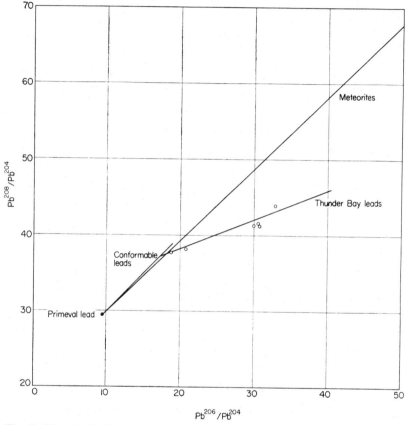

Fig. 2. The relationship between meteoritic lead isotope abundances, conformable lead isotope abundances and abundances of a typical suite of anomalous leads. The anomalous lead analyses show the largest scatter, and for clarity only these analyses are shown in detail.

and indicates that processes producing large fractionations of uranium relative to lead, produced no significant fractionation between uranium and thorium. The figure gives an atomic ratio Th^{232}/U^{238} equal to 3·72, referred to the present time.

Our attention has recently been attracted to the hypotheses of R. L. STANTON regarding the formation of a type of lead ore deposit of a conformable nature (STANTON and RUSSELL, 1959; STANTON, 1958; STANTON, 1954a; STANTON, 1955b). These hypotheses, which will not be reviewed here in detail, deal with a classification of leads proposed by H. F. KING, and of which Broken Hill Australia is the type example. Briefly they suggest that such deposits result from the replacement of iron in iron sulphide precipitated in off-reef facies associated with seaboard

44

volcanism. The precipitation of the iron sulphide lenses is the result of the rapid increase of sulphate-reducing bacteria in places where sulphate is abundant because of volcanic activity. The lead is presumed to have accompanied the volcanic ash as water-soluble halides, to have migrated in solution upwards through the ash deposited at the ocean bottom as the result of subsequent compaction, either replacing iron in iron-sulphide lenses if such are encountered or being lost to the oceans. These leads may have been derived from considerable depths below the

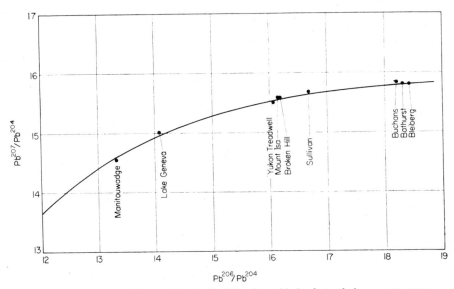

Fig. 3. Detail of Fig. 1, showing fit of conformable leads to their average curve.

surface by simple and short-term geological processes. In particular they may be of subcrustal origin and may represent mantle-lead except in the rare cases where the volcanic material passed through extremely radioactive materials (as may be the case, for example, at Vesuvius). STANTON and RUSSELL have shown that such leads fit a single growth curve within the limits of precision of the measurements and thus may have all been developed in a single homogeneous region. The scatter in their abundances about a single growth curve is an order of magnitude less than that obtained when vein leads are included in the plot. The supposition is that lead migrates through some thickness of pre-existing rock to form *vein* leads and therefore has had at least the *possibility* of becoming anomalous by picking up excess radiogenic leads from these rocks. Examples of these conformable leads are plotted in Figs. 1 and 2 and also in expanded scale in Figs. 3 and 4.

The remarkable fact that the Pb^{207}/Pb^{204} and Pb^{206}/Pb^{204} fit a single growth curve extremely well, suggests the possibility that all these leads were produced in a single homogeneous region. However, the excellent regularity of the Pb^{208}/Pb^{204} vs. Pb^{206}/Pb^{204} graph (Fig. 4) is even more remarkable. This shows that the source materials for these leads were not only uniform in uranium relative to lead, but also in thorium relative to lead. The present ratio of thorium relative to

45

uranium can easily be calculated from this graph and gives a value of 3·74. The details of this calculation are given below.

Anomalous leads were first interpreted numerically by Russell *et al.* (1954) who studied samples from the Sudbury, Ontario, district. In the paper cited it was suggested that anomalous leads represented mixtures of single "ordinary" or non-anomalous lead with variable amounts of a single radiogenic lead, and thus Pb^{206}/Pb^{204} and Pb^{207}/Pb^{204} should be related linearly. Since then this type of

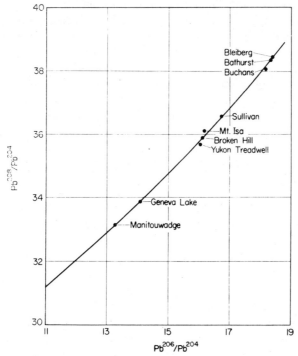

Fig. 4. Detail of Fig. 2, showing fit of conformable leads to their average curve.

relationship has been convincingly demonstrated for anomalous leads from the Thunder Bay region of Ontario (Farquhar and Russell, 1957). Analysis of anomalous leads show that the apparent ratio Th^{232}/U^{238} is extremely variable in the rocks producing the anomalous radiogenic component (e.g. 5·4 for Sudbury leads and 1·8 for Thunder Bay leads).

The following simple assumptions are sufficient to explain these features of observed lead isotope abundances. The condensation of primordial material to form parent bodies from which meteorites were derived, as well as the subsequent chemical and physical alterations resulting in the variety of meteorites observed today took place under conditions such that uranium existed in the quadrivalent state. In this oxidation state uranium compounds are very similar to thorium compounds and processes producing the observed variations in the lead/uranium ratios might produce little effect on the thorium/uranium ratio. The earth also condensed from similar material and the gross separation into core, mantle and crust also occurred under reducing conditions so that the thorium/uranium ratios

46

153

in its various parts were nearly identical to that ratio in meteorites. It was only in surface parts of the crust where oxidizing conditions developed that variations in the thorium/uranium ratio resulted. Conformable leads developed in regions of the earth below that at which oxidizing conditions prevail, and hence have isotope ratios produced in a uniform material with a thorium/uranium ratio equal to that of meteorites. Any addition to such leads of radiogenic lead produced in crustal rocks yields anomalous leads that may or may not be easily recognized as such.

The simplest, but certainly not the only method for satisfying the above assumptions is by asserting that surface leads are produced below the Moho and brought to the surface during orogenesis. This assumption is untenable if it is held that the uranium/lead ratio of the mantle is grossly different from the average for surface rocks, and this is often stated to be the case. However, there seems to be no real evidence for this statement, which now seems much less probable in view of PATTERSON's analyses of meteorites. VINOGRADOV (1955) has also presented evidence supporting his assertion that the lead/uranium ratio in meteorites is not as high as generally believed, while UREY (1956) has suggested that lead may be lithophile.

If these assumptions are true the concept of isochrons as applied to terrestrial leads is meaningless, although the zero isochron has significance in discussing meteoritic leads.

A PROCEDURE FOR DATING CONFORMABLE GALENAS

In line with the ideas outlined in the above paragraphs we have developed a method for dating galenas based on the following assumptions.

(1) The meteoritic lead isotope analyses are to be interpreted as PATTERSON suggests.

(2) All terrestrial lead develops isotope ratios by the accumulation of radiogenic lead in the mantle which is uniform in uranium, thorium and lead.

(3) Conformable leads as considered (and selected) by STANTON provide samples of mantle lead.

(4) All other lead ores, and particularly vein leads may have resulted from the migration of lead through crustal rocks of variable radioactivity and variable thorium/uranium ratios, and therefore have had the opportunity of becoming anomalous leads by picking up radiogenic lead developed in surface rocks.

(5) The observed scatter of the abundance of all ore leads about a single growth curve is solely the result of the inclusion of a large number of anomalous leads that have not been recognized as such.

After adopting these assumptions it is very easy to calculate the unique growth curve for lead in the earth's mantle. One end of the Pb^{207}/Pb^{204} vs. Pb^{206}/Pb^{204} growth curve lies at the point (9·50, 10·36) obtained by averaging Henbury and Canyon Diablo troilite lead isotope ratios. The other end of the same growth curve must lie on the zero isochron as defined by the meteoritic leads. Accepting PATTERSON's parameters for this isochron there results the equation

$$\frac{Pb^{206}}{Pb^{204}} = 1 \cdot 678 \frac{Pb^{207}}{Pb^{204}} - 7 \cdot 87$$

47

Table 2. Comparisons of galena ages by various procedures
(All samples shown are conformable leads except Bleiberg, which is included to
show that Houterman's B-type anomalous leads do not necessarily
give "anomalous" ages by this method)

Location			HOLMES HOU-TER-MANS (million years)	RUSSELL, FARQUHAR, CUMMING			RUSSELL, STANTON, FARQUHAR		
x	y	z		t_{206} (million years)	t_{208} (million years)	Ave (million years)	t_{6-7} (million years)	t_{8-4} million years)	Ave (million years)
Bleiberg, Austria (Mean of 2 samples			380	230	320	280	200	210	210
18·40	15·79	38·44							
Bathurst, New Brunswick (Mean of 5 samples)			470	280	370	330	280	270	270
18·31	15·80	38·33							
Buchans, Newfoundland			570	390	510	450	390	430	410
18·18	15·81	38·05							
Sullivan Mine, B.C. North Star Mine, B.C.			1460	1150	1200	1170	1340	1250	1290
16·68	15·67	36·59							
Mt. Isa, Australia (Mean of 10 samples)			1740	1390	1420	1410	1620	1520	1570
16·17	15·58	36·10							
Broken Hill, Australia (Mean of 17 samples			1750	1410	1520	1460	1640	1630	1630
16·14	15·58	35·88							
Yukon–Treadwell Mine Sudbury, Ontario			1740	1440	1620	1530	1640	1750	1690
16·06	15·50	35·66							
Geneva Lake, Ontario (Mean of 2 samples)			2790	2320	2320	2320	2710	2570	2640
14·08	15·02	34·10							
Manitouwadge, Ontario			3040	2630	2730	2680	2970	3060	3010
13·32	14·54	33·15							

which modern mantle lead must satisfy. These conditions completely specify the growth curve except for a single parameter representing the present value of the U^{235}/Pb^{204} ratio. This can be chosen if the isotope ratio of *any* point on the curve is known (except of course the primeval lead composition which has already been specified). Table 2 lists the conformable leads discussed by STANTON and RUSSELL. The final parameter for the growth curves was specified by insisting that the curve pass through the abundances of lead from the conformable deposit at Bathurst, New Brunswick. This choice was made because more analyses were available for this deposit than for the others, because it is a relatively young deposit and therefore it specifies the curve more precisely than an older sample, and because this deposit was studied personally by STANTON and therefore was identified positively by him as the type to which his hypotheses apply. In fact

48

the curve obtained fits all the samples well and this particular choice makes little difference to the equation for the curve, which in parametric form is

$$
\left.
\begin{aligned}
\mathrm{Pb}^{206}/\mathrm{Pb}^{204} &= 18.72 - 137\!\cdot\!8 \times 0\!\cdot\!0659(e^{\lambda t} - 1) \\
\mathrm{Pb}^{207}/\mathrm{Pb}^{204} &= 15\!\cdot\!82 - 0\!\cdot\!0659(e^{\lambda' t} - 1) \\[1ex]
\lambda &= 0\!\cdot\!1537 \times 10^{-9}\ \text{years}^{-1} \\
\lambda' &= 0\!\cdot\!9722 \times 10^{-9}\ \text{years}^{-1}
\end{aligned}
\right\} \tag{1}
$$

This procedure for calculating the parameters of equations (1) does not require that the ages of the deposits be known. Having obtained these parameters it is quite straightforward to find an equation for $\mathrm{Pb}^{208}/\mathrm{Pb}^{204}$ consistent with the observed isotope ratio for the conformable leads and equations (1). This is

$$
\mathrm{Pb}^{208}/\mathrm{Pb}^{204} = 38\!\cdot\!80 - 34\!\cdot\!2(e^{\lambda'' t} - 1)
$$
$$
\lambda'' = 0\!\cdot\!0499 \times 10^{-9}\ \text{years}^{-1} \tag{2}
$$

Figs. 3 and 4 illustrate the excellent fit of the conformable leads to these curves.

Equations (1) and (2) can then be used to estimate values for the age t of a galena of the conformable type. In principle it is possible to calculate t from the ratio of any two lead isotopes, but we have chosen $\mathrm{Pb}^{206}/\mathrm{Pb}^{207}$ (which age we call t_{6-7}) and from the value $\mathrm{Pb}^{208}/\mathrm{Pb}^{204}$ (which age we call t_{8-4}). Tables 3 and 4 give values of t_{6-7} and t_{8-4} as functions of the corresponding isotope ratio.

Ages calculated for the conformable leads studied are given in Table 2, where they are compared with values obtained by other methods. While this paper will not enter a detailed discussion of the age values, it is worthwhile noting that the Bathurst ores are overlain by Pennsylvanian sediments, suggesting a minimum age of 230 million years and that the oldest rocks nearby are Ordovician, suggesting an upper limit of 350–400 million years. Here our method seems to give a very satisfactory result. On the other hand the value for Sullivan seems older than previous geological estimates suggest as do in fact ages estimated for that deposit by all methods. In this regard it is interesting to note that the Coeur d'Alene district, Idaho, has given a number of well authenticated ages of similar magnitude.

A final point is that the ages obtained can be shifted somewhat up or down by assuming a slightly different zero isochron. We have preferred to use the zero isochron chosen by PATTERSON, as an impartial choice, but this may not be the best one. In particular if PATTERSON's zero isochron had been chosen by HOUTER-MANS in setting up his "model age" graphs, he would have obtained ages much less satisfactory than he has obtained with an arbitrarily selected zero isochron.

VEIN AND ANOMALOUS LEADS

It has already been pointed out that anomalous leads altered in crustal rocks will not conform to this simple theory and that all vein leads have had the possibility of becoming anomalous. The procedure for calculating age information about these leads varies somewhat from case to case.

In cases where a detailed study of anomalous leads has been made it may be possible to recognize the straight line relationships displayed by their abundances as was done, for example, in the case of Thunder Bay leads (FARQUHAR and

4

49

Table 3. The apparent age of a conformable lead ore, of the type described in the text, tabulated for various values of the Pb^{206}/Pb^{207} ratio. The age values are given in millions of years.

	0·000	0·002	0·004	0·006	0·008	0·010	0·012	0·014	0·016	0·018
0·90	3180	3154	3126	3099	3072	3045	3019	2994	2968	2943
0·92	2919	2895	2870	2846	2822	2789	2775	2751	2728	2705
0·94	2683	2660	2637	2615	2592	2570	2548	2525	2502	2479
0·96	2457	2435	2413	2390	2368	2346	2324	2301	2279	2258
0·98	2237	2215	2194	2172	2150	2129	2107	2085	2063	2042
1·00	2021	2000	1978	1957	1936	1915	1894	1873	1851	1830
1·02	1809	1788	1767	1746	1725	1704	1683	1662	1640	1619
1·04	1598	1577	1556	1535	1514	1493	1471	1450	1428	1406
1·06	1385	1363	1341	1320	1298	1277	1256	1234	1212	1190
1·08	1268	1146	1125	1103	1081	1059	1037	1015	993	971
1·10	949	927	905	883	861	839	817	795	773	751
1·12	729	707	685	662	640	618	596	573	551	528
1·14	505	483	460	437	414	391	368	345	323	300
1·16	276	253	229	206	183	159	136	112	89	65
1·18	41	18	−6	−30	−54	−78	−102	−126	−150	−174

Table 4. The apparent age of a conformable lead ore, of the type described in the text, tabulated for various values of the Pb^{208}/Pb^{204} ratio. The age values given are in millions of years.

	0·00	0·10	0·20	0·30	0·40	0·50	0·60	0·70	0·80	0·90
30	4583	4536	4489	4442	4395	4348	4300	4253	4205	4158
31	4111	4063	4015	3967	3919	3871	3823	3774	3726	3677
32	3628	3579	3530	3481	3432	3382	3333	3283	3233	3183
33	3133	3083	3033	2982	2931	2880	2829	2778	2727	2675
34	2626	2574	2523	2471	2419	2370	2315	2263	2210	2158
35	2105	2052	1999	1946	1993	1839	1786	1733	1679	1625
36	1571	1517	1463	1409	1355	1301	1246	1191	1136	1081
37	1026	970	914	858	802	746	690	634	577	520
38	463	386	329	291	233	175	117	59	0	−59
39	−117	−176	−235	−295	−355	−415	−475	−535	−596	−657

Russell, 1957) and Broken Hill leads (Russell *et al.*, 1957). The procedures then used are well described in the second paper which shows that the slope of the line along which the analyses follow can be used to give an upper limit to the time of deposition of the lead ores and also an upper limit to the age of the surface rocks producing the anomalous additions of radiogenic lead.

In particular cases where the line can be determined precisely either in the case of:

$$\frac{Pb^{207}}{Pb^{204}} \text{ vs. } \frac{Pb^{206}}{Pb^{204}} \text{ or } \frac{Pb^{208}}{Pb^{204}} \text{ vs. } \frac{Pb^{206}}{Pb^{204}}$$

50

157

it may be possible to extrapolate the line and get a good intersection with the conformable lead growth curve and hence determine the time of derivation of the lead from the earth's mantle with good precision.

In the case of a vein lead it may often be very difficult to determine whether it is anomalous or not and in some cases where a group of such leads is studied from the same location, the variation in abundances may be so small that it is impossible to determine a best straight line for them. In this case only approximate age information can be obtained. This is done by noting that the minimum possible slope for a line representing anomalous leads is 0·0459 and the maximum probable slope is 0·804. The first figure comes from the present ratio in which lead-207 and lead-206 are being produced. The second figure represents the ratio of the rates of production of these isotopes 3500 million years ago. We have at present no evidence that rocks older than this age have been preserved. By making use of these limiting values for the slope and extrapolating from the anomalous lead abundances to the conformable lead growth curve, one can obtain a minimum age for the time at which this lead was derived from the earth's mantle. In some cases this can be determined rather precisely while in other cases experimental error leads to a very uncertain extrapolated value.

CONCLUSIONS

This paper has re-examined the proposal for dating galenas put forward 5 years ago by the present writers. In view of the considerable advances made in studying and interpreting the isotope ratios of common lead, modifications and improvements of the older procedure are suggested. It is suggested that conformable leads represent samples of lead extracted from the mantle as a result of rather simple geological processes and that their isotope ratios accurately specify the time at which the lead was derived from the mantle which in this case is nearly equal to the age of the sedimentary beds with which the conformable leads are associated. In the case of vein or anomalous leads these two times might be quite different and procedures have been suggested for putting limits on the time of derivation of the lead from the earth's mantle and for the time of emplacement of the lead in the form of lead minerals.

REFERENCES

ALPHER R. A. and HERMAN R. C. (1951) The primeval lead isotopic abundances and the age of the earth's crust. *Phys. Rev.* **84,** 1111–1114.

CAHEN L., EBERHARDT P., GEISS J., HOUTERMANS F. G., JEDWAB J. and SIGNER P. (1958) On a correlation between the common lead model age and the trace element content of galenas. *Geochim. et Cosmochim. Acta* **14,** 134–149.

EBERHARDT P., GEISS J. and HOUTERMANS F. G. (1955) Isotopic ratios of ordinary leads and their significance. *Z. Phys.* **134,** 91–102.

FARQUHAR R. M. and RUSSELL R. D. (1957) Anomalous leads from the upper Great Lakes region of Ontario. *Trans. Amer. Geophys. Un.* **38,** 552–556.

GERLING E. K. (1942) Age of the earth according to radioactivity data. *C.R. Acad. Sci. USSR* **34,** 259–261.

HOLMES A. (1946) An estimate of the age of the earth. *Nature, Lond,* **157,** 680–684.

HOUTERMANS F. G. (1946) The isotope ratios in natural lead and the age of uranium. *Naturwissenschaften* **33,** 185–186; addendum, *Ibid.* 219.

51

Patterson C. C. (1956) Age of Meteorites and the earth. *Geochim. et Cosmochim. Acta* **10**, 230.

Russell R. D., Farquhar R. M., Cumming G. L. and Wilson J. T. (1954) Dating galenas by means of their isotopic constitutions. *Trans. Amer. Geophys. Un.* **35**, 301–309.

Russell R. D., Farquhar R. M. and Hawley J. E. (1957) Isotopic analyses of leads from Broken Hill, Australia. *Trans. Amer. Geophys. Un.* **38**, 557–565.

Stanton R. L. (1955a) Lower Paleozoic mineralization near Bathurst, New South Wales. *Econ. 'Geol.* **50**, 681–714.

Stanton R. L. (1955b) The genetic relationship between limestone volcanic rocks and certain ore deposits. *Aust. J. Sci.* **17**, 173–175.

Stanton R. L. (1958) Abundances of copper, zinc and lead in some sulphide deposits. *J. Geol.* **66**, 484–502.

Stanton R. L. and Russell R. D. (1959) Anomalous leads and the emplacement of lead sulphide ores. *Econ. Geol.* **54**, 558–607.

Urey H. C. (1956) Nuclear processes in geologic settings. *Nat. Acad. of Sci.—Nat. Res. Council Publ.* 400, p. 78.

Vinogradov A. P. (1955) *Lead Isotopes and their Geochemical Significance.* U.S.S.R. Academy of Science Press.

52

Isotopic Uranium–Lead Ages

VIII

Editor's Comments on Papers 19–22

The availability of isotopically enriched materials for peaceful purposes following the hostilities of 1939–45, coupled with further advances in mass spectrometry, opened the door to a vast field of analytical possibilities. In geochronology, the measurement of radiometric ages based on separate quantitative analyses of both uranium and thorium isotopes and of each of the three radiogenic isotopes of lead became at last a reality. By mixing isotopically enriched ''spikes'' of suitable composition with the extracted elements, very low concentrations of uranium, thorium and lead could be measured very precisely using a mass spectrometer. When such analyses were combined with analyses of the relative abundances of the lead isotopes in any sample, three indedpendent isotopic ages could be calculated, for any sample containing uranium and thorium, from a measurement of the time-dependent ratios Pb-206/U-238, Pb-207/U-235 and Pb-208/Th-232. For any given sample these three uranium–thorium–lead isotopic ages should all agree, within the limits of experimental uncertainty, provided that the only changes occurring in the original proportions of uranium, thorium, and lead incorporated in the sample were changes associated with radioactive transmutations.

A fourth age could be calculated from the Pb-207/Pb-206 ratio, for this ratio of radiogenic lead isotopes derived from the decay of uranium increases sytematically with time due to the different decay rates of the parent U-235 and U-238 isotopes. If a uranium-bearing mineral had suffered partial loss of radiogenic lead subsequent to its formation, the uranium–lead isotope ratios could be severely affected, but such alteration would be much less likely to effect the lead isotope ratios, and Pb-207/Pb-206 ages would thus appear to be more reliable than uranium–lead isotopic ages.

Isotopic analysis also permitted the true proportions of the radiogenic decay products to be identified and precisely measured in any naturally occurring mixture of radiogenic and nonradiogenic isotopes. For instance, if a uranium-bearing mineral incorporated some nonradiogenic (common) lead at the time of crystallization, or if common lead were introduced as a contaminant during analytical processing, mass-spectrometric analysis of the resulting mixture would reveal the presence of the nonradiogenic lead isotope Pb-204. A suitable correction could then be applied to the total Pb-206, Pb-207 and Pb-208 concentrations relative to Pb-204 to obtain the true proportion of radiogenic lead, provided only that the isotopic composition of the contaminating nonradiogenic lead was known.

One of the first reported applications of the new techniques (Tilton et al., 1955) was achieved by a brilliant team of investigators including Claire Patterson and Harrison Brown of the California Institute of Technology, together with experts in isotope dilution techniques, Mark Inghram, Richard Hayden, and David Hess from the University of Chicago's Department of Physics and the Argonne National Laboratories. The group was headed by George Tilton of the Department of Terrestrial Magnetism at the Carnegie Institution in Washington, D.C., and included Esper Larson from the U.S. Geological Survey, who had earlier (Larsen et al., 1952) developed a chemical method dating accessory minerals such as zircon, apatite, and sphene separated from common igneous rocks by determining their total lead content by emission spectrography and their radioactivity (U + Th) by alpha-counting.

Tilton and his co-workers successfully demonstrated that the isotope dilution technique could be used to determine precisely as little as 0.01 ppm of uranium and thorium and 0.1 ppm of lead present in the common and accessory mineral constituents of a Precambrian (1000 million years old) granite. The isotopic composition of lead could be determined on a sample totalling only a few micrograms, but, in order to work successfully with such small quantities, great precautions had to be taken to reduce contamination levels to an absolute minimum. This was particularly difficult in the case of lead, as one of the major sources of lead contamination was (and still is) the presence of lead in the atmosphere, chiefly due to the combustion of gasoline fuels containing lead tetraethyl. A specially clean laboratory was reserved for the work, all lead fittings were eliminated, lead gaskets were replaced by teflon, drafts were minimized, and a positive pressure of filtered air was maintained in the laboratory at all time. All water and acid reagents were carefully filtered and distilled, and solid reagents were recrystallized.*

Tilton's results showed conclusively that when sufficient care was taken to reduce the levels of contamination, isotopic uranium–lead ages could be determined for the accessory minerals zircon† and sphene. Unfortunately, the errors associated with the lead data for apatite were too large to allow meaningful ages to be calculated, but

---

*With present-day methods, entire U–Pb isotopic analyses can be successfully undertaken on zircon samples totaling only 250 μg in weight containing only 0.08 μg of lead (Krogh, 1971).
†A short report of isotopic U–Pb dating of zircon had been published by Vinogradov et al., in 1952.

the investigation did show that a pronounced enrichment of uranium and thorium relative to lead occurred at the time of formation of the granite. Subsequently the granite behaved as a closed system with respect to uranium and its decay products, but as an open system with respect to thorium and its decay products. Laboratory tests at the California Institute of Technology reported later by Tilton (1956) showed that leaching by quite dilute (0.1 molar) acid was effective in removing substantial fractions of the radioactive elements, thorium being more easily removed than uranium. Pb-208/Th-232 ages were therefore suspect, but the accessory minerals, particularly zircon, when carefully and patiently separated from any of the common plutonic rocks, could be dated by measuring their uranium and uranium-derived lead isotope concentrations.

While Tilton and his team* were developing the analytical techniques necessary to extend the isotopic uranium–lead methods to the common accessory minerals, L. H. Ahrens, then at Oxford University's Department of Geology and Mineralogy, began studying uranium, thorium, and lead isotopic ratios reported from the oldest known radioactive minerals, samples of uraninite and monazite from Rhodesia, Manitoba, Madagascar, and the Transvaal. Pb-207/Pb-206 ages for these minerals ranged from 2680 to 2420 million years, but the uranium– and thorium–lead isotopic ages were all discordant. A consistent pattern of discordant ages (Pb-206/Pb-207 > Pb-207/U-235 > Pb-206/U-238 ≫ Pb-208/Th-232) indicated that radiogenic lead had not been retained quantitatively in each of the samples (Holmes, 1954). Ahrens noticed that samples having successively lower Pb-207/Pb-206 ages had increasingly discordant (lower) Pb-206/U-238 ages relative to their Pb-207/U-235 ages, due apparently to loss of a successively greater proportion of the total radiogenic lead generated. He then found that if Pb-207/Pb-206 and Pb-206/U-238 ages were plotted against the Pb-207/U-235 age for each sample, a well ordered pattern emerged (Ahrens, 1955a). The data points for six samples converged toward a point where both uranium–lead and lead–lead isotopic ages were concordant. Ahrens called this age the "convergent age" and suggested that all six samples had orginally crystallized at precisely the same time (2800 million years ago) but that each had subsequently suffered loss of a variable proportion of radiogenic lead.

The convergent array of data points suggested to Ahrens that the lead loss was controlled by some physical, possibly nuclear, process which had operated with varying degrees but at a uniform rate throughout the whole lifetime of the minerals. Preferential loss of one of the intermediate daughter elements (e.g., radon-222) in the decay series, as had been suggested by Wickman (1942) and Kulp et al. (1954) as a possible cause for internally discordant uranium–lead isotopic ages, could not account for the pattern of discordant ages observed by Ahrens, particularly the Pb-207/Pb-206 ages, which approached most closely the true age of crystallization. Moreover, experi-

---

*See also Tilton et al., 1957.

ments by Giletti and Kulp (1955) showed that the rate of diffusion of radon from primary uranium minerals was far too low to account for the recorded magnitude of the radiogenic lead deficiencies.

Developing his observations, Ahrens (1955b) subsequently found that a semi-logarithmic linear relationship existed between the Pb-207/U-235 ages and the log (Pb-206/U-238) ages for the suite of samples, which could be extrapolated to give the convergent age ($t_{207/235} = t_{206/238}$) by intersection with an "age equality" curve drawn to connect all points having identical Pb-207/U-235 and Pb-206/U-238 ages.

The regularities observed by Ahrens immediately attracted the attention of George Wetherill at the Department of Terrestrial Magnetism in Washington, D.C. Wetherill pointed out that partial loss of radiogenic lead by geochemical leaching or alteration during a short episode in the subsequent history of the mineral would explain the pattern of discordant ages observed (Paper 19). Using a ratio plot of Pb-206/U-238 against Pb-207/U-235, he showed that the data used by Ahrens defined straight lines (chords) which intersected the age equality curve, renamed "concordia" by Wetherill, at two points. The upper intersection defined the primary age of crystallization of all the samples falling on the chord, while the lower intersection indicated the time of episodic lead loss.

In June the following year (1956) Wetherill published a full discussion of his interpretation of the discordant U–Pb age patterns, detailing his graphical procedure for determining the primary age of crystallization and secondary age of episodic lead loss (Paper 20). Provided that a suite of coeval samples could be found that had all lost varying proportions of radiogenic lead during some subsequent episode in their history, the actual proportion of lead lost from any one sample could be determined from its position on the chord passing through all the sample points. This chord could be extrapolated back to intersect the concordia curve giving the time of formation of the mineral suite.

Here was a powerful new method for determining the primary age of crystallization from a discordant set of isotope age data, but controversy soon arose over the validity of the secondary age of episodic lead loss that was indicated by the lower extrapolated intersection of the linear array with the concordia curve. In 1960, following Nicolaysen's (1957) suggestion that lead loss might occur by continuous diffusion during the entire lifetime of any uranium-bearing mineral, Tilton demonstrated that continuous diffusion resulting in either loss of lead or gain of uranium could account equally well for the observed linearity of data points on Wetherill's concordia plot (Paper 21). Samples having identical primary crystallization ages which had lost lead by continuous diffusion during their entire lifetimes would all lie on a single diffusion curve which looked very similar to one of Wetherill's discordant chords, the only difference being that diffusion curves would always intersect the concordia curve at $t = 0$ (see Paper 21, Fig. 1). The position of a sample on any diffusion curve would be determined by the diffusion coefficient divided by the square of the effective diffusion radius (D/a^2). Where continuous diffusion was involved, Pb-207 would be removed preferentially

165

because the half-life of the Pb-207 parent (U-235) was shorter than the Pb-206 parent (U-238), and Pb-207 atoms, being older, would have more time to diffuse out of the crystal lattice.*

Whatever the mechanism responsible for partial loss of radiogenic lead from suites of coeval uranium-bearing minerals, a linear relationship existing between Pb-206/U-238 and Pb-207/U-235 ratios could be used to determine the original age of crystallization by extrapolation of the data back to a point on the concordia curve where ages calculated from both independent ratios were identical. The elaborate chemical procedures necessary for successful analysis had tended to discourage investigation of U–Pb isotopic systems, but with the recognition that the true age of crystallization could be determined from a discordant suite of samples came the realization that a powerful tool had been placed in the hands of the geochronologist—one worth all the analytical effort involved.

The final establishment of the isotopic uranium–lead method as one of the most important available for determining primary crystallization ages resulted from the work of Leon Silver at the California Institute of Technology's Division of Geological Sciences in Pasadena. Working with Sarah Deutsch, Silver discovered that zircons separated from a single 250-lb. block of Precambrian granodiorite less than 2 cubic feet in volume had a significant range of uranium concentrations, the smaller zircon crystals being more radioactive. Not only did the uranium concentrations vary, but the separated zircons all gave discordant uranium–lead isotope ages indicative of partial lead loss that varied systematically with the size of the zircon fraction analyzed. Smaller zircons had lost proportionally more radiogenic lead than the larger zircons. Thus Silver showed that it was not necessary to collect uranium-bearing minerals from widely different rocks in order to obtain a linear array on the concordia plot. There was sufficient variability in the zircon population separated from a single rock sample to provide the data necessary to obtain the primary age of crystallization from discordant results by extrapolation to concordia (Paper 22; see also Silver, 1963).

In the case of metamorphic systems, Silver's work pointed the way to determining the original crystallization age in rocks that have suffered successive recrystallizations under metamorphic conditions, for zircons are most resistant to recrystallization and although they may suffer partial radiogenic lead loss during metamorphism, complete loss rarely occurs. Suites of zircons of different sizes from a single rock could provide the geochronologist with an eye to see through subsequent metamorphic events to the time of original crystallization, and U-Pb zircon analyses now provide some of the best isotopic evidence for the true antiquity of successively metamorphosed and reactivated Precambrian cratonic areas (see, e.g., Goldich et al., 1970).

For further discussions of a discordant uranium–lead isotopic ages the reader is referred to Stieff and Stern (1961), Wetherill (1963), and Pidgeon et al. (1966).

---

*Wasserburg (1963) later suggested that the diffusion coefficient might be related to the integrated radiation damage.

The interpretation of zircon ages has been reviewed by Catanzaro (1968), and systematics in systems involving Pb-208/Th-232 and the uranium–lead isotope ratios have been extensively discussed by Steiger and Wasserburg (1966, 1969) and Wasserburg and Steiger (1969).

Selected Bibliography

Ahrens, L. H., (1955a). The convergent lead ages of the oldest monazites and uraninites (Rhodesia, Manitoba, Madagascar, and Transvaal). *Geochim. Cosmochim. Acta*, **7**, 294–300.

Ahrens, L. H., (1955b). Implications of the Rhodesia age pattern. *Geochim. Cosmochim. Acta*, **8**, 1–15.

Catanzaro, E. J. (1968). The interpretation of zircon ages. *In* "Radiometric Dating for Geologists" (E. I. Hamilton and R. M. Farquhar, eds.), pp. 225–258. Interscience, London.

Giletti, B. J., and Kulp, J. L. (1955). Radon leakage from radioactive minerals. *Amer. Mineral*, **40**, 481.

Goldich, S. S., Hedge, C. E., and Stern, T. W. (1970). Age of the Morton and Montevideo gneisses and related rocks, Southwestern Minnesota. *Geol. Soc. Amer. Bull.*, **81**, 3671–3696.

Holmes, A. (1954). The oldest dated minerals of the Rhodesian Shield. *Nature*, **173**, 612.

Krogh, T. E. (1971). A low contamination method for decomposition of zircon and extraction of U and Pb for isotopic age determinations. *Ann. Rept. Geophys. Lab. Washington, D.C.*, pp. 258–266.

Kulp, J. L., Bate, G. L., and Broecker, W. S. (1954). Present status of the lead method of age determination. *Amer. J. Sci.*, **252**, 346.

Larsen, E. S., Keevil, N. B., and Harrison, H. C. (1952). Method for determining the age of igneous rocks using the accessory minerals. *Geol. Soc. Amer. Bull.*, **63**, 1045–1052.

Nicolaysen, L. O. (1957). Solid diffusion in radioactive minerals and the measurement of absolute age. *Geochim. Cosmochim. Acta*, **11**, 41–59.

Pidgeon, R. T., O'Neil, J. R., and Silver, L. T. (1966). Uranium and lead isotopic stability in a metamict zircon under experimental hydrothermal conditions. *Science*, **154**, 1538–1540.

Silver, L. T. (1963). The use of cogenetic uranium–lead isotope systems in zircons in geochronology. *In* "Radioactive Dating," pp. 279–288. International Atomic Energy Agency, Vienna.

Steiger, R. H., and Wasserburg, G. J. (1966). Systematics in the Pb-208/Th-232, Pb-207/U-235, and Pb-206/U-238 systems. *J. Geophys. Res.*, **71**, 6065–6090.

Steiger, R. H., and Wasserburg, G. J. (1969). Comparitive U–Th–Pb systematics in 2.7 × 10⁹ yr plutons of different geologic histories. *Geochim. Cosmochim. Acta*, **33**, 1213–1232.

Stieff, L. R.,and Stern, T. W. (1961). Graphic and algebraic solutions of the discordant lead–uranium age problem. *Geochim. Cosmochim. Acta*, **22**, 176–199.

Tilton, G. R. (1956). The interpretation of lead-age discrepancies by acid-washing experiments. *Trans. Amer. Geophys. Union*, **27**, 224–230.

Tilton, G. R., Davis, G. L., Wetherill, G. W., and Aldrich, L. T. (1957). Isotopic ages of zircon from granites and pegmatites. *Trans. Amer. Geophys. Union*, **38**, 360–371.

Tilton, G. R., Patterson, C., Brown, H., Inghram, M., Hayden, R., Hess, D., and Larsen, E. (1955). Isotopic composition and distribution of lead, uranium, and thorium in a Precambrian granite. *Geol. Soc. Amer. Bull.*, **66**, 1131–1134, 1139–1148.

Vinogradov, A. P., Zadorozhnyi, I. K., and Zykor, S. I. (1952). Isotopic composition of lead and the age of the earth. *Dokl. Akad. Nauk SSSR*, **85**, 1107–1110.

Wasserburg, G. J. (1963). Diffusion processes in lead–uranium systems. *J. Geophys. Res.*, **68**, 4823–4846.

Wasserburg, G. J., and Steiger, R. H. (1969). Systematics in the Th–U–Pb systems and multiphase assemblages. *In* "Radioactive Dating Methods of Low-lead Counting," pp. 331–347. International Atomic Energy Agency, Vienna.

Wetherill, G. W. (1963). Discordant uranium–lead ages. 2. Discordant ages resulting from diffusion of lead and uranium. *J. Geophys. Res.*, **68**, 2957–2965.

Wickman, F. E. (1942). On the emanating power and the measurement of geological time. *Geol. Fören. Förh. Stockholm*, **64**, 465–475.

19

GEOCHEMICAL NOTES

An interpretation of the Rhodesia and Witwatersrand age patterns

GEORGE W. WETHERILL

Department of Terrestrial Magnetism, Carnegie Institution of Washington, Washington 15, D.C.

(*Received 5 November* 1955)

IN two recent papers, AHRENS (1955a, 1955b) reported some interesting regularities in the discordant uranium-lead ages of uraninite and thucholite from the Witwatersrand, and in some very ancient monazite and uraninite samples from the Canadian and Rhodesian shields. He found that, when he plotted the isotopic analyses of these samples on D/P

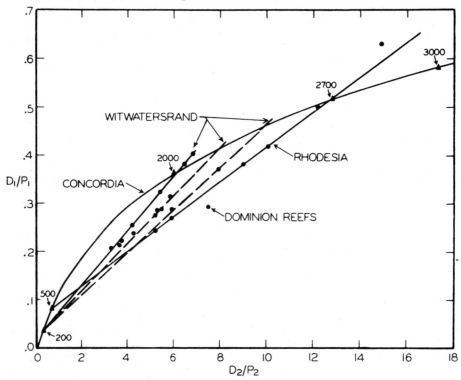

Fig. 1. D/P diagram showing the Witwatersrand and Rhodesia regularities found by AHRENS. D_1 and D_2 are the mole concentrations of the daughter isotopes, Pb^{206} and Pb^{207} respectively, while P_1 and P_2 refer to the mole concentrations of the parent isotopes, U^{238} and U^{235}. The curve marked "concordia" is the locus of all points such that $t_{U\text{-}235} = t_{U\text{-}238}$.

diagrams (Fig. 1), most of the points lay on straight lines. AHRENS felt that "a well-ordered age array suggests that physical causes, rather than chemical, have controlled lead loss," and refers to a private communication by MAYNE that "this opens up a possibility that some other, not properly understood, nuclear process could conceivably be the cause." In this note it is shown that there is at least one simple geochemical process which will

290

produce these age patterns. In Fig. 1 the experimental points are those discussed by AHRENS. The solid straight lines through these points are theoretical curves calculated by assuming that the true age of the Rhodesia samples is 2700 million years and varying amounts of lead were lost (or uranium was gained) 500 million years ago, and by assuming that the Witwatersrand samples are 2040 million years old (AHRENS, 1955b) and have lost lead or uranium 200 million years ago. The dashed straight lines are similar curves calculated for ages 2250 and 2450 million years. In the case of the monazite samples, one might expect that the assumption of a given Pb^{206} and Pb^{207} loss would imply a proportionate loss of Pb^{208}, and therefore, that the same correction for lead loss would bring all three ages into agreement. However, TILTON (1955) has shown that the thorium-lead is not homogeneously mixed with uranium-lead in monazite crystals, and large fractionations occur during leaching. Therefore, there should not be this simple relationship between the thorium-lead and the uranium-lead ages.

The excellent fit of the Witwatersrand main regularity to the theoretical curve is good evidence that these widely separated samples underwent a similar geological history, and is in agreement with the interpretation that they were formed 2040 ± 50 million years ago and suffered loss of lead because of some alteration or leaching process 200 ± 150 million years ago. The minor regularities are fitted less well by the dashed curves. If the minor regularities are real, they could therefore be interpreted as two groups of minerals having true ages of 2250 and 2450 million years and which have also lost varying amounts of Pb 200 million years ago. A similar statement can be made for the Rhodesia and Madagascar samples. It is rather doubtful that the Huron Claim samples can actually be identified with those from Rhodesia on the basis of this evidence, especially in view of the fact that only one of them falls on the theoretical curve, and this might well be coincidental.

Thus it seems that the regularities found by AHRENS can be more simply explained by chemical leaching or alteration at a single episode in geological history rather than "control of lead loss by physical processes which perhaps have operated at varying degrees but a uniform rate ever since the minerals were formed," as suggested by AHRENS.

The more difficult question of the uniqueness of the interpretation requires the results of a more general theoretical discussion, which is presently being prepared. It may merely be stated that aside from the possibility of the regularities being coincidental, there are a very limited number of similar geochemical processes which will produce regularities of this sort. All of these involve episodic uranium-lead fractionations at the times mentioned. Plausible arguments can be given for expecting that the interpretation given here is the correct one, although the available data are not sufficient to settle the question finally. In the case of the Rhodesia samples, there is good independent evidence from Rb-Sr and K-A determinations on Bikita lepidolite (WETHERILL, ALDRICH, and DAVIS, 1955) that the interpretation is correct in this case.

The properties of the other type of diagram used by AHRENS, the semilog plot, are similar to those of the D/P diagram, but this diagram is less useful for recent times, when lines that are straight on the D/P diagram become highly curved. For example, less of constant lead-lead age are straight lines passing through the origin on the D/P diagram, while these lines go to -00 on the semilog plot.

If this recent lead loss is accepted as the physical basis of the convergent age patterns of AHRENS, it is clear that the convergent age of a single sample has little meaning. Consider the Dominion Reefs uraninite for which AHRENS tentatively determines a convergent age of 2930 million years. If this sample had a true age of 2930 million years and lost lead 500 million years ago, it would have the lead-uranium ratios found. However, if the true age were 2730 million years and a single episode of lead loss took place in modern times, the same lead-uranium ratios would be found. Actually there are infinite number of possibilities.

291

169

It is only when a large number of samples are available, as in the case of the Witwatersrand main regularity, that a plausible argument can be made concerning the true age of the samples and their probable geological history.

The discovery of these regularities and their possible explanation raises the rather exciting possibility that it may be possible to use discordant uranium-lead ages to determine some aspects of the common geologic history of large areas, even entire geological provinces.

REFERENCES

AHRENS L. H. (1955a) *Geochim. et Cosmochim., Acta* **8,** 1.

AHRENS L. H. (1955b) *Geochim. et Cosmochim. Acta* **7,** 294.

TILTON G. R. (1955) Paper presented at Pennsylvania State University conference on nuclear geophysics.

WETHERILL G. W., ALDRICH L. T., and DAVIS G. L. (1955) *Phys. Rev.* **98,** 250.

Discordant Uranium-Lead Ages, I   20

George W. Wetherill

Abstract—A graphical procedure is described for rapid calculation of discordant uranium-lead ages resulting from multiple episodes of uranium-lead fractionation. A proof of the validity of this graphical procedure is given. The graphical procedure is extended to permit the calculation of the effect of the presence of primary radiogenic lead and of constant loss of intermediate daughter products.

Introduction—The element uranium has two long-lived isotopes, U^{238} and U^{235}, the final decay products of which are Pb^{206} and Pb^{207}, respectively. Measurement of the uranium concentration and the concentration of radiogenic Pb^{206} and Pb^{207} in a chemical system (such as a mineral) containing uranium permits the calculation of two uranium-lead ages, the U^{238}-Pb^{206} age, T_1 and the U^{235}-Pb^{207} age, T_2 calculated from the equations of radioactive decay

$$
\left.
\begin{aligned}
T_1 &= \frac{1}{\lambda_{U^{238}}} \ln\left(\frac{Pb^{206}}{U^{238}} + 1\right) \\
T_2 &= \frac{1}{\lambda_{U^{235}}} \ln\left(\frac{Pb^{207}}{U^{235}} + 1\right)
\end{aligned}
\right\} \quad (1)
$$

In this discussion it will be convenient to designate the U^{238}-Pb^{206} decay by superscript or subscript 1, and the U^{235}-Pb^{207} decay by the subscript 2. Thus, equations (1) become

$$
\left.
\begin{aligned}
T_1 &= \frac{1}{\lambda_1} \ln\left(\frac{D_1}{P_1} + 1\right) \\
T_2 &= \frac{1}{\lambda_2} \ln\left(\frac{D_2}{P_2} + 1\right)
\end{aligned}
\right\} \quad (1a)
$$

where the λ's are the decay constants, and D and P refer to the daughter and parent concentrations, respectively.

Where these two ages are found to be equal to one another, the ages are said to be 'concordant.' When they are unequal, they are said to be 'discordant.'

These two calculated ages will be equal to one another and to the true age of the mineral if the following assumptions are fulfilled: (a) There have been no gains or losses of uranium or lead during the time since the formation of the system. (b) There have been no gains or losses of intermediate members of the radioactive decay scheme, for example, radon, or ionium. (c) Proper corrections have been made for the initial concentration of Pb^{206} and Pb^{207}. (d) The chemical analyses have been properly performed and the correct decay constants λ_1 and λ_2 have been used. When these

assumptions have been fulfilled, the ages will be concordant; when they are not fulfilled, the ages will be either discordant or 'accidentally' concordant.

These papers will discuss discordance and accidental concordance arising from failure of assumptions (a), (b), and (c), which may be considered 'intrinsic discordance' as opposed to 'technical' discordance resulting from failure of assumption (d).

In this first paper a graphical scheme will be presented for the calculation of the effects of failure of assumption (a) at discreet episodes in the history of the mineral, and a discussion will be given of the effects of failure of assumptions (b) and (c), within the framework of this graphical method. In a subsequent publication this graphical procedure will be applied to problems of geochronology, namely, the interpretation of regularities in a group of discordant ages such as those by *Ahrens* [1955] as well as the inference of the true age of a group of cogenetic minerals giving discordant ages, even when no regularities are present. This graphical scheme will make use of a diagram (Fig. 1) in which the mole ratio Pb^{206}/U^{238} (D_1/P_1) is plotted as the ordinate and the mole ratio Pb^{207}/U^{235} (D_2/P_2) as the abscissa. In the case of concordant ages, for every age $\tau_0 = T_1 = T_2$ there will correspond unique values of D_1/P_1 and D_2/P_2 defined by the equations

$$
\left.
\begin{aligned}
\frac{D_1}{P_1} &= e^{\lambda_1 \tau_0} - 1 \\
\frac{D_2}{P_2} &= e^{\lambda_2 \tau_0} - 1
\end{aligned}
\right\} \quad (2)
$$

The locus of these values for $0 < \tau_0 < \infty$ is the curve marked 'concordia' on Figure 1.

Description of the graphical procedure—The procedure will be described without proof. The proof of its validity is found in the following section. Consider a sample of uranium mineral (or more generally a chemical system containing uranium) having a true age τ_0. If the assumptions

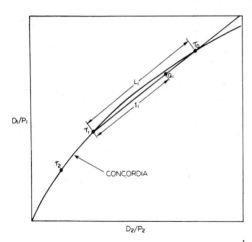

D_1/P_1

D_2/P_2

FIG. 1 – D/P diagram illustrating the graphical procedure for calculating the uranium-lead ages resulting from a given history of uranium-lead fractionation.

(a), (b), (c), and (d) are valid, the point characteristic of this mineral sample will lie on concordia at τ_0. If the mineral has lost lead or uranium or gained uranium during a geologically brief episode at a time τ_1 years ago, the position of the point (Q_1) characteristic of this mineral may be found by the following procedure: (1) Draw a straight line passing through the point on the curve 'concordia' corresponding to a true age τ_0 and that corresponding to a true age τ_1. (2) Defining the length of the straight segment $\overline{\tau_0\tau_1} = L_1$, measure off a distance along this segment of length $l_1 = R_1L_1$ from τ_1. R_1 is the ratio by which D_1/P_1 and D_2/P_2 changed at time τ_1. That is

$$R_1 = (D/P)_{\text{immediately after loss}} /$$

$$(D/P)_{\text{immediately before loss}}$$

The end of this segment is the desired point Q_1. In the case of lead loss, it is assumed that R is the same for both the lead isotopes, that is, that the lead which is lost has the same isotopic composition as the total lead which was present in the total lead which was present in the mineral. This is not a very severe restriction for lead loss, but it must be noted that lead addition will not in general fulfill this condition, and therefore the graphical procedure has been limited to losses of lead or uranium or additions of uranium. This case where R is unequal for D_1/P_1 and D_2/P_2 is discussed in the section on *Extensions of the graphical procedure*. For example, if half the lead were lost with no loss of uranium, the $R_1 = \frac{1}{2}$. If, on the

other hand, $\frac{1}{3}$ of the uranium were lost, $R_1 = \frac{3}{2}$, and $l_1 = \frac{3}{2}L_1$.

From the procedure given above, it is seen that regardless of how much uranium or lead is lost during this episode, the resulting point (Q_1) will lie on this straight line. From the coordinates of the point (D_1/P_1, D_2/P_2) found by this procedure, the discordant uranium-lead ages can be calculated by use of Eq. (2). Thus the effect of a single episode of uranium-lead fractionation has been determined.

For the special case $R_1 = 1$, that is, no fractionation, the result is that the point Q_1 characterizing the mineral remains on the curve 'concordia' at τ_0. If, on the other hand, all the lead were lost at τ_1, perhaps due to a remineralization with exclusion of lead, then $R_1 = 0$ and $l_1 = 0$ and the point characterizing the sample will lie on the curve 'concordia' at τ_1. In this case the age of the sample can be considered more properly to be τ_1.

For the case of multiple episodes of uranium-lead fractionation, the resulting point Q_n can be found by extension of this same procedure. Assume a second fractionation R_2 occurred at a later time τ_2. The effect of this is found by the following procedure: (1) Draw a straight line between the point Q_1 (found by the above procedure for the first fractionation) and the point τ_2 on concordia. (2) Defining the length of the straight segment $\overline{Q\tau_2} = L_2$, the desired point Q_2 will lie at a distance $l_2 = R_2L_2$ from τ_2 on concordia. For a third fractionation at τ_3, the procedure is repeated with $\overline{Q_2\tau_3} = L_3$ and $l_3 = R_3L_3$ giving point Q_3. For the case of n fractionations, the procedure is repeated n times, finally resulting in a point Q_n. From the coordinates of this point, the discordant ages T_1 and T_2 can be calculated. These are the ages that would be measured for a mineral with a true age τ_0 that has undergone n fractionations R_i, ($i = 1, 2, \cdots n$) at times τ_1 in the past.

This construction is illustrated by the following example, illustrated by Figure 2. Consider a uranium mineral having a true age of 1350 million years. 900 million years ago the uranium and lead within this mineral were partially separated. At this time 17 pct of the lead within the mineral was lost, while at the same time 50 pct of the uranium was lost. As a consequence of this fractionation, the ratios D_1/P_1 and D_2/P_2 change by a factor 1.65 and $R_1 = 1.65$. In recent times (essentially zero million years ago) a second fractionation occurred, resulting in the loss of 27 pct of the lead present in the mineral, with a loss of

322   GEORGE W. WETHERILL

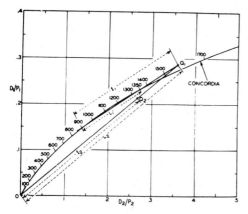

FIG. 2 – D/P diagram illustrating a numerical example
of the graphical procedure

only one per cent of the uranium. Hence R_2 = 0.74.

Following the graphical procedure described above, a straight line is drawn through the points 900 and 1350 on the curve 'concordia.' The point Q_1 is found by measuring the separation of these two points and measuring off a distance 1.65 times this separation along this line from the point 900. Q_2 is found by drawing a similar line between 0 and Q_1, and measuring off the proper distance from 0. This distance is 0.74 times the distance between 0 and Q_1.

Since this hypothetical mineral underwent two fractionations, it is characterized by the point Q_2. From the coordinates of this point (D_1/P_1 = 0.214, and D_2/P_2 = 2.64) the ages T_1 = 1260 $m \ y$ and T_2 = 1330 $m \ y$ are calculated from (1a). An exact analytic calculation of this example gives T_1 = 1250 and T_2 = 1330 $m \ y$. The accuracy of this graphical procedure depends only on the care with which the construction is made.

The effect of a continuous process can be approximated by graphically calculating the continuous process as a sum of episodic fractionations with the interval between episodes small.

Proof of the graphical procedure—The graphical procedure outlined above will be proved to be correct by showing that the coordinates of the resulting point Qn corresponds to the coordinates D_1/P_1 and D_2/P_2 which would be found by an analytic calculation of the effect of n fractionations $R_i(i = 1, 2, \cdots n)$ at times τ_1 in the past. In the following discussion the symbol t will be used to indicate time increasing in the usual sense and having the value zero at the time of mineralization. The symbol τ_1 will be used to indicate the time of an event measured back from the present.

This analytic expression will be obtained by solution of the equations of radioactive decay generalized to include the effects of gains or losses of parent and daughter isotopes. These are

$$\left.\begin{array}{l} \dfrac{dP}{dt} = -\lambda P + G_P \, P \\[2mm] \dfrac{dD}{dt} = \lambda P + G_D \, D \end{array}\right\} \quad (3)$$

where P and D refer to the concentrations of parent and radiogenic daughter isotopes, λ is the decay constant $= ln \ 2/\text{half-life}$, G_P and G_D are arbitrary functions of times representing gains, $(G > 0)$ or losses $(G < 0)$ of parent and daughter isotopes, respectively. These equations may be solved by elementary methods with the result

$$\frac{D}{P} = \lambda e^{F(\tau_0)} \int_0^{\tau_0} e^{-F} \, dt \qquad (4)$$

where

$$F = \lambda t + \int (G_D - G_P) \, dt,$$

and τ_0 is the true age of the mineral. (An equation essentially the same as (4) has also been derived by F. Wickman and his discussion of it will be published in the report of the Pennsylvania State University Conference on Nuclear Geophysics, September, 1955).

It may be seen by inspection of the form of F in (4) that a gain of parent represented by a function G_P will have exactly the same effect on the ratio D/P as a loss of daughter with $G_D = -G_P$. It will therefore be impossible to distinguish between loss of parent and gain of daughter or between gain of parent and loss of daughter by the effect on the ratio D/P. Without any loss of generality the expression $(G_D - G_P)$ can be replaced by an arbitrary function of time G, which will represent the net effect of gains or losses of parent and daughter. In this work where 'loss of daughter' is used, it will be understood that the possibility of gain of parent is also implied, and similarly for loss of parent.

By assuming different forms of G, the effects of different kinds of fractionation processes can be calculated. For the case of n episodic fractionations at times τ_i, $i = 1, \cdots n$ the function G will be given by

$$G = \sum_{i=1}^{n} a_i \, \delta[t - (\tau_0 - \tau_i)] \qquad (5)$$

where $\delta[t - (\tau_0 - \tau_i)]$ is the Dirac δ-function [*Dirac*, 1949] defined by

$$\left.\begin{array}{l} \delta(x) = 0 \quad x \neq 0 \\[2mm] \displaystyle\int_{-a}^{a} \delta(x) = 1 \quad a > 0 \end{array}\right\} \qquad (6)$$

Thus the δ-function can be visualized as a function which is zero everywhere except in the immediate neighborhood of one point. Similarly, G will be zero except at the moments τ_i. The difficulties of mathematical rigor associated with the use of (6) can be resolved in the usual way by replacing the δ-function by a gaussian and taking the standard deviation to be arbitrarily small. Thus the fractionation can be thought of as taking place over a period of time which is geologically short, for example, a million years.

By substitution of G (5) into (4) and integration, one obtains

$$\frac{D}{P} = \sum_{j=1}^{n} e^{\sum_{i=j}^{n} a_i} [e^{\lambda \tau_{j-1}} - e^{\lambda \tau_j}] + e^{\lambda \tau_n} - 1 \qquad (7)$$

The differential equations (3) can be combined in the form

$$\frac{d}{dt}\left(\frac{D}{P}\right) = \lambda + \left(\frac{D}{P}\right)(G + \lambda) \qquad (8)$$

or

$$\frac{d}{dt}\ln\left(\frac{D}{P}\right) = \lambda \frac{P}{D} + G + \lambda \qquad (9)$$

This equation will be integrated over a short period of time (2ϵ) including one of the episodes τ_i.

$$\int_{\tau_0 - (\tau_i - \epsilon)}^{\tau_0 - (\tau_i - \epsilon)} \frac{d}{dt}\ln\left(\frac{D}{P}\right) dt = \lambda \int_{\tau_0 - (\tau_i + \epsilon)}^{\tau_0 - (\tau_i - \epsilon)} \frac{P}{D} dt$$

$$+ \int_{\tau_0 - (\tau_i + \epsilon)}^{\tau_0 - (\tau_i - \epsilon)} \Sigma a_i \delta(t - (\tau_0 - \tau_i)) dt$$

$$+ \lambda t \Big|_{\tau_0 - (\tau_i + \epsilon)}^{\tau_0 - (\tau_i - \epsilon)} \qquad (10)$$

or

$$\ln \frac{\left(\dfrac{D}{P}\right)_{\tau_i - \epsilon}}{\left(\dfrac{D}{P}\right)_{\tau_i + \epsilon}} = \lambda \int_{\tau_0 - (\tau_i + \epsilon)}^{\tau_0 - (\tau_i - \epsilon)} \frac{P}{D} dt + a_i + 2\epsilon\lambda \qquad (11)$$

as $\epsilon \to 0$

$$\ln \frac{\left(\dfrac{D}{P}\right)_{\text{after loss}}}{\left(\dfrac{D}{P}\right)_{\text{before loss}}} \to a_i \qquad (12)$$

then

$$\frac{\left(\dfrac{D}{P}\right)_{\text{immediately after loss}}}{\left(\dfrac{D}{P}\right)_{\text{immediately before loss}}} = e^{a_i} \qquad (13)$$

or in accordance with the definition of R,

$$R_i = e^{a_i} \qquad (14)$$

(The proof of (14) has been materially simplified as a result of a suggestion by G. Wasserburg.)

Eq. (7) can now be rewritten as

$$\frac{D}{P} = \sum_{j=1}^{n} [e^{\lambda \tau_{j-1}} - e^{\lambda \tau_j}] \prod_{i=j}^{n} R_i + (e^{\lambda \tau_n} - 1) \qquad (15)$$

This is the general expression for the ratio D/P resulting from the decay of parent in a mineral of true age τ_0 which has undergone n fractionations, R_i at times τ_i. For a uranium mineral there will be two such expressions, one giving D_1/P_1 and the other D_2/P_2.

Using this general expression, the graphical procedure will be proved by induction. (a) It will first be shown that the point $(D_1/P_1, D_2/P_2)$ found by the graphical procedure agrees with the result of the analytic calculation for the case $n = 1$. (b) It then will be shown that if the construction is valid for $n = m$, it will also be valid for $n = m + 1$.

Demonstration of (a) and (b) above will prove the construction valid for any number of fractionations.

Regarding (a), by substituting $n = 1$ into (15) we obtain

$$\left.\begin{array}{l} \dfrac{D_1}{P_1} = R_1[e^{\lambda_1 \tau_0} - e^{\lambda_1 \tau_1}] + e^{\lambda_1 \tau_1} - 1 \\[4mm] \dfrac{D_2}{P_2} = R_1[e^{\lambda_2 \tau_0} - e^{\lambda_2 \tau_1}] + e^{\lambda_2 \tau_1} - 1 \end{array}\right\} \qquad (16)$$

These are the parametric equations for the locus of points representing minerals having a true age τ_0 and which underwent a fractionation R_1 at a time τ_1 in the past. According to the graphical construction, this locus should be a straight line passing through τ_0 and τ_1 on the curve 'concordia.'

That it passes through τ_0 can be seen by letting the parameter $R_1 = 1$, then

$$\left.\begin{array}{l} \dfrac{D_1}{P_1} = e^{\lambda_1 \tau_0} - 1 \\[4mm] \dfrac{D_2}{P_2} = e^{\lambda_2 \tau_0} - 1 \end{array}\right\} \qquad (17)$$

which are the coordinates of τ_0 on the curve 'concordia' as given by (2).

Similarly when the parameter $R_1 = 0$

$$\left.\begin{aligned}\frac{D_1}{P_1} &= e^{\lambda_1 \tau_1} - 1\\[6pt]\frac{D_2}{P_2} &= e^{\lambda_2 \tau_1} - 1\end{aligned}\right\} \quad (18)$$

which are the coordinates of τ_1 on concordia. By differentiating each of (16) with respect to the parameter R_1 and dividing, we obtain the slope of the locus

$$\frac{d\left(\dfrac{D_1}{P_1}\right)}{d\left(\dfrac{D_2}{P_2}\right)} = \frac{e^{\lambda_1 \tau_0} - e^{\lambda_1 \tau_1}}{e^{\lambda_2 \tau_0} - e^{\lambda_2 \tau_1}} \quad (19)$$

which is independent of R_1, that is of the position of the point on the locus. The slope is therefore the same at all points on the locus, that is the line is straight.

According to the graphical construction, the distance along this line (l) between τ_1 on concordia and the point characterizing a mineral which underwent a fractionation R_1 at τ_1 is given by $l_1 = R_1 L_1$ where L_1 is the length of the segment $\overline{\tau_0 \tau_1}$. The length can be found by use of the coordinates of τ_0 and τ_1 on concordia from equations (17) and (18) and the Pythogorean theorem to be

$$L_1{}^2 = [e^{\lambda_1 \tau_0} - e^{\lambda_1 \tau_1}]^2 + [e^{\lambda_2 \tau_0} - e^{\lambda_2 \tau_1}]^2 \quad (20)$$

By using (16) and (18) and the Pythogorean theorem we obtain similarly

$$l_1{}^2 = R_1{}^2[e^{\lambda_1 \tau_0} - e^{\lambda_1 \tau_1}]^2 + R_1{}^2[e^{\lambda_2 \tau_0} - e^{\lambda_2 \tau_1}]^2 \quad (21)$$

and by comparing (20) and (21) we obtain

$$l_1 = R_1 L_1 \text{ thus completing the proof for } n = 1. \quad (22)$$

Regarding (b) if it is assumed that the graphical construction is valid for $n = m$, it will now be shown that it is valid for $n = m + 1$.

The coordinates of a point which has undergone $n = m + 1$ fractionations will be

$$\left.\begin{aligned}\frac{D_1}{P_1} &= \sum_{j=1}^{m+1} [e^{\lambda_1 \tau_{j-1}} - e^{\lambda_1 \tau_j}] \prod_{i=j}^{m+1} R_i + (e^{\lambda_1 \tau_{m+1}} - 1)\\[6pt]\frac{D_2}{P_2} &= \sum_{j=1}^{m+1} [e^{\lambda_2 \tau_{j-1}} - e^{\lambda_2 \tau_j}] \prod_{i=j}^{m+1} R_i + (e^{\lambda_2 \tau_{m+1}} - 1)\end{aligned}\right\}$$
$$(23)$$

For the case $R_{m+1} = 1$, that is, no fractionation at stage $(m + 1)$ the point will have the coordinates

$$\begin{aligned}\frac{D_1}{P_1} &= \sum_{j=1}^{m} [e^{\lambda_1 \tau_{j-1}} - c^{\lambda_1 \tau_j}] \prod_{i=1}^{m} R_i\\[4pt]&\quad + (e^{\lambda_1 \tau_m} - e^{\lambda_1 \tau_{m+1}}) + (e^{\lambda_1 \tau_{m+1}} - 1) \quad (24)\\[6pt]&= \sum_{j=1}^{m} [e^{\lambda_1 \tau_{j-1}} - e^{\lambda_1 \tau_j}] \prod_{i=j}^{m} R_i + (e^{\lambda_1 \tau_m} - 1)\end{aligned}$$

and similarly

$$\frac{D_2}{P_2} = \sum_{j=1}^{m} [e^{\lambda_2 \tau_{j-1}} - e^{\lambda_2 \tau_j}] \prod_{i=j}^{m} R_i + (e^{\lambda_2 \tau_m} - 1)$$

which according to (15) are the coordinates of a point Q_m which has undergone m fractionations R_i at times τ_i. Thus the locus of a point which has undergone $m + 1$ fractions R_i at times τ_i passes through this point. For the case $R_{m+1} = 0$, we get the point

$$\frac{D_1}{P_1} = e^{\lambda_1 \tau_{m+1}} - 1 \quad (25)$$

and

$$\frac{D_2}{P_2} = e^{\lambda_2 \tau_{m+1}} - 1, \quad \text{that is, the point } \tau_{m+1} \text{ on}$$

concordia

By differentiating (23) with respect to R_{m+1} and taking their ratio we obtain

$$\frac{d\left(\dfrac{D_1}{P_1}\right)}{d\left(\dfrac{D_2}{P_2}\right)}$$

$$= \frac{\displaystyle\sum_{j=1}^{m} [e^{\lambda_1 \tau_{j-1}} - e^{\lambda_1 \tau_j}] \prod_{i=j}^{m} R_i + (e^{\lambda_1 \tau_m} - e^{\lambda_1 \tau_{m+1}})}{\displaystyle\sum_{j=1}^{m} [e^{\lambda_2 \tau_{j-1}} - e^{\lambda_2 \tau_j}] \prod_{i=j}^{m} R_i + (e^{\lambda_2 \tau_m} - e^{\lambda_2 \tau_{m+1}})}$$
$$(26)$$

which is independent of R_{m+1}. Thus the locus of points corresponding to minerals which have undergone varying fractionations R_{m+1} at time τ_{m+1} is a straight line passing through Q_m and τ_{m+1} on concordia, as given by the graphical procedure.

The distance L_{m+1} between Q_m and τ_{m+1} on concordia will be (by use of the Pythogorean theorem)

$$\begin{aligned}L^2{}_{m+1} &= \left\{\sum_{j=1}^{m} [e^{\lambda_1 \tau_{j-1}} - e^{\lambda_1 \tau_j}] \prod_{i=j}^{m} R_i\right.\\[4pt]&\quad \left. + (e^{\lambda_1 \tau_m} - 1) - (e^{\lambda_1 \tau_{m+1}} - 1)\right\}^2\\[6pt]&\quad + \left\{\sum_{j=1}^{m} [e^{\lambda_2 \tau_{j-1}} - e^{\lambda_2 \tau_j}] \prod_{i=j}^{m} R_i\right.\\[4pt]&\quad \left. + (e^{\lambda_2 \tau_m} - 1) - (e^{\lambda_2 \tau_{m+1}} - 1)\right\}^2 \quad (27)\end{aligned}$$

collecting terms

$$L^2_{m+1} = \left\{ \sum_{j=1}^{m+1} [e^{\lambda_1 \tau_{j-1}} - e^{\lambda_1 \tau_j}] \frac{\prod\limits_{i=j}^{m+1} R_i}{R_{m+1}} \right\}^2$$

$$+ \left\{ \sum_{j=1}^{m+1} [e^{\lambda_2 \tau_{j-1}} - e^{\lambda_2 \tau_j}] \frac{\prod\limits_{i=j}^{m+1} R_i}{R_{m+1}} \right\}^2 \quad (28)$$

The distance l_{m+1} between τ_{m+1} on the curve 'concordia' and the point characteristic of a mineral which has undergone $m + 1$ fractionations R_i at times τ_i will be

$$l^2_{m+1} = \left\{ \sum_{j=1}^{m+1} [e^{\lambda_1 \tau_{j-1}} - e^{\lambda_1 \tau_j}] \prod_{i=j}^{m+1} R_i \right\}^2$$

$$+ \left\{ \sum_{j=1}^{m+1} [e^{\lambda_2 \tau_{j-1}} - e^{\lambda_2 \tau_j}] \prod_{i=j}^{m+1} R_i \right\}^2 \quad (29)$$

By comparison of (28) and (29)

$$l^2_{m+1} = R^2_{m+1} L^2_{m+1}$$

Thus the locus of points characteristic of minerals which have undergone m fractionations R_i at times τ_i and an additional fractionation R_{m+1} at time τ_{m+1} will be a straight line passing through the point Q_m and the point τ_{m+1} on concordia. For a given fractionation R_{m+1}, the distance of the point from τ_{m+1} on concordia will be equal to $R_{m+1} L_{m+1}$ where L_{m+1} is the distance from Q_m to τ_{m+1} on concordia. By hypothesis the point Q_m (which represents the point characteristic of minerals which have undergone m fractionations R_i, $i = 1, 2, \cdots m$ at times τ_i) is given correctly by the graphical procedure. It has been shown above that the analytical calculation is in agreement with the graphical procedure for finding Q_{m+1} from Q_m. Therefore, the point Q_{m+1} found by the analytical calculation will be the same point found by the graphical procedure, thus completing the proof.

Extensions of the graphical procedure—In the foregoing discussion a procedure has been demonstrated for graphically calculating the discordant uranium-lead ages which will result when a mineral undergoes a series of episodes of uranium-lead fractionation (failure of assumption a). A brief discussion will now be given of the calculation of the effects of failure of assumptions (b), and (c) as well as the effect of unequal fractionation of Pb206 and Pb207.

Loss of intermediate decay products—For the equilibrium case (implicit in (1)) the loss of an intermediate decay product is equivalent to the

decay constant λ for the growth of daughter being less than the decay constant λ for the decay of parent (3).

Therefore, (3) can be replaced by

$$\left. \begin{aligned} \frac{dP}{dt} &= -\lambda P + G_P P \\ \frac{dD}{dt} &= \lambda \Lambda(t) P + G_D D \end{aligned} \right\} \quad (30)$$

where $\Lambda(t)$ represents the fraction of intermediate decay product which is retained.

These equations can be integrated to give

$$\frac{D}{P} = \lambda e^{F(\tau_0)} \int_0^{\tau_0} \Lambda(t) e^{-F} \, dt \quad (31)$$

It may be expected that Λ will be different for the two decay systems, U238 and U235.

For the case of constant loss of intermediate decay product the function Λ can be taken outside the integral. In this case the value of D/P will be decreased by a factor Λ. For combined multiple fractionation and constant intermediate product loss, the resulting values of D_1/P_1 and D_2/P_2 can be found by first applying the graphical procedure and finally multiplying D_1/P_1 by Λ_1, and D_2/P_2 by Λ_2.

Presence of primary radiogenic lead—Eq. (3) can be generalized to include the effects of primary radiogenic lead with the result

$$\frac{D}{P} = e^{F(\tau_0)} \left[\left(\frac{D}{P} \right)_0 e^{-F(0)} + \int_0^{\tau_0} \lambda e^{-F} \, dt \right] \quad (32)$$

where $(D/P)_0$ is the initial radiogenic daughter to parent ratio.

For the case of multiple episodic fractionations

$$\left(\frac{D}{P} \right) \sum_{j=1}^{n} [e^{\lambda \tau_{j-1}} - e^{\lambda \tau_j}] \prod_{i=j}^{n} R_i + (e^{\lambda \tau_n} - 1)$$

$$+ \left(\frac{D}{P} \right)_0 e^{\lambda \tau_0} \prod_{i=1}^{n} R_i \quad (33)$$

which differs from (15) only in the last term. Therefore, the effect of primary radiogenic daughter can be found by applying the graphical procedure and finally adding the term

$$\left(\frac{D}{P} \right)_0 e^{\lambda \tau_0} \prod_{i=1}^{n} R_i \quad \text{to} \quad \frac{D_1}{P_1} \quad \text{and} \quad \frac{D_2}{P_2} \quad (34)$$

By a proof along the lines of that in *Proof of the graphical procedure*, it can also be shown that the graphical calculation is valid for the case of primary radiogenic daughter if the starting point for the graphical procedure is taken to the co-ordinates of τ_0 on concordia augmented by

$$\left(\frac{D}{P}\right)_0 e^{\lambda \tau_0}$$

The effect of unequal fractionation of Pb^{206} and Pb^{207}—As a consequence of an earlier process of uranium-lead fractionation, the uranium in a mineral may be displaced from the lead. The radiogenic lead resulting from the decay of this uranium will then be in the vicinity of the uranium and the two lead isotopes will not be homogeneously distributed within the mineral. Any subsequent fractionation might then cause D_1/P_1 and D_2/P_2 to change by different factors.

In this case it can be shown that the points Q_{m+1} resulting from varying amounts of fractionation of this type will not lie on a straight line between Q_m and $\tau_{.n+1}$ on concordia but will lie on a curved line between these two points. The co-ordinate D_1/P_1 for Q_{m+1} will be

$$(e^{\lambda_1 \tau_{m+1}} - 1) +$$

$$R'_{m+1}\left[\left(\frac{D_1}{P_1}\right)_{Q_m} - (e^{\lambda_1 \tau_{m+1}} - 1)\right] \quad (35)$$

where R'_{m+1} is the factor by which D_1/P_1 is changed. A similar expression will give (D_2/P_2) for Q_{m+1}.

Concluding remarks—A graphical procedure has been demonstrated for calculating the effects of failure of assumptions (a), (b), and (c). By use of this procedure the discordant ages resulting from a given history can be calculated uniquely. However, if the discordant ages are given, it is not possible to infer the history uniquely. This graphical procedure will be found useful, however, in inferring the possible histories of a given geological unit and in some cases the probable history. These applications and their limitations will be discussed in a subsequent publication.

REFERENCES

AHRENS, L. H., The convergent lead ages of the oldest monazites and uraninites, *Geochim. et Cosmochim. Acta*, **7**, 294, 1955.

DIRAC, P. M. J., The principles of quantum mechanics, 3rd ed., p. 58, Oxford University Press, 1949.

Department of Terrestrial Magnetism, Carnegie Institution of Washington, Washington 15, D. C.

(Communicated manuscript received February 9, 1956; open for formal discussion until November 1, 1956.)

21

Volume Diffusion as a Mechanism for Discordant Lead Ages

G. R. TILTON

Geophysical Laboratory, Carnegie Institution of Washington
Washington 8, D. C.

Abstract. The discordant U-Pb ages of sixteen minerals having Pb^{207}/Pb^{206} ages of 2400 to 2700 m.y. closely fit a 2800 to 600 m.y. chord when placed on a Pb^{207}/U^{235}-Pb^{206}/U^{238} diagram. These minerals are from the continents of North America, Africa, Europe, Australia, and Asia. If the data are interpreted as evidence of an episode of loss of lead 600 m.y. ago from minerals that were crystallized 2800 m.y. ago, it is strange that the same time of loss is indicated for all four continents. Moreover, most of the samples are from shield areas where no evidence has been found for metamorphic events 600 m.y. ago.

As an alternative explanation, lead may be considered to diffuse continuously from crystals at a rate governed by a diffusion coefficient (D), the effective radius (a), and the concentration gradient. Calculations of the present-day Pb^{206}/U^{238} and Pb^{207}/U^{235} ratios as a function of the parameter D/a^2 yield ratios that lie on a 2800 to 600 m.y. chord on a Pb^{206}/U^{238}-Pb^{207}/U^{235} diagram for losses of up to two-thirds of the lead produced in the sample.

This method is capable of explaining the regularity of the age patterns from one continent to another and does away with the need for episodic losses of lead simultaneously in all the areas 600 m.y. ago. Other examples have been found of suites of minerals with discordant ages which fit this hypothesis; they include minerals with apparent ages of 1900, 1700, and 1100 m.y.

The behavior of uranium-lead ages in several zircons from metamorphic rocks is indicative of an activation energy for diffusion of lead of less than 10 kcal/mol over a temperature range of 50° to 400°–600°C. This inference, when considered with available data for the diffusion of argon from micas, is capable of explaining why uranium-lead ages measured from zircon may be either greater or less than the potassium-argon ages of coexisting mica.

INTRODUCTION

Since the first isotopic lead ages were published by *Nier* [1939], a problem has existed concerning lack of agreement between the Pb^{206}–U^{238} and Pb^{207}–U^{235} ages obtained on many mineral samples. In most, if not all, cases it can be assumed that minerals having discordant uranium-lead ages have not been closed systems with respect to uranium or lead or both since they were first formed. The cause of the age discordances is a matter of some importance since, in principle, such ages contain information about the post-crystallization history of a mineral that would not be obtained from concordant lead ages. The present paper deals with the problem of interpreting discordant ages found in minerals from rocks taken largely from Precambrian shield areas.

Certain regularities in the discordant uranium-lead ages of monazites from Rhodesia and Madagascar and of uraninite and thucholite from the Witwatersrand were discussed by *Ahrens* [1955a, 1955b]. Ahrens devised several methods of plot-

ting the ages, and he believed that the regularities in the discordant age patterns suggested 'control of lead loss by physical processes which perhaps have operated at varying degrees but at a uniform rate ever since the minerals were formed.' *Wetherill* [1956a, 1956b] then showed that the regularities noted by Ahrens could be explained by a simple geochemical process. For example, the Rhodesian age pattern could be produced by minerals with an age of 2700 m.y. that lost different fractions of their lead 500 m.y. ago. Wetherill developed a graphical procedure that permits rapid determination of the age and time of lead removal for suites of cogenetic minerals that have experienced a single episode of lead loss. The procedure is as follows: on a plot of Pb^{207}/U^{235} against Pb^{206}/U^{238}, a curve 'concordia,' is drawn representing the locus of all points such that $t(Pb^{207}/U^{235}) = t(Pb^{206}/U^{238})$. Samples having an age t_1 that lost lead at t_2 will lie on a chord connecting t_1 and t_2, their position along the chord depending on the amount of lead that was lost.

2933

Russell and Ahrens [1957] studied four suites of minerals with ages from 1000 to 2700 m.y. They noted that, if a single episode of removal of lead is assumed to be the cause of the age discordances, the time of lead removal appears to be a function of the age of a given suite of minerals. The younger a suite of minerals, the more recent the time of loss of lead appears to be. The authors suggested that a physical rather than a chemical process seems to be required to explain this regularity. The process should be one that causes a greater percentage loss of Pb^{207} than of Pb^{206}. Russell and Ahrens postulated that the differential losses might be related to differences in recoil energies from α-particle emission of various members of the two uranium decay series or to some 'wholly unforeseen process.' Serious questions can be raised about the samples used by Russell and Ahrens. Of four mineral suites, two are from the Goldfields District in Saskatchewan and one is from the Witwatersrand. As was mentioned by the authors, it is recognized that the Goldfields pitchblendes represent more than one generation. The isotopic data indicate more than one episode of lead

loss—one about 1200 and one about 200 m.y. ago [*Eckelmann and Kulp*, 1956; *Aldrich and Wetherill*, 1958]. The episode of loss at 1200 m.y. does not fit the authors' proposed regularities. For the Witwatersrand samples, the spread of points on a Pb^{207}/U^{235}–$Pb^{206}U/^{238}$ diagram [*Wetherill*, 1956a] indicates a complex history that involves more than one age of sample or time of lead loss, or both. The test of possible simple regularities in discordant lead ages should be limited to areas having simpler geological histories.

Since these earlier papers were written, the number of minerals having Pb^{207}–Pb^{206} ages of about 2600 m.y. has been considerably increased. Currently available results are given in Table 1, which includes only those minerals having Pb^{207}–Pb^{206} ages $\geq$ 2300 m.y. Minerals with still lower Pb^{207}–Pb^{206} ages that appear to belong to this group have been omitted to avoid possible erroneous inclusion of younger minerals.

If the samples from Table 1 are plotted on a Pb^{207}/U^{235}–Pb^{206}/U^{238} diagram, an array of points is obtained that corresponds to the case of a group of minerals having ages of 2600 to 2800

TABLE 1. Minerals Having Pb^{207}-Pb^{206} Ages of 2300–2800 m.y.*

Location	Mineral	Age, million years				Reference
		$\dfrac{Pb^{206}}{U^{238}}$	$\dfrac{Pb^{207}}{U^{235}}$	$\dfrac{Pb^{207}}{Pb^{206}}$	$\dfrac{Pb^{208}}{Th^{232}}$	
Mätäsvaara, Finland	Zircon	2520	2650	2770	2750	Tilton and Kouvo, unpublished
Heinävaara, Finland	Zircon	2250	2240	2620	1280	Tilton and Kouvo, unpublished
Koli, Finland	Zircon	1890	2270	2650	1790	Tilton and Kouvo, unpublished
Huhtilampi, Finland	Zircon	1820	2240	2640	1820	Tilton and Kouvo, unpublished
Sotkuma, Finland	Zircon	1810	2150	2470	1760	Tilton and Kouvo, unpublished
Saksagan, Repikhovo, Ukraine	Orthite	2420	2610	2780	2630	Vinogradov, 1956
Huron Claim, Manitoba	Uraninite	1560	1980	2480	1280	Nier, 1939
Dickinson Co., Michigan	Zircon	1740	2150	2570	1250	G. L. Davis, unpublished
Dickinson Co., Michigan	Zircon	1710	2100	2510	1300	G. L. Davis, unpublished
Cooke City, Montana	Uraninite	2600	2650	2700	...	Gast and others, 1958
Ebonite Claims, S. Rhodesia	Monazite	2640	2670	2700	2640	Holmes, 1954
Jack Tin Claims, S. Rhodesia	Monazite	2210	2460	2660	1940	Holmes, 1954
Irumi Hills, N. Rhodesia	Monazite	1990	2330	2640	1380	Holmes, 1954
Antsirabi, Madagascar	Monazite	1370	1850	2450	610	Holmes, 1954
Yadiur, India	Monazite	1410	1820	2330	1800	Holmes, 1955
Woodstock, W. Australia	Tanteuxenite	1900	2360	2790	2590	Greenhalgh and Jeffery, 1959

* Minerals having Pb^{206}-U^{238} ages that are greater than their Pb^{207}-Pb^{206} ages have been excluded from this table for reasons stated in the text.

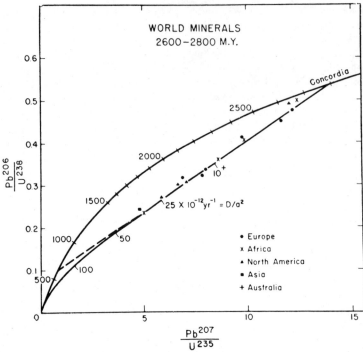

Fig. 1. Parent-daughter ratios for minerals having Pb^{207}–Pb^{206} ages of 2300 to 2800 m.y. The curve through the points is calculated for loss of lead by continuous diffusion from spheres of radius a having an age of 2800 m.y.

m.y. that lost varying amounts of lead with respect to uranium 500 to 600 m.y. ago (see Fig. 1). If the discordances are to be explained by a single episode of lead loss, the time of loss was about the same from one continent to another, and negligible losses occurred from 2700 to 600 m.y. ago. This would seem to be a rather remarkable coincidence.

A further difficulty arises from the fact that the North American and European minerals are from shield areas where age-determination studies have not, as yet, given evidence for igneous activity or metamorphism 500 to 600 m.y. ago. Minerals having ages in this interval do occur in South Africa [*Holmes and Cahen*, 1957], but it is not known to what extent the uranium/lead ratios in the Rhodesian minerals in Table 1 were affected by this episode of mineral crystallization.

VOLUME DIFFUSION

Mathematical treatment. If ages of 2600 to 2800 m.y. are accepted for the samples shown in Table 1, any mechanism postulated to explain the discordant ages must be one that produces a greater loss of radiogenic Pb^{207} than of Pb^{206}. Episodic loss of lead at some time in the past is one such mechanism. Diffusion also satisfies this requirement in at least a qualitative way. Owing to the shorter half-life of U^{235}, Pb^{207} has a greater mean age than the Pb^{206} in a mineral and thus has had a longer time to diffuse out.

A quantitative solution for the loss of a radiogenic daughter product by diffusion can be obtained for certain idealized cases. The assumptions used here are:

1. The mineral crystals are spheres of effective radius a. The value of a may be determined by grain boundaries or by lattice imperfections and need not be related directly to the physical size or shape of the grains.

2. Uranium is distributed uniformly within the spheres.

3. Diffusion of uranium and intermediate daughter products is negligible compared with that of lead.

4. The diffusion coefficient, D, is constant over the time the mineral has existed.

5. The diffusion of lead is governed by Fick's law, so that the differential equation for the change of lead concentration with time for any radial volume element in the sphere may be written

$$\frac{\partial C}{\partial t} = D\left(\frac{\partial^2 C}{\partial r^2} + \frac{2}{r}\frac{\partial C}{\partial r}\right) + \lambda N_0 e^{-\lambda t'} \quad (1)$$

where

C = atom concentration of lead daughter product.

N_0 = initial concentration of uranium parent atoms.

λ = decay constant of the uranium parent.

t' = time measured from the beginning of the diffusion process.

The boundary conditions are:

$$C = 0 \qquad t = 0$$

$$C = 0 \qquad \text{all } t \qquad r = a$$

A solution for equation 1 can be derived in a simple way from the equation for the diffusion of a component out of a sphere, for uniform initial distribution [Barrer, 1951, pp. 28–29]:

$$\frac{\overline{C}}{C_0} = \frac{6}{\pi^2} \sum_{n=1}^{\infty} \frac{e^{-n^2 \pi^2 D t / a^2}}{n^2} \quad (2)$$

where C_0 is the initial concentration and $\overline{C}$ is the average concentration after an elapsed time t. By radioactive decay uranium has continually generated lead so that, at time t',

$$C_0(t') = \lambda N_0 e^{-\lambda t'} \quad (3)$$

The average concentration of lead at the present time as the result of such a process is

$$\overline{C} = \frac{6}{\pi^2} \int_0^t \sum_{n=1}^{\infty} \frac{\lambda N_0 e^{-\lambda t' - n^2 \pi^2 D(t-t')/a^2}}{n^2} \, dt' \quad (4)$$

where t is the age of the mineral.

Equation 4 integrates to

$$\frac{\overline{C}}{N} = \frac{6}{\pi^2} \sum_{n=1}^{\infty} \frac{\lambda(e^{\lambda t - n^2 \pi^2 D t / a^2} - 1)}{n^2(\lambda - n^2 \pi^2 \, D/a^2)} \quad (5)$$

where N is the concentration of uranium (atoms) at present.

Equation 5 is equivalent to the solution given first by *Wasserburg* [1954]. *Nicolaysen* [1957] published an extensive table of values for equation 5. His tables were used in preparing the figures in the present paper.

Solutions of the diffusion equation for the sphere. Figure 1 shows a quantitative comparison between the discordant age pattern expected for hypothetical 2800 m.y.-old minerals that have lost lead by the diffusion model given above and the discordant ages of the minerals in Table 1. The curve labeled with values of D/a^2 is the locus of points which represent all possible samples that conform to the diffusion model. For example, the point at 10×10^{-12} yr^{-1} gives the Pb^{207}/U^{235} and Pb^{206}/U^{238} values that exist today in a mineral having an age of 2800 m.y. and a value of 10×10^{-12} for D/a^2 over the past 2800 m.y. For values of D/a^2 up to approximately 50×10^{-12} yr^{-1}, diffusion losses produce a group of samples that closely fit a 2800 to 600 m.y. chord. A group of minerals having an age of 2800 m.y. that lost different proportions of lead 600 m.y. ago, this being the only time at which any loss of lead occurred, would fit the same chord. Thus a process for loss of lead that took place continuously since the time these minerals crystallized can produce the same pattern of discordant ages as a process involving an episodic loss of lead. Because of this, the time of apparent episodic loss of lead found in Figure 1 need not have any particular geologic significance.

This explanation offers a solution to the problems that arise from an interpretation based on episodic lead loss. The discordant ages in Finland and Michigan can be explained without resort to a metamorphic episode 500 to 600 m.y. ago, for which no evidence is present. As was mentioned above, minerals with ages of 450 to 600 m.y. do occur in southern Africa [Holmes and Cahen, 1957], so that episodic loss is a possible explanation for the discordant age pattern of the African monazites. The diffusion mechanism for loss of lead helps to explain the linear relationship of samples from one continent to another in Figure 1. The diffusion model requires only that the samples be of approximately the same age, whereas the episodic-loss mechanism requires all the samples to be of the same age and all to have lost lead 500 to 600 m.y. ago.

The D/a^2 curve ('discordia') shown in Figure

TABLE 2. Uranium-Lead Ages from Belomoria, Karelia

Mineral	Age, million years			Reference
	$\dfrac{Pb^{206}}{U^{238}}$	$\dfrac{Pb^{207}}{U^{235}}$	$\dfrac{Pb^{207}}{U^{206}}$	
Uraninite	1760	1800	1870	Starik, 1956
Uraninite	1660	1760	1880	Starik, 1956
Uraninite	1680	1740	1820	Starik, 1956
Uraninite	1050	1300	1800	Starik, 1956
Uraninite	1150	1380	1780	Starik, 1956
Monazite	1250	1450	1800	Zhirov, 1958
Monazite	1220	1440	1760	Zhirov, 1958
Thucholite	1480	1655	1900	Zhirov, 1958
Thucholite	1250	1460	1850	Zhirov, 1958
Thucholite	1120	1340	1770	Zhirov, 1958
Thucholite	850	1170	1810	Zhirov, 1958

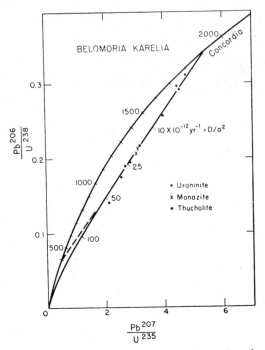

Fig. 2. Parent-daughter ratios for minerals from Belomoria, Karelia, compared with calculated ratios for loss of lead by continuous diffusion.

1 is intended only as an example of a solution by the diffusion mechanism, not as a least-squares fit of the data. Since the samples are unlikely to be of exactly one age, any such fit would be meaningless.

Nicolaysen [1957] published values for the ages of a number of samples after correcting the discordant ages for loss of lead by diffusion. His graphical method of solution differs from that used here but leads to the same solution for apparent age. He showed that the South African and Yadiur monazites have ages of 2600 to 2800 m.y. if diffusion according to the model discussed above is responsible for the discordant ages. His method does not show the lead-loss relationships illustrated by Figure 1.

A second example of a suite of minerals with discordant ages that fit the diffusion hypothesis is given in Table 2 and Figure 2. These minerals are from pegmatites in the White Sea region of Karelia. The fit of points to the D/a^2 curve in Figure 2 is substantially better than in Figure 1. This might be expected since the requirement that the minerals all have the same age is more likely to be true for the Karelian suite than for a group taken from various continents. In preparing Figure 2, several monazites having U^{235}-Pb^{206} ages greater than their Pb^{207}-Pb^{206} ages were rejected since such samples obviously do not conform to the basic assumptions of the diffusion calculations. The thucholite ages are probably equivalent to uraninite ages since

thucholites from Finland have been found to be uraninite embedded in an organic matrix (Kouvo, private communication).

Figure 3 compares the U-Pb ratios of a number of North American minerals having Pb^{207}-Pb^{206} ages of 1000 to 1150 m.y. against a D/a^2 curve for 1150-m.y.-old minerals. For minerals this young, 'discordia' lies closer to 'concordia' than in the two previous examples, and the time of apparent lead loss is closer to 0 m.y. so that interpretation of the data becomes more difficult. Clearly, however, the diffusion hypothesis is capable of explaining the discordant ages. This explanation has some merit because the majority of minerals in Figure 3 have U^{238}-Pb^{206} ages about 10 per cent lower than their Pb^{207}-Pb^{206} ages. If episodic loss of lead were the cause of the discordance, a more erratic spread of ages might be expected. In an earlier paper dealing with the isotopic zircon ages [*Tilton, Davis, Wetherill, and Aldrich,* 1957], the repeated finding of U^{238}-Pb^{206} ages that were 10 per cent lower than 207-206 ages led the authors to consider

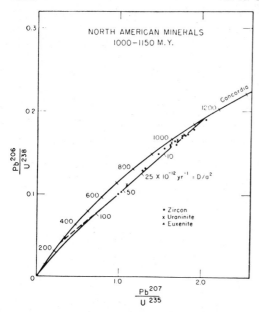

NORTH AMERICAN MINERALS
1000-1150 M.Y.

Concordia

$25 \times 10^{-12} \, yr^{-1} = D/a^2$

• Zircon
× Uraninite
▲ Euxenite

$\dfrac{Pb^{206}}{U^{238}}$

$\dfrac{Pb^{207}}{U^{235}}$

Fig. 3. Parent-daughter ratios for North American minerals with Pb^{207}-Pb^{206} ages of 1000 to 1150 m.y. The curve through the points is calculated for loss of lead by continuous diffusion.

the possibility that this difference might be due to loss of lead by faulty analytical procedures. Subsequent work has shown the analytical procedures to be reliable and age discordances real.

In western North Carolina and eastern Tennessee three zircons have discordant ages that do not fit the diffusion-loss hypothesis. These samples have age discordances that might be the result of an episodic loss of lead about 450 m.y. ago superimposed on diffusion losses. These zircons have been omitted from Figure 3.

A final diffusion comparison is made in Figure 4. These are zircon age determinations taken from the work of *Kouvo* [1958]. The line of best fit to the five points passes through the origin, but, until more analyses are obtained, a diffusion mechanism for lead loss cannot be excluded. The zircons are from a shield area where there is no evidence for metamorphism in recent time.

Effect of temperature changes. The diffusion calculations contain the implicit assumption, which would seem to be unrealistic, that temperature has remained constant over periods of billions of years. The effect of changes in tem-

perature on losses of lead by diffusion is difficult to evaluate. Equation 1 can be solved analytically for two cases: (*a*) no lead is present initially; (*b*) any such lead is homogeneously distributed. Calculations for temperature changes involve changing D after concentration gradients have already been established for lead as a result of previous diffusion. No analytical solutions exist for this problem.

Some information about the effect of temperature changes can be obtained from applications of equation 5. Consider the case of a mineral with an age of 2800 m.y. which lost lead as a result of an increase in the diffusion rate for a brief interval of time, e.g., 10 to 100 m.y., 1800 m.y. ago. At all other times let lead be lost at a rate governed by a particular value for D/a^2, say 10×10^{-12} yr^{-1}. The values 2800 and 1800 m.y. are chosen because the rubidium-strontium ages of biotite associated with several of the Finnish zircons in Table 1 are 1800 m.y. This with other zircon work indicates that 1800 to 1900 m.y. ago was a time of metamorphism in the area (Wetherill, Kuovo, Tilton, and Gast, paper to be submitted to the *Journal of Geology*). The concentrations of Pb^{206} and Pb^{207} that existed 1800 m.y. ago in a mineral which now has an age of 2800 m.y. can be calculated from equation 5. Of the lead present at that time, part is to be lost by heating, the loss being governed by Fick's law with the concentration of lead kept zero at the boundary of the crystal, then more lead is to be lost at a rate governed by $D/a^2 = 10 \times 10^{-12}$ yr^{-1} from 1800 m.y. to the present.

Two limiting cases are obvious: as the loss at 1800 m.y. approaches zero, the Pb^{207}/U^{235} and Pb^{206}/U^{238} ratios approach the point 10 on the D/a^2 curve for a 2800-m.y.-old mineral in Figure 1. As lead loss approaches 100 per cent, the ratios approach the point 10 on the D/a^2 curve for an 1800-m.y.-old mineral. For intermediate cases, the problem reduces to following the change in isotopic composition of lead as a function of the amount of lead lost by diffusion after lead has been generated in a mineral for 1000 m.y. This can be done by calculating from equation 5 the Pb^{207} and Pb^{206} at 1800 m.y. Next the amounts at successive times such as 800 and 0 m.y. ago are calculated. The radiogenic lead formed from 1800 m.y. ago to the time under consideration can be calculated and subtracted

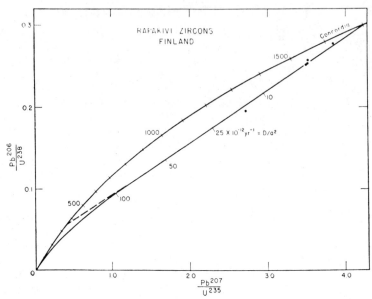

Fig. 4. Parent-daughter ratios for zircon from Finnish rapakivi granites compared with calculated ratios for loss of lead by continuous diffusion.

from the total lead present at that time. The difference represents what remains of the lead that was present 1800 m.y. ago. As the loss of lead approaches 100 per cent, the isotopic composition of the residual lead approaches that of lead generated from 2800 to 1800 m.y. ago in a mineral that experienced no loss of lead, as can be seen from equation 2. For a very long elapsed time after generation of the lead, the fraction of lead lost will be almost independent of the time interval in which it was generated. The Pb^{207}/Pb^{206} ratios 1800 m.y. ago in minerals that are now 2800 m.y. old were 0.3122 for $D/a^2 = 0$, 0.3061 for $D/a^2 = 10 \times 10^{-12}$ yr^{-1}, 0.3031 for $D/a^2 = 20 \times 10^{-12}$ yr^{-1}, and 0.3005 for $D/a^2 = 30 \times 10^{-12}$ yr^{-1}.[1]

These calculations show that the Pb^{207}/Pb^{206} ratios of the last three cases change in a regular manner toward 0.3122 as lead is lost by further diffusion. Since the ratios do not differ greatly from 0.3122 to start with, it can be assumed as an approximation that Pb^{207} and Pb^{206} are lost in amounts that are essentially proportional to their isotopic abundances at 1800 m.y. The resulting array of discordant ages

can be plotted as in Figure 5. Loss of half of the lead contained in the minerals at 1800 m.y. would yield points that plot approximately, but not exactly, at the mid-points of the lines connecting the D/a^2 curves. (The points will actually be displaced toward the 2800-m.y. curve.) Loss of 20 to 30 per cent of lead 1800 m.y. ago would not produce a serious departure from the D/a^2 curve for 2800-m.y.-old minerals. Such losses are quite within the spread noted for the five Finnish zircons, for example. Any episodic loss of lead 2500 to 2800 or 0 to 1000 m.y. ago would be even more difficult to detect.

The method illustrated in Figure 5 can be used to predict trends resulting from changes in the value of the diffusion coefficient, but it does not give information on the amount of lead existing in a mineral at a particular time that will be lost for a given change in D. More elaborate procedures for solving the diffusion equation after concentration gradients have been established by diffusion are required.

Diffusion from crystals with nonspherical shapes. The foregoing discussions have dealt with diffusion losses from crystals of spherical shape. Since this is an idealization not realized in nature, it is worth while to examine the effect of changes in crystal shape. The calculations for

---

[1] Decay constants used: U^{238}: 1.54×10^{-10} yr^{-1}; U^{235}: 9.72×10^{-10} yr^{-1}.

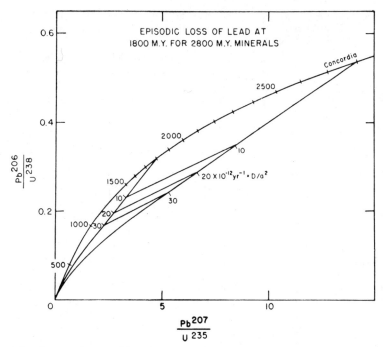

Fig. 5. Parent-daughter ratios resulting from episodic loss of lead 1800 m.y. ago superimposed on continuous loss of lead by diffusion for hypothetical minerals having an age of 2800 m.y.

the sphere can also be made for the infinite cylinder, that is, a cylinder in which diffusion takes place in the radial direction only. If the same assumptions outlined above for the sphere are used for the cylinder an equation for the average lead content of a cylinder of radius a may be derived by the same method used for the sphere, obtaining

$$\frac{\overline{C}}{N} = 4 \sum_{n=1}^{\infty} \frac{\lambda(e^{\lambda t - \xi_n^2 Dt/a^2} - 1)}{\xi_n^2 (\lambda - \xi_n^2 D/a^2)} \qquad (6)$$

The values ξ_n are the roots of the equation $J_0(x) = 0$, where $J_0(x)$ is the Bessel function of the first kind of zero order. The analogous solution for the infinite plate is

$$\frac{\overline{C}}{N} = \frac{8}{\pi^2}$$

$$\cdot \sum_{n=0}^{\infty} \frac{\lambda(e^{\lambda t - (2n+1)^2 \pi^2 Dt/a^2} - 1)}{(2n+1)^2 [\lambda - (2n+1)^2 \pi^2 D/a^2]} \qquad (7)$$

where a is the thickness of the plate.

Some ages resulting from diffusion losses from an infinite cylinder and an infinite plate are

given in Table 3. The calculations are for crystals having an age of 2800 m.y. If the results given in Table 3 are plotted on a diagram such as Figure 1, the points define D/a^2 curves that are indistinguishable from the one given in Figure 1 on the scale of the drawing. The only difference is that points corresponding to a particular value of D/a^2 plot at different positions along the curve for different shapes. For a given thickness or diameter and value of D, the sphere will lose more lead than the infinite cylinder, which in turn will lose more lead than the infinite plate.

These calculations show that diffusion processes occurring in one, two, or three dimensions will produce essentially the same D/a^2 curve given in Figure 1. This indicates that the relationship depicted there is not likely to depend in any critical manner on the shape of the crystal units from which diffusion losses have occurred. For example, a zircon crystal may be approximated by a section of cylinder capped by two hemispheres. A rigorous solution to diffusion losses from such a body is obtained from

TABLE 3. Ages Resulting from Loss of Lead by Diffusion from Nonspherical Crystals of Age 2800 Million Years

$D/a^{2*} \times 10^{-12}$ yr^{-1}	Infinite Cylinder		Infinite Plate	
	$\dfrac{Pb^{206}}{U^{238}}$	$\dfrac{Pb^{207}}{U^{235}}$	$\dfrac{Pb^{206}}{U^{238}}$	$\dfrac{Pb^{207}}{U^{235}}$
	m.y.	m.y.	m.y.	m.y.
6.25	2300	2545	2310	2535
12.50	2135	2435	2095	2400
25.0	1885	2260	1780	2175
50.0	1525	1985	1340	1785
100.0	1090	1550	835	1295
200.0	635	950	...	...

* Note that a is the radius of the cylinder and the total thickness of the plate.

a combination of the infinite cylinder and the sphere calculations.

As was mentioned earlier, it is by no means certain that the physical size and shape of a crystal are the determining geometric factor for diffusion loss. In fact, some evidence exists to the contrary. *Nicolaysen, de Villiers, Burger, and Strelow* [1958] have reported uranium-lead ages for three different size fractions of zircon from Ottensville granite where mineral age studies indicate an age of about 2000 m.y. The U^{238}-Pb^{206} ages varied with grain size as follows:

40–100 mesh (average mesh opening 0.023 cm): 816 m.y.
140–170 mesh (average mesh opening 0.010 cm): 689 m.y.
170–325 mesh (average mesh opening 0.0066 cm): 642 m.y.

According to the authors, there is no evidence of any recent geologic process that might have caused the loss of lead. These results do not preclude the possibility that the effective radius for diffusion is of the order of the grain size, but they strongly suggest that the effective radius is not the physical radius.

Activation energy for diffusion. In several cases there is evidence to indicate that zircon has not lost appreciable lead during conditions of metamorphism. A good example occurs in the Baltimore gneiss at Baltimore, Maryland. Zircons from two samples of gneiss have isotopic lead ages which indicate that the minerals have lost only 10 to 20 per cent of their lead over the past 1150 m.y., the presumed age of the minerals [*Tilton, Wetherill, Davis, and Hopson*, 1958]. Such a small loss of lead is found although the gneiss occurs in the Piedmont, where metamorphism was presumably intense as a result of processes associated with the formation of the Appalachians. The biotite age of 300 m.y. does in fact appear to be related to such processes.

The five Finnish zircons in Table 1 have discordant age patterns that are better described by diffusion losses that took place over the entire time the minerals have existed, even though biotite samples taken from the same rocks as the zircons give ages of 1800 m.y. (Wetherill, Kouvo, Tilton, and Gast, paper to be submitted to the *Journal of Geology*). These observations indicate that the diffusion rates may be relatively insensitive to temperature; that is, the activation energy for diffusion is low. If it is assumed that the Finnish zircons were subjected to temperatures of 400° to 600°C for 10 to 100 m.y. 1800 m.y. ago and have been at a temperature of approximately 50°C for most of the remainder of the time, calculations using the relation

$$D(T) = D_0 \exp(-Q/RT)$$

permit an estimation of the activation energy for diffusion, Q. The samples from Heinävaara and Sotkuma show the most evidence of possible loss of lead at 1800 m.y., but even these samples appear to have lost no more than 30 per cent of their lead at that time. Such a loss at 400° to 600°C coupled with additional required losses at 50°C give values of 3 to 10 kcal/mol for the activation energy (in the temperature range 50° to 600°C). The fact that zircons from shield areas like southern Ontario appear to have lost about 10 per cent of their lead by diffusion again argues for a low value of the activation energy, since there is no evidence for metamorphism in this area in the last 1000 m.y.

Activation energies observed for volume diffusion in metals vary from 20 to 80 kcal/mol and average about 30 kcal/mol [*Jost*, 1952, pp. 234–237]. Limited data for diffusion in ionic crystals indicate that activation energies are somewhat lower than for metals [*Jost*, 1952, p. 199; *Barrer*, 1951, p. 274]. Values of 35 to 85 kcal/mol are quoted in the literature for the diffusion of argon from micas [*Gerling and*

Morozova, 1957; *Amirkanov, Brandt, Bartnit-skii, Gasanov, and Gurvich*, 1959; *Fechtig, Gent-ner, and Zähringer*, 1960]. By comparison, the estimated activation energy for lead diffusion is surprisingly low, but qualitative considerations indicate the possibility of the lead value.

Diffusion in a lattice is generally considered to depend on the presence of vacancies. Then the term Q in the equation

$$D(T) = D_0 \exp(-Q/RT)$$

can be composed of two parts, E and U [*Jost*, 1952, p. 135]. E is the energy required to form 1 mol of vacancies in the lattice, and U is the energy barrier to be surmounted by a diffusing particle in moving from the vicinity of a vacancy. Since crystals are imperfect, they contain some vacancies as a result of impurities and imperfections. These are temperature-independent vacancies. Thus a plot of $\log D$ against $1/T$ may show a break in slope (knee) at a temperature that depends on the number of temperature-independent vacancies present. Ideally (for a simple, one-component lattice), the high-temperature branch is described by

$$D(T) = D_0 \exp(-[U + E]/RT)$$

and the lower branch by

$$D'(T) = D_0' \exp(-U/RT)$$

Examples fitting this hypothesis are the diffusion of sodium in NaCl and NaBr [*Mapother, Crooks, and Maurer*, 1950]. *Fechtig, Gentner, and Zähringer* [1960] show examples for the diffusion of argon in feldspar, anorthite, augite, and margarite, where knees in the curves occur at temperatures of 150° to 350°C. *Amirkanov, Brandt, Bartnitskii, Gasanov, and Gurvich* [1959] find an increase in the temperature dependence of D for argon diffusion from a muscovite at 600°C.

The effect of temperature-independent vacancies in crystals has also been illustrated from conductivity data. Since summaries of these data are given by *Jost* [1952, chapter 4] and by *Barrer* [1951, chapter 6], only a brief discussion will be presented here. Of particular interest are the experiments of *Koch and Wagner* [1937], who have shown that the addition of 0.1 per cent of CdCl$_2$ to AgCl decreases the temperature dependence of conductivity compared with that

of pure AgCl below a temperature of 300°C. They also show that, for a plot of $\log \sigma$ against $1/T$, the curve for AgBr plus 0.5 per cent PbBr$_2$ lies above that for AgBr plus 0.3 per cent PbBr$_2$ in the low-temperature region. The change to lower dependence on temperature occurs at approximately 300°C for 0.5 per cent PbBr$_2$ and 280°C for 0.3 per cent PbBr$_2$. This is expected, since 0.5 per cent PbBr$_2$ would create more vacancies in the lattice than 0.3 per cent PbBr$_2$ if electrical neutrality is to be maintained. At high temperatures the energy barrier for ion migration is found to be 25 kcal/mol for AgCl and 20 kcal/mol for AgBr. The corresponding values at low temperatures are 6.5 kcal/mol for AgCl plus CdCl$_2$ and 8.2 kcal/mol for AgBr plus PbBr$_2$. These experiments clearly demonstrate the importance of temperature-independent vacancies on the mobility of ions in a crystal lattice.

The activation energy for lead diffusion in zircon was estimated over a temperature range of 50° to 600°C. It is possible that temperature-independence vacancies control diffusion in this range, hence the low value of activation energy. Radiation damage would be expected to contribute to the supply of temperature-independent vacancies. However, if it controlled the supply, the diffusion coefficient would increase as the amount of bombardment increased. It seems unlikely that the relationships shown in Figures 1 to 4 would result from such a process.

Solubility is still another factor that may lower the value of the activation energy. For diffusion in lead, it is observed that silver and gold, which have low solubilities, diffuse with lower activation energies (and with faster rates at low temperatures) than metals such as bismuth and tin, which have greater solubilities. These relationships are illustrated by *Barrer* [1951, p. 288]. A related fact is that activation energies for diffusion in metals are higher for self-diffusion than for foreign metals in all the examples given by *Jost* [1952, pp. 234–237]. These data indicate that species with low stabilities in a lattice diffuse more readily than those with high stabilities. Lead has a low solubility in zircon, this being one of the principal reasons why zircon has been chosen over other minerals for determining ages by the lead methods.

TABLE 4. Comparison of Zircon and Biotite Ages* from Two Rocks

Rock	Mineral	Age, million years					
		$\dfrac{Pb^{206}}{U^{238}}$	$\dfrac{Pb^{207}}{U^{235}}$	$\dfrac{Pb^{207}}{Pb^{206}}$	$\dfrac{Pb^{208}}{Th^{232}}$	$\dfrac{Rb}{Sr}$	$\dfrac{K}{A}$
Baltimore gneiss	Zircon	1040	1070	1120	940		
	Biotite					305	340
Pikes Peak granite	Zircon	624	707	980	313		
	Biotite					1020	980

$* \ \lambda(Rb^{87}) = 1.39 \times 10^{-11} \ yr^{-1}$

$\lambda_{\epsilon}(K^{40}) = 0.583 \times 10^{-10} \ yr^{-1}$

$\lambda_{\beta}(K^{40}) = 4.72 \times 10^{-10} \ yr^{-1}$

It is possible that the lead ages could be reconciled with an activation energy of zero. In this case recoil and/or collision phenomena might be considered as the cause for the loss of lead. *Russell and Ahrens* [1957] proposed several mechanisms whereby the loss of greater proportions of Pb^{207} than Pb^{206} might be explained in terms of differences in recoil energies of various intermediate daughter products in the two uranium decay series. It is difficult to understand how such processes could produce losses of radiogenic lead of 10 to 50 per cent as required for most of the minerals in Table 1. Alpha decay imparts energies of the order of 100,000 electron volts to the parent recoil atom. The range-energy relationship for fission fragments reported by *Bøggild* [1941] shows that such an atom would have a range of $\sim 4 \times 10^{-6}$ cm in a mineral. After emission of 8 α particles, the average U^{238} nucleus would have migrated about 1×10^{-5} cm. If zircon crystals are considered to be spheres of radius 4×10^{-3} cm, corresponding to a size that would pass a 200-mesh sieve, it is found that only a few tenths of a per cent of lead can escape from the crystals by recoil.

If recoil energy is used only to transport the lead atoms to grain boundaries in order to permit escape of a greater proportion, the loss will be controlled by grain boundary diffusion. This in itself should be capable of explaining the discordant age patterns, since diffusion along grain boundaries would be essentially two-dimensional and ought to be somewhat analogous to the case of the infinite cylinder discussed earlier. Such a process would then give the same shape of 'discordia' curves derived for volume diffusion from the sphere.

Another mechanism having zero activation energy is to assume that the probability that a lead atom will be ejected from a crystal is proportional to the total bombardment in the crystal after the atom is formed. It does not seem possible to remove any substantial amount of lead by this means, and calculations show that, if such a process did control lead loss, the loss of Pb^{207} would be weighted too heavily to fit the discordant age patterns shown in Figures 1 to 4. The discordia curves appear to have unique characteristics that cannot be reproduced by simple models involving solely collision or recoil phenomena.

Diffusion may explain a difficulty concerning the behavior of the ages given by coexisting mica and zircon. The problem is illustrated by two examples in Table 4, taken from the work of *Tilton, Davis, Wetherill, and Aldrich* [1957] and *Tilton, Wetherill, Davis, and Hopson* [1958]. At Pikes Peak zircon has lost lead under conditions that do not appear to have affected the argon or strontium contents of biotite appreciably. In the Baltimore gneiss, biotite either crystallized or lost strontium and argon under conditions that did not seriously affect the lead in the zircon. In unmetamorphosed rocks, the uranium-lead ages of zircon are equal to or less than the rubidium-strontium and potassium-argon ages of micas. Here diffusion rates might be controlled mainly by the number of temperature-independent vacancies in the two mineral lattices. The results for the Baltimore gneiss can be explained by assuming that the temperature dependence in the interval 50° to 600°C is greater for the diffusion of argon and strontium in the biotite than for lead in the zircon. The curves for log D as a function of $1/T$ given by *Fechtig, Gentner, and Zähringer* [1960] show breaks in slope at temperatures of 150° to 350°C. Above these temperatures D increases sharply with T, presumably because temperature-dependent vacancies begin to predominate over temperature-independent vacancies in the lattice. D has a value of $\sim 10^{-15}$ cm²/sec at 500°C for the four minerals studied. In 10^7 years, argon atoms would migrate ~ 0.8 cm at this temperature. If the diffusion of lead in zircon is controlled by temperature-independent vacancies up to some temperature above 500° to 600°C, it should be possible to remove argon from biotite without affecting the lead content of the zircon to any great extent. Since there are but a few cases in the literature for which the rubidium-strontium age differs very greatly from the potassium-argon age of a mica, it appears that strontium diffusion is somewhat similar to argon diffusion in micas.

CONCLUSIONS

It has been shown that volume diffusion offers a promising explanation to several problems that have arisen in geochronology. It explains regularities that exist in the discordant uranium-lead ages of minerals from several continents. It does away with the need for episodic loss of lead in several cases where no evidence exists for metamorphic events at the time required. Moreover, it appears that volume diffusion will play a prominent part in resolving problems that have arisen from comparisons of uranium-lead ages of zircon with rubidium-strontium and potassium-argon ages of coexisting micas in granites and gneisses.

A fortunate aspect of the diffusion hypothesis is that it is amenable to rather rigorous testing even in the absence of laboratory determinations of the diffusion rates. Once the age of a group of rocks can be established by concordant uranium-lead ages or by other methods, the pattern of the discordant lead ages that will result from diffusion is determined. The importance of diffusion can then be judged from the number of cases for which discordant ages will fit the theoretical patterns. It would be particularly convincing to show that suites of minerals of two different ages from the same area fit two separate discordia curves for loss of lead by diffusion. It would be difficult to invoke an episodic-loss mechanism in such a case.

Acknowledgments. I am grateful to my colleagues, S. P. Clark, Jr., G. L. Davis, T. C. Hoering, and G. W. Wetherill, for helpful discussions and criticisms concerning various phases of this work.

REFERENCES

Ahrens, L. H., The convergent lead ages of the oldest monazites and uraninites (Rhodesia, Manitoba, Madagascar, Transvaal), *Geochim. et Cosmochim. Acta, 7,* 294–300, 1955a.

Ahrens, L. H., Implications of the Rhodesia age pattern, *Geochim. et Cosmochim. Acta, 8,* 1–15, 1955b.

Aldrich, L. T., and G. W. Wetherill, Geochronology by radioactive decay, *Ann. Rev. Nuclear Sci., 8,* 257–298, 1958.

Amirkanov, K. I., S. B. Brandt, E. I. Bartnitskii, S. A. Gasanov, and V. S. Gurvich, On the mechanism of losses of radiogenic argon in micas, *Izvest. Akad. Nauk SSSR, Ser. Geol.,* 1959.

Barrer, Richard M., *Diffusion in and through Solids,* University Press, Cambridge, 1951.

Bøggild, J. K., Range-velocity relation for fission fragments in helium, *Phys. Rev., 60,* 827–830, 1941.

Eckelmann, W. R., and J. L. Kulp, Uranium-lead method of age determination, 1, Lake Athabaska problem, *Bull. Geol. Soc. Am., 67,* 35–54, 1956.

Fechtig, H., W. Gentner, and J. Zähringer, Diffusionsverluste von Argon in Mineralien und ihre Auswirkung auf die Kalium-argon Altersbestimmung, *Geochim. et Cosmochim. Acta, 19,* 70–79, 1960.

Gast, P. W., J. L. Kulp, and L. E. Long, Absolute age of early Precambrian rocks in the Bighorn Basin of Wyoming, Montana and southeastern Manitoba, *Trans. Am. Geophys. Union, 39,* 322–334, 1958.

Gerling, E. K., and I. M. Morozova, Determination of the activation energy of argon isolation from micas, *Geokhimiya,* 304–311, 1957.

Greenhalgh, D., and P. M. Jeffery, A contribution to the Precambrian chronology of Australia, *Geochim. et Cosmochim. Acta, 16,* 39–57, 1959.

Holmes, A. H., The oldest dated minerals of the Rhodesian shield, *Nature, 173,* 612, 1954.

Holmes, A. H., Dating the Precambrian of Peninsular India and Ceylon, *Proc. Geol. Assoc. Can., 7,* 81–106, 1955.

Holmes, A. H., and Lucien Cahen, *Géochronologie africaine, Mém. acad. roy. sci. coloniales,* 169 pp., 1957.

Jost, W., *Diffusion in Solids, Liquids, Gases,* Academic Press, New York, 1952.

Koch, E., and C. Wagner, Der Mechanisimus der Ionenleitung in festen Salzen auf Grund von Fehlordnungsvorstellungen, I. Z., *physik. Chem., B, 38,* 295–325, 1937.

Kouvo, O., Radioactive age of some Finnish Precambrian minerals, *Geol. Comm. Finland, Bull. 182,* 70 pp., 1958.

Mapother, D., H. N. Crooks, and Robert Maurer, Self diffusion of sodium in sodium chloride and sodium bromide, *J. Chem. Phys., 18,* 1231–1236, 1950.

Nicolaysen, L. O., Solid diffusion in radioactive minerals and the measurement of absolute age, *Geochim. et Cosmochim. Acta, 11,* 41–59, 1957.

Nicolaysen, L. O., J. W. L. de Villiers, A. J. Burger, and F. W. E. Strelow, New measurements relating to the absolute age of the Transvaal system and of the Bushveld igneous complex, *Trans. Geol. Soc. S. Africa, 61,* 137–163, 1958.

Nier, A. O., The isotopic constitution of radiogenic leads and the measurement of geological time, II, *Phys. Rev., 55,* 153–163, 1939.

Russell, R. D., and L. H. Ahrens, Additional regularities among discordant lead-uranium ages, *Geochim. et Cosmochim. Acta, 11,* 213–218, 1957.

Starik, I. E., The role of secondary processes in age determination by radiometric methods, *Geokhimiya,* 18–29, 1956.

Tilton, G. R., G. L. Davis, G. W. Wetherill, and L. T. Aldrich, Isotopic ages of zircon from granites and pegmatites, *Trans. Am. Geophys. Union, 38,* 360–371, 1957.

Tilton, G. R., G. W. Wetherill, G. L. Davis, and C. A. Hopson, Ages of minerals from the Baltimore gneiss near Baltimore, Maryland, *Bull. Geol. Soc. Am., 69,* 1469–1474, 1958.

Vinogradov, A. P., Comparison of data on the age of rocks obtained by different methods and geological conclusions, *Geokhimiya,* 3–17, 1956.

Wasserburg, G. J., Argon[40]: potassium[40] dating, in *Nuclear Geology,* edited by H. Faul, John Wiley & Sons, New York, pp. 341–349, 1954.

Wetherill, G. W., An interpretation of the Rhodesia and Witwatersrand age patterns, *Geochim. et Cosmochim. Acta, 9,* 290–292, 1956a.

Wetherill, G. W., Discordant uranium-lead ages, *Trans. Am. Geophys. Union, 37,* 320–326, 1956b.

Zhirov, K. K., Data summarized in E. K. Gerling, The influence of metamorphism on the results of age determination according to lead, *Geokhimiya,* 288–295, 1958.

(Manuscript received June 17, 1960.)

Reprinted from *J. Geol.*, **71**, 721–729, 747–758 (1963)

22

URANIUM-LEAD ISOTOPIC VARIATIONS IN ZIRCONS: A CASE STUDY[1]

LEON T. SILVER[2] AND SARAH DEUTSCH[3]

ABSTRACT

Zircons in a single 250-pound block of Precambrian Johnny Lyon granodiorite from the Dragoon Quadrangle in Cochise County, Arizona, have been concentrated with special attention to yield and nature of impurities. Morphology, zoning, color, inclusions, size distribution, radioactivity, refractive indexes, cell dimensions, and other properties have been compared with the isotopic properties in the U-Pb system (and to a less precise degree in the Th-Pb system). It has been observed: (1) Uranothorite impurities, while less than 1 per cent in abundance, contribute much more than 50 per cent of the activity in conventionally prepared concentrates. The presence of uranothorite drastically affects the apparent ages in the zirons. (2) The uranothorite can be satisfactorily removed by an appropriate acid-washing procedure. (3) The uranothorite-free zircon concentrates are not homogeneous and show systematic variations in radioactivity and various isotopic properties as a function of average crystal size. (4) Individual zircons show internal variation in radioactivity and may differ in specific activity from other individuals by as much as an order of magnitude. (5) It is possible to strip outer layers experimentally from an aggregate of zircons to determine variations in composite internal isotopic properties. (6) The family of uranium-lead systems distinguished in the inhomogeneous zircon suite may be utilized to establish patterns of isotopic ratio variations that yield much more useful geochronological information than any single system. (7) All systems in this rock appear to have formed 1,655 million years ago and to have been profoundly disturbed by an event 90 million years ago. There is no evidence of any type of disturbance other than this simple episodic pattern. (8) While the mechanism of disturbance is not directly established, it is evident that radioactivity and radiation damage strongly influence susceptibility of the systems to disturbance. (9) It is possible to offer reasonable explanations for some of the puzzling discrepancies between uranium-lead and thorium-lead ages determined on a single mineral concentrate. (10) Recognition of the existence of *families of uranium-lead systems* among the variable members of a single mineral species, or in associated mineral species, in a single typical granitic rock provides a powerful tool for investigation of the processes and conditions that have influenced the age-dating systems. (11) The systematic variations in U and Th provide interesting information on the role of some trace elements and accessory minerals in the crystallization history of the rock. (12) The Johnny Lyon granodiorite is the oldest igneous rock dated in Arizona thus far and places a minimum age of $1,655 \pm 20$ million years on the orogeny called Mazatzal Revolution.

INTRODUCTION

In geochronological investigations of rock systems, methods involving isotope chemistry and mass spectrometry have evolved as precise and useful techniques for analysis of a number of natural radioactivity relationships. A vast body of data on parent-daughter ratios in the systems K^{40}/Ar^{40}, Rb^{87}/Sr^{87}, U^{238}/Pb^{206}, U^{235}/Pb^{207} and Th^{232}/Pb^{208} has been carefully determined by these techniques yielding geochronological information of far-reaching importance to studies of the earth. At the same time, it has become apparent also that numerous factors in the

geological setting and history of the analyzed minerals bear directly on the age interpretations made from the carefully measured isotopic relationships. Apparent ages calculated from several minerals in the same host rock commonly have been in significant disagreement. It is to be hoped that some of these contradictory "ages" may be reconciled, in part, by combining careful mineralogical and petrological studies of the analyzed phases with the isotopic measurements to achieve a better understanding of the geological factors. Few such attempts have been made.

Among the several naturally radioactive isotopes used in geochronology, the constant chemical coupling of U^{238} and U^{235} with their chemically similar decay products Pb^{206} and Pb^{207} introduces the fewest assumptions required for a comparison of apparent ages. It is fortunate too that Th^{232}, a similar actinide element commonly associated with uranium,

[1] Contribution No. 1070, Division of Geological Sciences, California Institute of Technology, Pasadena, California. Manuscript received February 21, 1962; revised May 25, 1962.

[2] Division of Geological Sciences, California Institute of Technology, Pasadena, California.

[3] Present address: University of Brussels, Belgium.

721

191

also decays to a stable lead nuclide, Pb[208]. Thus it is often possible to compare ages derived from three systems in the same analyses.

Larsen *et al.* (1952) suggested that one of the most attractive phases in common rock systems for studying these three isotopic decay relations was zircon. They proposed the so-called lead-α method, or Larsen method, a non-isotopic method of age determination. Tilton, Patterson, *et al.* (1955) and, in particular, Tilton *et al.* (1957, and numerous

FIG. 1.—Index map of Arizona showing location of Johnny Lyon Hills area in Cochise County.

other publications) demonstrated that precise and significant isotopic measurements could be made on the zircons of many different rocks. Tilton's work more than any other to date has defined the applicability and problems of U-Th-Pb systems in zircons in age-determination studies. The present report is an exploration of some of these problems.

The uranium-lead systems and, with less precision, the thorium-lead systems have been examined in zircon concentrates from a single block of granitic rock. Variations in isotopic properties have been compared with concentrate purity and homogeneity, radio-

activity, size, zoning, and other characteristics. A number of important regularities have been observed. These lead to the conclusion that within these $1\frac{1}{2}$ cubic feet of granite, there exist a large number of uranium-thorium-lead systems which, when studied as a family, can elucidate some types of apparent age contradictions and, indeed, provide some significant additional information on the history of the rock. They also raise some questions about accepted assumptions in age interpretations.

GEOLOGIC SETTING OF THE ROCK

The rock sample was obtained from a large Precambrian granodiorite pluton in the Johnny Lyon Hills in the Dragoon Quadrangle, Cochise County, in southeastern Arizona (figs. 1, 2). Detailed geologic studies have been made of the pluton and its setting by the United States Geological Survey. Preliminary accounts are available (Silver, 1955; Cooper, 1959), and a full report will be published soon (Cooper and Silver, forthcoming). The regional setting is shown on the geologic map of Cochise County (Arizona Bureau of Mines, 1959). The pluton is a post-tectonic intrusive in its Precambrian setting but it has been strongly modified locally by younger deformational, intrusive, and alteration processes, many of which are clearly Mesozoic or Cenozoic. This particular sample (L-312) was selected to be as free as possible from detectable metamorphic effects, as well as from the extensive weathering effects characteristic of this arid region.

The rock is a medium- to coarse-grained, somewhat porphyritic, hornblende-biotite granodiorite. In hand specimen, coarse white plagioclase is set in gray quartz, scattered pink potash feldspar, dark green platy biotite, greenish-black prismatic hornblende and, more rarely, euhedral brown sphene. A few very thin epidote-quartz veinlets, a fraction of a millimeter in thickness, transect the block.

In thin section, the rock is hypidiomorphic-inequigranular with a slightly seriate porphyritic texture. Tabular and zoned

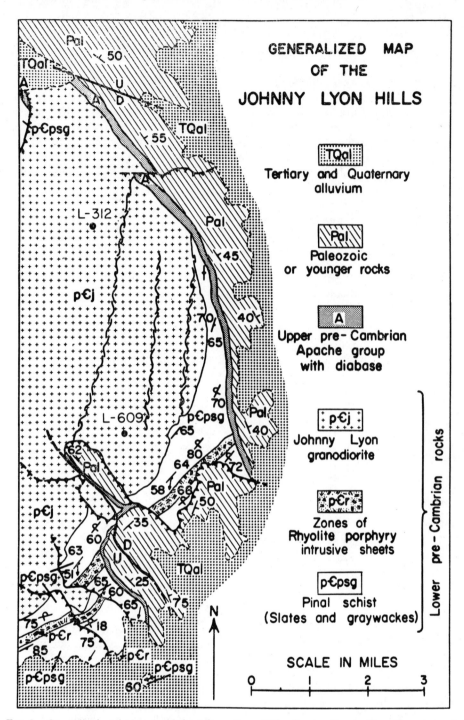

FIG. 2.—Generalized geologic map of the Johnny Lyon Hills, showing the stratigraphic and structural relationships of Johnny Lyon granodiorite to both older and younger rocks. Samples L-312 and L-609 were collected to be as free as possible of effects of the three major shear zones and other alteration features in the intrusive.

plagioclase crystals vary in composition from An_{35} at the cores to An_{20} at the rims. Microcline-microperthite is anhedral, texturally late, and bordered by myrmekite fringes on the plagioclase. Quartz is scattered in anhedral aggregates. The biotite and hornblende are subhedral to euhedral. Accessary minerals include magnetite, apatite, sphene, zircon, allanite, and uranothorite in sequence of decreasing abundance. Chemical, normative, and modal analyses are given in table 1. As in every other speci-

TABLE 1A

JOHNNY LYON GRANODIORITE*

CHEMICAL ANALYSIS

	L-312† (Weight Per Cent)	Average of 80 Granodiorites‡ (Weight Per Cent)
SiO_2	68.89	66.13
TiO_2	0.45	0.51
Al_2O_3	15.02	15.50
Fe_2O_3	1.40	1.62
FeO	1.89	2.70
MnO	0.08	0.07
MgO	1.43	1.73
CaO	3.44	3.70
Na_2O	3.84	3.55
K_2O	3.14	3.17
H_2O comb	0.88	0.89
P_2O_5	0.17	0.17
Others	n.d.	0.07
Total	100.68	99.90

* (L-312) NE 1/4, SW 1/4, Sec. 20, T. 14 S., R. 21 E.

† W. J. Blake, analyst.

‡ Johannsen, Vol. II (1932), 344.

men of this pluton we have examined, the major minerals show some alteration to sericite, chlorite, and epidote group minerals. Although not shown in the modal analysis, these minerals, principally replacing plagioclase and biotite, constitute about 3–5 per cent of the rock. We have no petrographic criteria for labeling these minerals deuteric or hydrothermal. We have simply accepted them as rather typical attributes of a rock of this composition.

MINERAL SEPARATIONS

The original reservoir of samples collected in the field weighed approximately 250 pounds. Almost all of the zircons used in this study came from a representative 60-pound aliquot of the original sample. It was necessary, however, to process some additional

TABLE 1B

JOHNNY LYON GRANODIORITE

MODAL ANALYSIS

(Point Count)

	L-312* (Weight Per Cent)
Quartz	25.84
Plagioclase	45.38
Myrmekite	1.93
Microcline-perthite	14.20
Biotite	6.44
Hornblende	4.06
Magnetite	1.34
Sphene	0.56
Other accessories	0.25
Total	100.00
Normative Analysis Salic:	
Quartz	25.02
Orthoclase	18.35
Albite	32.49
Anorthite	14.46
Total	90.32
Femic:	
Diopside	2.22
Hypersthene	3.99
Magnetite	2.09
Ilmenite	0.91
Total	9.21

C.I.P.W. Class I, 4, 3, 4

* Based upon 3,734 points on 13 sq. cm. of thin sections.

rock to replenish supplies of certain size fractions of zircon. Careful comparisons of physical and isotopic chemical characteristics of these new supplies with the old showed no measurable differences between them.

The principal aliquot was crushed in a

Dodge-type jaw crusher and then fed to a Braun UA-type disk pulverizer to produce a product of which more than 95 per cent passed through a 30-mesh sieve. The pulverizer product was carefully sized on a sequence of screens of 30, 50, 100, 200, 300 and 400 mesh.

Each of the size fractions was then processed through heavy liquids of density 2.96 (s-tetrabromoethane) and 3.33 (diiodomethane) at rates controlled to give maximum heavy mineral yields. The heavy min-

and minimum yield efficiency are plotted. *Curve I* shows the weight per cent of total crushed rock in each size fraction. *Curve II* shows the fraction (weight per cent) of the total zircon yield obtained from each size fraction of rock. *Curve III* shows the minimum yield efficiency for the recovery of zircon from each size fraction assuming all of the zirconium in the size fraction to be in the form of zircon.

From a comparison of *Curves I* and *II*, it is apparent that a large fraction of the zircon

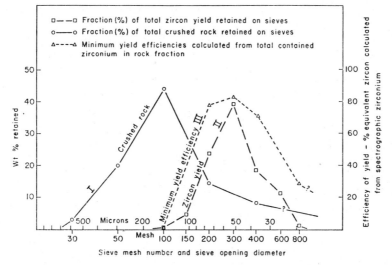

FIG. 3.—Curves relating zircon yields and yield efficiencies to over-all size distribution and sized fractions of crushed Johnny Lyon granodiorite sample L-312.

eral suites were then processed on a Frantz "Isodynamic" electromagnetic separator with emphasis given to maximum yields of zircons.

To measure the efficiency of the zircon extractions, the zirconium content of each of the size fractions of the rock was determined by emission spectroscopy prior to the mineral separations. A maximum possible value of zircon, liberated and unliberated, present in the size fraction was calculated, recognizing that some zirconium is to be found in other phases than zircon.

In figure 3, three curves relating crushed particle size distribution, zircon recovery,

in the rock tends to be liberated before the rock is completely reduced to particles of the average zircon grain size. This is fortunate, inasmuch as the zircon generally is freed in its characteristic original grain shape, and this tendency facilitates the recovery of a more representative sample of the zircon in the rock. Examination of *Curve III*, however, indicates that while extraction efficiency for liberated zircons is quite high (as in the retained 200-, 300-, and 400-mesh size fractions), a considerable part of the zirconium is tied up in material coarser than 100 mesh. All of this coarse material was reground to pass 100 mesh. It was then processed to recover the zircons as before. Although this involved reprocessing nearly 70 per cent of the total material, the over-all yield was increased by less than 10 per cent. It is concluded that the zircon size fractions used for this study (excepting the passed 400-mesh material) are reasonably representative of each of their size classes in the original rock sample.

The size classes of zircon to which all subsequent discussions will refer are given in table 2.

TABLE 2

Fraction Retained on Sieve Mesh No.	Prism Diameter Range (μ)	Average Prism (Diameter (μ))
R200..............	75–150	90
R300..............	54–75	64
R400..............	37–54	47
Passed 400........	20–37	33

TABLE 3

Mineral	No. of Grains
Zircon................	2,447
Uranothorite..........	17
Quartz and feldspar....	5
Apatite..............	2
Sphene and clinozoisite (?)......	2
	2,473

The zircons coarser than 150 μ are too few for analysis. The zircons finer than 20 μ prism diameter probably are considerably more abundant, but problems of systematic recovery prevented their use.

PHYSICAL PROPERTIES OF THE ZIRCON CONCENTRATES

Purity.—Each of the zircon concentrates obtained in this study was at least 99 per cent pure. The purity was determined by making permanent grain mounts in a high index mounting medium (Aerochlor no. 4665, $n = 1.65$) of several thousand grains. Grain counts of 2,000 or more grains were then taken. The retained 200-mesh zircon concentrate contained the greatest percentage of non-zircon grains. Table 3 gives the results of a count of nearly 2,500 grains in this size fraction.

The general spectrum of mineral impurities other than zircon was similar for each of the other concentrates, but the relative abundance of uranothorite declined in the finer sizes.

For several different purposes, 100-mg. aliquots of the R200-, R300-, and R400-mesh fractions were hand-picked to observable 99.99+ per cent purity. The removed mineral grains were also mounted and studied. They generally confirmed the mineralogy already determined.

For most of the analytical work, aliquots were not hand-picked. It was determined that a 1-hour wash in hot concentrated nitric acid would effectively remove all uranothorite and apatite. This was employed as a standard preparation procedure and will be discussed in a later section. The remaining impurities were minerals whose concentrations of uranium, thorium, and lead were such that their low abundance in the concentrates could not affect the analytical data within assigned limits of error.

Zircons.—The individual zircon grains are typically euhedral, pale lavender to lavender-brown, visibly zoned, transparent except for minute inclusions of foreign crystals and growth cavities.

Individual grains with excellent euhedral form comprise more than 50 per cent of the over-all zircon population (pls. 1, *e*, *f*, 2, *e*). It is readily apparent that most of the other grains were broken during the crushing pre-

---

PLATE 1

a–d, R200 fraction zircons displaying characteristic habits, zoning, and inclusions. Linear magnification ×220. *e*, R200 fraction zircons. ×100. *f*, R300 fraction zircons. ×100. *g*, Autoradiograph of R200 fraction of ground zircons, displaying α-particle tracks, zoning, and shadow zone. ×300. *h*, Same as 1, *g*. ×800.

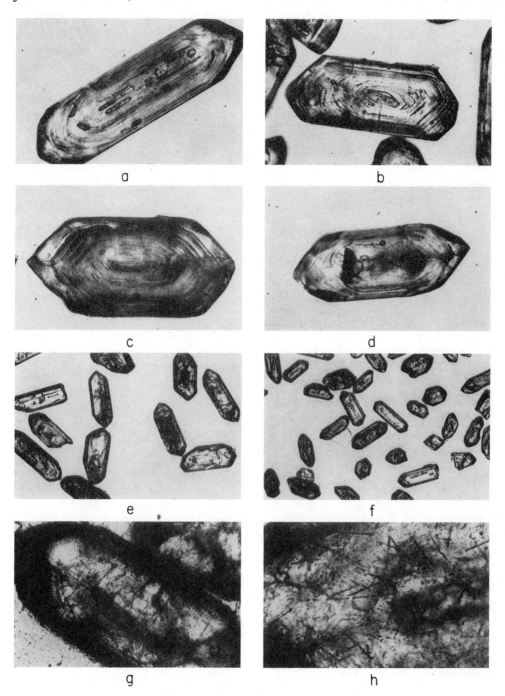

R200 fraction zircons displaying characteristic habits, zoning, and inclusions

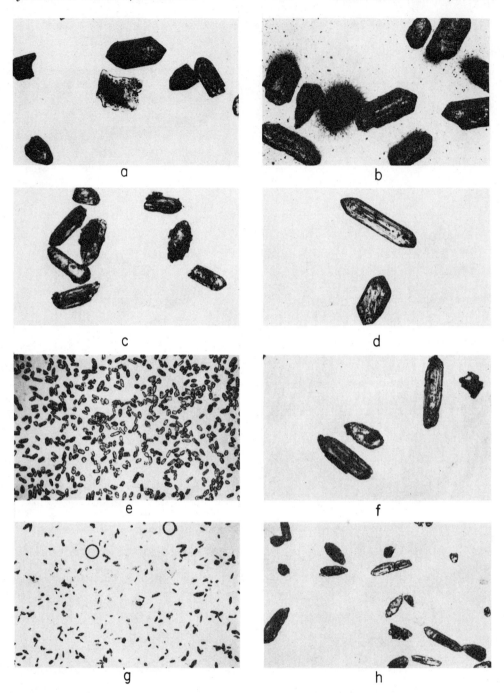

Photomicrographs showing enlargements of different Zircons

paratory to mineral separation. In general, the percentage of complete grains increases in the finer size fractions to more than 80 per cent of the grains in the passed 400-mesh size. Typically, the crystal forms present include two prominent prisms, a ditetragonal dipyramid and a tetragonal dipyramid (see pl. 1, *a–f*). The basal pinacoid and another dipyramid are more uncommon. The average length-to-width ratio for each fraction is in the range 2.3–2.5.

The color appears to vary in intensity among the grains but does not change appreciably in hue. Individually, lavender with a slight brown tint is predominant; in the aggregate the concentrates generally have a distinct brownish cast. Within a few individual grains there appears to be a slight color zoning with deeper intensities toward the center. This may sometimes be more apparent than real because the abundance of inclusions is usually greatest in the centers, and this may affect the light absorption in this region.

The zircons are distinctly zoned in most cases (see pls. 1, *a–f*, 2, *d*). The zoning is physically apparent under microscopic examination in the form of surfaces of optical discontinuity within the grains. From these surfaces Becke lines are generated that follow, in general, the typical forms of the external crystal faces. The Becke line behavior indicates that the higher index of refraction is more commonly in the crystal interior. In some grains, however, reversals in apparent index of refraction were noted. Some of the zones are marked by fine, dusty, opaque (?) inclusions. Although the external crystal forms are usually represented among the

zone forms, it is not uncommon that the relative development of the forms is changed drastically from within to without.

Only rarely are the individual zircon grains free from some type of inclusion or other imperfection. Inclusions fall into the following categories:

1. Stubby to slender prisms, usually less than 10 μ in diameter, and with length-to-width ratios of 2:12. They are distinctly lower in index than the host zircon with a very low birefringence invariably masked by that of the zircon. These characteristics plus terminations that are characteristically "rounded" basal pinacoids suggest apatite. The prisms may be aligned parallel to the zones or, equally as common, randomly oriented.

2. Equant, anhedral forms with very strong negative relief against the zircon. Indexes are apparently much lower than the grain-mount medium containing the zircon. Low or no birefringence. These may be feldspar or structural cavities. No evidence of a fluid phase was observed.

3. High index, euhedral grains, 1–5 μ in diameter, with indexes very close to that of the host. Forms and length-to-width ratios suggest these are probably other zircon grains. This is much less common than types 1 and 2.

4. Dark-brown zircon grains, 10–40 μ in diameter, with apparently rounded outlines overgrown by typical zoned zircon. Sometimes these zircon inclusions are surrounded by a halo of radiating fractures in the host. Only one or two of these are observed in a grain mount of several thousand grains.

5. Fine, opaque (?) "powder" usually barely resolvable with an oil immersion lens. Rarely this "powder" completely clouds a zircon core.

6. Red-brown plates, 1–5 μ in diameter. These

PLATE 2

a, Fragment of uranothorite in unwashed zircon concentrate. ×100. *b*, Autoradiograph of unwashed concentrate showing dense cloud of α-tracks over a uranothorite grain. ×100. *c*, Zircon residual (R200 fraction) from which outer 55 per cent of volume has been stripped by partial fusion. ×100. *d*, Zircon crystal aliquot (R200f raction) used for stripping experiment. Compare forms with 2, *c*, and 2, *f*. ×100. *e*, Zircon concentrate (R300 fraction) showing typical form and purity. ×8. *f*, Zircon residues of same experiment as 2, *c*. Compare residual form with internal zoning and with forms in 2, *d*. *g*, Residue of aliquot of zircon concentrate in 2, *e*, from which 99.7 per cent has been dissolved. Note common prismatic form. ×8. *h*, Enlargement of several grains from residue in 2, *g*, showing persistent prismatic forms. ×100.

appear to be an iron oxide of some form. Quite uncommon.

7. Tubes (see pl. 2, *d*), spheres, ellipsoids, isotropic with very strong negative relief, some of which can be clearly observed to be cavities. Most of these are geometrically oriented relative to the host zircon structure. None have been observed to have a fluid phase in them. Rarely these may have diameters equal to one-third that of the host.

8. Black, opaque, anhedral to subhedral, magnetic inclusions 1–3 μ in diameter.

9. Fractures radiating from nuclei of type 4 or more commonly from what appear to be simply darker zonal cores. Occasionally a brown film is observed coating some of these fractures. This type of fracture is observed in much less than 1 per cent of the zircons.

Types 1, 2, and 7 are by far the commonest types of inclusions. In considering these

TABLE 4

Grain Size Fraction	a_o	c_o
R200	6.607 ± 0.002	6.009 ± 0.002
R300	$6.608 \pm .002$	$6.012 \pm .002$
R400	$6.608 \pm .002$	$6.012 \pm .002$
P400	$6.610 \pm .002$	$6.010 \pm .002$
R200 (fusion residue)*	6.604 ± 0.002	6.006 ± 0.005

* This sample was held in sodium tetraborate flux for one hour at 1100° C.

zircons as chemical systems, the inclusions cannot be ignored. They will be discussed again in the section on nuclear emulsion studies.

Index of refraction measurements were made on all size fractions of zircon. High-index liquids prepared by R. P. Cargille Laboratories after the descriptions of Meyerowitz and Larsen (1951) were used with a sodium light source. The oils were checked for stability by comparison with standard zircons of known indexes.

In all size fractions, a range of index properties was observed. This range is almost identical for all size fractions with $n_o = 1.900$–1.920 (± 0.005) and $n_e = 1.950$–1.970 (± 0.005). In general, most of the zircons fall into the range $n_o = 1.910$–1.920

and $n_e = 1.955$–1.970 with a few darker individuals showing the lowest values. Observations of indexes of zoned fragments indicated slight variations. The intervals in the refractive indexes of immersion oils did not permit a refined determination of these differences.

Comparison of these index determinations with the range of published values (n_o from 1.920 down to 1.810) for zircon indexes indicates only minor reduction of the indexes as a result of radiation damage.

X-ray powder diffraction studies were made on several of the size fractions of the zircon. A Norelco X-ray source equipped with goniometer spectrometer was employed. Using filtered Cu radiation ($K\alpha = 1.5405$Å), uniformly ground powders of zircon mixed with NaCl for an internal standard were scanned at 1/8° per minute speed. Peaks used for the determination of cell dimensions were {200}, {400}, {112}, {312}, {321}, and {331}. Peak shapes were all well defined. A very slight suggestion of asymmetry toward lower 2θ values is the only indication of peak broadening. The multiple solutions for cell dimensions agreed within the indicated limits of error in table 4. The c_o/a_o ratio is distinctly larger than those reported by Holland and Gottfried (1955, p. 294) for both Ceylon gem zircons and granitic accessory zircons. It is clear, however, that the cell dimensions are not unusually expanded by radiation damage. Within the limits of assigned error the c_o values of all unheated zircon fractions are the same. Using the curves of Holland and Gottfried (1955) or Fairbairn and Hurley (1957), the c_0 value indicates a modest radiation dosage of about 1.0×10^{15} α-particles/mg.

The heated sample of the retained 200 fraction is apparently a partly or completely annealed zircon with distinctly reduced cell dimensions, but it preserves the distinctively larger c_o/a_o. This suggests that this ratio may be an original characteristic for these zircons. The application of existing curves for c_0 versus radiation damage thus may give unduly large values for the radiation damage retained by these crystals.

An explanation of this unusual c_o/a_o ratio possibly may be found in the zirconium/hafnium ratio. X-ray fluorescence comparisons of peak intensity ratios for Zr and Hf indicate the Johnny Lyon granodiorite zircons have the lowest Zr/Hf ratio of a dozen granitic zircons examined. These are preliminary results which have not been calibrated on an absolute basis.

Uranothorite.—The most abundant mineral impurity in the zircon concentration is uranothorite (see pl. 2, *a*, *b*). Although its abundance ranges from only 0.3 to 0.7 per cent in the various size fractions, the contribution by uranothorite to the radioactivity of each size fraction is as great or greater than that of the zircon. Its physical properties are quite variable. The individual particles show no crystal faces and are bounded by conchoidal to irregular fracture surfaces. It ranges in color from yellow to orange to red to brown and black even within single grains. Generally translucent, it may contain a disseminated opaque dust. The luster varies from glossy to dull. It is optically isotropic and index of refraction measurements yielded a number of values ranging from 1.702 to 1.720 for the sodium *D* line. The specific gravity appears to range between 5 and 6 and averages 5.3.

X-ray diffraction powder photographs yielded no reflections. After heating to about 600° C. in a reducing flame of a Bunsen burner, an aggregate of about twenty grains yielded a weak diffraction pattern most readily interpreted as a mixture of huttonite and thorite reflections.

One milligram of hand-picked uranothorite grains was mixed in a matrix of 25 mg. mixed ferric and aluminum oxide, and examined by emission spectroscopy. Thorium, 1–2 per cent, uranium, 0.2 per cent, silicon about 1 per cent, and traces of zirconium and lead were the only elements observed in addition to those of the matrix. This recalculates to about 25–50 per cent Th, 5 per cent uranium, about 25 per cent silicon, and

the results confirm an approximate composition of a member of the thorite family.

It is quite possible that in addition to thorite or uranothorite, which is clearly in a highly metamict state, related and equally metamict materials, such as thorogummite and huttonite, may be present among the grains assigned to this mineral in the concentrates.

Other impurities.—Quartz and the feldspars, epidote group minerals, apatite, and sphene may also be present as impurities. Apatite is removed in the hot-acid-wash treatment along with the uranothorite. The other minerals have been examined for their concentrations of uranium, thorium, and lead. It is clear that their combined contributions to the concentrates at their abundance levels is negligible, compared to other sources of error.

URANIUM-LEAD EQUILIBRIA IN THE ZIRCONS AND URANOTHORITE

The daughter-parent atom ratios, Pb^{206}/U^{238} and Pb^{207}/U^{235}, have been calculated for all of the zircon size fractions. In addition to appearing in the tables, they are presented in a series of daughter-parent diagrams (figs. 11, 12) of the type proposed and mathematically justified by Wetherill (1956). Pb^{206}/U^{238} is plotted relative to Pb^{207}/U^{235} for each system. The graphs show the well-known "Concordia" curve which is the locus for all values of these ratios that give a Pb^{206}/U^{238} age in agreement with the Pb^{207}/U^{235} age in the same system. The ages (T_{238}, T_{235}) are calculated from the relations

$$T_{238} = \frac{1}{\lambda U^{238}} \ln\left(\frac{Pb^{206}}{U^{238}} + 1\right)$$

$$T_{235} = \frac{1}{\lambda U^{235}} \ln\left(\frac{Pb^{207}}{U^{235}} + 1\right).$$

If $T_{238} = T_{235}$ in any system, it is called a concordant system; if they are unequal it is called discordant.

Aldrich and Wetherill have stated (1958, p. 267, 268) that the two ages of a mineral system must be in agreement within 3 per cent if either age is to be accepted as the true age. This implies that analytical accuracy must meet these limits to successfully establish truly concordant natural systems. Isotope-dilution techniques are capable of this accuracy on many favorable natural systems. In general, we feel the accuracy of our results meets such requirements, and the observed departures from equilibria are genuine natural relations rather than laboratory-induced effects.

In figure 11, the data from the four zircon size fractions and the two uranothorite washes are plotted. It is possible to draw a straight line through these points in which no point deviates from the line by more than 0.4 per cent of its value. None of the points fall on the Concordia curve, however. The line generated by them intersects the Concordia curve at points corresponding to ages of about 90 and 1,655 million years.

Mass spectrometric errors do not allow us to expect a better fit to a straight line. However, because the line passes so near to the origin of the graph, it is pertinent to make the following observation. Any analytical errors in the uranium or lead concentrations will move the plotted points along lines passing through their true position and the origin of the graph. It is clear that small shifts in the reported concentration values would not move the points very far from the chord. Therefore this fit in itself is no test of analytical precision. However, profound concentration errors would be required to disturb this apparent relationship seriously, or to move points from some distinctly different position into this chord.

Errors of position on the Concordia plot may be introduced in the determination of lead isotopic composition. These may be the result of mass spectrometric errors, or incorrect compensation of Pb^{206} and Pb^{207} introduced into the system with Pb^{204} by non-radioactive processes. Measurement errors in observed Pb^{206}/Pb^{207} are about 0.3 per cent in this data. This in itself is negligible. Further, these errors, if random, would have no systematic effect on a family of points. The method of correction for non-radiogenic Pb^{206} and Pb^{207} in these systems has already been described. In most cases, the samples have had a sufficiently high Pb^{206}/Pb^{204} ratio so that almost any normal error in the selection of an appropriate common lead would have an effect less than the mass spectrometry error. This is apparent in the very small range of variation observed in the Pb^{207r}/Pb^{206r} ratio for all zircon samples. The uranothorite washes have the lowest Pb^{206}/Pb^{204} values and these must be more suspect. The only argument offered here is that considering the two drastically different values ($Pb^{206}/Pb^{204} = 141.4$ and 65.9) and the size of the necessary corrections, their agreement is excellent. Finally, it must be suggested that a family of U-Pb systems that is not initially concordant and that is disturbed by handling generally would not provide a straight line distribution.

Figure 12 is a similar Concordia plot to which has been transferred the chord derived from figure 11. The only point duplicated is that representing the total R200-mesh zircon. The points obtained from the steps I and II of stripping-experiment A appear on opposite sides of this point, but still on or very near the chord. The two steps of experiment B plot on either side of the total zircon point, and also fall on the chord. The larger spread of the experiment A points reflects the greater degree of isolation of the highly discordant outer zone in step A-I.

Also plotted on this curve is an average value for the intermediate zone of R200-mesh zircon model. This is apparently the most concordant U-Pb system found in this group of experiments. It falls on the general chord.

Accepting the rectilinear quality of the data, and excepting analytical precision, common lead corrections, and decay constants from the remaining discussion, we may inquire into its origin and significance. Wetherill (1956) derived and Aldrich and Wetherill (1958) have discussed hypothetical models for data yielding such chords. For systems in the region of the Concordia plot where we observe these samples, it could be hypothesized that (1) the group originated 1,655 million years free of radiogenic lead; (2) its evolving uranium-lead systems were essentially undisturbed until 90 million years ago; (3) an episode of profound dis-

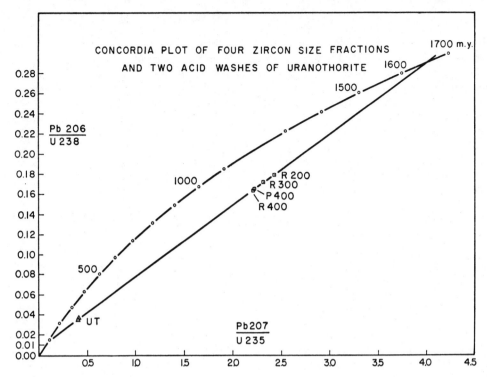

FIG. 11.—Graph of Pb^{206}/U^{238} vs. $Pb^{207}/U235$ for the four zircon size fractions and two uranothorite washes (UT) showing chordlike relation to the "Concordia" curve.

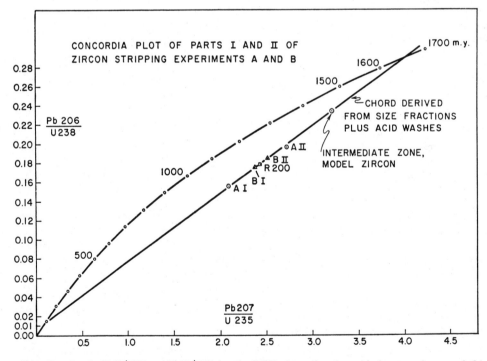

FIG. 12.—Graph Pb^{206}/U^{238} vs. Pb^{207}/U^{235} for the R200 zircon fraction and the several zones of this fraction isolated in the stripping experiments. Chord shown was transferred from fig. 11.

turbance characterized by loss of lead and/ or gain of uranium occurred in a brief interval of geologic time 90 million years ago; (4) the uranium-lead systems subsequently have not been disturbed further. Aldrich and Wetherill (1958) also observed that a similar chord could be obtained from minerals originating at a younger point in time, which incorporated a characteristic older radiogenic lead from external sources. We have dismissed the latter possibility in this case on the grounds that it is quite improbable that such a radiogenic lead would be attracted to these minerals in such systematic relation to other properties such as grain size, uranium concentration, and eTh/U ratio. Further, there is no evidence of such a primary radiogenic lead in some of the non-radioactive minerals in the host rock.

Each intersection of the chord with the Concordia curve represents the characteristic daughter-parent ratios of systems evolved without disturbance from the indicated point in time until the present. The chord between may be generated by any process which could mix the modern products of two such systems prior to analysis. Wetherill's rigorous proof, for example, may be represented as follows: (1) an original uranium system originating free of radiogenic lead at T_1 is allowed to evolve without disturbance until T_2. At T_2, a fraction of the radiogenic lead is lost, leaving a similar fraction of residual uranium, U_n, with no accumulated daughter. The lead, Pb_o, attributed to the other fraction of residual uranium, U_o, is still undisturbed. (Or, at T_2, a new fraction of uranium, U_n, is added to the system, which has no accumulated daughter, while the indigenous uranium, U_o, has its entire accumulated radiogenic lead, Pb_o, undisturbed.) From T_2 to the time of analysis both fractional systems are undisturbed and uranium continues to decay to radiogenic lead. Pb_o'/U_o' now represents the fractional system evolved from time T_1 to the present (the older intersection on Concordia) while Pb_n'/U_n' represents the present nature of the younger system initiated at time T_2 (the younger intersection). In

analysis, of course, it is the total mixture of these two systems, Pb_o'/U_o' and Pb_n'/U_n', that is measured. The chord is the locus of mixtures of two such systems where the present ratio $Pb_o'/U_o'/Pb_n'/U_n'$ varies from zero to infinity. Indeed, we may define a fraction of concordance (F) for a mixture Pb_x/U_x, which we analyze

$$F = \frac{Pb_x/U_x - Pb_n'/U_n'}{Pb_o'/U_o' - Pb_n'/U_n'}$$

for both uranium-lead isotope ratios. The "fraction of concordance" is a useful parameter for comparing disequilibrium relations in the family of systems. It may be obtained readily from the graphical parameters on the Concordia diagram.

If one considers each analyzed sample as a mineral concentrate (zircon) instead of a hypothetical mineral as in the Wetherill model treatment, then it is apparent that there are a number of additional situations which might provide a linear chord similar to the one we have obtained. Some geologically feasible concentrates might include:

1. Inherited, undisturbed zircon xenocrysts, 1,655 million years old, in a 90-million-year-old granite mixed with undisturbed zircons of the later generation. The mixing may be as old nuclei with overgrowths or as two distinct families of zircons.

2. A similar geologic setting where either the older or younger end members are not zircons, but impurities of minor relative abundance, but with major uranium and lead contribution (e.g., thorite, uraninite, xenotime, etc.).

3. A Precambrian granite that has an original generation of undisturbed (1,655-million-year-old) zircons mixed with a new generation of zircons originating from a metamorphic event (90 million years old).

4. A situation similar to 3 in which either one of the end members is a small but significant impurity in the concentrate.

5. Mixtures of two uranium-lead phases, either dissimilar zircons or zircons and impurities, whose different isotopic characters are determined solely by the same combination of older and younger events but which in neither case lie on the Concordia curve. In-

deed, any number of phases limited to sharing one or both parts of the same two-event history may be mixed to give analytical values that fall on a straight line between the extreme end members.

The previous list represents only some of many geologic possibilities. To these, of course, may be added non-geological "accidents," that is, the laboratory mixing of two unrelated uranium-lead systems. This is unlikely and unnecessary, but unfortunately, occasionally possible. We feel we have eliminated this consideration in this case.

Returning to the listed geological possibilities, the first two cases can be eliminated for the Johnny Lyon granodiorite on the grounds that clear-cut field relations demonstrate the Precambrian age of the intrusive (see fig. 2).

The third case is considered to be eliminated because of (1) the generally homogeneous physical characteristics of the zircon, (2) the lack of evidence of a significant metamorphic episode imposed on this sample of granite, (3) the evidence of the stripping experiments for disturbance throughout the zircon, as well as limiting the presence of 90-million-year overgrowth to something considerably less than 4 μ thick.

The fourth case is eliminated on the argument that the only significant radioactive impurity recognized in the autoradiographs of the unwashed zircon concentrates is uranothorite. The lead isotopic composition of the latter clearly labels it as an old system.

The fifth case might well apply to unwashed or inadequately washed mixtures of Johnny Lyon zircon and uranothorite. We have already presented our data on the acid elimination of uranothorite in our size-fraction samples. The stripping experiments show the pervasive discordant character *within* the zircon, where there is an inverse relation between degree of discordance and the Pb^{208}/Pb^{206} ratio (which we would expect to reflect the presence of the discordant uranothorite).

As long as the samples consist of large populations of grains, they are composite with respect to certain variable zircon characteristics. For example, the specific activity of individual grains in the same size fraction has been shown to range over an order of magnitude. Despite our inability to state whether a range or several distinct classes of Pb/U ratios exist within these zircon individuals, we consider it probable that each fraction we have analyzed as a whole, or stripped in stages, is a composite mixture in its isotopic character. To a considerable extent, the distribution of points along the linear chord must be derived from this mixing phenomenon. This does not alter the unique time significance of the chord intersections pointed out by Wetherill.

Recently, Tilton (1960) has revived a model of volume diffusion as a mechanism for discordant uranium-lead ages. Lead diffusion is considered continuous throughout the history of the radioactive crystals at rates governed by diffusion coefficients, effective crystal radii, and concentration gradients. This model had been invoked earlier by Nicolàysen (1957) in an interesting attempt at resolving discrepancies in the U-Pb ages of individual minerals. Nicolaysen recognized the value of testing continuous diffusion as a *possible* cause of disagreement on several samples from the same locality but did not cite any experimental investigations. Tilton's renewed interest, however, was directed toward the explanation of some patterns of discordant ages in assemblages of minerals with regional to intercontinental distributions. Using Nicolaysen's tables, Tilton calculated and presented curves on the Concordia plot that are the loci of Pb^{206}/U^{238} and Pb^{207}/U^{235} values, generated with diffusion loss, for families of uranium-lead systems. Members of each family originated at the same time and represent all values of D/a^2 for lead loss. No suites of analyses for minerals that were demonstrably cogenetic were presented, but Tilton suggested this mechanism might be generally applicable.

In figure 13, four curves representing the calculated loci of diffusion systems originat-

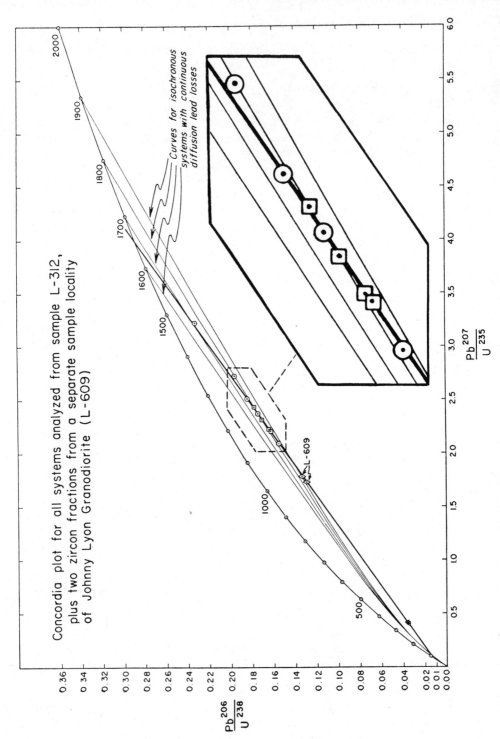

Fig. 13.—Concordia plot for all of the U-Pb systems analyzed from L-312 sample and for two zircon fractions from a locality, L-609, 3 miles south of L-312 in Johnny Lyon granodiorite (see fig. 2). Photographic enlargement (5×) shows the close fit of the data.

ing 1,600, 1,700, 1,800, and 1,900 million years ago are plotted. The complete suite of analytical data from the Johnny Lyon granodiorite L-312 sample is shown and the familiar chord is redrawn. In addition, the two points labeled L-609 represent two zircon fractions from an independent sample locality in the Johnny Lyon granodiorite about three miles away from L-312 (see fig. 2).

The L-312 zircon points in the inset have been enlarge five times to show the regularity of their fit to the chord. This set of data shows no obvious effects of continuous lead diffusion, despite the near parallelism of the curves in this region of the diagram. The L-609 points analyzed for an extended study (in preparation) of the same intrusive body simply confirm, by their positions on the chord, that they have been exposed to a similar episodic history.

EVIDENCE ON THE NATURE OF THE DISTURBANCE MECHANISMS

The recognition that there is a differential rather than a uniform response to a disturbing episode among various uranium-lead systems in the same rock is of great importance. It provides a basis for establishing informative isotopic patterns for rocks in many geologic situations. It suggests the possibilities of investigating the nature of disturbing processes as well as being utilized in the establishment of chronologies.

In this suite of data, one of the most significant clues to the mechanisms of disturbance lies in the general observation that the Pb^{206}/U^{238} ratio decreases as the U, eTh, and total radioactivity concentrations increase in the various fractions. This is shown in table 9 for the four size fractions and in table 11 for the stripping experiments. The only apparent exception to this is found in the calculated intermediate zone of the R200 composite zircon. The Pb^{206}/U^{238} ratio is an index of discordance when compared to the value of 0.2900 for Pb^{206}/U^{238} for a concordant 1,655-million-year-old system. In figure 14, Pb^{206}/U^{238} is plotted against equivalent uranium measured by α-particle scintillometry. (Uranium contributes 75–90 per cent of this activity, hence the similar correlation.) The coupling of the trend of increasing discordance with increasing radioactivity strongly suggests that the greater the total radiation flux in the history of these zircons, the greater their vulnerability to disturbance. It is worth noting that both L-609 samples are 50 per cent more radioactive (1,150 p.p.m. eU) and are significantly more discordant than any of the fractions in L-312. The L-312 uranothorite is 500 times more radioactive than the most radioactive zircon and is the most discordant system in the suite, although as a different mineral species its response to disturbance may not be directly analogous to that of the zircons. Surprisingly, the X-ray and index of refraction measurements have shown that the L-312 zircons now display the radiation damage to be expected in a young rock perhaps 100 million years old (Holland and Gottfried, 1955; Fairbairn and Hurley, 1957). Apparently, annealing of the zircon structure accompanied or followed the disturbance of the contained lead-uranium systems. The coupling of annealing and disturbance would explain some of the relations that have discouraged the use of radiation damage as a simple measure of age.

Several lines of evidence suggest that lead loss rather than uranium gain is the important change occurring in these disturbances. The positive correlation between radiogenic lead concentration and uranium concentration occurs in almost all fractions in increments several times too large to be explained by the accumulation of lead since a postulated gain of uranium 90 million years ago. It is difficult to accept the hypothesis that the uranium gain would be greatest where the lead concentrations were already highest, but this would be required by the greater discordances within the more radioactive systems. Mobility of lead during the 90-million-year-old episode is the only ready explanation (other than analytical error) that we have for the maximum in Pb^{206}/U^{238} ratio in the intermediate zone of the R200 composite zircon (fig. 10). The accumulated radiogenic lead attained a maximum concentration in the more radioactive outer parts of the zoned crystals prior to the disturbance. During the disturbing episode, diffusion of lead probably was partly influenced by the concentration gradients in the overall crystal system. For this reason, the maximum concentration, while being reduced, may also have broadened by diffusion both outward and toward the center. It may also have shifted laterally as well. Thus, despite a pervasive loss of lead

systematically higher at the outside, then the lost lead will not have a composition representative of the total lead prior to loss. It has been axiomatic that isotopic fractionation in the mass range of lead does not occur in nature. This process we describe would not produce fractionation of any originally homogeneous lead. It would take an inhomogeneous lead system and separate it into mobile and residual fractions in a non-representative manner. A very special loss process would be required to do otherwise in a sys-

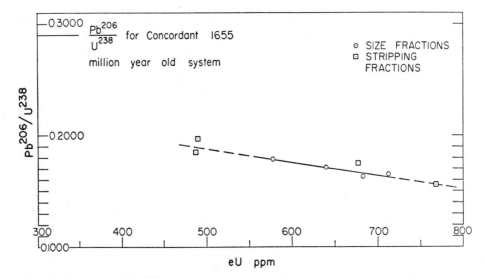

FIG. 14.—Variation of Pb^{206}/U^{238} as a function of total radioactivity of each fraction (eU = equivalent Uranium measured by α-particle scintillometer). Increasing departure from the concordant value of 0.2900 is a measure of increasing discordance in age with increasing radioactivity.

throughout the composite zircon, the intermediate zone lost less lead relatively than the neighboring zones.

Although we have not determined thorium precisely in these experiments, it is possible to draw some conclusions that have an important bearing on the commonly observed discrepancies between U-Pb and Th-Pb ages in the same mineral. The isotopic compositions of radiogenic lead in the stripping experiments show large variations in the Pb^{208}/Pb^{206} ratio in different parts of the composite crystal, while the Pb^{207}/Pb^{206} ratio is nearly uniform. If degree of lead loss is not uniform through the crystal but is

tem as inhomogeneous as this R200 fraction. If zonation in the Th/U ratio is a common characteristic in the minerals used in age dating, then the age discrepancies are not at all surprising.

Of course, some discrepancies between the U/Pb and Th/Pb ages simply represent the effect of minute quantities of highly discordant thorium-rich or uranium-rich impurities, such as thorite or uranite mixed with concentrates of more concordant zircons. A mixture of systems with distinctly different Th/U ratios and different degrees of disturbance will always give apparent ages in disagreement. This is illustrated by the data in table 14 on the R300 unwashed composite of zircon and uranothorite compared with the analyses of zircon and uranothorite separately. We have very little basis for understanding the extreme discordance in the uranothorite, other than to point out its characteristic extreme radiation damage as a condition that makes it susceptible to disturbance.

We have observed minerals of the thorite group in ten (about 20 per cent) of the granitic rocks we have recently examined. In mineral separation flow sheets it sometimes follows the zircon, monazite, or sphene, primarily because of variable magnetic susceptibility. It is also variable in color and form. Commonly, acid washing is

acid-wash studies on zircons as bearing as much on the nature of the minor minerals in the concentrates as upon the stability of the uranium-lead systems in the zircon, monazite, and other mineral species. More definitive studies of the acid mobility phenomenon in the light of the complete mineralogy of the mineral concentrates must be awaited.

Another suggested mechanism of discordance is the loss of intermediate daughter products, particularly radon (Giletti and Kulp, 1955). Loss of radon (Em^{222}) would provide a sequence of apparent ages $Pb^{206}/U^{238} < Pb^{207}/U^{235} < Pb^{207}/Pb^{206}$, which is the pattern we observe for our individual mineral concentrates. However, this mecha-

TABLE 14

SAMPLE	APPARENT AGES (MILLIONS OF YEARS)				AGE RATIO
	Pb^{206}/U^{238}	Pb^{207}/U^{235}	Pb^{207}/Pb^{206}	Pb^{208}/Th^{232}	$(Pb^{206}/U^{238})/(Pb^{208}/Th^{232})$
R300 zircon acid-washed.	1,030	1,230	1,610	1,390	0.74
R300 uranothorite wash no. 1................	230	350	1,280	245	0.93
R-300 unwashed ziron and uranothorite composite.	710	950	1,570	360	1.97

the only satisfactory method of separating thorite from the other minerals. Its solubility characteristics are also quite variable. The acid-washing experiments reported by Tilton, Patterson, et al. (1955), Starik (1956), and Tilton (1956) frequently have been cited as evidence on the nature of lead-uranium fractionation from the principal mineral of each concentrate. Unfortunately, documentation on the freedom of the concentrates from minor soluble minerals such as thorite is difficult to obtain, and certainly was not provided in these papers (see, e.g., Tilton, 1956, p. 230). We strongly suspect, on the basis of mineral separations on a similar sample from the same locality, that thorite may have been present in the sphene and zircon fractions of the Tory Hill, Essonville, granite cited by Tilton. We prefer to interpret the principal results of these older

nism operating by itself would require that in our different size fractions the Pb^{207}/Pb^{206} ages progressively rise as the Pb^{206}/U^{238} ages decrease. This is not the observed situation. The P400 zircon size fraction and the outer zone of the R200 composite crystal have the highest radioactivities and greatest surface areas. They could be expected to suffer greatest relative radon loss. The Pb^{207}/Pb^{206} ratios, although nearly constant, are, if anything, somewhat lower than in other samples. Even the extremely metamict uranothorite fails to confirm the pattern for radon loss. Thus, although the effect cannot be dismissed as entirely absent, radon loss alone cannot explain the major age patterns of the L-312 zircons. If continuous intermediate daughter loss is responsible for the linear patterns on Concordia it must involve isotopes in both decay chains, and must be governed by a special set of limiting conditions for which we can offer no ready phenomenological explanation at present.

The role of crystal size in influencing the mechanism of disturbance is not completely clear principally because more important factors, such as distribution of radioactivity and isotopic composition, are variable with grain size. However, the stripping experiments clearly showed striking isotopic differences in the outer 4-μ shell compared to the interior of the R200 zircon. One can explain many of the characteristics of the other size fractions with a similar 4-μ shell over relatively smaller cores. Thus some of the isotopic properties may vary nearly as a function of the ratio of surface area to mass, providing a dependence on grain size. To the extent that the surface "film" is most seriously discordant, it would appear that the rate effects for the disturbance mechanism are controlled to some degree by available surface area.

Russell and Ahrens (1957) have suggested loss of daughters by recoil of disintegrating atoms as a general mechanism of disturbance. A surface film would certainly be the site for maximum effect of such a phenomenon. It is possible, therefore that at least some of the depression of the Pb^{206}/U^{238} and Pb^{207}/U^{235} ratios in the outer zone reflects loss by recoil. The slight drop in Pb^{207}/Pb^{206} ratio in that zone is also compatible with greater loss of Pb^{207} because of the greater cumulative recoil energies (hence total range) of the U^{235} series over the U^{238} series. On the other hand, the directions of these effects are not unique to the recoil hypothesis. It is difficult to understand how a more general application of the recoil hypothesis alone would explain the correlation of disturbance with radioactivity in all of our fractions.

The pervasive disturbance throughout the R200 composite crystal and the evidence for the operation of lead loss must require some process of internal lead diffusion to have operated during the disturbing episode. The present lead concentration gradient in the crystal is well defined. The gradient be-

fore the 90-million-year event must have been similar in sign and much more strongly developed. Since the data indicates that the lead concentration reached a maximum near the outer margin, it is clear that diffusion controlled by this gradient alone could not have produced the extensive internal losses that have occurred. We conclude that smaller elements of structure within the crystal (in part, perhaps, defined by the radiation damage) were the effective domains with local concentration gradients in which conditions for diffusion were controlled.

It may be possible to make further deductions on the nature of a possible disturbance mechanism in the zircon crystals. In a broader study (in preparation) of zircons in the Johnny Lyon granodiorite, and associated Precambrian rocks, some geologic environmental factors can be introduced to provide some limits on conditions prevailing during the disturbance. Independent evidence is needed on the conditions prevailing in the alteration history of the Johnny Lyon granodiorite during late Mesozoic and early Cenozoic times.

SOME PETROLOGIC AND GEOLOGIC CONSIDERATIONS

Several results of this study have implications for our understanding of minor and trace element behavior in a crystallizing granitic melt. It is universally assumed in studies of accessory minerals in rocks that any given separate is both homogeneous and adequately representative of the respective mineral species in the rock. This is, of course, aside from the question of mineral concentrate purity. It is clear from this study that the zircons in the Johnny Lyon granodiorite are not homogeneous. Nor can we state, despite considerable painstaking efforts, that they are fully representative of all of the zircons in the rock. The basis of prism size that we have employed to fractionate arbitrarily the over-all zircon concentrate is not necessarily the most fundamental parameter to be used for this purpose, but it was convenient. We have found systematic trends in the zircon populations that strongly suggest that trace element variation in solid-solution-forming minor minerals is governed by the same petrological factors which influence major and minor element variations in the major minerals. In this instance, it appears that, on the average, zircons started to crystallize relatively early in the history of the melt and *grew over a very significant part of the total crystallization history.* As the residual melt was reduced in volume, U and Th were concentrated by exclusion from the major crystallizing phases. The partition coefficients between the zircons and their growth environment may also have changed but, in any event, the outer zones and finer sizes (later nucleation?) were markedly enriched in U and Th. That this is not a phenomenon unique to the Johnny Lyon granodiorite or to zircons will be documented in another paper (in preparation) on the distribution of uranium and thorium in igneous rocks. We see no reason to believe that such effects would be limited to uranium and thorium.

The evidence of profound disturbance in the uranium-thorium-lead systems has led us to re-examine the petrography and geologic setting of the sample for independent evidence of such an episode. We are faced with the question of whether the later chlorite-epidote-"sericite" alteration assemblage is a superimposed metamorphic (hydrothermal) effect or merely a deuteric suite. No obvious independent answer is available.

We have observed one condition that is suggestive. There is a surprisingly sparse development of pleochroic halos around zircons in the biotites of the L-312 sample. Deutsch *et al.* (1957) have shown that the optical density of a pleochroic halo is a function of the total radiation flux and the sensitivity of the biotite to irradiation. By comparing the pleochroic halos and radioactive sources in a rock of unknown age with those in an unaltered rock with similar biotites of known age, it is possible to estimate the age of the pleochroic halos.

Deutsch (1960) and Picciotto and Deutsch (1963) have evidence that the pleo-

chroic halos are quite sensitive to metamorphic effects.

We measured the optical density $\bar{D}$ of the halos in the L-312 biotite and the specific activity of the corresponding inclusions. In figure 15 the points are plotted on logarithmic plot of $\bar{D}$ as a function of specific α-activity measured by autoradiography. Also shown are three "isochrones" corresponding to the best-fitting line for data of the Elbe granite (30 million years old), the La Bresse granite (320 million years old), and a hypothetical granite of 1,000 million years age with biotite of similar sensitivity to that of the Elbe. Of six zircon inclusions observed in the biotite, four had no halos. Two zircons had very light halos with a mean density of 4 (in a range of 3 to 7 represented by the arrow). The data suggest an apparent age for the halos of 100 ± 100 million years. If the biotite has a normal sensitivity to radiation, it would appear that original pleochroic halos as radiation damage effects were completely erased by the younger event 90 million years ago. A more extensive study is required to evaluate the specific effects in this rock, but the method provides an interesting correlation with the evidence of annealing observed in the zircon crystal structures.

The application of this data to the regional geochronology has already been discussed in a preliminary report presented at the symposium, "Geochronology of Rock Systems," held by the New York Academy of Sciences in March, 1960 (Silver and Deutsch, 1961). The Johnny Lyon granodiorite is a post-tectonic intrusive whose age of crystallization places a younger limit on a major Precambrian orogenic event recorded in the rocks of the Dragoon Quadrangle (Cooper and Silver, 1954). The 1,655-million-year-old age of crystallization indicated in our analytical data is, therefore, a younger limit for this orogeny that has been correlated with the "Mazatzal Revolution" of southern and central Arizona. The recent paper by Giletti and Damon (1961, p. 642), placed the Mazatzal Revolution at somewhere between 1,200–1,550 million years based on a series of Rb^{87}–Sr^{87} dates from scattered exposures of Precambrian rocks in Arizona. Unfortunately, none of the work represented extended investigations at any one locality. Inasmuch as the authors have also recognized evidence of younger disturbances in their samples from the region, we believe that their data can only be interpreted as placing minimum limits on the age of their rock samples.

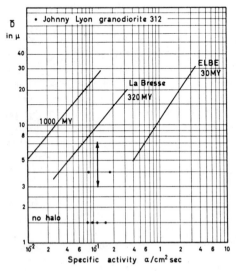

Fig. 15.—Graph showing optical density of pleochroic halos in biotite around zircon inclusions as a function of specific activity. Four zircons show no halo and two show halos to be expected in a 100-million-year granite when compared with effects in biotites of unaltered granites of known age (Deutsch et al., 1957).

The evidence for a major younger event (or events) sometime after the deposition of the Bisbee group (early to middle Cretaceous in age) is well documented in studies of the Dragoon Quadrangle and Cochise County. We are reluctant to interpret from our data that a single point in time covers all of the important events that occurred in the late Mesozoic and in much of the Cenozoic history of the region. We only conclude that some episode about 90 million years ago (late Cretaceous time) profoundly affected the mineral systems we have investigated. In subsequent work, we will report on broader manifestations of this younger event superimposed on Precambrian rocks.

ACKNOWLEDGMENTS.—The authors wish to acknowledge the continued critical advice and assistance of C. R. McKinney. J. Kawafuchi, D. Maynes, C. C. Patterson, and D. Ledent discussed analytical problems. H. P. Schwarcz and J. Bolinger assisted in mineral separations. V. Nenow constructed a very satisfactory furnace for control of fusion rates. Arthur Chodos provided emission spectrographic data on zirconium in the mineral separates. R. von Huene prepared the special ground zircon grain mounts for autoradiography studies. Professors Arden Albee and Barclay Kamb, colleagues at the California Institute of Technology, read the manuscript and discussed some significant points. This investigation was carried out as part of a program, "A Study of the Fundamental Geochemistry of Critical Materials," supported by the United States Atomic Energy Commission under contract AT(11-1)-208.

REFERENCES CITED

ALDRICH, L. T., and WETHERILL, G. W., 1958, Geochronology by radioactive decay: Ann. Rev. Nuclear Sci., v. 8, p. 257–298.

ARIZONA BUREAU OF MINES, 1959, Geologic map of Cochise County, Arizona: Tucson, Univ. of Arizona.

CHOW, T. J., and McKINNEY, C. R., 1958, Mass spectrometric determination of lead in manganese nodules: Anal. Chemistry, v. 30, p. 1499–1503.

—— and PATTERSON, C. C., 1959, Lead isotopes in manganese nodules: Geochim. et Cosmochim. Acta, v. 17, p. 21–31.

COOPER, J. R., 1959, Some geologic features of the Dragoon quadrangle, Arizona: Arizona Geol. Soc., Southern Arizona Guidebook II, p. 139–145.

—— and SILVER, L. T., 1954, Older Precambrian rocks of the Dragoon quadrangle, Cochise Co., Arizona: Geol. Soc. America Bull., v. 65, p. 1242.

—— —— in press, Geology and ore deposits of the Dragoon quadrangle, Cochise County, Arizona: U.S. Geol. Survey Prof. Paper 416.

DEUTSCH, S., 1960, Influence de lachaleur sur la coloration des halos pleochroiques dans la biolite: Il Nuov. Cim., v. 10, no. 16, p. 269–273.

——, KIEFFER, P., and PICCIOTTO, E., 1957,

Pleochroic halos and the artificial coloration of biolites by α particles: II Nuov. Cim., v. 10, no. 6, p. 797–810.

FAIRBAIRN, H. W., and HURLEY, P. M., 1957, Radiation damage in zircon and its relation to ages of Paleozoic igneous rocks in northern New England and adjacent Canada: Am. Geophys. Union Trans., v. 38, p. 99–107.

GILETTI, B. J., and DAMON, P. E., 1961, Rubidium-strontium ages of some basement rocks from Arizona and northeastern Mexico: Geol. Soc. America Bull., v. 72, p. 639–644.

—— and KULP, J. L., 1955, Radon leakage from radioactive minerals: Am. Mineralogist, v. 40, p. 481–496.

HOLLAND, H. D., and GOTTFRIED, D., 1955, The effect of nuclear radiation on the structure of zircon: Acta Crystallography, v. 8, p. 291–300.

JOHANNSEN, A., 1932, A descriptive petrography of the igneous rocks, v. 2, p. 344.

LARSEN, E. S., JR., KEEVIL, N. B., and HARRISON, H. C., 1952, Method for determining the age of igneous rocks using the accessory minerals: Geol. Soc. America Bull., v. 68, p. 181–238.

MEYEROWITZ, R., and LARSEN, E. S., JR., 1951, Immersion liquids of high refraction index: Am. Mineralogist, v. 36, p. 746–750.

NICOLAYSEN, L. D., 1957, Solid diffusion in radioactive minerals and the measurement of absolute age: Geochim. et Cosmochim. Acta, v. 11, p. 41–59.

PICCIOTTO, E., and DEUTSCH, S., 1963, Proceedings of the Varenna Summer Course, 1960.

RUSSELL, R. D., and AHRENS, L. H., 1957, Additional regularities among discordant lead-uranium ages: Geochim. et Cosmochim. Acta, v. 11, p. 213–218.

SILVER, L. T., 1955, Structure and petrology of the Johnny Lyon Hills area, Cochise County, Arizona: Doctoral dissertation, California Inst. of Technology, p. 1–407.

—— and DEUTSCH, SARAH, 1961, Uranium-lead method on zircons: New York Acad. Sciences Annals., v. 91, art. 2, p. 279–283.

——, McKINNEY, C. R., DEUTSCH, S., and BOLINGER, J., 1963, Precambrian age determinations in the western San Gabriel Mountains, Calif.: Jour. Geology, v. 71, no. 2, p. 196–214.

STARIK, I. E., 1956, The role of secondary processes in age determination by radiometric methods: Geochemistry, no. 5, p. 444–457.

TILTON, G. R., 1956, The interpretation of lead-age discrepancies by acid-washing experiments: Am. Geophys. Union Trans., v. 37, p. 224–230.

—— 1960, Volume diffusion as a mechanism for discordant lead ages: Jour. Geophys. Res., v. 65, p. 2933–2945.

——, DAVIS, G. L., WETHERILL, G. W., and ALDRICH, L. T., 1957, Isotopic ages of zircon from granites and pegmatites: Am. Geophys. Union Trans., v. 38, p. 360–371.

——, PATTERSON, C., BROWN, H., INGHRAM, M., HAYDEN, R., HESS, D., and LARSEN, E. S., JR., 1955, Isotopic composition and distribution of lead, uranium and thorium in a Precambrian granite: Geol. Soc. America Bull., 66, p. 1131–1148.

WETHERILL, G. W., 1956, Discordant uranium-lead ages: Am. Geophys. Union Trans., v. 37, p. 320–326

The Radioactivity of Rubidium

IX

Editor's Comments on Papers 23 and 24

Radioactivity associated with two of the alkali elements, rubidium and potassium, had been recognized for a number of years, ever since J. J. Thomson's (1905) experiments which showed that both rubidium and potassium alloys differed from many other light metals in emitting negatively charged particles. The weak beta activity associated with rubidium and potassium was confirmed by the researches of Campbell and Wood (Campbell and Wood, 1906; Campbell, 1908) and subsequent investigations indicated that the observed beta radiation was not due to traces of radioactive impurities but was definitely associated with the elements themselves.

In 1921, Aston investigated the mass-spectrum of the alkali elements and discovered the two isotopes of rubidium, Rb-85 and Rb-87, and two isotopes of potassium, K-39 and K-41. Five years later, Holmes and Lawson (1926) discussed the geologic significance of the radioactivity of rubidium and potassium in terms of their contribution to the internal heat of the earth, but by 1930 no transformation products of either rubidium or potassium had been detected. Since in both cases the emission of a beta particle was involved, production of radiogenic strontium for rubidium and radiogenic calcium from potassium was expected, but the difficulties of distinguishing between radiogenic and nonradiogenic strontium and calcium seemed insurmountable (see Rutherford, Chadwick, and Ellis, 1930, p. 543).

The isotopic abundances of rubidium and potassium were redetermined by Nier (1935, 1936), who discovered in the rare isotope K-40 the potassium mass spectrum. Nier suggested that the radioactivity associated with the alkali elements could best be explained by assuming that it was the Rb-87 isotope in the case of rubidium, and the K-40 isotope in the case of postassium, that was undergoing decay. The following year, Otto Hahn and his co-workers (Strassman, Walling, and Mattauch) at the Kaiser Wilhelm Institute for Chemistry in Berlin obtained from Prof. W. Papish at Cornell University about 3 kg of an ancient (2000 million years old) lepidolite mica from the Silver Leaf mines in Manitoba which contained more than 2 per cent rubidium. They separated the trace of strontium present in the mineral and showed it to be more than 99 per cent pure strontium-87 (Hahn et al., 1937; Mattauch, 1937). One month after the German analyses had been published, Hemmendinger and Smythe at the California Institute of Technology in Pasadena submitted the results of their work with a high-intensity mass spectrometer with which they had successfully separated the two isotopes of rubidium, and they showed that the beta activity was definitely associated with the Rb-87 isotope (Hemmendinger and Smythe, 1937).

Nier's suggestion had been confirmed, and it was clear that a new method of age determination could be based on the accumulation of radiogenic Sr-87 relative to the parent Rb-87. In Oslo, V. M. Goldschmidt (1937) suggested using the natural radioactivity of rubidium for age determination, and a discussion of the possibilities and advantages of the new method was presented in 1938 by Otto Hahn and Ernst Walling (Paper 23). Hahn and Walling compared the Rb–Sr method with age determination methods based on the accumulation of radiogenic lead and helium. Unlike lead, radiogenic strontium is derived from only one parent isotope, and the complexities introduced by the presence of thorium-derived lead in uranium minerals, together with the problem of helium leakage, both particularly severe in the case of very old minerals, could be avoided by using the "strontium method" for determining radiometric ages. For these reasons, the method seemed most appropriate to use for dating the ancient Precambrian shield areas.

There was now, however, little time to think about problems related to the history of the earth. In the last weeks of 1938, Otto Hahn split the nucleus of the uranium atom and for the next seven years the energies of most nuclear scientists were directed towards the future of the earth, not its past. Hahn did manage to continue his investigations into the possibilities inherent in Rb–Sr geochronology during the war years, but little progress was made until 1949 when L. H. Ahrens published the results of an extensive investigation into Rb–Sr dating which he had begun in South Africa at the University of Witwatersrand's Government Metallurgical Laboratory in Johannesburg and continued in the United States at MIT (Paper 24).

In his report Ahrens concluded that reliable age determinatiions could definitely be obtained by the Rb–Sr method. Using a spectrochemical procedure which was rapid if not particularly precise, he successfully analyzed thirty lepidolite micas for rubidium and strontium and calculated their Rb–Sr ages, which turned out to be in all respects acceptable. The oldest calculated Rb–Sr ages (2100–2350 million years) came from southeastern Manitoba, and these were essentially concordant with the Pb-207/Pb-206 ages Nier had obtained from the same area nearly ten years previously. Ahrens suggested that the Rb–Sr dating method could be usefully extended to minerals such as biotite and muscovite, which have much lower rubidium contents than lepidolite micas, provided than an isotopic analysis could be performed to correct for the presence of primary (nonradiogenic) strontium incorporated in the minerals at the time of crystallization.

The problem of determining the very small concentrations of rubidium and strontium encountered in the common rock-forming minerals was overcome at the Carnegie Institute's Geophysical Laboratory and the Department of Terrestrial Magnetism in Washington, D.C., by Aldrich and his colleagues, who used ion-exchange columns to separate the various alkali and alkaline earth elements, and stable isotope-dilution techniques for quantitative analysis of the strontium and rubidium (Aldrich et al., 1953; Davis and Aldrich, 1953).

There still remains the problem of the half-life of Rb-87, which is extremely difficult to determine accurately because of the large number of low-energy beta particles emitted by the disintegrating parent atoms. Such particles are prone to scattering and absorption effects. Two values for the half-life differing by more than 6 per cent are currently employed by geochronologists. The higher value, 5.0×10^{10} yr, was determined by Aldrich and his colleagues in 1956 by comparing measured radiogenic Sr-87/Rb-87 ratios in micas and potassium feldspars from six different rock units from which concordant isotopic uranium–lead ages had been obtained. A lower value, $(4.7 \pm 0.1) \times 10^{10}$ yr, was determined by Flynn and Glendenin (1959) by direct counting of the beta particle emission using a special liquid scintillation technique designed to minimize loss of the low-energy particles. More recently, McMullen and others (1966) at McMaster University, Ontario, allowed a kilogram of initially pure rubidium salt to accumulate radiogenic Sr-87 over a period of seven years, and determined a value of $(4.72 \pm 0.04) \times 10^{10}$ for the half-life of rubidium-87. This value agrees with Flynn and Glendinin's, and supports the contention of Kulp and Engels (1963) who found superior agreement between Rb–Sr and K–Ar ages when the 4.7×10^{10}-year half-life was used for rubidium decay.

Selected Bibliography

Aldrich, L. T., Doak, J. B., and Davis, G. L. (1953). The use of ion-exchange columns in mineral analysis for age determination. *Amer. J. Sci.*, **251**, 377–387.

Aldrich, L. T., Wetherill, G. W., Tilton, G. R., and Davis, G. L. (1956). Half-life of Rb-87. *Phys. Rev.*, **103**, 1045–1047.

Aston, F. W. (1921). The mass spectra of the alkali elements. *Phil. Mag., Ser. 6*, **42**, 430–441.

Campbell, N. R. (1908). The radioactivity of potassium, with special reference to solutions of its salts. *Proc. Cambridge Phil. Soc.*, **14**, 557–567.

Davis, G. L., and Aldrich, L. T. (1953). Determination of the age of lepidolites by the method of isotope dilution. *Geol. Soc. Amer. Bull.*, **64**, 379–380.

Flynn, K. F., and Glendenin, L. E. (1959). Half-life and beta spectrum of Rb-87. *Phys. Rev.*, **116**, 744–748.

Goldschmidt, V. M. (1937). Geochemische Verteilungsgesetz der Elemente, IX: Die Mengenverhältnisse der Elemente und der Atom-arten. *Skrift. Norske Vid.-Akad., Math.-Nat. Kl.*, No. 4, p. 140.

Hahn, O., Strassmann, F., Walling, E. (1937). Herstellung wägbarer Mengen des Strontiumisotops 87 als Umwandlungsprodukt des Rubidiums aus einem Kanadischen Glimmer: *Naturwissenschaft*, **25**, 189.

Hemmendinger, A., and Smythe, W. R. (1937). The radioactive isotope of rubidium. *Phys. Rev.*, **51**, 1052–1053.

Holmes, A., and Lawson, R. W. (1926). The radioactivity of potassium and its geological significance. *Phil. Mag., Ser. 7, 2*, 1218–1233.

Kulp, J. L., and Engels, J. (1963). Discordances in K–Ar and Rb–Sr isotopic ages. *In* "Radioactive Dating," pp. 219–238. International Atomic Energy Agency, Vienna.

Mattauch, J. (1937). Das Paar Rb-87/Sr-97 und die Isobarenregel: *Naturwissenschaft,* **25**, 189–191.

McMullen, C. C., Fritze, K., and Tomlinson, R. H. (1966). The half-life of rubidium-87. *Can. J. Phys.,* **44**, 3033–3038.

Nier, A. O. (1935). Evidence for the existence of an isotope of potassium of mass 40. *Phys. Rev.,* **48**, 283–284.

Nier, A. O. (1936). The isotopic constitution of rubidium, zinc, and argon. *Phys. Rev.,* **49**, 272.

Rutherford, E., Chadwick, J., and Ellis, C. D. (1930). "Radiations from Radioactive Substances." 588 pp. The University Press, Cambridge.

Thomson, J. J. (1905). On the emission of negative corpuscles by the alkali metals. *Phil. Mag., Ser. 6, 10*, 584–590.

The Possibility of Determining
the Geological Age of
Rubidium-Containing Minerals and Rocks†

OTTO HAHN and ERNST WALLING

The most reliable methods for the determination of the geologic ages of minerals, and hence the geologic periods in which these minerals were formed, are based on radioactive processes. The two primary radio-elements, uranium and thorium, are transformed at a very accurately known rate into uranium-lead having an atomic weight of 206 and thorium-lead having an atomic weight of 208. Thus, if the content of uranium and the content of lead in a well preserved uranium mineral are known, and if the content of lead-206 in the mixed-element lead is established by an atomic weight determination, it is possible to find the duration of the lead formation process with great accuracy. The same is true in principle of thorium minerals, but primary pure thorium minerals that have not changed since their formation are practically unknown. If there are minerals that contain both uranium and thorium, the determinations become very uncertain, since it is then difficult to decide whether the mineral might not originally have contained ordinary lead, which is a mixture of uranium-lead and thorium-lead. In such a case, greater certainty can be achieved only by recording a mass spectrum.

A phenomenon that accompanies the transformation of uranium and thorium is the formation of helium, which remains in the dense minerals if there is not too much of it and can also be used for age determinations. Geologically old, pure uranium minerals can generally be determined better by the lead method, while geologically later or uranium-poor minerals and also iron meteorites[1] can be determined better by the helium method.

Strontium formed by radioactive decay of rubidium was recently separated quantitatively from a rubidium-rich Canadian mica by Strassmann and Walling[2] in the present authors' institute. A mass-spectroscopic investigation of this strontium by Mattauch[3]

†Translated from Über die Möglichkeit Geologischer Altersbestimmungen Rubidiumhaltiger Mineralien und Gesteine, *Z. Anorg. Allgem. Chem.*, **236**, 78–82 (1938).

showed unequivocally that the strontium isotope formed has the atomic weight 87; the β-emitting component of the mixed element rubidium is thus the rubidium isotope having the atomic weight 87. The mass-spectroscopic investigation by Mattauch also showed that ordinary strontium not formed in the mineral is present in the preparation in a content of less than 3 per mil in comparison with the strontium formed in the mineral. In the meantime, the purity of the strontium separated from the mineral was also demonstrated by an optical method by Kopfermann and Heyden.[4] The chemical atomic weight of this strontium must naturally differ considerably from that of ordinary strontium. Mattauch calculated a value of 86.94, whereas the atomic weight determinations for normal strontium gave 87.63.

These results show that the mica used for the investigation contained practically no strontium at the time of its formation. The geologic age of this mica is fortunately also very accurately known from determinations of the geological ages of minerals that occur with the mica in the same deposit.[5] According to a private communication from Prof. Lane, the Chairman of the American committee on the measurement of geological time from radioactive data, the most probable age of the mica is 1975 million years. If this age is taken as certain for the mica, the half-life of the rubidium can be determined with greater certainty than before.[2] The values found for this constant were 2.3×10^{11} years for the mixed-element rubidium and 6.3×10^{10} years for the 27 per cent rubidium 87 present in the mixed element.

Now that the half-life of rubidium has been placed on a sound basis in this way and the mass-spectroscopic findings by Mattauch have shown that the content of ordinary strontium in a mica can be negligibly small, we are faced with the question of whether the radioactivity of rubidium, i.e., its transformation into strontium 87, can be used for the determination of geological ages. The question also arises of whether this "strontium method" can offer any advantages over the radioactive methods introduced before. Both questions can be answered in the affirmative.

Regarding the first question, if the rubidium and the strontium content of a mineral are known and if the excess of strontium-87 with respect to the normal composition of the mixed-element strontium is determined by mass spectroscopy, the duration of the transformation process, i.e., the age of the mineral, is then found directly from the transformation rate of rubidium, which is known from the above determination.

The quantitative determination of the rubidium content by chemical methods is still rather uncertain and troublesome because of the very low rubidium content in the mineral in most cases. The most reliable and most convenient of the methods known at present have been applied by Prof. Geilmann and Dr. Strassmann-Heckter to the above-mentioned mica. The results will be reported shortly. The optical determination of the rubidium content is probably faster and presumably not much less acurate once the possibility of accurate comparison has been provided by standard preparations. Thus Goldschmidt, Bauer, and Witte[6] in a detailed report on the occurrence of the alkali metals in minerals and rocks, gave a survey of the rubidium (and cesium) content of more than 30 minerals, in which the individual determinations were mostly carried out by three analyses using only a few milligrams of material.

As to the quantitative determination of the strontium content, here again there is a possibility of obtaining satisfactory results by spectral methods with much smaller quantities of materials than are required for the chemical separation.[7]

The weight-based separation of the quantity required for the mass-spectroscopic investigation could then be achieved, e.g., by the Strassmann-Walling method, but there would be no need to place any value on a quantitative yield.

The convenient feature of the strontium method is that radiogenic strontium is one of the rarer isotopes in the ordinary mixed element, since the mixed element contains only 7.5 per cent ^{87}Sr. Unless present in an excessively high content, therefore, ordinary strontium will hardly affect the accuracy of the mass-spectroscopic result in the case of geologically old minerals containing rubidium, where the quantities of strontium-87 formed are not inconsiderable.

The number of parameters required is the same in the "strontium method" as in the lead method for pure uranium minerals, the determination of the atomic weight of the lead formed from uranium being replaced by the mass-spectroscopic determination of strontium. A reliable determination should generally be possible with 20–30 mg of strontium salt. (The quantities of lead required for the atomic weight determinations are larger).

With regard to the second question, i.e., whether the "strontium method" promises any advantage over the methods used previously, it is possible to say the following. In the minerals containing rubidium, there is always only one possible transformation process, namely the transformation of rubidium-87 into strontium-87. If strontium-88, which is normally by far the most abundant isotope, is present as well as strontium-87, this is clear evidence of the admixture of ordinary strontium. The quantity of strontium-88 found then allows a reliable correction for the nonradiogenic fraction of the strontium-87. In the case of uranium minerals, on the other hand, thorium is a factor that must always be considered, since this also gives a lead isotope (as well as helium) that occurs in ordinary lead. The presence of thorium thus makes the determination very uncertain.

Since the rubidium content in the micas is never more than a few per cent and the quantity of strontium formed even in long geologic periods constitutes only a very small part of the total mineral, the strontium formation causes no change in the structure and negligible damage to the mineral. Moreover, rubidium decays by β emission, so that no helium is formed, whereas the formation of each atom of uranium-lead and thorium-lead is accompanied by the simultaneous formation of eight and of six atoms, respectively, of helium. Geologically old uranium and thorium minerals therefore contain many cubic centimeters of helium per gram of mineral under great pressure, which certainly does help [the] durability [of the helium fraction]. If one considers the state of a uranium mineral in which 10–20 weight per cent of uranium has been converted into lead, in which millions of years of high-energy α rays have loosened the structure before they became lodged as helium gas, it is surprising that such minerals can be extremely well preserved and that they can still be used at all for accurate age determinations.[8]

The main rubidium-containing minerals are certain types of mica (lepidolites) and feldspars. There are evidently many such minerals that contain up to a few per cent rubidium. Examples are the above-mentioned mica from Canada and various potash feldspars from various granite pegmatites.[6] Most mica and feldspars are geologically very old. The micas in particular have little tendency toward transformation; on the contrary, other silicates are often transformed into micas, whereas the latter are seldom transformed into other silicates.[9] The well crystallized or compactly crystalline micas found today have thus presumably undergone no transformation since their formation, whereas such a transformation is not unusual for uranium minerals and nearly always occurs for thorium minerals. The "strontium method" should therefore be particularly suitable for the determination of geologically very old mineral deposits, where uranium-lead determinations can be used only with reservations, thorium minerals are totally ruled out, and the helium method fails.

The "strontium method" thus seems the most promising method where one is particularly interested in age determinations for earlier geologic periods. It is not impossible, in the view of the authors, that minerals will then be found whose geologic ages are even greater than the 2000 million years of lepidolite, which gave rise to this work, and which is perhaps the oldest known at present.

References

1. F. Paneth, *Naturwissenschaft,* **19**, 164 (1931); F. Paneth and W. D. Urry *Z. Phys. Chem., Sec. A,* **152**, 127 (1931).
2. F. Strassmann and E. Walling, *Ber. Deut. Chem. Ges.* **71**, 1 (1938); O. Hahn, F. Strassman, and E. Walling, *Naturwissenschaft,* **25**, 189 (1937).
3. J. Mattauch, *Naturwissenschaft,* **25**, 189 (1937); **25**, 738 (1937).
4. M. Heyden and H. Kopfermann, *A. Phys.,* in press.
5. F. Strassmann and E. Walling, loc. cit.: F. S. Delury and H. V. Ellsworth, *Amer. Mineral.,* **16**, 596 (1931); Ref. in "Report of the Committee on the Measurement of Geologic Time 1932," p. 5; F. Hecht and E. Kroupa, *Z. Anal. Chem.,* **106,** 98 (1936).
6. V. M. Goldschmidt, H. Bauer, and H. Witte, *Göttinger Nachrichten, Sect. 4,* 39 (1934).
7. W. Noll, *Chem. Erd.,* **8**, 507 (1934).
8. V. M. Goldschmidt and L. Thomassen, *Videns. Kap. Scr., Mat.-Naturw. Kl.,* II, 44 (1923).
9. C. Doelter, *Handbuch der Mineralchemie* II, 2, p. 440.

BULLETIN OF THE GEOLOGICAL SOCIETY OF AMERICA

VOL. 60, PP. 217-266, 2 PLS. 10 FIGS.   FEBRUARY 1949

24

MEASURING GEOLOGIC TIME BY THE STRONTIUM METHOD

BY LOUIS AHRENS

ABSTRACT

The strontium method for measuring geological age has been examined in each of its aspects, and the general conclusion is made that evidence is now sufficient to show this method as capable of providing reliable age measurements.

Of the total strontium in the earth's crust, 0.5–1.0% is considered to be radiogenic, almost all of which is concentrated in potassium and cesium minerals. Lepidolite contains the highest proportion of radiogenic strontium, in which mineral it usually predominates completely. Amazonite, pollucite, hydrothermal pegmatitic microcline, zinnwaldite, some specimens of lithium-rich muscovite, and probably also rhodizite and lithium-rich biotite commonly contain a high proportion of radiogenic strontium; age measurements may be made on most of these minerals, if not all. Age determinations on other minerals are also possible, and the strontium method could probably be extended to granite biotites, which would increase its scope very considerably. With lepidolite, a reasonably reliable age can be obtained in most instances without an isotope analysis, although one is desirable. An isotope analysis is imperative for age determinations on all other minerals. For a quantitative isotope analysis of strontium, the mass spectrograph is superior in sensitivity and accuracy to other methods; a quantitative analysis can be made on as little as 0.3 mg. of strontium salt, and a conveniently low proportion of radiogenic strontium can be determined in the presence of an excess of common strontium.

In all, 32 strontium age determinations are known, 30 of which have been made according to a spectrochemical procedure outlined in this paper. The procedure is rapid—10–15 determinations (without isotope analyses) may be made in quadruplicate in 3 days—but lacks precision, since age reproducibility is usually only within about ±10–15%, even when carried out in quadruplicate. A more accurate procedure for determining Sr/Rb ratios is urgently needed to increase the usefulness of the strontium method; investigation of other spectrochemical methods is recommended.

Where comparisons were possible, strontium ages tally reasonably well with lead and helium (magnetite) ages, with few exceptions.

The strontium method should be of particular value in pre-Cambrian time and is probably superior to the lead and helium methods for dating very ancient rocks. The extreme antiquity of pegmatites from southeast Manitoba has been established beyond reasonable doubt; their age (2100×10^6 years) is considered to be greater than any other region on the earth's crust on which sufficient data are available. The method is, however, apparently not so suitable for dating relatively young specimens, and even under the most favourable circumstances, that is, if lepidolite is available, it seems unlikely that an age of less than 50×10^6 years could be measured successfully.

Lead, helium, and strontium methods have been compared.

INTRODUCTION

Some 10 years ago, Goldschmidt (1937a, p. 140) and Hahn and Walling (1938) suggested that the natural beta decay of rubidium to strontium could be utilized for determining the geological age of a mineral. During the decade since, considerable investigation has been carried out on the use of the strontium method, and some speculation, a fair proportion sounding a note of pessimism, has been made on the scope and limitations of the method.

In this paper the strontium age method is examined in some detail in each of its aspects and several new age determinations and other data are given, in addition to an account of some recent investigations on methods of analysis. It is hoped that after perusing this paper the interested reader will have cause to feel that within a reasonable period this elegant method for measuring geological time should provide reliable and accurate age data, particularly on pre-Cambrian time.

It is hoped further that active interest in this method will be stimulated: in each of the several aspects of the strontium method, only preliminary investigation has been made, and there is ample room for a tremendous amount of research.

A preliminary report of a portion of the work described in this paper has been given by Ahrens (1947a).

ACKNOWLEDGMENTS

Much of the earlier research described in this paper was carried out at the Government Metallurgical Laboratory, University of the Witwatersrand, Johannesburg, and thanks are due to the Director, Prof. L. Taverner, for permission to use laboratory equipment for spare time research. Mr. J. Levin, also of the Government Metallurgical Laboratory, kindly prepared some very pure specimens of lepidolite by a mineral dressing procedure. Grateful acknowledgment is also made to staff members of the Geology and Physics departments of the Massachusetts Institute of Technology where most of the later research was carried out. Thanks are extended to Prof. W. J. Mead and Prof. H. W. Fairbairn of the Dept. of Geology, for generously providing facilities and all necessary operational requirements; and to Dean G. R. Harrison and Dr. R. McNally (Jr.) of the Physics Department, also for providing facilities and for making useful suggestions for the recording of isotope effects. It is a pleasure also, to acknowledge the ready co-operation of several persons in South Africa, America, and elsewhere in providing specimens.

A major portion of the research described in this paper was carried out when I was a Research Fellow of the South African Council for Scientific and Industrial Research; grateful acknowledgment is made to this Council for having provided the necessary opportunity for this research.

NATURAL RADIOACTIVE DECAY OF RUBIDIUM

Like its alkali metal homologue, potassium, rubidium is beta active, and it decays into a stable isotope of strontium. Rubidium consists of two isotopes, Rb^{87} and Rb^{85}, the relative abundances of which are 27.2 and 72.8%, respectively; according to Hahn, Strassmann, and Walling (1937) and Hemmendinger and Smythe (1937), Rb^{87} alone is radioactive, whence the above transmutation may be written as:

$$Rb^{87} \rightarrow Sr^{87} + beta \ particle$$

For convenience, radiogenic strontium will be written thus, $Sr^{87}*$, so as not to confuse ordinary Sr^{87} found in common strontium.

The radioactive disintegration of any radioactive element may be expressed as

$$N = N_0 e^{-\lambda t} \dots\dots\dots\dots\dots\dots\dots\dots\dots\dots\dots (1)$$

where N_0 is the number of atoms originally present and N the number left after a time, t, has elapsed; λ is the decay constant. In the above equation, N_0 and N refer to atoms of Rb^{87} and λ its decay constant: the period of half life, T, is equal to $0.693/\lambda$.

Since, as will be shown later, the half life is very long in comparison to geologic periods, only an insignificant proportion of rubidium will have decayed even in the most ancient minerals (Fig. 1), and consequently one may regard the quantity of rubidium in a mineral as having remained constant throughout its lifetime. Equation (1) can then be simplified and written as follows,

$$N = N_0 - \lambda t N,$$
$$N_0 - N = \lambda t N \dots\dots\dots\dots\dots\dots\dots\dots\dots\dots (2)$$

$N_0 - N =$ the number of atoms of Rb^{87} that have decayed, which is equivalent to the number of strontium atoms that have formed, whence we may write,

$$\%Sr^{87}* = \%Rb^{87} \times t \times \lambda,$$

or

$$t \ (age) = \frac{\%Sr^{87}*}{\%Rb^{87}} \times 1/\lambda \ \dots\dots\dots\dots\dots\dots\dots\dots (3)$$

Figure 2 compares the use of the simplified equation (3) with equation (1). For a given amount of Rb^{87}, calculations were made using these two equations, of the quantities of Sr^{87} generated at various times and Figure 2 shows a plot of per cent variation vs. time. The amount of deviation corresponds to the percentage deviation in the age measurement if the simplified equation is used; for most of geologic time the deviation is small, and the convenient simplified equation may therefore be used.

The accuracy of a determination of geologic age of a mineral by the strontium method will depend upon the accuracy with which λ, $\%Rb^{87}$, and $\%Sr^{87}*$ can be estimated. The experimental determination of Rb^{87} and $Sr^{87}*$ is discussed later.

Two relatively early determinations of λ (Hahn and Rothenbach, 1919; Mühlhof, 1930) are equal to 9.2×10^{-12} yr.$^{-1}$ and 6.0×10^{-12} yr.$^{-1}$ respectively. A more recent and indirect determination is reported by Strassmann and Walling (1938). From lepidolite from the Silver Leaf mine, Southeast Manitoba, Canada, strontium was extracted, determined quantitatively and analysed mass spectrographically. The analysis showed > 99.7% of the strontium to be radiogenic; the rubidium content of the lepidolite was also determined. The age of the lepidolite was con-

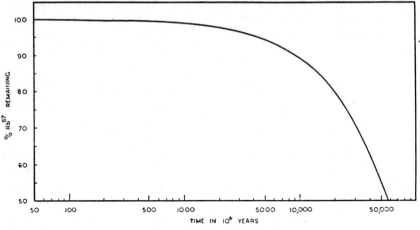

FIGURE 1.—*Decay of Rb*[87]

Showing %Rb[87] remaining after different periods have elapsed.

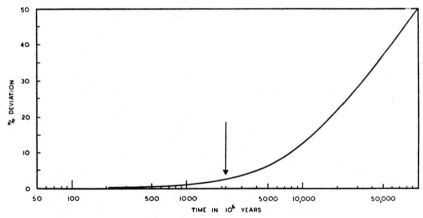

FIGURE 2.—*Percent deviation in computing ages by use of simplified equation*

(Arrow indicates age of oldest known mineral).

sidered to be equal to that of the near-by Huron Claim pegmatite which had been dated by lead age measurements. An age of 1975×10^6 years was used in this computation, and λ was calculated as 1.11×10^{-11} year^{-1} and T as 6.3×10^{10} years. Unfortunately this age is somewhat approximate, and according to Goodman and Evans (1940, p. 284–287) the value of λ may be in error by as much as 30–40%.

The above value of λ has been used for all strontium age determinations so far reported, but does not appear to be in serious error because a new value of 5.81×10^{10} years for the half life of Rb[87] has recently been determined by Eklund (1946) using a direct method; this value is in reasonably good agreement with the indirect determination and has been used for all the age determinations reported in this paper.[1]

Substitution of the new value for λ $(0.693/5.8 \times 10^{10}$ years $= 1.195 \times 10^{-11}$ year$^{-1} = \lambda)$ in equation (3) gives,

$$\text{Age} = \frac{\%Sr^{87*}}{\%Rb^{87}} \times 8.37 \times 10^{10} \text{ years} \dots\dots\dots\dots\dots\dots\dots(4)$$

The value of λ is believed to be known with an accuracy of about $\pm 15\%$, which is satisfactory for general purposes but not for work requiring a high precision. Within reasonable time however, according to Prof. Sanborne Brown (private communication) of the Department of Physics, Massachusetts Institute of Technology, an accuracy of about $\pm 5\%$ should be possible, which should suffice for almost all age requirements.

Rb^{87} and Sr^{87} are isobars, that is, they have the same mass (A), but differ in the number of protons (Z) in the nucleus. Rubidium has $Z = 37$, whereas strontium has $Z = 38$. According to a rule of nuclear stability, stable isobaric pairs which differ by one unit of Z should not exist. Very few exceptions are known. In this instance Rb^{87} is unstable, and Rb^{87} and Sr^{87} are thus not exceptional, although the rate of decay is very slow. Furthermore, nuclear theory indicates that the greater the difference in spin between parent and daughter nuclei, the slower the rate of decay. Rb^{87} has a spin of $3/2$ and Sr^{87} a spin of $9/2$: this large difference apparently accounts for the existence of Rb^{87} and its very slow rate of decay to Sr^{87}.

GEOCHEMISTRY OF RUBIDIUM IN IGNEOUS MINERALS

The abundance of rubidium in the earth's crust has been given by Goldschmidt (1937a) as about 0.03% by weight. Although this value may be subject to some revision because of more recent quantitative abundance measurements of rubidium in minerals and rocks, the magnitude is in all probability correct; rubidium is therefore not a relatively abundant element nor very rare, which of course is fortunate for the purposes of geological age measurement.

It may be well to review briefly the general geochemistry of rubidium in igneous minerals. Rubidium forms no minerals of its own, but tends to substitute for potassium in potassium minerals and for cesium in the cesium mineral, pollucite; rubidium may also be present in some specimens of beryl, but the nature of its location within this mineral is not clear. Since Rb^+ (radius $= 1.49$ A) is larger than K^+ (radius 1.33 A) for which ion it proxies during crystallization of minerals, rubidium tends to concentrate relative to potassium during the formation of potassium minerals, in the mother liquor fractions (residua), in accordance with the general geochemical rule (Goldschmidt, 1937b, p. 661) that when two ions of like charge but of slightly different size compete for a given lattice site within a growing crystal, the smaller ion will be accepted in preference because of its greater electrostatic attraction. This rule undoubtedly holds for the pair rubidium: potassium.

[1] Two separate determinations of the half life of Rb^{87} have just been reported by O. Haxel, F. G. Houtermans, and M. Kemmerich (Phys. Rev., vol. 74, p. 1886–1887, 1948). After considering both their determinations, these authors give $6.0 \pm 0.6 \times 10^{10}$ yr. as their best value: this value is in close agreement with the value used in this paper, and the use of the newer value would alter the strontium ages given only very slightly.

* * * * * * *

QUANTITATIVE DETERMINATION OF Sr/Rb RATIOS

GENERAL

The usefulness of the strontium method will depend, naturally, to a very large extent on the ease and accuracy with which quantities of rubidium from 2.5—about 0.05% and even less, and of strontium from 0.02–0.0002%, can be determined.

Chemical analysis has apparently been successful for estimating rubidium in lepidolites and in other minerals where the concentration of rubidium is less (Stevens and Schaller, 1942), but amounts of about 0.1% and less would be difficult to determine chemically.   Amounts of strontium less than 0.02% have been determined chemically by Strassmann (Strassmann and Walling 1938; Hahn *et al.*, 1943), but the procedure is somewhat lengthy, and considerable difficulty would be experienced in determining quantities of strontium as low as 0.001%, which amount is average for lepidolite of an intermediate age, namely about Ordovician.   In addition, a relatively large bulk of the mineral is required for the chemical analysis and some significant loss of strontium is likely during chemical manipulations.   Thus, in the ancient lepidolite from Manitoba, Canada, which is the most ancient thus far analyzed and which contains about 0.015% Sr, the recovery of strontium may be 10% too low (Strassmann and Walling, 1938).   In younger minerals and in minerals which contain smaller amounts of rubidium, greater loss of strontium is to be expected.

Both rubidium and strontium may be determined spectrochemically.   Rubidium has been determined by several different techniques (Goldschmidt, *et al.*, 1935; Tolmacev and Filippov, 1935; Rusanov and Vasil'ev, 1939; Adamson 1942; Ahrens, 1945a; and others) and no serious difficulties have been reported.   Using some refinements in technique, it should be possible to determine this element with a reproducibility of ±5.0%.

The determination of strontium in minerals and rocks has been, in most instances, by semi-quantitative methods.   Noll (1934) describes a method he developed for investigating the geochemistry of strontium.

Strock (1942) carried out investigations on the determination of strontium in some specimens of lepidolite analyzed by Stevens (1938).   Using a series of standards prepared from a base identical in composition to that of the mineral, excellent reproducibility is reported, using Sr 4607 with a lithium line as an internal standard line.   As lithium is used as an internal standard and because the lithium content of lepidolite varies markedly in different specimens, one would have to know the lithium content beforehand in order to use this method.

More recently, Hybbinette (1943) combined a preliminary chemical concentration of strontium (precipitation of strontium as oxalate in the presence of a large excess of calcium) followed by a later spectrochemical analysis of the strontium in the ignited precipitate, using calcium as an internal standard.   The method is interesting and the accuracy reported good, but a serious loss of strontium may be expected when the strontium content of the mineral is low. Hybbinette's statement that the chemical enrichment is required because a spectrochemical analysis cannot be made directly when the concentration of strontium is below 0.01% requires some comment, because strontium is a sensitive element spectrochemically and less than a tenth of this amount can invariably be detected and estimated.

DEVELOPMENT OF A SPECTROCHEMICAL PROCEDURE

Purpose.—The author has attempted to determine rubidium and strontium in a single operation. In this method to be described, an attempt is made to use self-internal standardization, that is, the ratio Sr/Rb is determined from the intensity ratio of a pair of strontium and rubidium lines.

Apparatus.—Two spectrographs have been at my disposal; a larger Hilger quartz spectrograph of the Government Metallurgical Laboratory, University of the Witwatersrang, Johannesburg, and a large Wadsworth type grating spectrograph (linear dispersion 2.5 A/mm) of the Cabot Spectrographic Laboratory, Department of Geology, Massachusetts Institute of Technology, Cambridge.

The material to be analyzed was placed in the anode, of purest available graphite or carbon, the dimensions of which were; external diameter $\frac{3}{16}''$, internal diameter $\frac{1}{8}''$ and depth $\frac{1}{4}''$. The cathode, of similar material, was $\frac{3}{16}''$ in diameter and was pointed.

A d.c. arc source (line voltage 220 volts) was used, and after a pre-arcing period of about 2 seconds at low amperage, the current was increased to 6–7 amps, and the specimen arced to completion. The spectra were recorded on Ilford Ordinary plates and also on Eastman Spectroscopic plates, Type 103-0, both of which were found suitable, and were developed under standardized conditions for 4 minutes at 18°C in Kodak D 19 developer. After fixing, washing, and drying, microphotometric measurements were made on Hilger nonrecording microphotometers, one at the Government Metallurgical Laboratory and one at the Dept. of Physics, Massachusetts Institute of Technology.

Only the purest electrodes were used (Hilger's and National Carbon) because more impure grades revealed the presence of strontium lines of considerable intensity.

A stepped sector of seven steps having a transmission factor of two was used to obtain line intensity ratios and to decrease the intensity of Rb 4202, which in lepidolite is very intense.

Selection of suitable analysis lines.—The most sensitive lines of rubidium are located in the red end of the spectrum, but these lines were disregarded for this work because at relatively large concentrations of rubidium; these ground state lines are prone to reverse excessively; furthermore, their location is not convenient. Rb 4202 and Rb 4215, less sensitive lines, can be used to determine concentrations of rubidium as low as 0.005%, which sensitivity is ample; but Rb 4215 could not be used because it is co-incident with Sr 4215 and the head of a CN band. Rb 4202 was therefore selected provisionally, and wave-length tables were examined for the presence of interfering lines (Table 2). Only lines listed as having an intensity greater than "50" are given.

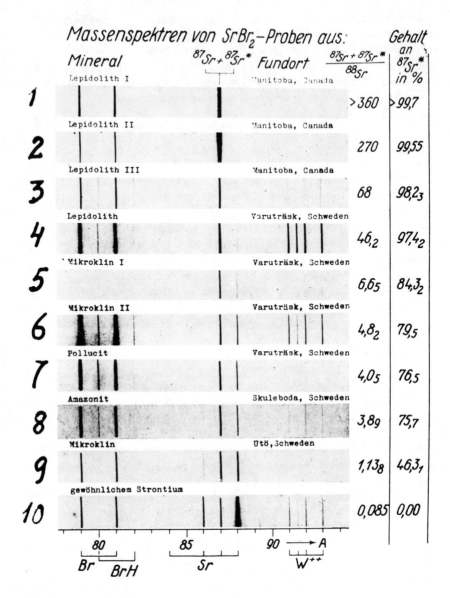

MASS SPECTRA (MATTAUCH, 1947) OF STRONTIUM IN SEVERAL RUBIDIUM-RICH MINERALS

*   *   *   *   *   *   *

DETERMINATION OF RADIOGENIC STRONTIUM IN TOTAL STRONTIUM

Common strontium consists of four isotopes, Sr^{88} (82.6%), Sr^{87} (7.02%), Sr^{86} (9.86%) and Sr^{84} (0.54%); all radiogenic strontium is of mass 87 (Sr^{87*}). Methods for estimating the different isotopes of an element depend essentially on the fact that each isotope is of slightly different mass, and the common procedure for making an isotope analysis is by means of the mass spectrograph. Relative abundance measurements on the different isotopes of strontium can be made with precision and ease, and, according to Mattauch (1947), an accurate quantitative analysis may be made on as little as 0.3 mg. of strontium salt, using his method of photographic recording. This is indeed good news for the strontium method because the concentration of strontium rarely exceeds 0.01% in the rubidium—rich minerals; but for this mass spectrographic sensitivity, the strontium method would be severely restricted and cramped in its application. Furthermore, according to a private communication from Dr. A. Keith Brewer, qualitative tests (Brewer, 1938) on the isotope composition of the strontium in lepidolites which he analyzed were made without any chemical pre-enrichment.

Plate 1, which is reproduced here with the kind permission of Prof. Mattauch, shows typical mass spectra of strontium isolated from several rubidium-rich minerals. In each of the mass spectra, Sr^{87} is either predominant or a major isotope. In many minerals, naturally, the proportion of radiogenic strontium is extremely small, and proportions of radiogenic strontium range from nearly 100% radiogenic strontium in some minerals, to negligible proportions, that is, say, less than 0.001% in others. An important limiting factor of the strontium method will be the lowest proportion of radiogenic strontium to nonradiogenic strontium that it will be possible to determine accurately with the mass spectrograph. According to Mattauch (1947), if Sr^{87*} is predominant, the mass spectrographic analysis is made by determining the intensity ratio of lines 87/88, since Sr^{88} produces the strongest line in common strontium; if the amount of common strontium approaches that of radiogenic strontium, the ratio of the lines 87/86 may be determined as well. If common strontium is predominant, it should be easy to detect a small quantity of Sr^{87*} since in common strontium the line-intensity ratio 87/86 is near unity and the accuracy in determining line-intensity ratios by the photometric method as used by Mattauch is greatest if the value of the intensity ratio is near unity. Even if the proportion of radiogenic strontium is 1%, the value of the intensity ratio 87/86 would increase 15% over the value for common strontium. In the absence of further data, 1–100% will be regarded tentatively as a *minimum* useful range in the proportion of radiogenic strontium in a mineral that can be accurately determined mass spectrographically.

It is fortunate that Sr^{87} in common strontium is not more abundant, because had this isotope been a major one the determination of small proportions of radiogenic strontium in the presence of an excess of nonradiogenic strontium would not have been so convenient. In this respect, it may be recalled that K^{40} is beta active, and decays, in part only, to Ca^{40}, and there has been some speculation as to whether this transmutation could be utilized for the determination of geological age, particularly in certain micas. But since Ca^{40} comprises about 97% of common calcium, correction is more difficult, and there seems at the moment to be little likelihood of developing a really satisfactory calcium method for determining geological age.

Two other methods of demonstrating the presence of different isotopes of an element are in the field of optical spectroscopy. First, in optical line spectra, many lines may be split into hyperfine components, that is, the line has a hyperfine structure which depends on two properties of the nucleus, its spin and its mass. Once the hyperfine components of a given line corresponding to particular isotopes have been identified, the relative abundance of the different isotopes may be measured by determining the relative intensities of some of the hyperfine components. Heyden and Kopfermann (1937) were able to show that strontium in a specimen of lepidolite analyzed by Mattauch was radiogenic. The method is sensitive, and Heyden and Kopfermann employed only a few mgs. of strontium salt, but quantitatively is probably not as suitable as the mass spectrograph.

Second, in molecular spectra it has been shown that a slight change in the mass of one of the components of the molecular emitters is capable of causing a measurable shift in certain band heads, due to a change in vibrational frequency. Provided a large dispersion is used and resolution sufficient, the presence of isotope band heads due to strontium isotopes Sr^{88}, Sr^{87} and Sr^{86} may be observed in certain emission molecular spectra produced by the emitter SrF (Ahrens, 1948b). Quantitative isotope abundance measurements on these three isotopes may be made with a fair degree of accuracy, though not as accurately as with the mass spectrograph, but the amount of strontium required for a quantitative isotope analysis of this type is rather excessive (at least 200 mgs.) because the band sequence that is used is relatively weak.

In addition to the above methods, the suggestion has been made by Prof. Robley D. Evans of the Physics Dept., M. I. T., in a private communication, that the nuclear excitation of Sr^{87} could perhaps be used for determining the abundance of this isotope. Sr^{87} alone among the strontium isotopes is converted into an isomeric metastable (excited) state when bombarded with suitable missiles. If the intensity of excitation is measured, the amount of Sr^{87} (total) could be determined after suitable calibration has been made with standards. The principle underlying this method is not one of a variation in mass, but of nuclear configuration. Some preliminary experiments have been carried out using a relatively weak source of X-radiation, but so far no response from Sr^{87} has been detected; it is hoped to continue further investigations using a more powerful source and different missiles.

* * * * * * *

EXTENDING THE SCOPE OF THE STRONTIUM METHOD

The usefulness of the strontium method would be increased and its scope extended if age determinations could be made on a greater variety of minerals and on minerals of wider distribution.   There is little doubt that lepidolite is suitable for age determinations, and it is highly likely that age determinations could also be made on hydrothermal microcline, amazonite, and pollucite, but all these minerals are somewhat rare.   Pegmatitic muscovite and biotite, in particular the lithium-rich varieties, are certainly worth investigating and since they are relatively common in pegmatite their use would be of considerable value.   They are, however, restricted to pegmatites.

The application of the strontium method to rocks would increase its usefulness tremendously.   At first thought, this possibility seemed rather remote because of a high concentration of common strontium relative to rubidium and radiogenic strontium in rocks.   A closer scrutiny of the available analytical data provides cause for optimism, however; certainly basic rocks may be ruled out, but in some granite types which invariably contain 0.02–0.1% Rb the proportion of common strontium may be sufficiently low to permit a satisfactory quantitative analysis for the determination of radiogenic strontium. One to 100% radiogenic strontium has been considered as a minimum, but tentative, useful range; for the purposes of this discussion, therefore, only those minerals or rocks which contain 1% or more radiogenic strontium will be considered.

Using analytical data published by Sahama (1945), the proportion of radiogenic strontium in some Finnish pre-Cambrian rocks was found (Ahrens, 1948a) to be quite considerable in some instances. Most of the granite and gneissic types were calculated to contain more than 1% and the proportion of radiogenic strontium in some was as high as 30%. Detailed data may be obtained from Ahrens (1948a), but the general conclusion may be made that the proportion of radiogenic strontium in these rocks should be sufficient to permit strontium age measurements.

The Caledonian rocks of Great Britain analyzed by Nockolds and Mitchell (1948) were found to contain an extremely low proportion of radiogenic strontium (0.01%) as a whole. In the biotite separated from the granite, the proportion of radiogenic strontium usually fell within 1–5%. Consequently it should be possible to determine the ages of these specimens by an analysis of the biotite fractions. These rocks are not very ancient (approximately 350×10^6 years) and hence the possibility of having a relatively high proportion of radiogenic strontium is less than in older rocks.

It would seem reasonable then that the strontium method could be applied to some granite-like rocks, in particular if the mica fraction could be separated, because the proportion of radiogenic strontium is likely to be at a maximum in the mica.

RADIOGENESIS OF STRONTIUM AND THE STABILITY OF THE SILICATE LATTICE

In silicate minerals where Rb^+ substitutes for K^+, it occupies a lattice site of high co-ordination. When Rb^{87} decays to Sr^{++}, an ion forms which (1) is slightly smaller than the parent ion, and (2) carries two charges instead of the single charge of the parent. The change in dimensions of the ions should produce negligible effects in the lattice because Sr^{++} is very similar in size to K^+ and only a relatively small number of Sr^{++} ions will be produced. The presence of a doubly charged ion in the place of the singly charged ion could cause some local electrostatic headaches, but it is very doubtful that any significant strain could develop within the lattice as a result of the formation of a limited number of doubly charged cations. Even in the most ancient minerals (2000×10^6 years), only about 0.6% of the total rubidium will have decayed, and in lepidolite, the mineral richest in rubidium and hence in radiogenic strontium, about 0.1–0.2% of the total number of K^+ sites will have been occupied by Sr^{++}: this proportion is extremely low and is undoubtedly insignificant. In the transmutation, $K^{40} \rightarrow Ca^{40}$, a similar situation obtains, and an extremely small proportion of K^+ sites are occupied by Ca^{++}.

Microscopic evidence has shown that in these ancient leipdolites, as well as in the younger, the mineral may be obtained in a fresh state of preservation and free from signs of internal strain. This condition may be contrasted with the internal strain and resultant poor state of uranium minerals, caused by the production of a large volume of helium, a large quantity of lead, and by the change of valence $U^{4+} \rightarrow U^{6+}$, on oxidation of UO_2 to UO_3: this change in valence results in a large change in size of the uranium ion. All these factors contribute to the production of shattering and to the metamict state of uranium minerals, which makes them very susceptible to losing some of their constituents; whereas UO_2 is relatively inert, UO_3 more reactive, forms stable uranyl compounds in the presence of acidic solutions, and in the presence of alkaline solutions may form uranates.   The shatter cracks in uranium minerals also form passage ways for possible losses of the gaseous products of decay—in particular, radon, whose half-life period is 3.85 days.

PRESENCE OF BARIUM AND OF CALCIUM AS A POSSIBLE AID FOR ESTABLISHING THE GENESIS OF STRONTIUM

If a mass spectrographic analysis of strontium in a mineral cannot be carried out, the concentration of either barium or calcium may sometimes serve as a useful aid in determining qualitatively, at least, whether the strontium is likely to be largely radio-genic.

It will be necessary to compare very briefly the geochemical characteristics of these three alkaline earths.   Chemically they are similar, but the radii of their ions are somewhat different: $Ba^{++} = 1.43$ A, $Sr^{++} = 1.27$ A, and $Ca^{++} = 1.06$ A. Sr^{++} is of intermediate size and in a sense one may regard the geochemistry of strontium as intermediate between that of barium and calcium.

Strontium is invariably associated with calcium in calcium-rich minerals, but it may be present to some limited extent in potassium minerals, with many of which small amounts of barium are frequently present.   Consider first the Ba:Sr association.   In general, strontium and barium tend to concentrate in the potassium minerals of earlier formation, and residual hydrothermal solutions should contain only very small quantities of these two elements; furthermore, because the radius of $Ba^{++} > Sr^{++}$, the general tendency will likely be for barium to concentrate relative to strontium in the residual differentiates.   For the Ca:Sr association, the reverse is probably true, and strontium tends to concentrate relative to calcium in the crystals of later formation.   In this manner, strontium can be considered to hold an intermediate 'geochemical' position between calcium and barium.

More than a trace of calcium is rarely found in late potassium hydrothermal minerals, but should the supply of calcium not be exhausted, some may enter these minerals and will undoubtedly carry some primary strontium with it.   The calcium contents of lepidolites are invariably very low, as evidenced by published chemical analyses, by Stevens (1938) for example, and also by personal observations on calcium spectra of the specimens so far analyzed.   In a few instances, however, the calcium lines have been found to be unusually strong and far in excess of an amount which could have been generated by the radioactivity of potassium.   In most, though not all, of these instances, the strontium ages have tended to be high.   In no instance where the calcium spectra have been extremely faint, has the strontium age appeared to be excessive.

Some observations have been made on the barium spectra on ten of the specimens of lepidolite referred to in Table 6.   In one specimen (No. 23) about 0.003% Ba was found, and in another (No. 24) about 0.0001–2 % Ba was found; in the eight other specimens (Ahrens, 1947a, no barium could be detected and is consequently less than about 0.0001 % in each case.   Specimens No. 23 and 24 have a greater apparent age than the other three specimens of lepidolite that have been analyzed from South West Africa. The spectra of the ten specimens were recorded on the quartz spectrograph and at the particular setting used observations could be made on the sensitive barium lines; the other spectra were recorded on the grating instrument, and the setting used did not include the wave-lengths of any very sensitive barium lines.

Naturally, this indirect evidence is of a very qualitative nature, but nevertheless I have found it of some value for the age determinations on the lepidolite specimens. It seems indeed likely that if the spectra of calcium are very faint and that if those of barium are absent, then the chances of a significant amount of contamination should be reduced to a minimum. Since calcium and barium, as well as strontium, are spectrally very sensitive, extremely small amounts of these elements can be detected. An experienced spectrochemist can readily estimate approximately the magnitudes of calcium and barium present by an inspection of the strengths of the lines.

BRIEF COMPARISON OF THE LEAD, HELIUM, AND STRONTIUM METHODS

EXPLANATION

As three methods which utilize the natural radioactivities of elements are now in a reasonable degree of development for the measurement of geological time, it seems appropriate to attempt a general comparison. Although factual data are still very scant, the brief discussion to be given is considered to be reasonably accurate and without bias and should portray, in a general way and rather tentatively, the relative merits of the lead, helium, and strontium methods.

APPLICATION

Lead method: Has been applied essentially to age measurements on pegmatites; satisfactory use of zircons would increase range of application very considerably.

Helium method: Suitable magnetite is relatively widely distributed and it appears that this method has potenialities of wide application.

Strontium method: So far confined to age measurements on pegmatites, but application to some igneous rocks possible, for example by determining the ages of biotite in granite.

USEFUL AGE RANGE

Lead method: Recent as well as pre-Cambrian. For very ancient ages, probably less reliable.

Helium method: Has been applied to fairly recent as well as pre-Cambrian measurements. In general, the more ancient pre-Cambrian measurements may be open to considerable hazard.

Strontium method: Not applicable to recent determinations, $<$ about 50–100 $\times$ 10^6 years. Best suited for pre-Cambrian and is probably the most reliable of the three methods for measuring very early pre-Cambrian time.

MINERALS GENERALLY USED AND STRUCTURE STABILITIES

Lead method: Uraninite, monazite, samarskite, and other strongly radioactive minerals. Lattice structure invariably disrupted and metamict condition common; serious consequences may result.

Helium method: Most satisfactory measurements thus far have been made on magnetite; satisfactory specimens can, according to present data, quite frequently be obtained.

Strontium method: Most work so far on lepidolite, which mineral can be obtained in an excellent condition. Other minerals such as hydrothermal pegmatitic microcline, amazonite, pollucite, Li-rich muscovite, which can very probably be used, can frequently be obtained in a satisfactory state.

POSSIBILITY OF A LOSS OF EITHER PARENT OR DAUGHTER ELEMENTS

Lead method: Chiefly as a result of poor state of highly radioactive minerals, chances of losses considerable; in particular, loss of U from minerals containing U as UO_3. Loss of Rn also possible.

Helium method: Since He is a gas, leakage is potentially great, but apparently He losses are insignificant in a high proportion of magnetites which show no evidence of recrystallization.

Strontium method: In lepidolite, and in the other minerals referred to above, parent and daughter elements should be tightly held by ionic bonding, and losses are considered quite negligible in good material. Unlike the helium method in particular, even a slight degree of recrystallization may have no serious consequences.

CORRECTION FOR THE PRESENCE OF SOME PRIMARY DAUGHTER ELEMENT

Lead method: Correction can usually be made for the presence of primary lead by means of the mass spectrograph.

Helium method: No direct correction can be applied to the entrapment of some He during the formation of a mineral, but indirect investigations thus far made have shown that, in the types investigated, little significant entrapment of He could be observed.

Strontium method: The presence of primary strontium can be corrected very easily on the mass spectrograph, even when the common strontium is in considerable excess; only about 0.3 mgs. of Sr salt are required.

METHOD, ACCURACY, AND SPEED OF ANALYSIS

Lead method: Pu, Th, and Pb usually analyzed by chemical methods, and unless these elements are present in very small quantities, accuracy is of a high order. Complete analysis of specimen rather tedious.

Helium method: Special He apparatus and suitable counting equipment. In most instances reproducibility about 10–15% per single determination. About 2–3 determinations may be made in a day.

Strontium method: Spectrochemical analysis probably best suited, speed considerable—10–15 age determinations can be made in quadruplicate in about 3 days—but the reproducibility thus far has been relatively poor; quadruplicate determinations can usually be reproduced to within 10–15%.

REFERENCES CITED

Adamson, O. J. (1942) *Minerals of the Varutrask pegmatite. XXXL. The feldspar group*, Geol. Fören. Förh., vol. 64, p. 19–54.

Ahrens, L. H. (1945a) *The relationship between thallium and rubidium in minerals of igneous origin*, Geol. Soc. S. Africa, Tr., vol. 48, p. 207–231.

———— (1945b) *Quantitative spectrochemical examination of the minor constituents in pollucite*, Am. Mineral., vol. 30, p. 616–622.

———— (1947a) *The determination of geological age by means of the natural radioactivity of rubidium: a report of preliminary investigation*, Geol. Soc. S. Africa, Tr., vol. L, p. 23–54.

———— (1947b) *Analyses of the minor constituents in pollucite*, Am. Mineral., vol. 32, p. 44–51.

———— (1948a) *The geochemistry of radiogenic strontium*, Mineral. Mag., vol. XXVIII, p. 277–295.

———— (1948b) *Molecular spectroscopic evidence of the existence of strontium isotopes Sr^{88}, Sr^{87} and Sr^{86}*, Phys. Rev., vol. 74, p. 74–77.

Berggren, T. (1940) *Minerals of the Varuträsk pegmatite. XV. Analyses of the mica minerals and their interpretation*, Geol. Fören. Förh., vol. 62, p. 182–193.

Bray, J. M. (1942) *Spectroscopic distribution of minor elements in igneous rocks from Jamestown, Colorado*, Geol. Soc. Am., Bull., vol. 53, p. 765–814.

Brewer, A. K. (1938) *A mass spectrographic determination of the isotope abundance and of the atomic weight of rubidium*, Am. Chem. Soc., Jour., vol. 60, p. 691–693.

Davis, C. W. (1926) *The composition and age of uranium minerals from Katanga, South Dakota and Utah*, Am. Jour. Sci., vol. 11, p. 201–217.

Eckel, E. B. (1933) *A new lepidolite deposit in Colorado*, Am. Ceram. Soc., Jour., vol. 16, no. 5, p. 239–245.

Eklund, S. (1946) *Studies in nuclear physics. Excitation by means of X-rays. Activity of Rb^{87}* Arkiv. Mat., Astron. Fysik., vol. A 33, no. 14.

Erämetsä, O., Sahama, Th. G., and Kanula, V. (1941) *Spektrographische Bestimmungen an Rubidium und Caesium in einigen finnischen Mineralen und Gesteinen*, Bull. Comm. Geol. Finland, no. 128, p. 80–86.

Goldschmidt, V. M. (1937a) *Geochemische Verteilungsgesetze der Elemente. IX. Die Mengenverhältnisse der Elemente und der Atom-Arten*, Skrift. Norske. Vid.-Akad., Math.-Nat. Kl., no. 4.

———— (1937b) *The principles of distribution of chemical elements in minerals and rocks*, Jour. Chem. Soc. (London), p. 655–673.

———— Bauer, H., and Witte, H. (1934) *Zur Geochemie der Alkali-Metalle. II.* Nachr. Ges, Wiss. Göttingen, Math.-Phys. Kl. 1, p. 39–55.

Gonyer, F. L. (1937) *Report of the Committee on the Measurement of Geologic Time*, Nat. Res. Council, p. 60.

Goodman, C., and Evans, R. D. (1941) *Age measurements by radioactivity*, Geol. Soc. Am., Bull., vol. 52, p. 491–544.

Hahn, O., and Rothenbach, M. (1919) *Über die Radioactivität des Rubidiums*, Physik. Z., vol. 20, p. 194–202.

——— and Walling, E. (1938) *Über die Möglichkeit geologischer Altersbestimmungen rubidiumhaltiger Mineralen und Gesteine*, Zeit. Anorg. Allgem. Chem., vol. 236, p. 78–82.

——— Strassmann, F., and Walling, E. (1937) *Herstellung wagbärer Mengen des Strontiumisotopes 87 als Umwandlungsprodukt des Rubidiums aus einem kanadischen Glimmer*, Naturwiss., vol. 25, p. 189.

———— Mattauch, J., and Ewald, H. (1943) *Geologische Altersbestimmungen mit dem Strontium methode*, Chem. Ztg., vol. 67, no. 5–6, p. 55–56.

Hamilton, G. N. G. (1938) *The geology of the country around Kubuta, Southern Swaziland*, Geol. Soc. S. Africa Tr., vol. 41, p. 41–48.

———— (1938) *Further notes on the geology of the country around Kubuta*, Geol. Soc. S. Africa Tr., vol. 41, p. 199–203.

Harrison, G. R. (1939) *M. I. T. wavelength tables*, John Wiley and Sons, Inc., New York.

Hemmendinger, A., and Smythe, W. R. (1937) *Radioactive isotope of rubidium*, Phys. Rev., vol. 51, p. 1052–1053.

Hendricks, S. B. (1939) *Polymorphism of the micas*, Am. Mineral., vol. 24, p. 729–771.

Hess, F. L., and Stevens, R. E. (1937) *Rare-alkali biotite from Kings Mountain, North Carolina*, Am. Mineral, vol. 22, p. 1040–1044.

Heyden, M., and Kopfermann, H. (1937) *Über die Kernspinänderung beim radioaktiven β-Zervallsprozess Rb 87 → Sr 87.*, Zeit. f. Phys., vol. 108, p. 232–243.

Holmes, A. (1934) *The Gordonia uraninite and the upper pre-Cambrian rocks of Southern Africa*, Am. Jour. Sci., vol. 27, p. 343–355.

(1947a) *The construction of a geological time-scale*, Geol. Soc. Glasgow, Tr., vol. 21, pt. 1, p. 117–152.

———— (1947b) *A revised estimate of the age of the earth*, Nature, vol. 159, p. 127–128.

Hurley, P. M., and Goodman, C. (1943) *Helium age measurement. 1. Preliminary magnetite index*, Geol. Soc. Am., Bull., vol. 54, p. 305–324.

Hybbinette, Anna-Greta. (1943) *Bestämmung av låga halter strontium i några svenska pegmatitmineral*, Svensk. Kem. Tid., vol. 55, p. 151–155.

Ishibashi, M., and Ishihara, T. (1942) *Chemical studies of micas. Chemical constituents of lepidolite and its meaning from the standpoint of geochemistry and chemical sources*, Chem. Soc. Japan, Jour., vol. 63, p. 767–780.

Khlopin, V. G., and Vladimirova, M. E. (1938) *Geological age of uraninites and monazites from the pegmatite veins of North Karelia*, Bull. Acad. Sci. U.R.S.S. Cl. Sci.-Mat. Nat., p. 499–508.

Lacroix, A. (1922) *Minéralogie de Madagascar*, vol. 1, 624 p.

Larsen, E. S., and Keevil, N. B. (1942) *The distribution of helium and radioactivity in rocks. III. Radioactivity and petrology of some Californian intrusives*, Am. Jour. Sci., vol. 240, p. 204–215.

Mattauch, J. (1947) *Stabile Isotope, ihre Messung und ihre Verwendung*, Angew. Chem., A., no. 2, p. 37–42.

Muench, O. B. (1938) *"Glorieta" monazite*, Am. Chem. Soc., Jour., vol. 60, p. 2661–2662.

Mühlhoff, W. (1930) *Aktivität von Kalium und Rubidium gemessen mit dem Elektronenzählrohr*, Ann. Phys., vol. 7, 205–224.

Nier, A. O. (1938) *The isotopic constitution of radiogenic leads and the measurement of geological time. II*, Phys. Rev., vol. 55, p. 153–163.

———— (1941) *The isotopic constitution of lead and the measurement of geological time. III*, Phys. Rev., vol. 60, p. 112–116.

Nockolds, S. R., and Mitchell, R. L. (1948) *The geochemistry of some Caledonian plutonic rocks: a study in the relationship between major and trace elements of igneous rocks and their minerals*, Royal Soc. Edinburgh, Tr., vol. 61, pt. II, no. 20, p. 533–575.

Noll, W. (1934) *Geochemie des Strontiums. Mit Bemerkungen zur Geochemie des Bariums*, Chem. d. Erde., vol. 8, p. 507–600.

Norman, J. E., and O'Mear, R. G. (1941) *Froth flotation and agglomerate tabling of micas*, U. S. Bur. Mines, R.I. 3558.

Pierce, W. C., and Nachtrieb, N. H. (1941) *Photometry in spectrochemical analysis*, Ind. Eng. Chem., Anal. Ed., vol. 13, p. 774–781.

Rollwagen, W. (1939) *Die physikalischen Erscheinungen der Bogenentladung in ihrer Bedeutung fü die spektralanalytischen Untersuchungsmethoden*, Spectrochem. Acta, vol. 1, p. 66–82.

Rusanov, A. K., and Vasil'ev, K. N. (1939) *Spectral analysis of minerals and solutions*, Zavodskaya Lab., vol. 8, p. 832–837.

Sahama, Th. G. (1945) *Spurenelemente der Gesteine im südlichen Finnisch-Lappland* Bull. Comm. Geol. Finl., no. 135.

Shimer, J. A. (1943) *Spectrographic analysis of New England granites and pegmatites*, Geol. Soc. Am. Bull., vol. 54, p. 1049–1066.

Stevens, R. E. (1938) *New analyses of lepidolite and their interpretation*, Am. Min., vol. 23, p. 607–628.

————, and Schaller, W. T. (1942) *The rare alkalies in micas*, Am. Mineral., vol. 27, p. 525–537.

Strassmann, F., and Walling, E. (1938) *Die Abscheidung des reinen Strontium-Isotops 87 aus einen alten rubidium-haltigen Lepidolith und die Halbwertszeit des Rubidiums*, Ber. Deut. Chem. Ges., vol. 71, Abt. B., p. 1–9.

Strock, L. W. (1942) *Blank and background effect on photographed spectral lines*, Jour. Optical Soc. Am. vol. 32, no. 2, p. 103–111.

Taylor, G. L. (1935) *Note on ages of granites in Black Hills of South Dakota region*, Am. Jour. Sci., vol. 29, p. 278–291.

Tolmacev, Y. M., and Filippov, A. N. (1935) *The presence of rare alkali metals in amazonite*, Acad. Sci. U.R.S.S., C. R., vol. 1, p. 321–323.

Wickman, F. E. (1943) *A graph for the calculation of the age of minerals according to the lead method*, Sver. Geol. Undersök., ser. C., no. 458.

Wright, J. F. (1932) *Geology and mineral deposits of a part of Southeastern Manitoba*, Geol. Survey Canada, Mem. 169, 150 p.

DEPARTMENT OF GEOLOGY, MASSACHUSETTS INSTITUTE OF TECHNOLOGY, CAMBRIDGE, MASS.
MANUSCRIPT RECEIVED BY THE SECRETARY OF THE SOCIETY, DECEMBER 29, 1947.

The Distribution of
Radiogenic Strontium

X

Editor's Comments on Papers 25–29

Throughout the 1950s rubidium–strontium dating as a feasible method applied to the common micas and potassium feldspars achieved ever increasing recognition. Mass-spectrometric techniques and isotope dilution were successfully used to determine precisely the small concentrations of rubidium and strontium involved, and with special care the method could be used on young Tertiary materials (Jäger, 1962). A correction for the presence of any nonradiogenic common strontium was applied on the basis of an isotopic analysis of the total strontium extracted from the sample. Common strontium, like common lead, consists of four stable isotopes, Sr-84, Sr-86, Sr-87, and Sr-88. As only Sr-87 is derived from the decay of rubidium, contamination of radiogenic strontium with common strontium is revealed by the presence of the two most abundant common isotopes, Sr-88 and Sr-86, in the mass-spectrum of the mixture. The accepted abundance of Sr-87 in common strontium was 7.0 atom per cent, corresponding to an Sr-87/Sr-86 ratio of 0.710.

At the end of the decade new results were published which led to a major advance in Rb–Sr geochronology. The story began in 1958, when G. D. L. Schreiner, working at the Bernard Price Institute of Geophysical Research in the University of the Witwatersrand, Johannesburg, South Africa, found that powdered whole-rock samples of granite were just as suitable for age determination by the Rb–Sr method as individual mineral separates. In his published report, Schreiner (1958) showed that there was no significant difference between the Rb–Sr ages determined on total-rock samples and those determined on mineral fractions (feldspar and biotite) separated from the Red granite in the Bushveld Complex. In 1959, Australians W. Compston and P. M. Jeffery repeated Schreiner's experiment using whole-rock and mineral separates from the Boya granite near Perth in Western Australia. Three samples of this granite were analyzed, two mineral separates (biotite and microcline), and a single powdered aliquot of the whole-rock. Compston and Jeffery were surprised to find that, contrary to

Schreiner's experience, whole-rock and mineral separates from the Boya granite gave highly discordant Rb–Sr ages (Paper 25).

Rb–Sr ages calculated for the three Boya granite samples were 650 million years (biotite), 1290 million years (microcline), and 2340 million years (total rock). They showed a pattern of increasing age with increasing common strontium concentration, if the accepted value for the common strontium-87 abundance was adopted. Measuring all their isotopic concentrations relative to Sr-86, Compston and Jeffery suggested that the mineral separates, which had a relatively high concentration of Rb-87, had lost a significant proportion of their radiogenic Sr-87 some time after crystallization. The radiogenic Sr-87 lost from these rubidium-rich minerals did not escape from the whole-rock system, but was simply redistributed among the other strontium isotopes, so that immediately after the disturbing event all mineral constituents throughout the entire whole-rock system had identical strontium-87 concentrations relative to common strontium-86. Each mineral now had an ''anomalous'' common strontium isotopic composition, for the Sr-87/Sr-86 ratio, irrespective of the rubidium content of the mineral, had increased above the accepted figure of 0.710 for ''normal'' common strontium. Subsequent additions of radiogenic Sr-87 then accumulated at a rate dependent upon the rubidium concentration in each mineral.

Compston and Jeffery showed (Paper 25, Fig. 1)* that by extrapolating the Sr-87/Sr-86 growth curves for their three analyzed samples backward with respect to time (measured according to the radiogenic-Sr-to-Rb ratio) a point of intersection was reached at 520 million years when all three samples had the same Sr-87/Sr-86 ratio (0.82). This age was interpreted as the time of strontium isotopic homogenization, when radiogenic strontium concentrated in the rubidium-rich minerals was redistributed among all the strontium-bearing minerals within the granite. Assuming that all the minerals containing strontium originally crystallized with Sr-87/Sr-86 ratios of 0.710, then the calculated whole-rock age (2430 million years) would be the age of primary crystallization.

The model proposed by Compston and Jeffery represented a most important advance in the interpretation of discordant Rb–Sr ages, and established the rubidium–strontium method on a par with the isotopic uranium–lead methods as one of the most powerful means available to geochronologists for the determination of primary crystallization ages in isotopically disturbed systems. The model was elaborated by the Australian workers in a number of publications (e.g., Compston et al., 1960; Compston and Jeffery, 1961; Riley and Compston, 1962), and it became clear that where metamorphosed systems were concerned, the age of primary crystallization could be determined only by measuring several whole-rock samples having varying Rb–Sr ratios selected from a single rock unit. The time of subsequent metamorphism and strontium isotopic homogenization could be obtained by analyzing a suite of minerals separates, provided that these also had a range of Rb–Sr contents.

---

*There is a drafting error in this figure. The legend on the abscissa should read ''Sr-87/Sr-86 common strontium.''

At MIT, where Ahrens had earlier urged the adoption of the Rb–Sr method for radiometric age determination, Compston and Jeffery's model provided an explanation for a large number of discordant Rb–Sr ages that had been obtained by Harold Fairbairn in association with P. M. Hurley and W. H. Pinson on coexisting biotite, potash feldspar, and whole-rock samples of plutonic igneous rocks collected near Sudbury, Ontario (Paper 26). Apparent Rb–Sr ages, assuming an initial Sr-87/Sr-86 ratio of 0.710 in all the samples, showed a wide range from 1065 million years to 2215 million years, but application of Compston and Jeffery's model indicated that only two real events were recorded in the existing isotopic data as indicated by intersection of the radiogenic Sr-87/S-86 growth lines. Fairbairn and his colleagues regarded the use of whole-rock analyses as marking a "milestone in the interpretation of discordant Rb–Sr data" (Paper 26, p. 254).

The graphical presentation of Rb–Sr isotopic data proposed by Compston and Jeffery was simplified by L. O. Nicolaysen of the Bernard Price Institute of Geophysical Research at the University of the Witwatersrand (Paper 27). Nicolaysen showed that a more effective way of presenting the data was to plot total measured Sr-87/Sr-86 ratios as ordinate against the measured Rb-87/Sr-86 ratio as abscissa. Such a plot had the distinct advantage that data obtained for each analyzed sample appeared as a single point, and all samples having the same initial common strontium ratio and the same age but with different Rb-87/Sr-86 ratios would define a line whose slope would be proportional to the age. The intercept on the ordinate of the line defines the value of the initial Sr-87/Sr-86 ratio. Compston and R. T. Pidgeon called such lines "isochrones" (see Paper 28, p. 273) following Houtermans' use of the word in conjunction with lead-isotope data (see Paper 15).*

In 1962, Compston and Pidgeon suggested that the Rb–Sr whole-rock dating method could be applied to argillaceous shales to determine the age of deposition, provided that a uniform Sr-87/Sr-86 ratio were established in the sediment at the time of deposition irrespective of variations in the rubidium concentration. Application of any radiometric age method to sedimentary rocks is complicated by the presence of inherited radiogenic daughter products in the detrital components of the rocks. Two of the three formations studied by Compston and Pidgeon gave isochron ages in good agreement with the known stratigraphic age, but the third (the Cardup shale) gave scattered results in excess of the stratigraphic age due apparently to the presence of inherited radiogenic strontium. The whole problem of inherited radiogenic strontium in sedimentary age determinations was further discussed by Whitney and Hurley (1964), who recommended that application of whole-rock Rb–Sr dating would be most useful for determining the age of ancient nonfossiliferous argillaceous sediments.

The whole-rock Rb–Sr method is currently enjoying widespread application to all manner of igneous, metamorphic, and sedimentary materials. In his review of the method presented to the symposium on radioactive dating held at Athens under

---

*It is now conventional to refer to any line connecting samples of equal age as an "isochron," dropping the "e."

the auspices of the International Atomic Energy Agency in November, 1962, Hurley and his colleagues at MIT summarized the advantages of using whole-rock analyses (Paper 29), pointing out that loss of radiogenic daughter products during an extended post-crystallization cooling history would further limit the usefulness of mineral separates for determining the true crystallization ages. If radiogenic Sr-87 lost from rubidium-rich mineral phases was redistributed among rubidium-poor phases and retained within the whole-rock system, the primary crystallization age could only be obtained by analyzing whole-rock samples.

Hurley also pointed out the petrogenetic significance of the initial Sr-87/Sr-86 ratio. This ratio could be used to distinguish between reworked, rubidium-rich, sialic material which would have a high initial Sr-87/Sr-86 ratio at the time of emplacement, and differentiates from the rubidium-depleted mantle which would have lower initial Sr-87/86 ratio at the time of emplacement. In the case of igneous rocks, Rb–Sr isotopic analyses of a cogenetic suite of igneous differentiates, with different rubidium concentrations derived from a single magma source, would define a single isochron whose slope would be proportional to the time of intrusion. The intercept of the isochron on the Sr-87/Sr-86 ordinate would define the value of this ratio in the magma source region at the time of differentiation.

For further discussion of various aspects of the Rb–Sr method, the reader is referred to papers by Hurley et al. (1962), Faure and Hurley (1963), Lanphere et al. (1964), Arriens et al. (1966), Bence (1966) and Armstrong (1968). A short review of the method was published in 1964 by Moorbath.

Selected Bibliography

Armstrong, R. L. (1968). A model for the evolution of strontium and lead isotopes in a dynamic earth. *Rev. Geophys. Space Phys.,* **6,** 175–199.

Arriens, P. A., Brooks, C. Bofinger, V. M., and Compston, W. (1966). The discordance of mineral ages in granitic rocks resulting from the redistribution of rubidium and strontium. *J. Geophys. Res.,* **71,** 4981–4994.

Bence, A. E. (1966). The differentiation history of the earth by rubidium–strontium isotopic relationships. *In* "Variations in Isotopic Abundances of Strontium, Calcium, and Argon and Related Topics," 14th annual progress report, A.E.C. contract AT(30-1)-1381, p. 35–8. MIT, Cambridge, Massachusetts.

Compston, W., and Jeffery, P. M. (1961). Metamorphic chronology by the rubidium–strontium method. *Ann. N.Y. Acad. Sci.,* **91,** 185–191.

Compston, W., Jeffery, P. M., and Riley, G. H. (1960). Age of emplacement of granites. *Nature,* **186,** 702–703.

Faure, G., and Hurley, P. M. (1963). The isotopic composition of strontium in oceanic and continental basalts: Application to the origin of igneous rocks. *J. Petrol.,* **4,** 31–50.

Hurley, P. M., Hughes, H., Faure, G., Fairbairn, H. W., and Pinson, W. H. (1962). Radiogenic strontium-87 model of continent formation. *J. Geophys. Res.,* **63,** 5315–5334.

Jäger, E. (1962). Rb–Sr age determinations on micas and total rocks from the Alps. *J. Geophys. Res.,* **67,** 5293–5306.

Lanphere, M. A., Wasserburg, G. J., Albee, A. L., and Tilton, G. R. (1964). Redistribution of strontium and rubidium isotopes during metamorphism, World Beater Complex, Panamint Range, California. *In* "Isotopic and Cosmic Chemistry," 269–320. North-Holland, Amsterdam.

Moorbath, S. (1964). The rubidium–strontium method. *Quart. J. Geol. Soc. London,* **120s,** 87–99.

Riley, G. H., and Compston, W. (1962). Theoretical and technical aspects of Rb–Sr geochronology. *Geochim. Cosmochim. Acta,* **26,** 1255–1281.

Schreiner, G. D. L. (1958). Comparison of the Rb-87/Sr-87 age of the Red granite of the Bushveld complex from measurements on the total rock and separated mineral fractions. *Proc. Roy. Soc. Ser. A,* **245,** 112–117.

Whitney, P. R., and Hurley, P. M. (1964). The problem of inherited radiogenic strontium in sedimentary age determinations. *Geochim. Cosmochim. Acta,* **28,** 425–436.

Reprinted from *Nature*, **184**, 1792–1793 (1959)

25

Anomalous 'Common Strontium' in Granite

W. Compston and P. M. Jeffery

THE term 'common strontium' is customarily used to denote the element as found in minerals with high ratios of strontium to rubidium. These include strontium-rich minerals such as celestite and also many common rock-forming minerals containing strontium in the parts per million concentration range. All analyses reported so far, with the single exception of meteoritic strontium, indicate that the isotopic abundances of common strontium are constant to within a few per cent. In particular, no significant increase has been found in the world-wide strontium-87 abundance due to the continual addition of radiogenic strontium with time[1], nor has any case been previously reported of strontium-87 enrichment due to local mixing with radiogenic strontium. In this communication, we wish to cite an instance of local strontium-87 enrichment: two common strontium minerals isolated from a Precambrian granite at Boya near Perth, Western Australia, show strontium-87 abundances some 8 per cent higher than the 'normal' value, while the common strontium for the granite as a whole is apparently some 15 per cent higher.

The discovery was made through attempted strontium/rubidium age determinations on biotite, microcline and 'total rock'[2] samples of the granite using the normal figure for strontium-87 abundance (7·0 atom per cent = 0·71 strontium-87/strontium-86) to correct for common strontium. The average ages calculated on this basis, together with the calculated ratio of radiogenic to common strontium-87, are shown in Table 1.

Table 1. MINERAL AGES USING 0·71 AS STRONTIUM-87/STRONTIUM-86 IN COMMON STRONTIUM

Material	Radiogenic ^{87} Sr/ Common ^{87} Sr	Age (million years)
Biotite	1·75	650
Microcline	0·31	1290
Total rock	0·17	2430

For individual samples, this pattern of increasing age with increasing common strontium content, was reproducible to better than 10 per cent. Both the microcline and the total rock samples contain relatively large amounts of common strontium. As a result, their ages were very sensitive to the value assigned to the common strontium 87/86 ratio.

In a graph relating the ratio radiogenic strontium-87/ rubidium-87 to variations in the common strontium 87/86 ratio (Fig. 1), it was found that the three straight lines representing the three age determinations intersected approximately at a point, corresponding to an 87/86 ratio of 0·82. If this figure is used in correcting for common strontium, the biotite, microcline and total rock samples will give a concordant age of about 520 million years.

The presence of anomalous common strontium was confirmed by analyses of epidote and apatite concentrates from the original granite sample. The ratio strontium/rubidium in the epidote was about 45; no rubidium could be detected in the apatite. Both possessed strontium-87/strontium-86 ratios of 0·77. Even though this value is significantly lower than predicted for complete age concordance, it is sufficiently anomalous to indicate a partial mixing of radiogenic with common strontium.

We believe that the correct interpretation of the age data is that the granite crystallized at least

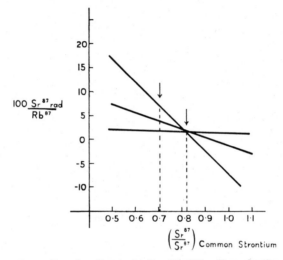

Fig. 1. The ratio radiogenic strontium-87/rubidium-87 as a function of common strontium strontium-87/strontium-86 ratio

2,400 million years ago (corresponding to the total rock age and using the 'normal' strontium-87/strontium-86 ratio for common strontium), but at about 520 million years ago it was metamorphosed. At this time, some or all of its minerals were reconstituted: radiogenic strontium left the biotite and potassium feldspar, to be partially or completely mixed with the original common strontium in other minerals. Apparently the granite as a whole remained a closed system with respect to rubidium and strontium throughout its life-time, including the period of metamorphism. Independent evidence for a 2,400 million year event was given by concordant muscovite and feldspar strontium/rubidium ages. These minerals occurred in a pegmatite dyke cutting the same granite at a place several miles distant. The biotite in the granite here also registers the 520 million year reconstitution. Presumably the much coarser grain size of the pegmatitic minerals was responsible for their apparent complete retention of radiogenic strontium during the metamorphism.

Geological evidence for the metamorphism cannot be seen in hand-specimens of the granite, but in thin section, strain quartz, moderately decomposed feldspar (epidote, sericite), and slightly chloritized biotite provide evidence for stress or reheating. Regionally there has been widespread injection of dolerite dykes, as yet undated. These are probably associated with the metamorphism of the granite, but they themselves cannot be directly responsible since we have lately found precisely the same discordant biotite and total rock ages in a granite situated about 100 miles distant where the occurrence of dolerite dykes is very rare.

It is evident from this study that the clearest distinction must be made in radioactive age determinations between mineral age and the age of original crystallization of the whole rock. An isolated determination by the strontium/rubidium method using the biotite from a granite yields the biotite age only, which is by no means necessarily the age of emplacement of the granite. Knowledge of the relationship between these two ages requires at least one further analysis: preferably another age determination using a total rock sample, or else a measurement of the common strontium strontium-87/strontium-86 ratio for the particular granite under investigation.

Details of the age determinations cited here will be published elsewhere. We are indebted to Dr. A. F. Wilson of the Department of Geology for advice and assistance, and to the Carnegie Institution of Washington for financial support.

W. COMPSTON
P. M. JEFFERY

Department of Physics,
University of Western Australia,
Nedlands.
Aug. 17.

[1] Gast, P. W., *Bull. Geol. Soc. Amer.*, **68**, 1449 (1955).
[2] Schreiner, G. D. L., *Proc. Roy. Soc.*, A, **245**, 112 (1958).

Printed in Great Britain by Merritt & Hatcher Ltd., London & High Wycombe

Geochimica et Cosmochimica Acta, 1961, Vol. 23, pp. 135 to 144. Pergamon Press Ltd. Printed in Northern Ireland

The relation of discordant Rb–Sr mineral and whole rock ages in an igneous rock to its time of crystallization and to the time of subsequent Sr^{87}/Sr^{86} metamorphism*

26

H. W. FAIRBAIRN, P. M. HURLEY and W. H. PINSON

Department of Geology and Geophysics, Massachusetts Institute of Technology,
Cambridge, Mass.

(*Received* 30 *June*, 1960)

Abstract—Interpretation of discordant Rb–Sr ages of coexisting biotite and K-feldspar in igneous rocks, mostly from Sudbury, Ontario, has been attempted using supplementary whole-rock ages. Following a model proposed by COMPSTON and JEFFERY, it is postulated that, if the igneous body is a closed system, and a post-crystallization thermal event interrupts the accumulation of Sr^{87} in biotite and K-feldspar, the whole-rock analysis will give the true age and, owing to diffusion of radiogenic Sr out of biotite and K-feldspar, the apparent ages of these two minerals would be less than the whole-rock age. The common intersection of the three radiogenic growth lines (Sr^{87}/Sr^{86} plotted against age) gives the time of metamorphism. For the majority of the twelve examples the model offers an apparently valid explanation of the discrepant ages in terms of known field relations and two orogenic events at 1·2 billion years and 1·6 billion years.

INTRODUCTION

OF THE various problems which beset isotopic age investigations none is more challenging than the discordant ages shown by coexisting minerals in certain igneous and metamorphic rocks. The disagreement is usually far in excess of possible analytical error and may be caused by redistribution of radiogenic isotopes as a result of reheating of the rock during a metamorphic episode. The particular aspect of this problem under consideration here is the disagreement between Rb–Sr ages of coexisting minerals in rocks of the deformed Pre-Cambrian belt in Ontario just north of Lake Huron. Comparison is also made with K–Ar ages on four of the samples.

The first clue to the riddle of discordant ages was disclosed by Carnegie Institution investigators (TILTON *et al.*, 1959b) in their recent study of biotite, muscovite, K-feldspar, and zircon in gneisses and pegmatites from the Maryland piedmont. They conclude that the gneissic rocks of the region originated at least as long ago as 1100 million years; in the Paleozoic, pegmatites (and probably granites) were intruded 450 million years ago, followed by a metamorphism at 300 million years shown by biotite.

A subsequent clue, marking a distinct advance, has been presented by COMPSTON and JEFFERY (1959), who demonstrated that the Rb–Sr age of a *whole-rock* sample of granite in Western Australia was greatly in excess of the Rb–Sr ages of the component biotite and K-feldspar. In explanation, they postulated that, if the granite were a closed system, and a post-crystallization thermal event interrupted the accumulation of radiogenic Sr^{87} in biotite and K-feldspar, the whole-rock analysis would give the true age and, owing to diffusion of Sr out of biotite and K-feldspar, the apparent ages of these two minerals would be less than the whole-rock age. In more detail the concept is as follows.

---

* MIT age studies No. 21.

135

The Model

Fig. 1 shows the increase* of Sr^{87}/Sr^{86} with time for a hypothetical whole-rock sample of granite and for its component biotite and K-feldspar. It is assumed that, at t_0 when crystallization was complete and the decay of Rb^{87} to Sr^{87} commenced, the initial value of Sr^{87}/Sr^{86} was 0·712. Gast (1960) has assembled abundant evidence in support of this value. As biotite normally has more Rb than K-feldspar and each has more than the whole rock, the accumulation of radiogenic Sr^{87} with time follows the same order. The corresponding radiogenic growth lines have highest slopes for biotite, and lowest for the whole rock. If the radioactive decay

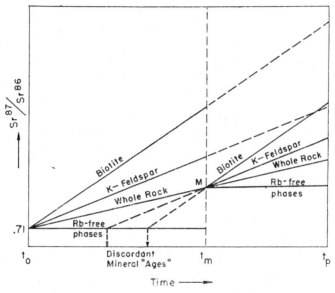

Fig. 1. Schematic diagram of radiogenic Sr growth lines for biotite, K-feldspar, and whole-rock fractions of a granitic rock, showing effect of a thermal episode at time t_m.

proceeded without any interruption to the present (t_p), the growth lines would project straight across the diagram from t_0 to t_p and, since they intersect at t_0 where the initial $Sr^{87}/Sr^{86} = 0·712$, we would have three concordant ages and no further problems to solve. This is essentially what Schreiner (1958) demonstrated for granite of the Bushveld complex in the first published investigation of whole-rock age. If, however, at time t_m in Fig. 1 a thermal disturbance intervenes at sufficiently high temperature, it may bring about a redistribution of radiogenic Sr^{87} amongst certain Rb-poor minerals without net loss or gain of Sr in the rock as a whole. Assuming a normal partition redistribution of this radiogenic Sr^{87} between the minerals, and no loss of Sr from the granite upon completion of this thermal episode, the build-up of radiogenic Sr^{87} will begin anew but with a higher initial value of Sr^{87}/Sr^{86} than the original 0·712 (M in the diagram). If Sr^{87}/Sr^{86} at t_p is measured for biotite, and the slope of the radiogenic growth line is computed, the apparent

---

* For readers unfamiliar with details of the Rb–Sr age method, Rb^{87} decays to Sr^{87}, and Sr^{86}/Sr^{88} is constant for a given mineral. Thus the ratio Sr^{87}/Sr^{86} increases with time for any mineral containing Rb.

136

age, based on an initial Sr^{87}/Sr^{86} of 0·712, would be at the intersection of the broken-line extension of the biotite growth line with the 0·712 line. Similarly for K-feldspar another "age" may be read from the diagram, somewhat larger than that for biotite. In contrast, the whole-rock age will be considerably larger, as its radiogenic growth line is flatter than the others and was assumed unchanged by the metamorphism at t_m.

Additional evidence of this metamorphic interruption in the post-crystallization history of the granite would be found in all minerals normally low in Rb or lacking it. These now would be expected to have values of Sr^{87}/Sr^{86} greater than the normal 0·712 (see M in the diagram). Their radiogenic growth lines have essentially zero slopes both before and after metamorphism but have different Sr^{87}/Sr^{86} values before and after the event.

In summary the diagram outlines an ideal case in which, assuming no gain or loss of Sr from the rock system, the whole-rock age is the true age and the trisection of the three radiogenic growth lines at M gives co-ordinates for the time of metamorphism t_m and the initial higher value of Sr^{87}/Sr^{86} in the post-metamorphic environment.

The validity of this model was successfully tested by COMPSTON and JEFFERY (1959) with a granite from Western Australia. An excellent trisection of radiogenic growth lines for biotite, K-feldspar and the whole rock indicated a metamorphic event at 520 million years and the age of the granite to be 2430 million years. Biotite and K-feldspar "ages" were 650 and 1290 million years, respectively.

The present writers have confirmed the above hypothesis in a number of examples from and near Sudbury, Ontario, now to be presented. Before this Australian work came to our attention interpretation of mineral ages in this region was unsatisfactory (FAIRBAIRN et al., 1960; TILTON et al., 1959a; WETHERILL et al., 1960), as no true age "signals" could be heard above the all-pervasive orogenic "noise". We therefore regard the whole-rock approach as a milestone in interpretation of discordant Rb–Sr data.

SCOPE AND METHODS OF INVESTIGATION

Although our investigations were not originally designed for the work which eventually was found necessary, a considerable body of evidence is at hand which we believe confirms the model illustrated by Fig. 1. The diagrams we have constructed give independent evidence of two orogenic periods already suspected along the north shore of Lake Huron and elsewhere and at the same time indicate approximately the true ages of the rocks. Although the precision of whole-rock age determinations will in general be inferior to biotite and K-feldspar (smaller Rb/Sr and consequently flatter slopes in the diagrams), there is no gainsaying the geological importance of the age so determined. If desired, the error may be reduced by replicate analysis. In the diagrams to be presented the limits of error are usually shown for the whole rock only; errors for the component minerals are smaller and have not in all cases been computed.

The data have been obtained by isotope dilution procedures to determine Rb and Sr. The ratio of Sr^{87} to Sr^{86} was obtained either by direct measurement or by computation from an isotope dilution analysis. The growth lines are located by

137

Table 1. Rb–Sr analytical data

Sample number	Rb (p.p.m. by wt.)	Common Sr (p.p.m. by wt.)	Radiogenic Sr^{87} (p.p.m. by wt.)	Rb^{87}/Sr^{87} (in sample)	Sr^{87}/Sr^{86} (in sample)	Age (million years) for $Sr^{87}/Sr^{86} = 0.712$
B 3086	1926	3·5	9·66	48·2	25·7 (C)	1190 ± 35
F 3086	553	17·0	3·20	36·0	2·63 (C)	1365 ± 40
R 3086	263	43·6	2·29	14·0	1·25 (M)	2065 ± 70
B 3087	2297	4·6	11·54	58·5	24·6 (C)	1195 ± 35
F 3087	630	17·8	4·57	30·8	3·32 (C)	1720 ± 55
B 3094	938	40·7	5·17	33·1	2·01 (C)	1310 ± 40
F 3094	260	55·0	2·18	12·3	1·11 (M)	1985 ± 60
R 3094	234	44·6	2·20	12·5	1·21 (M)	2215 ± 80
B 3200	474	18·2	2·39	25·8	2·93 (C)	1205 ± 40
R 3200	98·7	311	0·78	1·25	0·737 (M)	1870 ± 170
F 3418	186	99·6	1·45	6·25	0·863 (M)	1840 ± 90
R 3418	112	149	0·735	2·84	0·767 (M)	1685 ± 75
F 3424	326	119	2·02	9·55	0·883 (M)	1465 ± 70
R 3424	135	157	0·72	3·27	0·755 (M)	1270 ± 120
B 3479	976	45·8	5·31	32·5	1·90 (C)	1275 ± 40
F 3479	217	43·8	1·37	14·0	1·03 (M)	1490 ± 50
R 3479	193	92·9	1·68	6·71	0·897 (M)	2055 ± 110
F 3767	193	67·8	1·41	8·90	0·926 (M)	1700 ± 85
R 3767	189	161	1·69	4·17	0·815 (M)	2020 ± 120
F 3768	339	87·2	2·10	11·8	0·958 (M)	1470 ± 60
R 3768	160	119	1·27	4·56	0·821 (M)	1880 ± 85
B 3769	382	57·1	2·79	16·0	1·21 (C)	1730 ± 85
R 3769	94·1	369	0·715	1·01	0·732 (M)	1800 ± 225
F 3771	285	122	1·29	8·25	0·820 (M)	1065 ± 75
R 3771	104	295	0·732	1·38	0·737 (M)	1675 ± 185
F 3774	257	43·7	1·80	15·0	1·13 (M)	1655 ± 65
R 3774	172	136	1·56	4·41	0·829 (M)	2140 ± 95

Decay constant used $\lambda_{Rb} = 1.47 \times 10^{-11}$ year^{-1} (Flynn and Glendenin, 1959).

To compute Sr^{87}/Sr^{86} for any time t, given Sr^{87}/Sr^{86} and Rb^{87}/Sr^{87} for present time p, and where the fraction of initial Rb^{87} remaining at present time p is $Rb_p{}^{87}/Rb_t{}^{87}$ is

$$\left(\frac{Sr^{87}}{Sr^{86}}\right)_t = \left(\frac{Sr^{87}}{Sr^{86}}\right)_p \left[1 - \frac{(Rb^{87}/Sr^{87})_p}{(Rb^{87})_p/(Rb^{87})_t}\left(1 - \frac{Rb_p{}^{87}}{Rb_t{}^{87}}\right)\right]$$

$\dfrac{Rb_p{}^{87}}{Rb_t{}^{87}} = e^{-0.693 t/T}$ for any time t and a half-life T of 47 billion years corresponding to the above decay constant.

(M)—Measured directly on unspiked sample.
(C)—Computed from isotope dilution analysis.

138

computing Sr^{87}/Sr^{86} for specific times for a given value of Rb^{87}/Sr^{87}. The error limits of the lines (standard deviation of arithmetic means) are based on $\pm0\cdot3$ per cent error in measured Sr^{87}/Sr^{86}, $\pm0\cdot5$ per cent in computed Sr^{87}/Sr^{86} and about ±3 per cent error in the ratio Rb^{87}/Sr^{87}. The age error is then read directly from each diagram, and varies with the slope of the individual lines.

DISCUSSION OF EXAMPLES

· Figs. 2 and 3 show diagrams for two localities in the Creighton granite. Although Fig. 3 is incomplete (no whole rock analysis is available) the biotite and K-feldspar lines intersect at approximately the same time point as in Fig. 2. Following the interpretation given for the model (Fig. 1), the Creighton granite underwent a metamorphic episode approximately 1·2 billion years ago. The unusually large Sr^{87}/Sr^{86} for sample 3087 at the time of metamorphism (Fig. 3) reflects an unusual Rb content (0·2 per cent for biotite). Although comparison of whole-rock ages is not possible for the two localities it is notable that the two K-feldspar "ages" are discordant beyond possible analytical error. The two biotite "ages" show this less conspicuously, but each is clearly discordant with its coexisting K-feldspar. In the absence of metamorphism since crystallization of the granite it would be reasonable to expect concordant K-feldspar, biotite, and whole-rock ages for both samples.

Fig. 4 of the Murray granite shows a reasonably good trisection centering between 1·2 and 1·3 billion years and, because of high Rb/Sr, unusually high precision of the whole-rock age. From the standpoint of field interpretation it is significant that this granite and the Creighton granite have identical whole-rock ages (cf. Fig. 2). In contrast, their biotite and K-feldspar "ages" show confusing differences.

Fig. 5 for the Cutler granite, about 100 miles west of Sudbury, is thought to be another example of metamorphism at about 1·2 billion years. No K-feldspar data are at hand, but the sharply contrasting slopes of the biotite and whole-rock lines make for high precision of the intersection and the postulated time of metamorphism. Although the error in the whole-rock age is large, its geological significance far outweighs its deficiency as a physical quantity.

Fig. 6 lacks a biotite line and the intersection of the K-feldspar and whole-rock lines has an error parallelogram extending from 1·1 to 1·35 billion years. It seems reasonable to assume that this granite, as with the preceding examples, has been involved in the same metamorphism, at about 1·2 billion years.

The Copper Cliff rhyolite was sampled at two localities, with results as shown in Figs. 7 and 8. The single area of agreement is the whole-rock age (2165 and 2215 million years). The K-feldspar–whole-rock intersection in Fig. 8 at 1·52 billion years may be evidence of an orogeny around 1·6 billion years ago (FAIRBAIRN et al., 1960), as the error parallelogram extends just beyond this age. No biotite is available from this sample.

The K-feldspar line in Fig. 7 presents an unexplained anomaly, as there is no obvious reason for its slope and Sr^{87}/Sr^{86} ratio to be lower than that of the whole-rock line. Clearly it cannot be explained on the basis of the model of Fig. 1. As there is abundant quartz in the rock (presumably very low in Rb), redistribution of

139

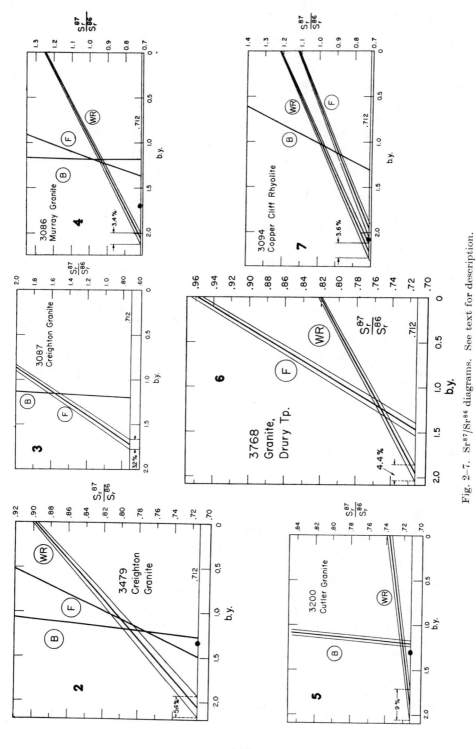

Fig. 2–7. Sr^{87}/Sr^{86} diagrams. See text for description.

140

257

Sr during metamorphism should show the normal relation of lowest slope and Sr^{87}/Sr^{86} for the whole-rock line.

The whole-rock age (1665 ± 185 million years) for the rhyolite in Fig. 9 is questionable as a true age according to the model, as it coincides with the 1·6–1·7 billion year orogeny everywhere evident in the region (FAIRBAIRN et al., 1960). There are also field reasons for supposing the rhyolite to be older than this. Absence of biotite in the rock unfortunately leaves the diagram incomplete.

Breccia in the Creighton granite gives the two intersecting lines of Fig. 10. The error parallelogram of the intersection is larger than might be desired but it seems reasonable to interpret it as evidence of the 1·6 billion year orogeny. As the whole-breccia age agrees with that of the unbrecciated Creighton granite the brecciation itself must be younger and for reasons given elsewhere (FAIRBAIRN et al., 1960) is believed to have developed 1·6–1·7 billion years ago.

Concordant biotite and whole-rock ages are shown in Fig. 11, with an intersection of the two lines almost exactly at the normal Sr^{87}/Sr^{86} value of 0·712. The contrast in precision of the two ages is clearly demonstrated. At this locality biotite apparently escaped the recrystallization so prevalent in the region.

Finally in Figs. 12 and 13 we have two examples of whole-rock and K-feldspar intersections at Sr^{87}/Sr^{86} values less than 0·712. In each case the "ages" calculated using 0·712 for the normal Sr correction are discordant. It may be significant that, although the data are derived from different rocks, the intersection is at approximately 0·70 for both, with the micropegmatite substantially older than the rhyolite. Unexplained, however, are the relative age relations indicated by these two diagrams compared with the field relations (FAIRBAIRN et al., 1960). The micropegmatite intrudes the Onaping volcanic formation of which the rhyolite (Fig. 12) is a part, thus disagreeing with the suggested laboratory conclusion.

The preceding example raises the question of the magnitude of departure from the normal value of Sr^{87}/Sr^{86} (=0·712) which can be ascribed to experimental error. It is implicit in this approach to dating of metamorphism that $Sr^{87}/Sr^{86} > 0·712$ upon termination of any intermediate thermal episode. In most of the examples presented here Sr^{87}/Sr^{86} at the critical metamorphic intersections is unquestionably larger than 0·712 with respect to any reasonable analytical error. Excluding Fig. 11, which shows no evidence of metamorphism, Fig. 5 is the only diagram with an intersection close to 0·712. It differs from 0·712 by about the same amount as the low intersection (0·70) shown by Figs. 12 and 13. Its error parallelogram, however, is well above the 0·712 level and suggests that the deviation from 0·712 is real. The parallelogram, moreover, is much smaller in terms of Sr^{87}/Sr^{86} error than are those of Figs. 12 and 13, which have upper limits in each case of approximately 0·712. This, together with the unlikelihood of an initial $Sr^{87}/Sr^{86} < 0·712$ suggests a non-metamorphic history for the two rocks, comparable with Fig. 11, and analytical error as the source of the anomalous intersections.

CORRELATION WITH K–Ar BIOTITE AGES

Four K–Ar biotite ages have been determined for comparison with the preceding Rb–Sr data. These are indicated at the appropriate point on the age scale in each

141

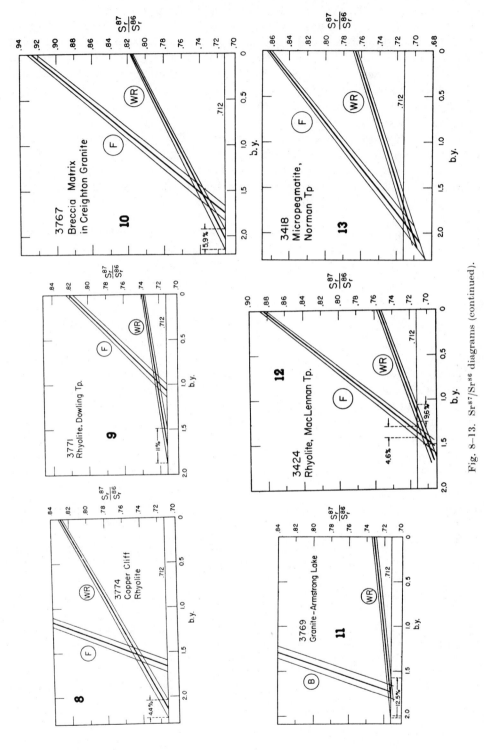

Fig. 8-13. Sr^{87}/Sr^{86} diagrams (continued).

142

figure by solid circles. The value in Fig. 2 for the Creighton granite at 1·33 billion years is close to the 1·2 billion-year orogeny under discussion. Biotite in the Murray granite (Fig. 4) on the other hand discloses evidence of both the 1·2 (Rb–Sr) and 1·6 billion years (K–Ar) orogenies. That is, the later orogeny had no apparent effect on Ar, in contrast with the Sr, which in this example is an ideal illustration of the model represented in Fig. 1.

Biotite in the Cutler granite (Fig. 5), with a K–Ar age of 1·28 billion years, compares favorably with the intersection at 1·2 billion years. On the other hand, the K–Ar age (2·07 billion years) of biotite in the Copper Cliff rhyolite (Fig. 7) closely approaches the Rb–Sr whole-rock age, indicating that Ar in this sample was not affected by later orogenic events. This is an unexplained feature of much interest for, although the Cutler and Copper Cliff localities are 100 miles apart, they are part of the same deformed belt which parallels the north shore of Lake Huron.

In summary these four K–Ar biotite ages show no consistent pattern with respect to the assumed times of orogeny and Rb–Sr whole-rock ages. The values obtained are obviously comparable with individual Rb–Sr ages but are unpredictable.

CONCLUSIONS

The radiogenic Sr model suggested by COMPSTON and JEFFERY stands up remarkably well as an explanation of most of the discordant mineral ages at Sudbury. Existing isotopic age evidence of two orogenies (at 1·2 and 1·6 billion years) is confirmed by the radiogenic growth line intersections of most of the examples discussed. Additional studies should of course be made, including isotopic lead ages on zircon. The present data are preliminary and should be amplified by selection of samples which permit successful age analysis by all three current methods, and should include determination of Sr^{87}/Sr^{86} for Rb-poor minerals.

Acknowledgements—The U.S. Atomic Energy Commission, Division of Research, through annual grants to P. M. HURLEY as supervisor, subsidized the investigation throughout. We are greatly indebted to J. E. THOMSON and the Ontario Department of Mines for generous assistance in collecting material. Two critical samples were provided by the geological staff, International Nickel Company of Canada, through the courtesy of F. ZURBRIGG. The co-operation of the Dept. of Mineralogy, Harvard University, and especially of JUDITH FRONDEL, made possible a rapid X-ray spectrographic survey of many samples. During the early stages of the work the experience of L. F. HERZOG was invaluable for the mass spectrometric measurements. To all these individuals and organizations, as well as to numerous and unnamed students and technical assistants, we extend our thanks.

REFERENCES

COMPSTON W. and JEFFERY P. M. (1959) Anomalous common strontium in granite. *Nature, Lond.* **184**, 1792.

FAIRBAIRN H. W., HURLEY P. M. and PINSON W. H. (1960) Mineral and rock ages at Sudbury-Blind River, Ontario. *Proc. Geol. Assoc. Canad.* In press.

FLYNN K. F. and GLENDENIN L. E. (1959) Half-Life and beta spectrum of Rb^{87}. *Phys. Rev.* **116**, 744.

GAST P. W. (1960) Limitations on the composition of the upper mantle. *J. Geophys. Res.* **65**, 1287.

Schreiner G. D. L. (1958) Comparison of the ^{87}Rb → ^{87}Sr ages of the red granite of the Bushveld complex from measurements on the total rock and separated mineral fractions. *Proc. Roy. Soc. A* **245,** 112.

Tilton G. R., Davis G. L., Wetherill G. W., Aldrich L. T. and Jäger E. (1959a) The ages of rocks and minerals. *Carn. Inst. Wash. Year Book* **58,** 171–178.

Tilton G. R., Davis G. L., Wetherill G. W., Aldrich L. T. and Jäger E. (1959b) Mineral ages in the Maryland piedmont. *Carn. Inst. Wash. Year Book* **58,** 171.

Wetherill G. W., Davis G. L. and Tilton G. R. (1960) Age measurements on minerals from the Cutler Batholith, Cutler, Ontario. *J. Geophys. Res.* **65,** 2461–66.

144

GRAPHIC INTERPRETATION OF DISCORDANT AGE MEASUREMENTS ON METAMORPHIC ROCKS

27

L. O. Nicolaysen

Bernard Price Institute of Geophysical Research, University of the Witwatersrand*
and National Physical Research Laboratory, C.S.I.R.

The first systematic age measurements based on the decay $Rb^{87} \rightarrow Sr^{87}$ were largely carried out on pegmatites and massive granites, and the ages obtained for coexisting mica and K-feldspar in these rocks were usually in good agreement (Aldrich and Wetherill, 1958, Table VIII). Recently published age measurements on certain metamorphic rocks of granitic composition have disclosed the following important feature: gross discordance between the mica age and the K-feldspar age (with the mica age very much less than the feldspar age) when these ages are calculated according to the following assumptions:

(1) There have been no changes in the content of Rb^{87} and Sr^{87} in the mineral since the time of formation except for the nuclear transformations.

(2) The content of nonradiogenic Sr^{87} in the mineral can be estimated by taking 0.71 times the measured proportion of Sr^{86} or 0.084 times the measured proportion of Sr^{88}; these factors result from the many Sr isotopic abundance measurements on terrestial minerals with high Sr/Rb ratios, which show constant values to within a few per cent and indicate that any world-wide increase in the Sr^{87} abundance over geological time has been less than 2 per cent. It is therefore assumed that the Sr incorporated in the mineral at the time of formation was typical of this world-wide "pool" of common strontium.

For the gneiss of the Baltimore, Md., area, Tilton *et al.* (1958) thus reported the presence of $\sim$300 million-year (m.y.)-old biotite in the form of unstrained crystals coexisting with $\sim$1160 m.y.-old K-feldspar in the form of irregular shaped microclines having delicate projections and interlocking contacts with other minerals. They conclude that the microcline and zircon probably record a 1000 to 1100 m.y.-old crystallization, while biotite records a second crystallization 300 to 350 m.y. ago.

Compston and Jeffery (1959) presented certain data for the Boya granite near Perth, Australia, indicating the coexistence of a 650 m.y.-old biotite and a 1290 m.y.-old microcline, if these assumptions are employed. However, these authors presented a graph of the assumed Sr^{87}/Sr^{86} ratio in the nonradiogenic component versus a parameter proportional to the calculated age. This plot showed that the biotite, microcline, and total rock samples give a concordant age close to 520 m.y. if the value 0.82 is used for the Sr^{87}/Sr^{86} of the primary component; the presence of anomalous Sr^{87}/Sr^{86} (with a ratio 0.77) was also reported for the apatite and epidote in this rock. The interpretation of these authors is that the granite crystallized at least 2400 m.y. ago, but was metamorphosed at about 520 m.y. ago. During this metamorphism the rock is believed to have remained a closed system.

My associates and I have studied the granite occurring 3 miles northeast of Mbabane in the Swaziland Protectorate in Africa. This granite is termed

* Permanent address.

198

"G5" in the Swaziland granite classification of Hunter. It shows the same general pattern when the above-mentioned assumptions are used. The total-rock sample and the sample of quartz plus total feldspar show ages close to 2600 m.y., the idiomorphic sphene crystals have a Pb^{207}-Pb^{206} age close to 2700 m.y., but two micaceous samples, consisting of purified biotite and chloritized biotite respectively, show ages of $\sim$1950 m.y.

L. T. Silver reported elsewhere in this monograph that measurements on metamorphics are now taking us on a new avenue toward some of the fundamental questions of petrogenesis. We fully endorse this optimistic approach. For example, if the Boya granite had shown apatite and epidote with Sr^{87}/Sr^{86} values equal to 0.82, this would have been an indication of an intense metamorphic process with very high effective mobility as far as the strontium is concerned. We now have the tools for distinguishing between times of formation of heterogenous chemical systems and times of partial and total fusion, and particularly for distinguishing between certain paragneisses and orthogneisses.

An important question that should challenge all workers in this field concerns the most fruitful and unambiguous way of displaying and analyzing the Rb-Sr analytical data on such metamorphic rocks. A plot of the parameters Sr^{87}/Sr^{86} versus Rb^{87}/Sr^{86} (both expressed as atomic ratios) has certain advantages: (1) for most K-feldspars and total rocks and for many metamorphic micas, the Sr^{87} and Sr^{86} isotopes are of approximately the same proportion, and their ratio can be measured rather precisely (standard deviation <0.5 per cent); and (2) this plot is amenable to study of the effect of relative movements of the two elements rubidium and strontium within the various minerals and rocks.

As FIGURE 1 indicates, any age calculation on a mineral or rock (represented by the point marked 1) consists effectively of drawing a line between 1 and a point on the vertical axis (representing the primary strontium isotope ratio in the mineral at the time of its formation); the slope of this line, that is, $\tan \alpha = (\Delta Sr^{87}/\Delta Rb^{87})$ is measured and substituted into the age equation:

$$t \text{ calculated } = \frac{\ln (1 + \tan \alpha)}{\lambda},$$

where λ is the decay constant of Rb^{87}.

An age may also be obtained by drawing a line between mineral component 1 and another mineral component 2 if we make the assumption that these two minerals formed at the same time and incorporated primary strontium of identical isotopic composition, represented by the point 0. In both cases we still make the assumption referred to earlier, that is, that no fractionation of strontium or rubidium has taken place in any of the minerals since their formation. FIGURE 2 illustrates a graph of this nature.

Examples of this type of plot are given in FIGURE 3 for the Mbabane (Swaziland G5) granite and in FIGURES 4 and 5 for the Baltimore gneiss data given by Tilton *et al.* (1958). The Towson, Phoenix, and Woodstock domelike occurrences of this gneiss are situated about 12 miles from each other. Tilton *et al.* have specifically interpreted the data on the Towson microcline as indi-

cating that the strontium incorporated in the Phoenix microcline was similar
to "common strontium" in its isotopic composition, in spite of the considerable
distance between the rock samples.

One feature is common to the three areas: the feldspar and biotite samples
have concordant ages if a particular "anomalous" and high value of the Sr^{87}/Sr^{86}

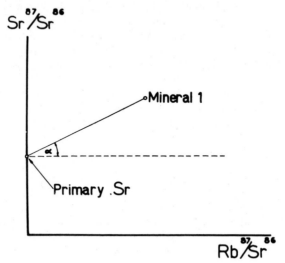

FIGURE 1.   Correction for nonradiogenic strontium.

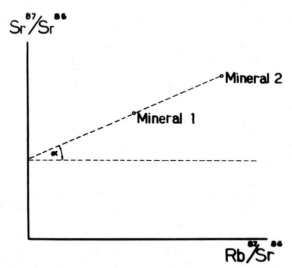

FIGURE 2.   Use of a mineral pair for establishing the nonradiogenic strontium component.

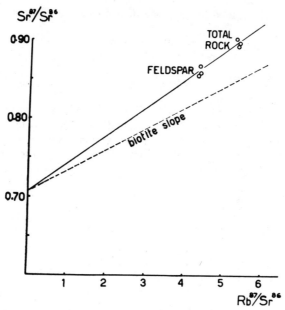

FIGURE 3.   Isotopic analyses of the granite northeast of Mbabane, Swaziland Protectorate, Africa.

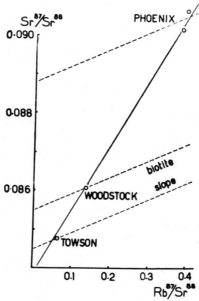

FIGURE 4.   Isotopic analyses of microclines from various domes of the Baltimore gneiss.

or Sr^{87}/Sr^{88} ratio is used for the nonradiogenic component as:

	Northeast of Mbabane	Phoenix Dome	Boya granite
"Concordant" age using anomalous strontium	$\sim$1950	$\sim$300	$\sim$520
Isotope ratio of the anomalous strontium	$Sr^{87}/Sr^{86} = 0.75$	$Sr^{87}/Sr^{88} = 0.089$	$Sr^{87}/Sr^{86} = 0.82$

This relationship is shown by the lines of lesser slope in FIGURE 4, the extension of straight lines joining the microcline and biotite points; if the bulk of the rubidium and strontium in the Phoenix Dome sample is contained in this pair of minerals, the "total-rock" point will fall close to the same line and the same type of "triple concordance" (shown for the Boya granite) will hold.

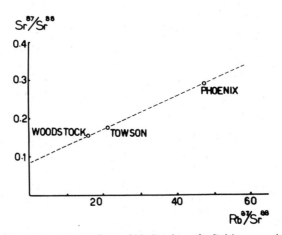

FIGURE 5.   Isotopic analyses of biotites from the Baltimore gneiss.

Several questions arise from an examination of the data for the Swaziland granite G5, the Boya granite, and the Baltimore gneiss:

(1) What critical evidence would indicate whether the feldspar has remained a closed system with respect to rubidium and strontium, or whether the metamorphism was intense enough to bring about an isotopic mixing of the strontium in all phases present so that it would be justifiable to speak of the microcline having an "age" of $\sim$1950 m.y. (Mbabane) or $\sim$300 m.y. (Phoenix Dome)?

(2) What was the destiny of the radiogenic Sr^{87} generated in biotite between the original crystallization of the rock and the time of metamorphism? This information may help to distinguish between micas that have suffered complete and incomplete expulsion of strontium during metamorphism.

(3) Did the particular whole-rock sample behave as a closed system or an open system with respect to rubidium and strontium during metamorphism?

Concerning the first question: it may be suggested for example, that further analyses of feldspar samples from different localities of the Baltimore gneiss may yield points that continue to lie along the steeper line in FIGURE 4; this would be very strong evidence that these feldspars all crystallized with an identical strontium as a primary component and that each feldspar has remained as a closed chemical system (with respect to rubidium and strontium), although there may have been slight changes in physical morphology. The slope of the line (equivalent to $\sim$1100 m.y.) would date the time of formation of the family of related microclines as a heterogenous chemical system. If isotopic mixing has taken place during metamorphism, these points would not be expected to have this regularity, and the presence of anomalously high Sr^{87}/Sr^{88} ratios in the apatite and epidote of these rocks would be predicted. As shown in FIGURE 4, the presence of rather different strontium isotopic compositions for the apatites of the Phoenix, Woodstock, and Towson domes would be expected.

As a first step towards understanding the next problem, consider a simplified granite in which just two phases (mineral 1 and mineral 2) contain the bulk of the rubidium and strontium in the rock. Let p_1 and p_2 be the weight proportions of these two minerals in the rock. Assume Rb and Sr are always present in sufficiently small proportions so that p_1 and p_2 are not affected by transfers of these elements. Consider internal fractionations alone, and that any Sr lost from phase 2 (for example, biotite) was gained by phase 1 (for example, total feldspar).

Let R_1, R_2 be their respective contents of Rb^{87}

A_1, A_2 be their respective contents of Sr^{87}

S_1, S_2 be their respective contents of Sr^{86}

at the instant before metamorphism. During metamorphism the points for mineral 1 and mineral 2 will move in a manner similar to that shown in FIGURE 6a. The positions of the mineral fractions at the following times are indicated by subscripts, noted here in parentheses: time of rock formation(s), instant before metamorphism (m), instant after metamorphism (n), and present day (measured). Mineral ages small compared to the half life of Rb^{87} ($\sim$50 $\times$ 10^9 yrs.) are illustrated; in this case the very slight curvature of the vertical "growth-lines" in FIGURE 6 can be neglected.

The following relations are clear:

(1) After any fractionations the points for 1, 2, and the total rock continue to lie on a straight line and will also lie on a straight line when measured at any particular time after the metamorphism.

(2) If only Rb moves between the two phases during metamorphism, the lengths of the vectors (V_1 and V_2) describing the movement of each point have the relation $V_1 p_1 S_1 = V_2 p_2 S_2$.

(3) If only Sr* moves (from phase 2 to phase 1) during metamorphism, as in FIGURE 6a, and the Sr^{87} content of 2 declines from A_2 to ($A_2 - q$), then the hori-

* This Sr is assumed to be a representative sample of the Sr in the biotite at the time of metamorphism.

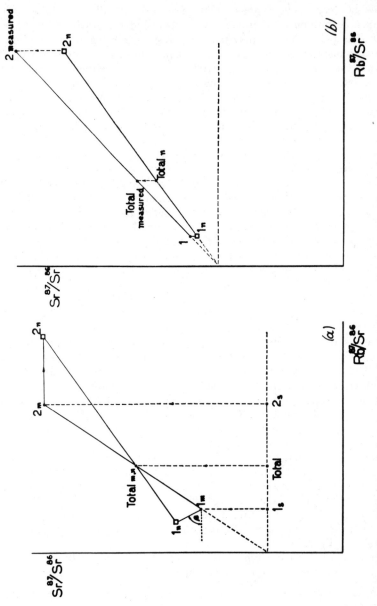

FIGURE 6. (a) Growth histories of the mineral pair from initial crystallization to the time of metamorphism and during metamorphism. (b) Growth histories of the mineral pair after metamorphism.

zontal vector describing the movement of 2 has a length $(R_2'S_2)|q\,(.1_2 - q)|$, and the direction of movement of 1 is defined by

$$\tan\beta = \frac{S_1.1_2 - S_2.1_1}{R_1S_2}.$$

that is, a function independent of q, p_1, or p_2.

Consider any metamorphosed granite in which there is a close approximation to a two-phase distribution of most of the rubidium and strontium, for example, some granites containing microcline microperthite and biotite; there will always be a "triple concordance" of the ages of the microcline, total rock and biotite, with a particular value of the Sr^{87}/Sr^{86} ratio used for the primary strontium component. However, as Compston and Jeffery have noted elsewhere in this monograph, taking this as the age of the metamorphic impress on the minerals is justified only if there was complete mixing of the strontium isotopes between the minerals during metamorphism. If there was merely a partial transfer of strontium from the biotite to the microcline, *then this "triply concordant" age would certainly overestimate the age of the metamorphism*. This possibility is clarified in FIGURE 6b.

When the bulk of the rubidium and strontium in the rock is contained in the two minerals plotted and there is a strong possibility that one phase (for example, feldspar) acted as a "soak" for strontium expelled from the other phase (for example, biotite), there is a particular need for caution in interpreting the data.

An application of the graphical procedure of FIGURE 2 occurs where a uniform biotite-bearing igneous rock has been metamorphosed. An array of points are plotted for two fractions, biotite and "the rest", as successively higher metamorphism is attained. For internal transfers the slopes of the lines (between the two fractions) are precisely defined by certain simple relationships between the weight proportion of biotite, the rubidium and strontium contents of the fractions, and the times of initial crystallization and metamorphism.

If good independent evidence is available (for example, from U-Th-Pb minerals) concerning the time of initial crystallization of a rock and the time of a subsequent metamorphism, this information imposes useful restraints on the interpretation of the Rb-Sr data. In particular, it allows a distinction between different ways in which expulsion of radiogenic Sr^{87} from the mica could have occurred.

Equivalent Graphs for the Uranium-Lead System

The axes in the graphs referred to may be transformed to U^{238}/Pb^{204} versus Pb^{206}/Pb^{204}. This plot is useful for study of fractionations in U and Th minerals. Consider a large uniform crystal of such a mineral, that has been subjected to chemical alteration in recent times. If several portions of the crystal are analyzed, the relative movement of the chemical components will be manifested in a distinct pattern of discordant Pb-U ages for each fraction. These data will give rise to an array of points on the graph of U^{238}/Pb^{204} versus Pb^{206}/Pb^{204}. The horizontal or vertical alignment of this array can allow a significant distinction between discordant ages due to: (1) movement of uranium with respect to primary lead; and (2) movement of radiogenic lead with respect to primary lead.

To assist in the interpretation of discordant age measurements it is important that workers should publish their detailed strontium or lead isotopic data in "raw" form. It is preferable to quote the strontium data as "Sr^{87}/Sr^{86}" or "Sr^{87}/Sr^{88}" in the mineral or rock sample, instead of "per cent radiogenic Sr." Modal analyses are very relevant to interpretation of the data.

Acknowledgment

The experimental data on the Swaziland granite occurring northeast of Mbabane were obtained in a joint investigation with my colleague A. J. Burger, to whom I am also indebted for fruitful discussions.

References

ALDRICH, L. T. & G. W. WETHERILL. 1958. Geochronology by radioactive decay. Ann. Rev. Nuclear Sci. **8:** 257–298. Ann. Reviews, Inc. Stanford.

COMPSTON, W. & P. M. JEFFERY. 1959. Anomalous "common Strontium" in granite. Nature. **184:** 1792.

TILTON, G. R., G. W. WETHERILL, G. L. DAVIS & C. A. HOPSON. 1958. Ages of minerals from the Baltimore gneiss near Baltimore, Maryland. Bull. Geol. Soc. Am. **69:** 1469–1474.

JOURNAL OF GEOPHYSICAL RESEARCH   VOLUME 67, No. 9   AUGUST 1962

Rubidium-Strontium Dating of Shales by the Total-Rock Method

W. COMPSTON AND R. T. PIDGEON

Department of Geophysics
Australian National University, Canberra

28

Abstract. Measurements of Rb^{87}, Sr^{86}, and Sr^{87}/Sr^{86} are reported for total-rock samples of three shale formations. These data allow calculation of the time of deposition for one formation known to be of Lower Silurian age and of either the deposition or metamorphism for another. The calculation is based on the model that each formation was once uniform in its ratio of Sr^{87}/Sr^{86} and that the observed distribution of Rb/Sr through each formation was established at that time. By considering the sources of error in our measurements, we believe that a small but definite range in Sr^{87}/Sr^{86} actually occurred, rather than complete uniformity. Nevertheless, excellent agreement is found between our calculated value and the current estimates for the absolute age of Lower Silurian. The data for the third formation indicate that its ratio of Sr^{87}/Sr^{86} was always highly nonuniform. Measurements of Sr^{87}/Sr^{86} for sea water from various localities are also reported. The variation in this ratio is very small, and the absolute values are significantly lower than those calculated for Sr^{87}/Sr^{86} in the shale formations at the times of their deposition. These observations show that the strontium fixed in the shales during deposition was not entirely precipitated from sea water: some must already have been enriched in radiogenic strontium relative to sea water. We therefore believe that a small amount of detrital mica or feldspar distributed irregularly through the shales must be responsible for the nonuniformity of their initial ratios of Sr^{87}/Sr^{86}.

1. *Introduction.* In studying the geochronology of two areas over the past two years, we had occasion to analyze samples of sedimentary rocks. Most of these were alkali-rich shales and metashales. Many possessed a high ratio of Rb/Sr, so that the radioactive decay of Rb^{87} produced large increases in the ratio of Sr^{87}/Sr^{86} during their lifetimes. We were led, therefore, to try calculations of the absolute age of the samples to make some assessment of the potential of such shales as dating material.

Results for two formations, the Cardup shale and the Cooma metamorphosed sediments are discussed in sections 3 and 4 and are shown to provide valuable information. They suggested that an attempt should be made to determine the absolute accuracy possible in shale dating, and a shale whose stratigraphic position is precisely known was chosen for this purpose. Results on this shale, the State Circle shale, were excellent and are described in section 5.

These results suggest that Rb-Sr dating of shales by the total-rock method can be a useful and accurate method of dating sediments. Detailed work on a wide range of sedimentary rocks is planned for the future.

2. *Measurement of isotopic ratios.* The mass spectrometer used in this work is a Metropolitan-Vickers type MS2-SG. It was modified to use a Cary vibrating-reed electrometer and to measure isotopic ratios by rapid switching of the magnetic field. Rubidium and strontium were analyzed as chlorides by the triple-filament technique. Full details of the chemical and mass-spectrometric procedures will be published later.

To compare correctly measurements of Sr^{87}/Sr^{86} made on this machine with those made elsewhere, we give our results to date on the standard strontium used by *Aldrich, Herzog, Doak, and Davis* [1953]:

$$Sr^{88}/Sr^{86} = 8.293 \pm 0.006$$

$$Sr^{87}/Sr^{86} = 0.7053 \pm 0.0003$$

The uncertainties quoted are the standard errors of the mean. Measurements of this standard approximately 1 per cent higher for each ratio were recently reported by *Faure and Hurley* [1962], and the measurements by these authors on the strontium from basic igneous rocks is also 1 per cent higher than unpublished observations here. On the other hand, our measurements on the Moore County achondrite agree exactly with those of P. W. Gast (personal

3493

TABLE 1. Rb-Sr Isotopic Analyses of the
Cardup Shale Samples

Specimen	Rb^{87}, $\mu m/g$	Sr^{86}, $\mu m/g$	$\dfrac{Rb^{87}}{Sr^{86}}$	$\dfrac{Sr^{87}}{Sr^{86}}$
Composite	0.799 ± 2	0.0209	38.2	1.207
white shale 1		0.0213	37.5	1.216
Dark band 2	0.599 ± 5	0.0193	31.0	1.109
		0.0197	30.4	1.130
White band 3	0.630 ± 3	0.0206	30.6	1.210
		0.0204	30.9	1.212
Black shale 4	0.918 ± 3	0.0160	57.3	1.304
		0.0166	55.3	1.286
Black shale 5	0.540 ± 2	0.01535	35.2	1.099

communication). These systematic differences appear to be characteristics of individual mass spectrometers: the true isotopic ratios cannot be determined until an absolute calibration of one machine has been made using known mixtures of the separated isotopes. Such a calibration of the A.N.U. machine is in progress. Determination of Sr^{86} and Rb^{87} are independent of systematic machine error if the 'spike' solutions have been calibrated against standard solutions of the natural elements.

The deviations of our Sr^{88}/Sr^{86} and Sr^{87}/Sr^{86} measurements from their mean values over a series of replicate observations appear to be quite independent of one another. Thus the Sr^{87}/Sr^{86} measurements of unspiked strontium *made on the A.N.U. machine* should not be normalized to a fixed value of Sr^{88}/Sr^{86}, in contrast to the evident validity of this precedure on the machines used by Faure and Hurley. This situation also is probably due to the characteristics of the individual machines.

3. *The Cardup shale.* Rubidium-strontium isotopic analyses of biotite from the 2450-m.y.-old Mundaring granite in Western Australia have suggested that the western margin of this mass may have experienced a mild regional metamorphism some 600 to 700 m.y. ago (Compston, Riley and Jeffery, in preparation). This was one of several possible explanations for the biotite apparent ages, and it requires, if correct, that the deposition of the overlying Cardup shale must have taken place later than this time, since metamorphic effects in the shale are almost nonexistent. It was in order to test this hypothesis of regional metamorphism that

measurements on total rock samples of shale were first undertaken. Subsequently, convincing evidence for a metamorphic event at 660 m.y. was obtained from another source so that this figure could be set as an upper limit to the time of deposition.

The Cardup shale can be traced for some 30 miles along the Darling scarp near Perth, Western Australia. It comprises black carbonaceous shale, white banded shale, sandstone, and minor conglomerate. The beds unconformably overlie 2450-m.y.-old gneiss and are themselves overlain by Quarternary gravels and sands. They dip 70° to the west and have a weakly developed fracture cleavage. They have been intruded by a spilitic-type sill and later by quartz dolerite dikes. Biotite developed in the contact zone of the latter gives an age of about 550 m.y., which sets a lower limit to the time of deposition.

The white shale is composed of fine fibrous sericite, quartz, and biotite, the latter becoming locally more abundant to produce the banded appearance. It is high in Si, Al, and K, very low in Ca, and is thought to be derived from illite-rich clays [*Prider*, 1941]. The biotite is believed to be detrital (Prider, personal communication). The black shale is composed of sericite with minor quartz and graphitic material. It is apparently free of biotite. Idioblastic tourmaline is present in both shale types. This is believed to be due to recrystallization of original shale constituents rather than to the later introduction of external material [*Prider*, 1941]. According to *Singh* [1958], the shales were formed by 'slow or aperiodic introduction of fine clastic material into neritic, possibly lower neritic waters.' Their tectonic environment is 'stable shelf facies.'

Except in the immediate vicinity of the dikes, petrological evidence suggests that the shale has remained a closed chemical system since alkali fixation was completed in the sericite, presumably very soon after deposition. Thus, except for the problem of inherited radiogenic strontium in the detrital biotite, it was expected that total-rock age determinations would yield values very close to the age of shale deposition.

The samples analyzed are shown in Table 1. Single dissolutions were made, Rb^{87} and Sr^{86} were determined by isotope dilution on separate aliquots of the sample solution, and the Sr^{87}/Sr^{86} ratio of the sample was calculated from the

TABLE 2. Cardup Shale Apparent Ages

Specimen	$10^2 \theta$	Age, m.y.
1	1.32	900
2	1.34	910
3	1.63	1110
4	1.04	710
5	1.10	750

measured isotope ratios of the mixtures of aliquots and Sr^{86} 'spike.'

To calculate the age, the following relationships were used:

$$t = (\text{age}) = 1/\lambda \ln (1 + Sr^{87}{}_* / Rb^{87})$$

and

$$Sr^{87}{}_* = Sr^{86}(R_p - R_i)$$

where λ denotes the Rb^{87} decay constant (here taken as 1.47×10^{-11}/year), $Sr^{87}{}_*$ the radiogenic strontium, and R_p and R_i the present-day and initial Sr^{87}/Sr^{86} ratios of the sample.

For Paleozoic minerals $\ln(1 + Sr^{87}{}_*/Rb^{87})$ very closely approximates to the ratio $Sr^{87}{}_*/Rb^{87}$, which will be here denoted as θ. Thus,

$$\lambda t = \theta = (Sr^{86}/Rb^{87})(R_p - R_i) \qquad (1)$$

To determine θ, it is usually sufficient to assume R_i to be 0.71, an average value for common Sr in most rocks. If this is done with the data for the Cardup shale, we obtain the apparent ages in Table 2. These should all be equal and less than 660 m.y. if the ages measured are to be the ages of deposition. However, only the black shale specimens give ages reasonably close to 660 m.y., the rest being greater, as if they had experienced a recent loss in Rb^{87} or a gain in radiogenic strontium at any time in their history.

Loss in Rb^{87} and gain in $Sr^{87}{}_*$ after diagenesis are unlikely on the grounds that the specimens show no independent evidence for being open systems. On the other hand, undecomposed detrital biotite is present, and some of the sericite also may be fine flakes of unaltered muscovite from the adjacent Archaean shield, or second-generation (or older) hydromicas formed during previous sedimentary cycles. Thus the most likely explanation of the high apparent ages is a variable excess of radiogenic strontium

which has been present in the specimens from the time of deposition.

This situation is best seen by using the graphical expression of data given by *Nicolaysen* [1961]. Equation 1 may be rearranged to give

$$R_p = R_i + \theta(Rb^{87}/Sr^{86}) \qquad (2)$$

which becomes the equation of a straight line if R_p and Rb^{87}/Sr^{86} are regarded as variables. The age is proportional to the gradient θ of such a line, and R_i is given by its intercept on the R_p axis.

Equation 2 was developed to handle a number of samples having the same (unknown) age and the same (unknown) R_i, for which the R_p-Rb^{87}/Sr^{86} points corresponding to the different samples will define a single straight line. This line constitutes an isochrone for such a group of samples. For the Cardup shale, it is known that the age of deposition is approximately the same for all samples and probably not greater than 660 m.y. This corresponds to an isochrone whose gradient θ is 0.0097. Figure 1 shows the R_p-Rb^{87}/Sr^{86} plot for these samples with isochrones drawn through sample points corresponding to this maximum value for the gradient. Had the samples (assumed 660 m.y old) all possessed the same R_i these isochrones would superimpose. Instead, the figure shows that each sample must have commenced with significantly different values for R_i, ranging from 0.735 to 0.90, or even higher if the gradient corresponding to the lower age limit of 550 m.y. is assumed. These Rb-Sr age determinations provide a measure of the variability of the initial Sr^{87}/Sr^{86} ratio throughout the shale rather than an accurate measure of age.

As suggested above, undecomposed detrital micas are probably responsible for this variability. The age of the detrital biotite would be 660 m.y. It follows, therefore, that much of the detrital mica must be sericite, lying in the apparently older sericite-bearing bands. This is consistent with the observations that the dark band, specimen 2, containing most detrital biotite, has a lower R_i value than the composite shale, and this in turn has a lower value than the most sericite-rich member.

The carbonaceous shale specimens appear least contaminated with inherited radiogenic strontium. This raises the important question

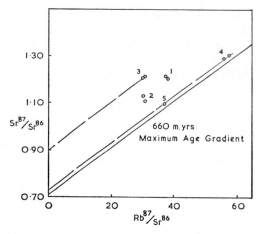

Fig. 1. Sr^{87}/Sr^{86} versus Rb^{87}/Sr^{86} for the Cardup shale total-rock samples.

whether carbonaceous shales in general possess an extremely low content of detrital mica.

4. *The Cooma metamorphosed sediments.* Analyses of the Cooma metasediments were made primarily to compare their ratios of Sr^{87}/Sr^{86} and Rb/Sr with those of an associated igneous body, the Cooma gneiss. We wished to see whether the value of the initial ratio of Sr^{87}/Sr^{86} for the gneiss was consistent with the theory that it was formed wholly from the metasediments by some process of palingenesis. This study will be reported later. Only the dating application of this data will be considered here.

The specimens analyzed were collected approximately 3 miles west of Cooma, N.S.W., from the Binjura beds [*Joplin,* 1942–1943]. The main rocks here are aluminous pelites consisting largely of chlorite, biotite, muscovite, and quartz, and psammites, containing small amounts of chlorite and mica, which are usually interbedded with the pelites. Tuffaceous sandstones and rocks gradational between the pelites and psammites are also present. The beds are tightly folded and, according to Joplin, have probably experienced two separate metamorphisms, not widely separated in time. The older of these was regional, producing chlorite and biotite schists, whereas the second is evident as zones of progressively higher metamorphic grade surrounding the Cooma gneiss, apparently a thermal effect associated with the emplacement or generation of this mass.

No fossils have been discovered, and so the precise stratigraphic position of the Binjura beds is not known. Unmetamorphosed siliceous shales nearby contain Upper Ordovician graptolites, but the mutual relationship of the two formations is obscure. However, *Browne* [1943] proposed an Upper Ordovician age, and Joplin (personal communication) a Middle or Lower Ordovician age, on the basis of interpretations of structural and chemical evidence. A definite minimum age limit of 395 m.y. is set by unpublished mineral and whole-rock rubidium-strontium analyses from the intrusive Cooma gneiss and associated pegmatites (Pidgeon and Compston, in preparation). This figure is essentially in agreement with the potassium-argon age reported by *Evernden and Richards* [1962] for biotite from the same body.

The analytical data for four samples are shown in Table 3. A fist-sized piece of each was crushed to pass 40 mesh, then split to about 10 grams. This was crushed further to pass 100 mesh, and half-gram portions were taken for analysis. Replicate measurements of R_p using unspiked aliquots have been made for most dissolutions. Comparison with the calculated R_p values shows that no great increase in experimental error would arise if only the latter were available. For this reason, sample 10B has been given equal weight with the rest. The data are plotted in Figure 2 as R_p-Rb^{87}/Sr^{86} points, and in Figure 3 as θ-R_i lines [*Riley and Compston,* 1962].

A maximum age of deposition for the Binjura beds can be set from Figure 2 on the following assumptions:

1. The samples analyzed have remained closed systems with respect to rubidium and strontium during metamorphisms.

2. Sample 3 was deposited with R_i equal to that of basic igneous rocks.

With regard to the first assumption, petrographic and chemical studies by *Joplin* [1942] suggest that the metasediments have retained their macrochemical compositions up to a higher metamorphic grade than these samples. This may also be true for their trace constituents, in particular rubidium and strontium. With regard to the second, basic igneous rock R_i has the lowest value known (excepting that of certain types of meteorite). It is 0.702 ± 0.001, as

TABLE 3. Rb-Sr Isotopic Analyses of the Cooma Metasediments Samples

Specimen		Rb^{87}, $\mu m/g$	Sr^{86}, $\mu m/g$	Rb^{87}/Sr^{86}	R_p (calc.)	R_p (meas.)
Quartz-chlorite- muscovite schist	3A	(0.619)	0.04769	(12.98)	0.794	0.7907
						0.7889
						0.7894
	3B	0.6081	0.04811	12.64	0.799	(0.795)
	3C	0.6080	0.04809	12.64	0.790	0.7903
						0.7892
	3D	0.6093	0.04815	12.65	0.785	0.7886
Quartz-epidote rock	7A	0.00607	0.2377	0.0255	0.713	0.7155
	7B					0.7161
	7C					0.7156
Quartz-biotite schist	9A	0.4937	0.1555	3.175	0.731	0.7301
	9B	0.5061	0.1617	3.130		0.7331
						0.7312
Quartz-muscovite- cordierite-biotite schist	10A	0.5673	0.06582	8.62	0.765	0.7629
						0.7626
	10B	0.5768	0.06628	8.70	0.764	
Averaged data	3C			12.64		0.7898
	3D			12.65		0.7886
	7A			0.0255		0.7157
	9A			3.175		0.7301
	9B			3.130		0.7321
	10A			8.62		0.7628
	10B			8.70		0.7640

measured on the A.N.U. machine. The line joining this value and the sample 3 point in Figure 2 therefore defines a maximum age isochrone, and lines of higher gradient drawn through the sample points have no physical meaning. (Sample 3 is considered because lines of different gradient drawn through this point will produce a greater range in the R_p intercept than corresponding lines drawn through the other points.) The value of this maximum age, 460 m.y., calculated from the slope of the isochrone, is consistent with the Middle or Lower Ordovician age proposed by Joplin. R_i for the different samples, found from Figure 2 by extending lines of the same maximum gradient through each point to the R_p-axis, ranges from 0.702 to 0.715.

Other lines drawn in Figure 2 are the minimum age isochrone for 395 m.y. and the least-squares fit to all the experimental points. If it may be assumed that R_i was the same for all samples, the physical meaning of the least-squares line is clear: it is an approximation to the isochrone on which all points will lie in the absence of experimental error. As such, its gradient would correspond to an event at 403 m.y., with 95 per cent confidence interval of 410 to 394 m.y., and the event itself could refer either to the original deposition of the sediments, if the samples have remained closed systems during metamorphism, or to the metamorphism itself if they have not. On the other hand, if there were a measurable variance in R_i, the chronological meaning of the least-squares gradient would be much less obvious. In this case, the statistical derivation of the gradient would amount to deliberate choice of the smallest variance in R_i which will fit the measured data. However, there is no a priori knowledge of R_i variance. For this reason the 'age' corresponding to the least-squares gradient merits no preference at present over that based upon other assumptions.

An analysis of variance in Figure 2 will reveal whether experimental error or R_i variance is responsible for the observed scatter of points. A simpler (if less objective) indication is given by the θ-R_i plot in Figure 3, which emphasizes

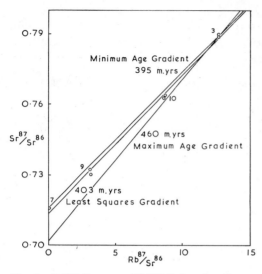

Fig. 2. Sr^{87}/Sr^{86} versus Rb^{87}/Sr^{86} for the Cooma total-rock samples.

how the value of 0.711 for R_i in sample 7 would produce an intersection of all lines apparently to within measurement reproducibility. The difference between 0.711 and the measured 0.716 seems quite beyond the limits of experimental error. It must be concluded that a significant R_i difference exists, at least between sample 7 and the rest, and therefore that the least-squares line shown in Figure 2 is not the best isochrone approximation. On the other hand, the least-squares fit to samples 9, 10, and 3 only, would be an excellent approximation, and this line yields an age of 416 m.y. with 95 per cent confidence interval of 425 to 407 m.y. This result, a Silurian age for the Binjura beds, does not conflict with known geological facts. Referring to the petrographic appendix, it is clear that there are valid geological grounds also for isolating sample 7: this sample was originally an impure sandstone, whereas the other three were shales.

5. *The State Circle shale.* This formation is a member of the Lower Silurian sequence in the Canberra District [*Öpik*, 1958]. It consists of noncalcareous sandy shale and black shale containing beds of fine-grained sandstone. It has a total thickness of about 200 feet. From its areal extent, Öpik suggests that this formation is a transgressive unit indicating a general deepening of the Silurian sea.

Specimens selected for analysis were from an occurrence of the State Circle shale at Etheridge Creek on the eastern slopes of Black Mountain, about two miles northwest of Canberra.

The age of the shale is upper Lower Silurian, zone 22, determined by the occurrence of the graptolites *Monograptus turriculatus, M. spiralis, M. exiguus.* On the time scale compiled by *Kulp* [1961] the absolute age of the formation would be about 420 m.y. or using that of *Evernden and Richards* [1962], about 430 m.y. For convenience, we have assumed in the following discussion that no age uncertainty exists in the Palaeozoic time scale beyond these limits. It is acknowledged, however, that the Silurian absolute age is the least well known.

Samples 12 and 13 were collected within a few feet of each other, whereas sample 17 was some 70 feet higher in the section. It is possible that the age of sample 17 may be 1 or 2 per cent less than the others.

The detailed and averaged analytical data are given in Table 4, and the averaged data are plotted in Figure 4. As before, only the measured R_p values are used. The averaged Rb^{87}, Sr^{86}, and R_p values for two or more sample dissolutions were used in this case, rather than individual points corresponding to the different dissolutions, because complete data were available for only two out of the eight dissolutions. 12C and 12D were small weight dissolutions spiked for rubidium only. Thus the experimental variance is not obvious from Figure 4, but it may be estimated from Table 4.

The two lines drawn through the sample-13

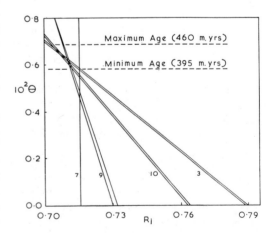

Fig. 3. $\theta(Sr^{87}{}_*/Rb^{87})$ versus R_i (initial Sr^{87}/Sr^{86}) for the Cooma total-rock samples.

TABLE 4. Rb-Sr Isotopic Analyses of the State Circle Shale Samples

Specimen	Rb^{87}, $\mu m/g$	Sr^{86}, $\mu m/g$	Rb^{87}/Sr^{86}	R_p (calc.)	R_p (meas.)
Shale 12A	0.6238	0.05458		0.7822	0.7868
	0.6231				
B				0.7885	0.7857
					0.7860
C	0.6232				
D	0.6198				
E		0.05446		0.7817	
Shale 13A		0.02578		0.841	0.8450
B	0.5332				
C	0.5353	0.02579		0.853	0.8465
Shale 17A	0.5788	0.03677	15.74	0.817	0.8160
B	0.5765	0.03603	16.00	0.8183	0.8181
Averaged data					
12	0.6225	0.05452	11.42		0.7862
13	0.5342	0.02578	20.72		0.8457
17	0.5776	0.0364	15.87		0.8170

point are isochrones corresponding to the limits of 420 and 430 m.y. Assuming that the deposition of all three samples occurred within a 2-m.y. interval, so that the isochrones through each point will be parallel, then R_i values required for equal ages range from limits of about 0.714 to 0.719. None of these samples can have commenced with basic igneous rock R_i, since this would yield an age greatly in excess of the lower limit. Instead, two alternatives seem to be possible: a measurable but small R_i variance among the samples, perhaps as much as the above limits, or no significant R_i variance, the deviations of the points from the least-squares fit being caused by experimental error only. At present, there are insufficient data to objectively distinguish between these alternatives. However, it is our impression that a significant R_i variance exists. In either case, it is definite that the average R_i for the formation will be about 0.714, and the least-squares lines could be an isochrone equivalent to an age of 433 m.y. in essential agreement with the estimate from *Evernden and Richards* [1961].

6. *Sea-water strontium.* It has been suggested that the presence of older detrital micas was responsible for the larger R_i variance in the Cardup shale. This explanation could also be applied to the Binjura beds and to the State Circle shale, but, since the apparent range in R_i is much smaller in these instances, it also ap-

peared possible that variations in the Sr^{87}/Sr^{86} ratio of sea water might have been responsible.

In theory, sea-water Sr^{87}/Sr^{86} may range from the minimum value 0.702 of basic igneous rocks up to an undefined maximum value in regions of the sea contaminated with radiogenic strontium from the weathering of ancient rubidium-rich rocks. Through the growth of authigenic minerals and by exchange of strontium in detrital clays, it seems likely that the strontium fixed in at least some shales should have the same isotopic composition as the environmental sea water. Granted such equilibration mecha-

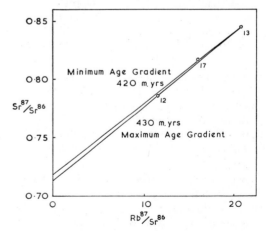

Fig. 4. Sr^{87}/Sr^{86} versus Rb^{87}/Sr^{86} for the State Circle shale total-rock samples.

TABLE 5.   Rb-Sr Isotopic Analyses of
Water Samples

Location	87/86	Rb, $\mu g/g$	Sr, $\mu g/g$
Rottnest Island, Perth, Western Australia	0.7065	0.117	7.76
Sorrento Beach, Perth, Western Australia	0.7068	0.118	7.78
Elwood Beach, Port Phillip Bay, Victoria	0.7088	0.112	7.51
Rosedale Beach, Batemans Bay, N.S.W.	0.708 ± 1	0.116	7.76
Mundaring Weir, Western Australia	0.723 ± 1	0.0073	0.088

nisms, any change in sea-water Sr^{87}/Sr^{86} with time or from place to place in the deposition site of a sediment would produce a variance in the sediment R_i. For this reason, analyses of five water samples were undertaken to see whether significant changes in sea-water Sr^{87}/Sr^{86} do in fact occur.

The results are shown in Table 5. The four sea-water samples show a striking uniformity in strontium concentration and isotopic composition, which furthermore are in agreement with results for Atlantic Ocean water found by P. J. Anderson, W. H. Pinson, L. F. Herzog, and R. F. Cormier (personal communication), after making allowance for the 1 per cent difference in ratio measurement mentioned earlier. No sea-water Sr^{87}/Sr^{86} ratio equals the lowest R_i value (0.714) inferred for the State Circle shale, or the least-squares value (0.711) for the three Binjura beds samples. It appears either that these sea-water samples are not representative of the depositional environments or that detrital minerals of higher Sr^{87}/Sr^{86} than typical seawater are responsible for the observed shale R_i.

One fresh-water sample from Mundaring weir was also analyzed. The watershed here is a 2450-m.y.-old granodiorite bathylith abundantly intruded by quartz-dolerite dikes. The Sr^{87}/Sr^{86} ratio measured has approximately the value expected, but in comparison with that of sea water the strontium concentration is extremely low. It is evident that no reasonable dilution of sea water by the water from Mundaring weir could measurably change the sea-water Sr^{87}/Sr^{86} ratio The over-all evidence from water analyses therefore suggests that the small R_i variance inferred

for the Binjura beds and the State Circle shale, as well as the gross variance present in the Cardup shale, must be due to older detrital mica or clays which have not exchanged their excess radiogenic strontium with sea water. Convincing evidence for the existence of such material elsewhere has been recently reported by *Hurley, Brooking, Pinson, Hart, and Fairbairn* [1961], in the form of 200-m.y. (and older) illites from the Mississippi River delta deposits.

7. Comparison with potassium-argon dating. Considerable work has been done on the evaluation of illite for K-Ar dating of sedimentary rocks, [*Evernden, Curtis, Obradovich, and Kistler*, 1961] and on the inverse problem of deducing the genesis and history of illites from their apparent K-Ar ages [*Hurley et al.*, 1961]. To consider the dating application only, the difficulties in the K-Ar method are the inheritance of radiogenic argon retained by older micas and illites and the diffusive loss of argon from extremely small shale components. Loss of daughter product from the whole rock seems much less likely for the Rb-Sr method, but the inheritance of radiogenic strontium locked in older micas is at least as likely as the inheritance of argon.

In contrast to the argon situation, however, there is some hope that approximate correction for inherited radiogenic strontium can be made, depending on the nature of the shale constituents and on the processes which determine their final rubidium and strontium distributions. If the shale as a whole is composed of a uniform mixture of detrital and authigenic components, its Sr^{87}/Sr^{86} will be anomalously high (relative to sea water) but *uniform*. If in addition the rubidium concentration of the shale independently varies from place to place, so that samples of different Rb/Sr ratios can be obtained for analysis, an isochrone could be defined by plotting the R_p-Rb^{87}/Sr^{86} points of the different samples. This isochrone would designate the time at which the various Rb/Sr ratios and the uniform Sr^{87}/Sr^{86} ratio were established. If this is also assumed to be the time of deposition or early diagenesis of the shale, then fixation of rubidium in clay mineral at this time is implied.

The small R_i variance and the excellent apparent age for the State Circle shale indicate that the above conditions have been closely approached in this formation. On the other hand, the Cardup shale samples show definite non-uniform mixing of authigenic and detrital components, the latter being of at least two different ages. (The case of nonuniform two-component mixtures has a simple θ-R_i pattern which is discussed by *Riley and Compston* [1962]. The Cardup shale data do not show such a pattern.)

To check for inherited radiogenic argon in the State Circle shale, potassium and argon analyses were made on sample 13. The results were:

Potassium	3.23 per cent
Ar^{40}/K^{40}	18.2×10^{-3}
Apparent age	289×10^6 years
Air correction	88.5 per cent

The high air correction and probable diffusive loss of argon minimize the significance of this young age. No further K-Ar work was done.

APPENDIX

Petrography of the Analyzed Rocks

The Binjura Beds Specimens

Estimates of metamorphic grade are based on the metamorphic facies proposed by *Turner and Verhoogen* [1960].

Specimen 3. Quartz-chlorite-muscovite schist. Specimen 3 is foliated schist consisting of alternating bands, approximately 0.6 mm wide, of (1) quartz and minor green chlorite and (2) chlorite and muscovite with minor quartz, graphite, and possibly iron ore. A few limonite stringers are present, suggesting weathering solutions. The metamorphic facies is considered to be greenschist or albite-epidote hornfels.

Specimen 7. Quartz-epidote rock. Specimen 7 is a blue-gray, fine-grained homogeneous rock consisting of approximately 60 per cent quartz, 38 per cent clinozoisite, 2 per cent iron ore, and a few subhedral crystals of zircon. Some iron ore is altered to limonite. *Joplin* [1942, p. 167] suggests that this may have been a psammite with a tuffaceous matrix. However, the high Sr^{87}/Sr^{86} ratio (Figure 3) indicates that this rock was possibly originally a graywacke or a contaminated sandstone.

Specimen 9. Quartz-muscovite-biotite schist. Specimen 9 is a light brown, strongly foliated schist with some biotite muscovite bands, ap-proximately 0.3 mm wide, in an approximately homogeneous matrix of quartz (approx. 70 per cent) biotite and muscovite (approx. 30 per cent), and some andalusite and chlorite. The micas are oriented in at least two foliation planes, indicating a complex directed pressure during metamorphism. The metamorphic facies is considered to be albite-epidote hornfels or hornblende hornfels.

Specimen 10. Quartz-muscovite-cordierite-biotite schist. Specimen 10 is a green foliated schist consisting of alternating bands, approximately 0.1 mm wide, of (1) muscovite and minor biotite and (2) quartz and minor muscovite and biotite. Pinitized porphyroblasts of cordierite with associated iron ore constitute approximately 10 per cent of the rock. The metamorphic facies is thought to be hornblende hornfels.

The State Circle Shale Specimens

Specimen 12. Shale. Specimen 12 is a pink, fine-grained, foliated shale. It consists of bands, approximately 0.1 mm wide, of (1) fine-grained semiopaque green-brown clay with minor sericite and (2) quartz with interstitial sericite clay and limonite. An orientation of clay particles at approximately 30° to the bedding plane was probably produced by a minor movement.

Specimen 13. Shale. Specimen 13 is yellow and weakly foliated. It consists of semiopaque, fine, green-brown clay with sericite, some quartz fragments, and iron ore. A little limonite occurs as a stain on the clays. Specimen 17 is very similar to this rock.

Acknowledgments. We wish to thank M. J. Vernon and V. M. Bofinger for assisting with several analyses, Dr. I. McDougall for the argon analysis, Dr. G. A. Joplin for a stimulating tour of the Cooma region and for commenting on the text, and Professor D. A. Brown for suggesting the use of the State Circle shale and showing us favorable sampling sites.

The analyses of the Cardup shale were made by W. Compston while he was at the Department of Physics, University of Western Australia.

REFERENCES

Aldrich, L. T., L. F. Herzog, J. B. Doak, and G. L. Davis, Variations in strontium isotope abundances in minerals, 1, Mass spectrometric analysis of mineral sources of strontium, *Trans. Am. Geophys. Union, 34,* 457–460, 1953.

Browne, W. R., The geology of the Cooma district, N.S.W., 1, *J. Proc. Roy. Soc., N. S. Wales, 48,* 172–222, 1914.

Browne, W. R., The geology of the Cooma district, N.S.W., 2, The country between Bunyan and Colinton, *J. Proc. Roy. Soc. N. S. Wales, 77,* 156–172, 1943.

Evernden, J. F., G. H. Curtis, J. Obradovich, and R. Kistler, Evaluation of glauconite and illite for dating sedimentary rocks by the potassium argon method, *Geochim. et Cosmochim. Acta, 23,* 78–99, 1961.

Evernden, J. F., and J. R. Richards, Potassium argon ages in Eastern Australia, *J. Geol. Soc. Australia,* in press, 1962.

Hurley, P. M., D. G. Brooking, W. H. Pinson, S. R. Hart, and H. W. Fairbairn, K-Ar age studies of Mississippi and other river sediments, *Bull. Geol. Soc. Am., 72,* 1807–1816, 1961.

Joplin, G. A., Petrological studies in the Ordovician of New South Wales, 1, The Cooma complex, *Proc. Linnean Soc. N. S. Wales, 67,* 156–196, 1942.

Joplin, G. A., Petrological studies in the Ordovician of New South Wales, 2, The northern ex- tension of the Cooma complex, *Proc. Linnean Soc. N. S. Wales, 68,* 159–183, 1943.

Kulp, J. L., Geologic time scale, *Science, 133,* 1105–1114, 1961.

Nicolaysen, L. O., Graphic interpretation of discordant age measurements on metamorphic rocks, *Ann. N. Y. Acad. Sci., 91,* 198–206, 1961.

Öpik, A. A., The geology of the Canberra city district, *Bull. 32,* Bureau of Mineral Resources, Australia, 1958.

Prider, R. T., The contact between the granite rocks and the Cardup series at Armadale, *J. Roy. Soc. W. Australia, 27,* 27–55, 1941.

Riley, G. H., and W. Compston, Theoretical and technical aspects of Rb-Sr geochronology, *Geochim. et Cosmochim. Acta,* in press, 1962.

Singh, J. S., The geology of the Darling scarp in the Mundijong area, Unpublished thesis, University of Western Australia, 1958.

Turner, F. J., and J. Verhoogen, *Igneous and Metamorphic Petrology,* McGraw-Hill Book Co., New York, 1960.

(Manuscript received March 14, 1962; revised May 4, 1962.)

Reprinted from *Radioactive Dating*, 201–217 (1963)

NEW APPROACHES TO GEOCHRONOLOGY BY STRONTIUM ISOTOPE VARIATIONS IN WHOLE ROCKS

P.M. HURLEY, H.W. FAIRBAIRN, G. FAURE, W.H. PINSON, Jr.
MASSACHUSETTS INSTITUTE OF TECHNOLOGY, CAMBRIDGE, MASS.

Abstract — Résumé — Аннотация — Resumen

29

NEW APPROACHES TO GEOCHRONOLOGY BY STRONTIUM ISOTOPE VARIATIONS IN WHOLE ROCKS. Conventional methods of geochronology using K-Ar, Rb-Sr, U-Pb relationships in separated mineral phases are seriously limited through the loss of daughter isotopes during the cooling history of the rock. Sr^{87}, lost from rubidium-rich mineral phases appears, however, to remain in the rock as a whole, so that whole-rock rubidium-strontium age measurements seem to be generally free of this limitation. It is essential with this procedure to know the initial Sr^{87}/Sr^{86} ratio in the system and, if the rock has any lithologic complexity, to know something of its history. These requirements have been studied and methods of satisfying them appear to be achievable.

The measurement of whole-rock ages and initial values of Sr^{87}/Sr^{86} in materials of the sialic crust leads to a determination of the genesis of the material if·the ratio of Sr^{87}/Sr^{86} in possible source materials is known. A study of this ratio in mantle source regions indicates values in almost all cases between 0.706 and 0.709 for a wide range of igneous rocks believed to be derived from these regions. Crustal materials including sediments rapidly develop higher ratios owing to a several-times enrichment in rubidium. It is therefore possible in the crystalline basement rocks of the continents to determine the proportion of reworked crustal material relative to new additions of sialic material from deep non-sialic source regions. So far it has been found that the major proportion of sialic basement is composed of new material with no previous sialic history; the minor proportion is reworked crustal material. It is possible to date the rise of the primary sial in most cases. With this greater insight into the genesis of igneous and metamorphic rocks it is possible to apply the whole-rock rubidium-strontium method more generally without determining the initial ratio of Sr^{87}/Sr^{86} each time by isochron plot, by utilizing the ubiquitous source region value of 0.708 ± 0.001.

A knowledge of the true age of intrusion at depth by this whole-rock method permits the interpretation of the generally lower ages of the separated minerals in terms of the subsequent cooling and uplift history of the rock body. The difference between the true age of intrusion and the ages shown by conventional methods on separated minerals is found to be as great as several hundred millions of years in some pre-Cambrian shield areas.

NOUVELLES MÉTHODES DE GÉOCHRONOLOGIE FONDÉES SUR L'ÉTUDE DES DIFFÉRENCES DE COMPOSITION ISOTOPIQUE DU STRONTIUM DANS LA ROCHE. L'emploi des méthodes classiques de géochronologie fondées sur les relations existant dans les couples d'éléments K-A, Rb-Sr et U-Pb dans des phases minérales distinctes est sérieusement limité par la perte de produits de filiation durant la période de refroidissement de la roche. ^{87}Sr perdu par les phases minérales riches en rubidium semble, cependant, demeurer dans l'ensemble de la roche, de telle sorte que les mesures d'âges de roches d'après Rb-Sr échappent généralement, semble-t-il, à cette limitation. Avec cette méthode, il est essentiel de connaître le rapport initial $^{87}Sr/^{86}Sr$ dans le système et, si la roche présente une certaine complexité lithologique, d'avoir quelques connaissances sur son évolution. Les auteurs ont étudié ces conditions et il leur a semblé qu'il était possible de les remplir.

Les mesures d'âges de roches et des valeurs initiales du rapport $^{87}Sr/^{86}Sr$ dans les roches du sial permet de déterminer la genèse des roches si l'on connaît le rapport $^{87}Sr/^{86}Sr$ dans les matières qui ont pu donner naissance à ces roches. Une étude de ce rapport dans les régions mères de recouvrement donne, dans la plupart des cas, des valeurs entre 0,706 et 0,709 pour toute une série de roches ignées dont on pense qu'elles proviennent de ces régions. Dans les roches du sial, y compris les sédiments, les rapports deviennent rapidement plus élevés du fait d'un enrichissement multiple en rubidium. Il est donc possible de déterminer dans le soubassement cristallin des continents la proportion des matières de la croûte qui ont été remaniées par rapport aux nouvelles additions de matières sialiques provenant de régions mères non sialiques profondes. On a constaté jusqu'à présent que la plus grande partie du soubassement sialique est composée de matières nouvelles n'ayant aucune ascendance sialique préalable; le reste du soubassement est constitué par des matières de la croûte remaniées. Dans la plupart des cas, il est possible de déterminer la date du soulèvement du sial primaire.

201

281

Avec ces connaissances accrues de la genèse des roches ignées et métamorphiques, il devient possible d'appliquer la méthode Rb-Sr à l'ensemble de la roche d'une manière plus générale, sans avoir à déterminer chaque fois le rapport initial $^{87}Sr/^{86}Sr$ par tracé isochrone, en prenant $0,708 \pm 0,001$ comme valeur de ce rapport pour la région mère ubiquiste.

Les connaissances obtenues par cette méthode sur l'âge véritable de l'intrusion en profondeur permettent d'interpréter les âges généralement inférieurs des minéraux pris séparément par rapport aux dates du refroidissement et du soulèvement ultérieurs de la masse rocheuse. La différence entre l'âge véritable de l'intrusion et les âges obtenus par les méthodes classiques pour les minéraux pris séparément semble être de l'ordre de plusieurs centaines de milliers d'années dans certains boucliers précambriens.

НОВЫЕ МЕТОДЫ ГЕОХРОНОЛОГИИ, ОСНОВАННЫЕ НА ВАРИАЦИЯХ ИЗОТОПОВ Sr В СКАЛЬНЫХ ПОРОДАХ. Обычные методы геохронологии, основанные на использовании отношений K/Ar, Rb/Sr, U/Pb в выделенных минеральных фазах, ограничены из-за потери дочерних изотопов во время охлаждения горной породы. Однако утерянный из богатой рубидием минеральной фазы Sr^{87} остается, по-видимому, полностью в скальной породе, так что измерения возраста цельной горной породы по отношению Rb/Sr кажутся свободными от этого ограничения. При использовании этого метода важно знать первоначальное отношение Sr^{87}/Sr^{86} в изучаемой системе. Если порода имеет сложную историю, то необходимо знать некоторые данные из ее истории. Выяснилось, что представляется возможным выработать методы, удовлетворяющие этим требованиям.

Измерение возраста цельных горных пород и первоначальных значений отношения Sr^{87}/S^{86} в породах сиалического щита ведет к определению происхождения этой породы, если отношение Sr^{87}/Sr^{86} в исходных образцах известно. Изучение этого отношения в зонах мантии, откуда брался исходный образец, дает почти во всех случаях значения между 0,706 и 0,709 для широкого диапазона вулканических горных пород, которые предположительно происходят из этих зон. В породах коры, включая осадочные отложения, эти отношения быстро возрастают благодаря многократному обогащению рубидием. Поэтому в кристаллических породах основания континентов можно определять отношение переработанных пород коры к новым добавкам сиалических материалов из глубоких несиалических зон, откуда брались исходные образцы. Установлено, что главная часть сиалического основания состоит из нового материала без сиалической предистории; меньшая доля состоит из преобразованных коровых материалов. В большинстве случаев представляется возможным датировать поднятие первичной сиали. Благодаря такому пониманию образования вулканических и метаморфических пород представляется возможным более широко применять рубидий-стронциевый метод цельных пород без определения каждый раз первоначального отношения Sr^{87}/Sr^{86} по изохронной диаграмме. При этом используется повсеместное значение $0,708 \pm 0,001$ для зоны, из которой брались исходные образцы.

Определение истинного времени интрузии в глубину при помощи этого метода дает возможность объяснить общее уменьшение возраста выделенных минералов в условиях последующего охлаждения и поднятия скаловой основы. Было обнаружено, что разница между истинным временем интрузии и временем, определенным обычными методами на выделенных минералах, может достигать в некоторых областях Докембрия нескольких сотен миллионов лет.

NUEVOS METODOS GEOCRONOLOGICOS BASADOS EN LAS VARIACIONES DE LOS ISÓTOPOS DEL ESTRONCIO EN ROCAS ENTERAS. La pérdida de isótopos descendientes en el curso del enfriamiento de las rocas restringe seriamente la utilidad de los métodos geocronológicos corrientes fundados en la determinación de las razones K/A, Rb/Sr y U/Pb en fases minerales separadas. Sin embargo, el ^{87}Sr desprendido de las fases minerales ricas en Rb permanece en la roca en conjunto, de modo que aquella limitación no rige, al parecer, para las determinaciones de edades por medio de la razón Rb/Sr en la roca entera. Para aplicar este método, es indispensable conocer el valor inicial de la razón $^{87}Sr/^{86}Sr$ en el sistema y, cuando la roca es litológicamente compleja, conocer hasta cierto punto el proceso de su formación. Los autores han estudiado estos requisitos y estiman posible hallar métodos que los satisfagan.

La medición de las edades de las rocas enteras y los valores iniciales de la razón $^{87}Sr/^{86}Sr$ en materiales de la corteza siálica permiten determinar la génesis del material siempre que se conozca la razón $^{87}Sr/^{86}Sr$ en los posibles materiales originales. El estudio de dicha razón en regiones primitivas del manto terrestre conduce en casi todos los casos a valores comprendidos entre 0,706 y 0,709 para una amplia serie de rocas ígneas, que se suponen procedentes de esas regiones. En los materiales de la corteza, incluso los sedimentos, los valores de dicha razón aumentan rápidamente debido a un repetido enriquecimiento en Rb. Por lo tanto, es posible determinar en las rocas cristalinas del basamento continental la proporción de material de corteza transformado con respecto a los nuevos aportes de material siálico proveniente de regiones primitivas no siálicas profundas. Hasta ahora se ha comprobado que la mayor proporción del basamento siálico se compone de materiales nuevos, sin antecedentes siálicos; la proporción menor corresponde a material de corteza transformado. En la mayor parte de los casos, es posible determinar la edad del ascenso del sial primitivo. Gracias a esta comprensión más cabal de la génesis de las rocas ígneas y metamórficas, se puede generalizar la aplicación del método Rb/Sr a la roca entera, sin necesidad de determinar cada vez la razón $^{87}Sr/^{86}Sr$ inicial con el diagrama isocrónico, utilizando para todas las regiones primitivas el valor 0,708 ± 0,001.

El conocimiento de la verdadera edad de la intrusión en profundidad, obtenido por este método de la roca entera, permite interpretar las edades, generalmente inferiores, de cada mineral, con referencia a la posterior historia del enfriamiento y ascenso de la masa rocosa. Se ha comprobado que la diferencia entre la verdadera edad de la intrusión y. las edades deducidas por los métodos corrientes de estudio de cada mineral es grande, alcanzando a varios cientos de miles de años en algunas zonas del escudo precámbrico.

INTRODUCTION

The principal weakness in standard methods of age dating by U-Pb, K-Ar, and Rb-Sr methods on separated minerals is the loss of the daughter element by diffusion or recrystallization since the initial time of emplacement of the rock mass as a geologic unit. In an ancient orogenic belt this emplacement may have occurred at great depth and appear at the surface today only after major uplift and erosion. Thus, in Precambrian rocks particularly, it is necessary to consider a period of time during rise and cooling in which the loss of daughter elements from parent-rich mineral phases is likely to occur.

In this report we are interested in the applicability of the whole-rock Rb-Sr method in general as an approach to age dating which can commonly avoid the above-mentioned weakness of the methods using separated minerals. Rubidium and strontium are lithophile elements that readily substitute for potassium and calcium in some of the commoner rock-forming minerals, and are not as prone to low-temperature hydrothermal migration as uranium, lead, and argon. It is therefore probable that this pair will be most suitable for this purpose, if measurements can be made accurately enough and the initial daughter abundance can be determined.

The application of this method requires a knowledge of the value and homogeneity of the Sr^{87}/Sr^{86} ratio in a rock mass at its time of emplacement, the means of testing whether there has been significant migration of the daughter or parent across the boundaries of the system represented commonly by several kilograms of whole-rock sample, and a Rb/Sr ratio large enough for the increment of Sr^{87} to be measured precisely enough.

As a corollary to the study of the initial Sr^{87}/Sr^{86} ratio in sialic rocks for whole-rock age dating, we find that this information throws light on the genesis of the rocks. In fact the origin of the sialic crust and the igneous rock types in it, the history of recycling and development of metamorphic rocks, and the development of whole-rock geochronology, are so closely interwoven that it is difficult to study the subjects separately. This report barely touches on these broader questions, as they are treated elsewhere (HURLEY et al. [13]).

Investigations of the Rb-Sr relationships in the whole-rocks for age dating, and for information on the metamorphic history of the rocks, have been reported by SCHREINER [19] , COMPSTON and JEFFERY [3, 4] ,

ALLSOPP [1], NICOLAYSEN [17], GAST [10,11], COMPSTON and PIDGEON [5], and by the group in the geochronology laboratory at the Massachusetts Institute of Technology. This group has been reporting progress annually in USAEC reports [15] which have been distributed to a large number of investigators, and also has reported in the following: FAIRBAIRN et al. [6,7], HART [12], FAURE and HURLEY [8], POWELL [18], BEALL et al. [2], HURLEY et al. [13], MOORBATH et al. [16], and WHITNEY [21].

INITIAL RATIO OF Sr^{87}/Sr^{86}

For whole-rock age dating the degree of homogeneity of the ratio Sr^{87}/Sr^{86} in a rock mass at time of emplacement is significant only in terms of the error in age determination it may cause. For example, a Precambrian quartz-mica schist is usually so enriched in Sr^{87} owing to its high Rb/Sr ratio that an assumed initial Sr^{87}/Sr^{86} ratio of 0.712 (i.e. average shale at time of deposition) will be within a precision error of the whole-rock measurement. On the other hand a very old calcium-rich gneiss may require a careful analysis of a variety of lithologic phases tested in an isochron plot (see below) before a judgement can be made on whether a suitable value for the initial Sr^{87}/Sr^{86} ratio can be determined.

In a recent survey of the Sr^{87}/Sr^{86} ratio in young oceanic and continental basalts, FAURE and HURLEY [8] found that the ratio Sr^{87}/Sr^{86} was 0.708 * with limits in 25 samples of + 0.002 and - 0.003. The standard deviation precision error in a single determination is estimated to be ± 0.002, so that in this sampling the limits are almost entirely accounted for by the precision of measurement. This suggests that the variation of Sr^{87}/Sr^{86} ratio in the source regions of basalt is universally very low, probably in the range 0.001 about the mean value of 0.708.

In another survey of alkaline rocks and rocks of the typical carbonatite suite, which are intrusive into much more ancient host rocks and are probably of subcrustal origin, POWELL [18] found an average Sr^{87}/Sr^{86} ratio (corrected to time of emplacement) of 0.7065 ± 0.0003. Again, in the samples analysed the precision error would account for most of the variation so that the homogeneity of the source region of these magmas is probably also within the range 0.001.

A number of measurements of the initial ratio of Sr^{87}/Sr^{86} in intrusive magmatic rocks and in a variety of effusive volcanics in this laboratory to date have shown values close to 0.708 and even granitoid rocks of doubtful origin have shown this ratio more commonly than any higher value and no value lower than 0.706. Also FAURE et al. [9] recently measured the ratio in mafic intrusive rocks such as the Duluth Gabbro, Sudbury norite, and Bushveld norite, which range in age through much of geologic time and have found no variation that is significant within their precision of measurement. A variation from 0.708 in young rocks to about 0.706 in the most ancient could, however, be compatible with their data.

---

* All values for isotopic ratios given in this report are based on a normalization to a single value of 0.1194 for the stable isotope ratio Sr^{86}/Sr^{88}.

Therefore, it is tentatively concluded that subsialic source regions (whether mantle or lower crust is not relevant), have had a ratio Sr^{87}/Sr^{86} throughout most of geologic time in the range 0.706 to 0.709 and that the assumption of a value of 0.708 ± 0.002 will most likely cover most derivatives from these regions.

In belts of mixed metasediments and igneous rocks in which the source materials were heterogeneous at time of emplacement the question of the initial ratio of Sr^{87}/Sr^{86} in any lithologic phase is more complex. The most simple and approximate assumption for rocks of this type is that the material in any sample may have three components of radiogenic Sr^{87} (designated by asterisk), or

$$*Sr^{87} \approx *Sr^{87}_1 + *Sr^{87}_2 + *Sr^{87}_3 + \lambda(Rb^{87}_1 t_1 + Rb^{87}_2 t_2 + Rb^{87}_3 t_3),$$

neglecting the decay of the parent. The subscript numerals refer to the three principal events in the history of the material: the average age of the provenance of the original sediment, t_1, the time of sedimentation, t_2, and the time of subsequent tectonism, intrusion, metasomatism, or metamorphism, t_3. The rubidium and strontium notation must be specified as follows, all in ppm:

Rb^{87}_1 = Rb^{87} in the sample that is due to allogenic component in sediments.

$*Sr^{87}_1$ = Initial radiogenic Sr^{87} in the allogenic sedimentary material at time t_1, as given by an assumed common strontium ratio of Sr^{87}/Sr^{86} = 0.708 (the value of this constant is arbitrary).

Rb^{87}_2 = Rb^{87} in sample that is caused by cation fixation or adsorption or other authigenic process from the marine environment at t_2.

$*Sr^{87}_2$ = Radiogenic Sr^{87} (in excess of Sr^{87} allowed for in the assumed common strontium isotope ratio of 0.708) deposited by precipitation or other authigenic process at t_2.

Rb^{87}_3 = Rb^{87} in that component of the sample representing age t_3. This can be negative if rubidium was removed at that time.

$*Sr^{87}_3$ = Radiogenic Sr^{87} (in excess of Sr^{87} allowed for in the assumed common strontium isotope ratio of 0.708) added at t_3. A loss of strontium at t_3 is considered to be unlikely, so that it is assumed that there cannot be a negative value of this parameter.

The total measured Rb^{87} in the sample is:

$$Rb^{87} = Rb^{87}_1 + Rb^{87}_2 + Rb^{87}_3 .$$

Solutions for these various parameters can be approached by selective sampling. In most complex sections there are usually some clearly detrital materials (e.g. conglomerate boulders) which can be used to find t_1 by isochron plot (see below) on selected samples. Similarly, t_3 can generally be found easily by isochron plot on material that is intrusive. In these cases, the parameters in the equations above that do not pertain to these ages, approach an insignificantly low value.

The value of t_2 is the principal, and most difficult, remaining objective. In a study of the Rb-Sr relationships in marine shales P. R. WHITNEY (M.I.T. Ann. Rept. NYO 3943) finds that an approximation to the age of sedimentation can be achieved in cases where there has been a considerable increase in the Rb/Sr ratio due to the process of rubidium fixation and adsorption. A whole-rock isochron plot on non-limy, pelitic material will yield an apparent age value that includes both $*Sr_1^{87}$ and $*Sr_5^{87}$. The apparent age includes a mixture of t_1 and t_2, but the mixture is not a random one because the Rb_5^{87} addition in non-limy fractions represents a high Rb/Sr ratio, relative to that in the purely detrital material. Therefore, the points at the high Rb/Sr end of the plot give a slope that tends to represent t_2. By knowing t_1 from the selective sampling mentioned above Whitney has been able to correct for the inherited $*Sr_1^{87}$ and approach the age t_2. The result is compared favourably with the true age of sedimentation known from stratigraphy. However the finding of t_2 is beset with difficulties and may be possible only in unusually favourable cases.

COMPSTON and PIDGEON [5] have also investigated shales along essentially the same lines and have found that the initial ratio of Sr^{87}/Sr^{86} can be higher than in sea water and variable. They arrive at approximately the same conclusion as stated above, namely that these rocks can be dated in some cases, at least approximately, if the radiogenic Sr^{87} in the detrital component is not too large a proportion of the total.

With this background it is seen that initial ratios of Sr^{87}/Sr^{86} in mixed rocks will be variable. For example in the schists and gneisses of eastern Connecticut, BROOKINS (M.I.T. Ann. Rept. NYO 3943) has found the apparent initial ratio in an isochron plot of metasediments to be higher than that in the igneous-looking rocks. The terms "apparent isochron" and "apparent initial ratio" are used for mixed rocks because it is possible to get a meaningless value of age or initial ratio under some circumstances. For example, if a homogeneous Rb-rich material of older age A is mixed in all proportions with a homogeneous younger material of less, but not zero Rb content, of age B, the average isochron plot will have a greater slope and therefore indicate a greater age than either A or B, and the apparent initial ratio will be lower than that for A and B. In this case the plot will appear to have an upward concave. Conversely, if A is younger and B is older, the apparent isochron plot will show too young an age, too high an initial ratio, and will be concave downward.

Despite these complexities, it has so far been found that the apparent initial ratios of basement crystalline rocks are surprisingly uniform. This suggests that the apparent values are probably fairly close to the true values. Also, the values are commonly close to 0.708, which indicates that average sialic basement was dominated at its origin by new or fairly recent additions of sial from subsialic source regions. A study of this effect has been reported by HURLEY et al. [13].

From information to date it appears likely that an assumed range for the Sr^{87}/Sr^{86} initial ratio of 0.707 - 0.709 will turn out to be correct for most homogeneous granitoid rocks of the basement complex, and most volcanic rocks. This means that, if the rock is at least as sialic in composition as granodiorite, the error due to the assumption of 0.708 will be less than 100 Myr (million years). Any error of 0.001 in assumed initial ratio will be equal to 25/r Myr in the age value, where r is the ratio Rb/Sr, so that a granitic rock with r = 1.0 would have an age error of 25 Myr on this basis. This suggests that, in the future a single whole-rock Rb-Sr analysis will be meaningful on the major proportion of homogeneous crystalline rocks of Precambrian age, without an isochron plot, using the assumed initial ratio value of 0.708. This statement, of course, will need much more data to establish it, but it is predicted on the basis of information now available. The broader range 0.706 - 0.710 mentioned previously should cover most basement gneisses that are less homogeneous than the restricted group above.

EXAMPLE CASE-PROGRESS REPORT ON SUDBURY DISTRICT, ONTARIO

The age of the Precambrian basement rocks in the vicinity of Sudbury, Ontario, have been studied by FAIRBAIRN et al. [7] and WETHERILL et al. [20]. Figure 1 shows a summary of the age measurements on separated minerals by the former. It is evident that the history of the region was complex and that there were at least three major geologic events: the original orogeny that formed the basement, the intrusion of the igneous rocks of the Sudbury basin, and the Grenville orogeny. However, age measurements on micas

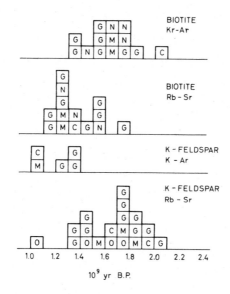

Fig. 1

Age measurements on separated minerals in the Sudbury District, Ontario. (From FAIRBAIRN et al. [7]

G = granitic rocks,

N = norite,

C = Copper Cliff "rhyolite",

M = micropegmatite,

O = Onaping volcanics.

(Decay constants: K^{40}, $\lambda = 5.30 \times 10^{-10}$ yr^{-1},

$\lambda_e = 0.585 \times 10^{-10}$ yr^{-1} ;

Rb^{87} , $\lambda = 1.47 \times 10^{-11} yr^{-1}$)

and feldspar by K-Ar and Rb-Sr methods showed an almost uniform distribution of age values ranging from Grenville at about 1000 Myr, to the highest
values at about 2200 Myr, and the geologic history could not be outlined except on weak grounds and by using geological relationships.

Following these investigations FAIRBAIRN et al. [6] and FAURE et al. [9]
made whole-rock Rb-Sr analyses on the original basement rocks, and on the
rocks of the nickel irruptive. A much more detailed investigation has also
been started by T.E. Krogh, who is analysing different lithologic types along
the Grenville front.

A plot of Sr^{87}/Sr^{86} against Rb^{87}/Sr^{86} in whole-rock samples of the basement granitic rocks is shown in Fig. 2. These points fall on a typical isochron with an initial Sr^8/Sr^{86} ratio of about 0.708, and a slope equivalent

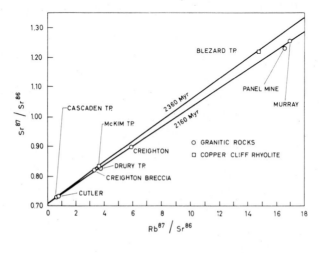

Fig. 2

Whole-rock Rb-Sr analyses on granitic basement rocks and Copper Cliff "rhyolite", in Sudbury District
(Myr = million years)

Isochron age values based on Rb^{87} decay constant = $1.47 \times 10^{-11} yr^{-1}$

to an age of 2160 Myr if a value of $1.47 \times 10^{-11} yr^{-1}$ is used for the Rb decay
constant. An isochron on 2 samples of the Copper Cliff Rhyolite shows a
slightly older age of 2360 Myr, which is compatible with the field relations.
Faure has found an excellent isochron on the rocks of the nickel irruptive
showing an age of 1650 ± 50 Myr (M.I.T. Ann. Rept. NYO 3943).

These are lucid age measurements in a complex area that previously
yielded diffuse answers, and demonstrate the effectiveness of the whole-rock
Rb-Sr method. Similar results have been found for the basement rocks of
New England (M.I.T. Ann. Rept. NYO 3943) and of the Cape Smith-Wakeham
Bay Belt and Labrador (M.I.T. Ann. Rept. NYO 3942). In these cases the
isochron age values on whole-rocks are again greater and more uniform
than the age values on separated minerals.

POSSIBLE CAUSES OF THE LOWER AGE VALUES IN SEPARATED MINERALS

To date, work in this laboratory has indicated that a whole-rock iso-chron study of a crystalline rock-mass in a typical basement complex yields a higher age value than is found on separated minerals from the same samples. The age difference is commonly one or more hundreds of millions of years.

It is well known that different minerals and different methods will yield different age values, and it is commonly stated that these differences are due to diffusion loss of daughter isotopes during thermal events or periods of metamorphism that occur after the original rocks were emplaced. We wish to propose that the general case is not a thermal event or subsequent period of metamorphism, but simply the cooling history of the belt. After an orogenic uplift is essentially completed, with base-level peneplanation and restoration of gravitational equilibrium, the rocks exposed at the surface probably represent original depths of emplacement up to 20 km or more. The geological relationships mapped in such a belt (intrusives, contacts, etc.) occurred at depth, and the dating of these events must therefore allow for the fact that the region in which the relationship took place was commonly at elevated temperatures for a long period of time.

The relaxation of an orogeny involves a rate of erosion and a rate of gravitational adjustment to this erosion, or uplift. It cannot be stated which of these is the controlling process in any single case, but it is probable that both are roughly exponential. Therefore, if the ultimate depth of erosion is Z_0, the erosion depth represented by the surface rocks at any time can be approximately stated as

$$Z = Z_0 (1 - e^{-0.693t/\tau_{\frac{1}{2}}}),$$

where t is the age of the rocks in Myr, and $\tau_{1/2}$ is the half-life of the mountain system.

In a rough plot of mean elevation <u>versus</u> age for major orogenic belts it is seen that $\tau_{1/2}$ is about 100 Myr. If Z_0 is, say, 20 km, then most Precambrian shield areas that are base-levelled have exposed crystalline rocks that remained at a depth greater than 10 km for 100 Myr. During this time the uplift was at its greatest rate so that abnormal geothermal gradients would be expected. The temperature gradient $\delta T/\delta Z$ can be approximated by the sum of two terms, one constant and one transient. The constant gradient is probably close to 20°C/km (using normal values of earth heat-flow and conductivities) and the transient gradient due to the uplift of the rock decreases from a value of about 15°C/km early in the uplift to zero. Therefore, over the time of interest a total gradient of about 35°C/km is reasonable. This would mean that the rocks sampled today in Precambrian areas probably remained at temperatures above 350°C for 100 Myr and above 150°C for 200 Myr.

The diffusion of Sr^{87} and Ar^{40} from mineral structures is probably ex-pressed approximately by the relationship

$$D/a^2 = D_0/a^2 \exp(-E/RT)$$

where a is the radius of an assumed spherical region in which the diffusion occurs. Experimental work reported in the literature to date suggests values for the activation energy E that range from 30 to 70 kcal/mole. There is a sharp break between loss and retention of daughter isotope for D/a^2 in the range of values between 10^{-17} and 10^{-19}. This involves a small range in temperature (less than 50°C), so that the minerals would lose essentially all of their radiogenic Sr^{87} or Ar^{40} during the cooling time down to a certain temperature, which is controlled by the value of E, and then rather quickly start to retain all of the daughter product. The temperatures at which this change occurs are not known, both because of uncertainty in the value of E for any mineral and because there may be other processes involved. Recently HART (oral communication, NRC Subcommittee on Nuclear Geophysics Symposium 1962) has reported losses of strontium and argon from minerals in a contact zone under temperature conditions that would be equivalent to very low values for E (lower than could be compatible with laboratory tests). HURLEY et al. have reported [14] a similar effect in the very low K-Ar and Rb-Sr ages (< 5 Myr) in biotite gneisses of probable Jurassic age in the centre of the uplift of the Alps of New Zealand.

It is therefore too early to make definite statements but we consider it probable that an error of 100 Myr on the low side may be expected in age measurements on separated Precambrian minerals because of diffusion or other loss during cooling, and that this error may even be as great as several hundred million years if some of the recent indications of low-temperature loss are found to be common.

Applying this concept to the Sudbury case, if the discrepancy between the whole-rock ages and mineral ages was due to cooling and uplift we would have a diffusion history as indicated in Fig. 3. However, in the Sudbury district the magnitude of the discrepancy is so large that we feel it may be due in part to an unusual history because of its location between two major orogenic regions. It is possible that this hinge belt may have remained at depth as the positive block to the north rose and the negative Huronian trough to the south sank. Finally, after the completion of the Penokean orogeny, the Sudbury basement rocks were elevated along with it.

Similar observations in New England lead us to the same conclusions. Whole-rock isochron values have been found to exceed separate mineral age values by 100 to 200 Myr in the several cases tested. Again the K-Ar and Rb-Sr values on micas are in rather close agreement, giving the appearance of concordance, and yet frequently the mica, in both intrusive and host rocks, yields the same results. We tentatively conclude that this is due to uplift and cooling, so that rocks of obviously different geologic age show the same mineral age values, and that the true age of emplacement at depth is more likely to be given by the whole-rock isochrons. The age patterns in the New England province show a smooth variation from one age level to another regardless of lithologic type, and fit readily the concept of uplift and cooling as the controlling factor in mineral age values.

Petrogenesis, and the Time of Differentiation of the Sialic Crust

As mentioned above, the source regions of basalt have developed a ratio of Sr^{87}/Sr^{86} only as great as about 0.708 during geologic history, and the

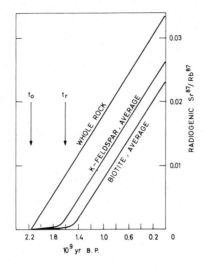

Fig. 3

Retention and development of radiogenic Sr^{87}/Rb^{87} in whole rocks and separate minerals in basement granitic rocks of Sudbury District, Ontario, estimated from analytical results and probable history of cooling and uplift

growth since 3 billion years ago appears to have been within the limits 0.704 - 0.708. The same range seems to cover the source regions of sialic rocks from present data. On the other hand average sialic crustal material starting at, say, 0.706 at 3 billion years would now have a ratio of about 0.74, if general information on the rubidium and strontium contents of crustal rocks is reasonably correct.

If the sialic crust was, to a great extent separated from mafic source regions at an early time in earth history and was reworked with only small additions ever since, it should show an Sr^{87}/Sr^{86} ratio commonly from 0.73 to 0.74 as an average, but probably with large scatter. If, on the other hand, the continental sial has been growing by continuous major additions from subsialic source regions, it should show a progressive decrease from about 0.74 to 0.71 with the decreasing age of the basement rocks. The latter is found to be the case (HURLEY et al. [13]). This implies that a large proportion of the material in typical sialic basement is derived from freshly separated sial that has risen from deep-source regions. It also implies that when a good isochron plot on an old pluton, or the direct measurement of Sr^{87}/Sr^{86} (corrected for slight increments due to time since emplacement) on a young pluton, show an initial ratio of 0.708, the material of the pluton is not contaminated with significant amounts of pre-existing sialic material. The same applies to intrusive or volcanic rocks.

Petrogenetic studies of this kind have so far favoured the concepts of continental growth and continuing additions to primary sial. Even some paragneisses are dominated by new additions of material which have had no pre-history of association with rubidium more than in the source regions of basalt. It is not determinable if the additions were in the form of true sialic magmas, or water-rich granitizing liquors. But, if metasomatic, the fact that the strontium (and by geochemical analogy the calcium) is new and that the pre-existing strontium has been removed must be accounted for. This is not true of schists.

The test of initial Sr^{87}/Sr^{86} ratio is being applied to a wide variety of rock types. Results to date suggest that igneous-looking plutonic and hypa-byssal rocks, acid volcanics, alkaline rocks and the unusual rocks of the carbonatite kindred, at least for the most part, are primary derivatives from subsialic source regions.

Much more work is needed, of course, and the whole concept may be faulty in some way that we have not yet appreciated. Therefore, this report is a statement of progress, with predictions that are tentative.

CONCLUSIONS

Whole-rock Rb-Sr age measurement techniques offer promise of mini-mizing the problem of loss of daughter isotope that is the principal drawback of the methods on separated minerals. Owing to low Rb-Sr ratios in whole-rock samples compared to those in separated mineral phases, it is necessary to know the initial ratio of Sr^{87}/Sr^{86} in the rock at the time of formation. In relatively homogeneous bodies of rock, or where it is known from geo-logical reasoning that the rock system developed from a homogeneous source material, the initial ratio can be found by isochron plot. However it is be-coming increasingly evident that most igneous rocks had an initial ratio so close to 0.708 that this value can be assumed if a resulting error of $25/r$ Myr from this assumption is acceptable. The Rb/Sr ratio, r, ranges from about 0.25 for granodiorites to several times this in granites, so the above error would be less than 100 Myr in a large proportion of the basement rocks of interest.

With isochron plots the method appears to be useful for dating strati-graphic sections containing differentiated volcanics, basement complexes (if certain tests are favourable) and even marine shales under limited conditions.

Information on the origin of igneous-looking rocks may be derived from the initial ratio of Sr^{87}/Sr^{86}. Many rocks from plutons to extrusives have shown initial ratios close to 0.708, which is the average found for modern basalts, and it is interpreted that these rocks are composed dominantly of material that is a primary derivative from subsialic source regions.

ACKNOWLEDGEMENTS

Unpublished analytical data that have been helpful in the development of some of the interpretations in the report have been kindly made available to us by several members of the laboratory staff. We wish to thank in par-ticular J. L. Powell, P. R. Whitney, D. G. Brookins, M. L. Bottino, G. H. Beall, E. Beiser, and T. E. Krogh, and also S. Moorbath who has been a visitor for a year. We appreciate very much the helpful advice and criticism of H. Hughes. The research has been made possible by the support of the Division of Research of the United States Atomic Energy Commission.

REFERENCES

[1]   ALLSOPP, H. L. , Rb-Sr age measurements on total rock and separated mineral fractions from the Old
        Granite of the Central Transvaal, J. Geophys. Res. 66 (1961) 1499-1500.

[2]   BEALL, G.H. , HURLEY, P.M. , FAIRBAIRN, H.W. and PINSON, W.H. , (Abstract) Comparison of K-Ar
        and whole-rock Rb-Sr dating in New Quebec and Labrador, Amer. Geophys. Un. 43d Ann. Meeting
        Program (1962) 64.

[3]   COMPSTON, W. and JEFFERY, P.M. , Anomalous common strontium in granite, Nature (Lond.) 184 (1959)
        1792.

[4]   COMPSTON, W. and JEFFERY, P.M. , Metamorphic chronology by the rubidium-strontium method,
        Annals of the N.Y. Acad. Sci. 91 Art. 2 (1961) 185-191.

[5]   COMPSTON, W. and PIDGEON, R.T. , Rubidium-strontium dating of shales by the total-rock method,
        J. Geophys. Res. 67 (1962) 3493.

[6]   FAIRBAIRN, H.W. , HURLEY, P.M. and PINSON, W.H. , The relation of discordant Rb-Sr mineral and
        whole-rock ages in an igneous rock to its time of crystallization and to the time of subsequent Sr^{87}/Sr^{86}
        metamorphism, Geochim Acta 23 (1960a) 135-144.

[7]   FAIRBAIRN, H.W. et al. , Mineral and rock ages at Sudbury-Blind River, Ontario, Proc. Geol. Assoc.
        of Canada, 12 (1960b) 41-66.

[8]   FAURE, G. and HURLEY, P.M. , (in press). The isotopic composition of strontium in oceanic and conti-
        nental basalts: application to the origin of igneous rocks, J. Petrology.

[9]   FAURE, G. , HURLEY, P.M. , FAIRBAIRN, H.W. and PINSON, W.H. , (Abstract) Isotopic compositions
        of strontium in continental basic intrusives, Amer. Geophys. Un. Program 43rd Ann. Meeting (1962) 66.

[10]   GAST, P.W. , Limitations on the composition of the upper mantle, J. Geophys. Res. 65 (1960) 1287.

[11]   GAST, P.W. , The rubidium-strontium method, Annals of the N.Y. Acad. Sci. 91 Art. 2 (1961) 181-184.

[12]   HART, S.R. , Mineral ages and metamorphism, Annals of the N.Y. Acad. Sci. 91 Art. 2 (1961) 192-197.

[13]   HURLEY, P.M. , HUGHES, H. , FAURE, G. , FAIRBAIRN, H.W. and PINSON, W.H. , Radiogenic Sr^{87}
        model of continent formation, J. Geophys. Res. (In Press).

[14]   HURLEY, P.M. , HUGHES, H. , PINSON, W.H. and FAIRBAIRN, H.W. , Radiogenic argon and strontium
        diffusion parameters in biotite at low temperatures obtained from Alpine Fault uplift in New Zealand,
        Geochim. Acta, 26 (1961) 67-80.

[15]   M.I.T. Geochronology Reports (HURLEY, P.M. , Ed. ,) Variations in isotopic abundances of Sr, Ca, and
        Ar, U.S. Atomic Energy Commission Reports NYO 3941-3943 (1960-1962).

[16]   MOORBATH, S. , HURLEY, P.M. , FAURE, G. , FAIRBAIRN, H.W. and PINSON, W.H. (Abstract),
        Evidence for the origin of mineralized Tertiary intrusives in southwestern states from strontium isotope
        ratios, Amer. Geophys. Un. Program 43d Meeting, (1962) 66.

[17]   NICOLAYSEN, L.O. , Graphic interpretation of discordant age measurements on metamorphic rocks,
        Annals of the N.Y. Acad. Sci. 91 Art. 2 (1961) 198-206.

[18]   POWELL, J. L. , The strontium isotope composition and origin of carbonatites, Ph.D. Thesis, M.I.T.
        (1962).

[19]   SCHREINER, G.D.L. , Comparison of the ^{87}Rb ^{87}Sr ages of the red granite of the Bushveld complex
        from measurements on the total rock and separated mineral fractions, Proc. Roy. Soc. A245 (1958) 112.

[20]   WETHERILL, G.W. , DAVIS, G.L. and TILTON, G.R. , Age measurements on minerals from the Cutler
        batholith, Cutler, Ontario, J. Geophys. Res. 65. (1960) 2461-2466.

[21]   WHITNEY, P.R. , The rubidium-strontium geochronology of argillaceous sediments, Ph.D. Thesis,
        M.I.T. (1962).

DISCUSSION

P. HAHN-WEINHEIMER: I tested your Rb/Sr isochron method with
granite from the Black Forest and found that three specimens, collected from
different places in this granite pluton, lay on one straight line. I was thus
able to test the homogeneity of the pluton and found an initial ratio of
0.710 ± 0.001 and an age of 200 Myr. I consider that in our case also the
method worked very well.

M. GRÜNENFELDER: Have you any Rb/Sr ages for feldspars, specifically alorites and oligoclases, in those rocks (e.g. the Appalachian rocks you referred to) where you found low Rb/Sr ages in biotites but high total rock ages?

P.M. HURLEY: Work on this point has still to be completed. We have measured a large number of feldspars from various places, the nearest to the Appalachians being the multiple batholith of Nova Scotia. In this case, however, the same age of 360 Myr was found for feldspar, biotite and mica. Initial work on the isochron of the whole rock gives a concordant result. We cannot say yet whether feldspars will give approximately the same age. However, we feel from work on feldspars in other areas that although feldspars commonly have a higher rubidium-strontium content than biotite and muscovite, they reach a magnitude considerably short of the isochron age.

M. GRÜNENFELDER: What is your strongest argument in postulating that you are losing daughter elements and are not losing parent elements in systems like biotite or feldspars?

P.M. HURLEY: We feel that in a massive, multiplex, granite pluton with variations in its lithologic phases, i.e. in the Rb/Sr ratio, and where there has been some alkaline metasomatism, causing a gain in alkalis, or where there has been a loss of alkali from the whole system by some sort of metamorphism - remembering that we are talking about a large system - you will find variations between the centre and the edge, or between one lithologic phase and another. Consequently, the homogeneity of our results, which give a plot along a single isochron, leads us to believe that there is no overall gain or loss of either parent or daughter in this large system. This is a megasystem, compared to the individual mineral grain. We believe that the homogeneity of the isochron plot constitutes the best support for our hypothesis. We know from many laboratory experiments that argon is lost by diffusion.

L.T. SILVER: At least three questions are immediately provoked by your report. Firstly, how large a sample have you found necessary to obtain a closed system in each case? Secondly, have you found any systems which are not homogeneous, do not fall on a single isochron and therefore do not concord with your simple model? Lastly, it is fairly obvious that the "age" derived from these isochrons is related to an important event - a time when all members of the system had a uniform strontium isotopic composition. However, this may have one significance for a rhyolite and quite another for a high-rank metamorphic gneiss. What do you consider to be the geological significance of this time of initial uniformity of isotopic composition?

P.M. HURLEY: We were surprised to find that the sample size did not make much difference. A handful of any homogeneous rock seems to be about as good as a 50-lb sample. The answer to the second question is that this situation occurs where we are definitely dealing with metasediments, particularly in areas of schists, such as the Collins Hill formation, which is adjacent to the Glastonbury and Monson gneisses in Connecticut, is definitely sedimentary in origin, and has received no particular addition of material except probably by metamorphism. In such cases, we find considerable scatter. Not only that, but on passing an isochron as well as possible through the scattered points, in several cases it did not fall on a low ratio of 709 or 708, but reached 720 or more. The host rocks, in which we found this, were the pegmatites in the Connecticut area. Very good isochrons also show a high initial ratio, indicating that these materials have evolved, not from newly-derived subsialic material, but in some way from formations very similar to that of Collins Hill, which is a metasediment. We have been surprised at the fact that most of the gneisses tested seemed to be more homogeneous than foreseen. We expected a variation, but they tend to be homogeneous and to show a low initial ratio.

Your third question raises a much more complex point into which I did not want to delve because it would require several lectures. The subject is covered in a paper which is to appear in the Journal of Geophysical Research in December. These isochrons you refer to are associated with the development of a variety of systems derived from a homogeneous unit at one point in time. Where the rock is a granite and its initial ratio falls at the basalt level, we feel convinced that the mass of granite magma or whatever it was became chemically separated and then developed these separate lithologic phases. The work on the gneisses appeared so similar to that on granites, particularly in granite gneisses and orthogneisses, that we can see no difference, but we have never found very much difference in a paragneiss, or in rather banded gneisses, before reaching mixtures of schist and gneiss. We have finally concluded that a surprisingly large percentage of the material of the crystal and crust is derived from subsialic source regions, including the material of many paragneisses, i.e. the alkaline material. Whatever the method of emplacement in these bodies, the system was greatly influenced by the addition of new material of a sialic or alkali-rich nature at the time, so that inhomogeneities before that, inherited in any sedimentary material, were relatively small. This could be due to the fact either that the abundance of the sedimentary material played only a small part in the manufacture of the gneiss, or that the age was low. We have reached the conclusion that, generally speaking, the age of the sediment in material is not great and this accounts for the very small influence of inherited strontium-87 in metasedimentary rocks of gneissic type.

J.L. KULP: I should like to comment on just two aspects of this complex subject. First, the Glastonbury situation. You showed a whole-rock age of about 360, whereas the ages obtained by individual minerals on the pegmatites - quite a few of them by your own laboratory - are concordant at about 250. I think that this is an excellent illustration of the need for paying attention to the individual mineral measurements, as well as to the whole-rock measurement, because clearly we obtain different kinds of information from these two sources. The whole-rock age would appear to represent the last time, or possibly the only time, the gneiss was homogenized, and the mineral ages the time of pegmatite formation. Both of these events are of interest geologically and it is significant that they can be dated by these different methods. I think we must exercise a certain amount of care when interpreting the extrapolation in most of your whole-rock isochrons; they did indeed appear to extrapolate below 710. You suggested 708. Nevertheless, 1000 million years ago the value of the oceans was about 710 and the difference is very small. Depending on the ratio of rubidium to strontium in the mega-system and on the time interval between a primary basement and a secondary metamorphism, you might or might not be able to see this small difference and I wonder if the Connecticut gneiss may not have had a still earlier history.

Finally, I would like to say something about the time interval between the completion of metamorphism and the effective beginning of the mineral clocks. In your paper, you suggested qualitatively how you arrived at 100 million years as the average time for an orogeny to cool off. I think we have at least two places where we have enough data to start measuring this. Kovo's work in Finland showed 1850 - 1900 for the extrapolated concordant-zircon ages. The highest biotite age, as you know, is only about 1780. Catanzaro, in our laboratory, has recently finished a similar study in the Little Belt mountains of Montana, where the extrapolated zircon age is about 1900 - 1950, a concordant monzonite shows 1920 - 1930, but the highest biotite potassium/argon age we can obtain is about 1800; other ages range down from that, depending on the perfection of the mineral. The last case of a different type is the Spruce Pine pegmatite situation. These Spruce Pine pegmatites in Western Carolina give concordant Rb/Sr and K/Ar ages of about 330 to 335 on micas; the error is quite small. The U/Pb determinations and the Rb/Sr on feldspar, however, give about 355 and are concordant. This I think is an illustration of possibly a 10 - 20-million year cooling period, where there is a super-imposed metamorphism which only produced pegmatites, and was not a primary one. I should like to hear your comments on the possible interpre-tation of the Connecticut Valley and on the examples of this cooling history; do you think these are reasonable and have you any other specific examples?

P.M. HURLEY: I have a slide showing an isochron on the Connecticut pegmatite minerals which comes out on a line as straight as any that we have found, giving an age of 260 or 270 on the pegmatites, although the initial value, about 0.72, is very high. This checks very well with the initial ratio, the ratio, let us say, calculated for the Collins Hill formation as a whole at the time of pegmatite formation. The inherited material that could go into a system should be homogenized to yield a homogeneous isochron plot. If the material were homogenized, then its strontium-87 abundance would re-flect its pre-history; it would not be 0.708. If it did not have a different value, then either its pre-history was very short, or its Rb/Sr ratio was very low. In either case this has very little bearing on its subsequent history as a unit, in which we are measuring our whole range of Rb/Sr ratios after homogenization. You can take any material of the sial that you wish, any sedimentary basin, mix it up, make it homogeneous and then, from that point on, generate a variety of systems that will give you an isochron indi-cating the time that it was homogenized. However, the initial ratio would not be 0.708. On the basis of our measurements so far we believe it to be impossible that a continental sialic crust was generated almost entirely about 3000 million years ago and has been reworked ever since. The measurements in fact show just the opposite; that the crust has been continuously evolving, because in all the basement rocks we have measured the initial 87/86 ratio is constantly low at the time that the age gives.

Regarding your second question, I think you were not talking about the same cooling time. We do not really know what alterations occur in biotite and muscovite, nor do we know the average value of some nebulous D/A^2 factor needed to produce a coefficient over some diffusion radius squared value and we do not know the activation energies very well. It could well turn out, as has been found by us and others, that these activation energies are sometimes down at the level of 30 kcal/mole; in which case, a temperature of 150-200°C is sufficient to cause continuing and substantial loss of argon and radiogenic strontium from mineral grains, and such temperatures could well have existed in these rocks for a period of 1 - 200 million years, taking any normal hypothesis for the time required for the relaxation of a mountain system. I think that is the time we are talking about in cooling. The rock has to form at a deep level, in a hot environment, and the whole system has to rise and erode before the cooling takes place. This is not the same as the cooling in, say, the Sierra Nevada.

A. VINOGRADOV (Chairman): You said that you had been able, using your isochron method, to determine the age on mantle material. However, I thought it was still impossible to determine the age of basic and ultrabasic rocks, using presently available radioactivity methods. In the case of contemporary basalts we have obtained ages of the order of hundreds of millions of years. Could you give us a definite example of the age of similar rocks compared to the age established by geologists? Your paper raises the question "What is rock age?", because your statements seem to mean that the ages of individual minerals and those of complete rocks are not the same.

P.M. HURLEY: We are unable to measure the age of volcanic rocks. In the instances where we are able to determine the age, the volcanic rocks are differentiated rocks. They must contain a fairly good range from basalt up to, say, some acid lava, such as a rhyolite or trachyte, so that we can obtain an Rb/Sr ratio from more or less zero up to some value at which the radiogenic strontium can be considered as significantly measurable. In the examples given, the rocks - in the younger cases - were mostly an association of basalts up to rhyolite; in the older cases, a range from basalt up to an intermediate volcanic could be used on the plot, where these rocks occurred together in the same sequence. The method cannot be applied to any rock that does not have a sufficiently high Rb/Sr ratio to yield enough radiogenic strontium-87 for significant measurements above the zero value. In the instances given, we have found only that the volcanic rocks appear to have about the same age as that established by geological methods. An isochron plot has to be made because the zero value is not known. If you do not know the initial value, but merely assume a zero value of 0.708, then a single calculation on a rhyolite, for example, could be completely wrong. A plot from acid to basic rocks is needed to provide the slope giving the age. Assumptions regarding the common strontium composition at the time of origin are dangerous, because although in most cases we have found a low initial ratio around 0.708, in some cases the value has been as high as 0.720.

Radiogenic Argon

XI

Editor's Comments on Papers 30–34

Probably the most versatile of all the radiometric methods of age determination, the potassium–argon method, like the related rubidium–strontium method, had its origins in investigations related to the radioactivity of the alkali elements. In 1935, Nier found the isotope K-40 whose existence had been predicted independently by Klemperer (1935) and Newman and Walke (1935) to account for the observed beta activity of potassium. Two years later, Symthe and Hemmendinger (1937) with their high-intensity mass spectrometer (see also p. 217) demonstrated that all the radioactivity of potassium was indeed associated with the K-40 isotope.

Radioactive disintegration of K-40 by beta emission would result in the production of an isotope of calcium, Ca-40, but there was little hope of using a potassium-40/calcium-40 method for determining the age of geologic materials because calcium, unlike strontium, is a common constituent in nearly all rocks and the most abundant isotope of common calcium happens unfortunately to be Ca-40 (96.97 atom per cent). The chance of detecting a small radiogenic increment of Ca-40 relative to the other common calcium isotopes seemed most unlikely.* Rutherford in his classic text "Radiations from Radioactive Substances," mentioned that in 1928 Kolhörster had observed the emission of gamma radiation from potassium (see Rutherford et al., 1930, p. 544), but there was no suggestion that this would in any way alter the nature of the decay product of potassium disintegration.

The first step toward the establishment of a reliable method of age determination based on the radioactive disintegration of potassium was taken in 1937 at the Kaiser Wilhelm Institute of Physics in Berlin by C. F. von Weizsäcker, the son of the German

---

*In 1951, Ahrens discussed the feasibility of the calcium method for age determination and concluded that it might be useful for Precambrian lepidolite micas. Coleman (1971) has recently reported calcium dates on a number of pegmatitic micas.

Undersecretary of State (Paper 30). Von Weizsäcker took note of the fact that the inert gas argon is the third most abundant gas in the atmosphere, making up nearly 1 per cent of the total by volume, far more abundant than all the other inert gases (helium, neon, krypton, and xenon) put together. Furthermore, the isotopic composition of atmospheric argon is dominated (99.6 per cent) by the isotope Ar-40,* having the same mass number as radioactive potassium. With these facts in hand, von Weizäcker suggested that a significant proportion of the radioactive K-40 atoms might be decaying, not by beta emission, but by a process involving capture of one of the orbital electrons surrounding the nucleus. Such a process would result in the production of Ar-40, which would account for the predominance of this isotope in the atmosphere.†

The process of electron capture had been considered as a possible mode of decay on purely theoretical grounds by C. Møller in 1937, but if it was occurring in the case of K-40 disintegration, then not one, but two stable daughter products would be generated, Ca-40 by beta emission and Ar-40 by electron-capture. In any given unit of time, a constant proportion of the disintegrating K-40 atoms would be converted into Ar-40, and the possibility of finding an excess of this isotope in an old potassium mineral seemed much more likely than the possibility of ever detecting the presence of radiogenic calcium-40.

There was strong support for Weizsäcker's suggestion, particularly if one considered Köhlhorster's earlier observation of the gamma activity of potassium, which could not in any way be associated with the process of beta emission and was, therefore, indicative of a second mode of decay. Experimental confirmation of the production of Ar-40 from K-40 by electron capture was reported by Thompson and Rowlands (1943) in England, and the search for systematic variations with age in the argon content of potassium-bearing minerals began.

Argon had originally been discovered in the terrestrial atmosphere by Sir William Ramsay, and its presence in rocks had been recognized by several investigators, most notably Lord Rayleigh (R. J. Strutt) and E. S. Shepherd, but no one had noticed or even suspected that there might exist a systematic increase in the argon content of potassium-bearing minerals with age. The initial results of analyses designed to confirm such a trend were disappointing, for without a mass spectrometer it was virtually impossible to distinguish between radiogenic argon produced in situ by radioactive decay of potassium and contaminating atmospheric argon.

Success came first to L. T. Aldrich and A. O. Nier, working in the Department of Physics at the University of Minnesota with a high sensitivity mass spectrometer and large samples (~ 400 g). Aldrich and Nier found that the isotopic ratio Ar-40/Ar-36 in argon extracted from four different potassium minerals was several times higher than the atmospheric value (Paper 31). This was positive proof that a significant enrichment of radiogenic argon-40 could be found in such minerals, and clearly pointed the way toward a new method for determining the age of geologic materials.

---

*The commission on inorganic nomenclature adopted the symbol Ar instead of A for argon in 1957.
†Shillibeer and Russell (1955) and Damon and Kulp (1958) subsequently discussed the predominance of Ar-40 in the atmosphere in relationship to the evolution of the atmosphere and the age of the earth.

Among the first applications of the K–Ar method to terrestrial materials were those reported by Smits and Gentner (Paper 32) in Germany, and Gerling and his associates (1952) in Russia. Gerling successfully extracted radiogenic argon from silicates by prolonged heating and found the calculated ages to be in good agreement with previously determined uranium–lead and helium ages. Smits and Gentner's paper, reproduced here (Paper 32) is quite typical of much of the early work. The methods of gas analysis adopted by the German workers were based on techniques developed by Paneth for helium age determinations. Argon and other gases were released from potassium-bearing salts by dissolving the minerals in water. The argon was purified and measured volumetrically on a McLeod gauge. A separate isotopic analysis of the gas was used to determine the concentration of Ar-40 relative to Ar-36 in order to correct the total volume of argon measured for any atmospheric contamination. The concentration of Ar-40 in the air relative to Ar-36 according to Smits and Gentner's measurement was 293 ± 10.* Atmospheric contamination determined from the Ar-40/Ar-36 ratios ranged from 30 per cent to less than 6 per cent of the total measured volume of argon in the analyzed samples, allowing precise ages to be determined from the radiogenic component.

By the mid-1950s, the foundations of the K–Ar method were firmly established. The isotopic abundance of potassium-40 (0.0119 ± 0.0001 atom per cent) was redetermined by Nier in 1950,† and although the branch decay of potassium necessitated the determination of two decay constants, one for beta decay (λ_β) and one for electron capture (λ_e), satisfactory measurements of the specific beta and gamma activities were established by 1957 (see Smith, 1964, for a review, and Beckinsale and Gale, 1969, for a reappraisal of the K-40 decay constants). Major analytical advances for the determination of radiogenic argon came through the application of isotope-dilution techniques, together with the development of ultra-high-vacuum technology (Alpert, 1953) and high-sensitivity mass spectrometers (Reynolds, 1956). Use of a direct-fusion radio-frequency induction furnace for the liberation of argon from silicate minerals was introduced by Carr and Kulp (1955, 1957).

The K-Ar method had enormous potential, for potassium, unlike all the other radioactive elements, is one of the eight most widely distributed and abundant elements in the earth's crust, occurring in most feldspars and micas, and its measurement posed few serious analytical problems. L. H. Ahrens, writing from the Department of Geology and Mineralogy at Oxford University, spoke for most geochronologists when he said,

> No other method, perhaps, shows indications of greater promise than does the argon method. It is too early yet to attempt a definite prediction,

---

*The currently accepted figure is 295.5 ± 0.1, determined by Nier (1950).
†In 1935, Nier had obtained a value of 0.011 ± 0.001 atom per cent for the isotopic abundance of K-40.

but available data must be regarded as very encouraging when due allowance is given to the usual allotment of teething troubles associated with the development of a new method. [Ahrens, 1956, p. 55.]

It was hardly surprising to see research directed toward development of the new method start up almost simultaneously at several institutions in the United States, at the University of Toronto (Department of Physics) in Canada, and at a number of European laboratories. The main question in everybody's mind was whether or not radiogenic argon-40 was retained quantitatively in all potassium minerals.

One of the more comprehensive reports on potassium–argon dating to be published in the mid-1950s was that of G. J. Wasserburg (Paper 33). Working on the problems associated with the new method for his PhD at the University of Chicago, in association with R. J. Hayden from the Argonne National Laboratory, Wasserburg was confident that radiogenic argon would be quantitatively retained in potassium minerals because of the negligible radiation demage produced by potassium decay and because the much larger atomic radius of argon compared to helium would inhibit loss by diffusion. Using an isotope-dilution technique for their argon analyses (a distinct advantage over the older total-volume method), Wasserburg and Hayden successfully measured Ar-40/K-40 ratios in 10 samples of feldspar. Replicate analyses of different-size fractions of one of the feldspars (Paper 33, Table I) agreed within experimental error (2.3 per cent), and there was no real evidence for loss of radiogenic argon due to diffusion in any of the potash feldspars. Although the decay constant for potassium-40 disintegration was reasonably well known (Wasserburg used a value within 4 per cent of the currently accepted figure) the actual proportion of disintegrating K-40 atoms generating argon rather than calcium was more uncertain.* Using isotopic U–Pb ages determined on coexisting uraninites as an independent measure of the age of their analyzed feldspar samples, Wasserburg and Hayden found the branching ratio for electon capture and beta emission to be 0.085 (± 10 per cent), much lower than previously published values. In conjunction with a total decay constant of 5.5×10^{-10} yr^{-1}, calculated K–Ar ages for the feldspars were then in good agreement with the U–Pb ages of the coexisting uraninites.

A few months after Wasserburg and Hayden's report appeared, a group working at the Carnegie Institution in Washington, D.C., headed by George Wetherill, published the results of a study comparing Ar-40/K-40 ratios of coexisting feldspars and micas separated from the same rocks (Paper 34). In each of seven cases studied, the feldspar yielded a lower Ar-40/K-40 ratio than the coexisting mica. This was the so-called "mica–feldspar discrepancy," the first indication that potash feldspars might be unreliable for K–Ar dating. Discrepancies between Ar-40/K-40 ratios determined on coexisting micas and potash feldspars, first reported by Wetherill and his colleagues, were also found by geochronologists working at Berkeley (Folinsbee et al., 1956) and

---

*This proportion is determined by the ratio of the two separate decay constants involved, λ_β and λ_e, and is referred to as the "branching ratio."

were subsequently confirmed by a number of workers including Wasserburg himself (Wasserburg et al., 1956; Goldich et al., 1957). Experimental studies of argon diffusion in mica and feldspar by Reynolds (1957) showed the argon did indeed diffuse much more readily from feldspar at elevated temperatures, and extrapolation of the experimental data suggested that significant loss of argon from feldspar would occur even at low temperatures as a result of continuous diffusion over geologically long periods of time.

This was a great disappointment, for potash feldspars constitute one of the most common groups of minerals found in plutonic igneous and metamorphic rocks, and expectations were high that they could be dated by the K–Ar method. However, the mica group is almost as abundant, and reliable K–Ar ages could be obtained from both muscovite and biotite. The K–Ar method was also found to give reasonably reliable ages on authigenic minerals such as glauconite and sylvite (Lipson, 1958) and could thus be used on sedimentary rocks to date the time of deposition.*

By the end of the 1950s, there was little doubt that the K–Ar method was over its teething troubles, and was fast becoming established as one of the best methods available for the determination of radiometric ages on a wide range of geologic materials. The full potential of the method was to be realized during the following decade.

For further information pertaining to the history and fundamentals of K–Ar dating, the reader is referred to two excellent texts on the subject, one compiled by Schaeffer and Zahringer (1966), and one by Dalrymple and Lanphere (1969). The effects of sample purity and inhomogeneity in K–Ar dating have recently been discussed by Engels and Ingamells (1970) and Engels (1971).

Selected Bibliography

Ahrens, L. H. (1951). The feasibility of a calcium method for the determination of geological age. *Geochim. Cosmochim. Acta,* **1,** 312–316.

Ahrens, L. H. (1956). Radioactive methods for determining geological age. *In* "Physics and Chemistry of the Earth" (L. H. Ahrens, K. Rankama, and S. K. Runcorn, eds.), Vol. 1, pp. 44–67. Pergamon Press, New York.

Alpert, D. (1953). New developments in the production and measurement of ultra-high vacuum. *J. Appl. Phys.,* **25,** 860.

Beckinsale, R. D., and Gale, N. H. (1969). A reappraisal of the decay constants and branching ratio of ^{40}K. *Earth Planet. Sci. Lett.,* **6,** 289–294.

Carr, D. R., and Kulp, J. L. (1955). Use of A-37 to determine argon behavior in vacuum systems. *Rev. Sci. Instr.,* **26,** 379–381.

Carr, D. R., and Kulp, J. L. (1957). Potassium–argon method of geochronometry. *Geol. Soc. Amer. Bull.,* **68,** 763–784.

Coleman, M. L. (1971). Potassium–calcium dates from pegmatitic micas. *Earth Planet. Sci. Lett.,* **12,** 399–405.

Dalrymple, G. B., and Lanphere, M. A. (1969). "Potassium–argon dating," 258 pp. W. H. Freeman and Co., San Francisco, California.

---

*See Hurley (1966) for a more recent review on K–Ar dating of sediments.

Damon, P. E., and Kulp, J. L. (1958). Inert gases and the evolution of the atmosphere. *Geochim. Cosmochim. Acta., ***13**, 280–292.

Engels, J. C. (1971). Effects of sample purity on discordant mineral ages found in K–Ar dating. *J. Geol.,* **79**, 609–616.

Engels, J. C., and Ingamells, C. O. (1970). Effect of sample inhomogeneity in K–Ar dating. *Geochim. Cosmochim. Acta.,* **34**, 1007–1017.

Folinsbee, R. E., Lipson, J., and Reynolds, J. H. (1956). Potassium–argon dating. *Geochim. Cosmochim. Acta.,* **10**, 66–68.

Gerling, E. K., Ermolin, G. M., Baranovskaya, N. V., and Titov, N. E. (1952). First experience with the application of the argon method for the determination of the age of minerals. *Dokl. Akad. Nauk. SSSR,* **86**, 593–596.

Goldich, S. S., Boadsgaard, H., and Nier, A. O. (1957). Investigations in A^{40}/K^{40} dating. *Trans. Amer. Geophys. Union,* **38**, 547–551.

Hurley, P. M. (1966). K–Ar dating of sediments. *In* "Potassium–Dating" (O. A. Schaeffer and J. Zähringer, eds.), pp. 134–151. Springer-Verlag, New York.

Klemperer, O. (1935). On the radioactivity of potassium and rubidium. *Proc. Roy. Soc., Ser. A,* **148**, 638–648.

Lipson, J. (1958). Potassium–argon dating of sedimentary rocks. *Geol. Soc. Amer. Bull.,* **69**, 137–150.

Møller, C. (1937). On the capture of orbital electrons by nuclei. *Phys. Rev.,* **51**, 84–85.

Newman, F. H., and Walke, H. J. (1935). The radioactivity of potassium and rubidium. *Phil. Mag., Ser. 7,* **19**, 767–773.

Nier, A. O. (1935). Evidence for the existence of an isotope of potassium of mass 40. *Phys. Rv.,* **48**, 283–284.

Nier, A. O. (1950). A redetermination of the relative abundances of the isotopes of carbon, nitrogen, oxygen, argon, and potassium. *Phys. Rev.,* **77**, 789–793.

Reynolds, J. H. (1956). High sensitivity mass spectrometer for noble gas analysis. *Rev. Sci. Instr.,* **27**, 928–934.

Reynolds, J. H. (1957). Comparative study of argon content and argon diffusion in mica and feldspar. *Geochim. Cosmochim. Acta,* **12**, 177–184.

Rutherford, E., Chadwick, J., and Ellis, C. D. (1930). "Radiations from Radioactive Substances," 588 pp. The University Press, Cambridge.

Schaeffer, O. A., and Zahringer, J. (eds.) (1966). "Potassium–Argon Dating," 234 pp. Springer-Verlag, New York.

Shillibeer, H. A., and Russell, R. D. (1955). The argon-40 content of the atmosphere. *Geochim. Cosmochim. Acta.,* **8**, 16–21.

Smith, A. G. (1964). Potassium–argon decay constants and age tables. *Quart. J. Geol. Soc. Lond.,* **1205**, 129–141.

Smythe, W. R., and Hemmendinger, A. (1937). The radioactive isotope of potassium. *Phys. Rev.,* **51**, 178–182.

Thompson, F. C., and Rowlands, S. (1943). Dual decay of potassium. *Nature,* **152**, 103.

Wasserburg, G. J., Hayden, R. J., and Jensen, K. J. (1956). A^{40}–K^{40} dating of igneous rocks and sediments. *Geochim. Cosmochim. Acta.,* **10**, 153–165.

30

The Possibility of a Dual
Beta Decay of Potassium†

C. F. VON WEIZSÄCKER

The radioactivity of potassium is probably due to the rare isotope of mass 40. This $^{40}_{19}$K has two stable isobars: $^{40}_{20}$Ca and $^{40}_{18}$A. The transition to ^{40}Ca, in which negative electrons are emitted, explains the normal radioactivity of potassium. However, in light of energy considerations, the transition to ^{40}A must also be possible, at least if the neutrino has a rest mass of zero, since two isobars whose charges differ by only one unit cannot then be stable at the same time.[1] It may be concluded from the empirical stability of ^{40}A that its transition to ^{40}K with emission of an electron is energetically impossible; it then follows directly that the reversal of this process, i.e., the conversion of ^{40}K into ^{40}A with absorption of an electron of the atomic shell, is energetically possible, even if the energy difference between ^{40}K and ^{40}A is not sufficient for the production of a positron. According to recent calculations by Møller[2], the a priori probability of this absorption is by no means small in relation to that of a normal β emission. The empirical existence of three pairs of immediately adjacent and apparently stable isobars makes our argument appear uncertain[3]. However, it is possible that in these cases one of the two nuclei is a very long-lived β emitter, which satisfies the Sargent equation and therefore emits only electrons whose ranges are undetectably small. However, it seems worthwhile to look for empirical indications of the occurrence of the transition ^{40}K $\rightarrow$ ^{40}A.

Positrons have not been found in Kβ radiation; 7 per cent of the number in which the normal β electrons occur should still have been detectable.[4] However, there is no reason to assume a large participation of the positron emission in comparison with the electron absorption in the transition to ^{40}A. It is therefore more promising to look for the argon formed in the decay. In particular, reasons can be offered for the assumption that practically all the ^{40}A content of the atmosphere is the result of the transformation of potassium in the earth's crust.

---

†Translated from Über die Möglichkeit eines dualen β-Zerfalls von Kalium, *Physik. Z.*, **38**, 623–624.

It is well known that the abundance of the chemical elements exhibits a fairly regular and (apart from the predominance of even atomic numbers) on the whole smooth trend with the atomic number, which is probably connected with the manner of their formation in the interior of the stars. All inert gases are a few powers of ten rarer than their neighboring elements, but, when compared with one another, they show the same trend in their abundances. Only argon is almost 1000 times as abundant as one would have expected from an interpolation based on this pattern. However, if one writes the abundances of its isotopes separately, it is found that the excess is due exclusively to the isotope ^{40}A. If the logarithm of the number of atoms in the accessible part of the earth divided by a suitable constant ($\log A$) is taken as 10 for oxygen, the values in the first column of the table are found for the inert gases[5]. Separation into isotopes is shown in the table only for argon, since the isotopes of the other inert gases do not differ much from one another in their abundances.

Inert gas	$\log A$	Even-numbered neighboring elements				$\log \alpha$
		Lighter	$\log A$	Heavier	$\log A$	
$^{4}_{2}He$	2.8					
$_{10}Ne$	1.9	O	10.0	Mg	9.4	7.8
$^{36}_{18}A$	1.9					7.0
$^{38}_{18}A$	1.1	S	8.4	Ca	8.3	
$^{40}_{18}A$	4.4					4.0
$_{36}Kr$	0.9–1	Se	4.7	Sr	5.5	5.2
$_{54}Xe$	0.9–2	Te	2.4	Ba	5.3	5.0

The rarity of the inert gases on the earth must be connected in some manner with processes related to the formation of the earth and with the fact that the inert gases do not form chemical bonds. The parallelism of the tendency of their abundances with that of the other elements suggests that even without detailed knowledge of this process, one could determine, purely empirically, a factor α by which their abundances have been reduced. Because of the relatively smooth variation of the abundance with the atomic number, a presumably satisfactory approximation to the "true" abundance of an inert gas should be given by an average of the abundances of its two even-numbered neighboring elements. The abundances given in the table for these elements are taken from meteorite analyses,[6] which probably approximate more closely to the average composition of the earth than the abundances in the accessible crust of the earth, which represents the result of a chemical separation. The values of α calculated in this way again show a regular variation from which only ^{40}A deviates. There is no doubt, however, that the same α must be chosen for all three isotopes of argon, since the escape of the inert gases, no matter how it happened, cannot have caused such an effective isotope separation, of which, moreover, the other inert gases show no trace. If the value determined for ^{40}A is chosen, then ^{36}A and ^{38}A become rarer than their neighboring isotopes by a factor of 1000; if the value for ^{36}A and ^{38}A, which agrees with those for the other inert gases, is chosen, and if it is assumed that all the ^{40}A found on earth at present was there from the outset, then ^{40}A becomes more abundant in the cosmos than oxygen. According to Page,[7] on the other hand, neon is rarer than oxygen by a factor of about 20, and argon is rarer than neon by a factor of about 7, in planetary nebula. The difficulty is avoided if one assumes that terrestrial ^{40}A was formed from ^{40}K only after the loss of the inert gases.

On this assumption, it is possible to estimate the branching ratio between the decay into ^{40}Ca and into ^{40}A. However, the result of the estimate also depends on an estimate of the percentage of the argon produced in the earth that was retained in the rock and the percentage that entered the atmosphere. Only the abundances of the inert gases in the atmosphere are known and therefore used in the table. On the other hand, it is usual to give the abundances of the other elements (i.e., including that of the oxygen, according to which the table is normalized) for a surface layer of the earth to 16 km deep. If we idealize the process by the assumption that all the argon formed in a layer up to d km deep reached the atmosphere, while all that produced in deeper layers was retained, and if we represent by x the ratio of the number of argon atoms now present in the atmosphere to the number of ^{40}K atoms that have decayed in a layer up to 16 km deep since the loss of the inert gases, then the fraction of ^{40}K nuclei that chooses decay into ^{40}A is $16x/d$. From the intensity of the Kβ radiation and the relative abundance of the radioactive isotope ^{40}K, a half-life of about 2×10^9 yr is found for the latter. If the loss of the inert gases is placed at the beginning of the earth's history, the quantity of ^{40}K that has decayed since then must be approximately equal to the quantity present now. The present value of log A for potassium in the earth's crust is 8.3, so that for ^{40}K, log $A \approx 4.5$, and $x \approx \frac{1}{3}$. This value is naturally uncertain by a factor of the order of 2 to 10, since neither the geochemical abundance of potassium nor the half-life of ^{40}K is accurately known and an error by a factor of f in the latter leads to an error by a factor of 2^f in the value of x. It is even more difficult to determine d. If we arbitrarily take $d = 16$, we find that approximately every third ^{40}K nucleus chooses decay into argon. It should then be possible to detect argon in old potassium minerals. A mineral 500 million years old that has not lost any [radiogenic] argon should have an argon content of about 1/100 of the potassium content.

The quantities of argon produced from potassium are thus very small, and could not appreciably change the cosmic abundance of argon. The fact that they are appreciable on earth and thus provide evidence of the dual decay of ^{40}K is due only to the abnormal rarity of the inert gases on the earth, which is an accident from the standpoint of nuclear physics. If our assumption is correct, the transposition of the atomic weights of K and A, which was so disturbing in the earlier theories of the periodic system, is simultaneously explained as a secondary effect.

Confirmation of the assumption that ^{40}K exhibits dual decay with abnormally long lifetimes on both sides would also be of some interest in nuclear physics. This long lifetime could then be regarded with certainty, not as a special property of the transition ^{40}K $\rightarrow$ ^{40}Ca, but as a property of the nucleus ^{40}K itself. Since according to general rules ^{40}A and ^{40}Ca must have zero spin, it could be explained, e.g., simply by Gamow's assumption that ^{40}K has a very high nuclear spin, so that the β decay, in which an electron and a neutrino must assume a high angular momentum, is to a close approximation forbidden.

References

1. G. Beck, *Z. Phys.*, **91**, 379 (1935).
2. C. Møller, *Phys. Rev.*, **51**, 84 (1937).
3. K. T. Bainbridge and E. B. Jordan, *Phys. Rev.*, **50**, 282 (1936).
4. Prof. von Hévésy, personal communication.
5. From G. Berg, "Das Vorkommen der chemischen Elemente auf der Erde." Leipzig 1932.
6. I. and W. Noddack, *Svenck Kem. Tids.*, **46**, 173 (1934).
7. G. Page, *Nature* **138**, 503 (1936).

PHYSICAL REVIEW   VOLUME 74, NUMBER 8   OCTOBER 15, 1948

Argon 40 in Potassium Minerals

L. T. ALDRICH AND ALFRED O. NIER

Department of Physics, University of Minnesota, Minneapolis, Minnesota

(Received July 8, 1948)

An investigation has been made of the isotopic composition of the argon from four potassium minerals. In each case a high A^{40}/A^{36} ratio compared to that of atmospheric argon is observed showing directly that K^{40} decays to both Ca^{40} and A^{40}. From the absolute amounts of radiogenic A^{40} and K^{40} in the minerals a lower limit on the branching ratio λ_K/λ_β can be made. If the half-life, 7×10^8 yrs., is assumed for $K^{40} \rightarrow Ca^{40}$, λ_K/λ_β must be at least 0.02. The possibility of using this method for measuring geological age is suggested.

IT was suggested by von Weizsäcker[1] in 1937 that the abnormally high abundance of A^{40} in argon might be explained by assuming that K^{40} not only decays to Ca^{40} through β-emission but also to A^{40} through K-capture. Indirect evidence to support this view has been obtained by a number of investigators.[2-6] Bleuler and Gabriel[7] studied the X-radiation emitted during the decay process of potassium and concluded that there are 1.9 times as many K-captures occurring as there are β-rays emitted. Suess[8] was unsuccessful in his search for argon in sylvine and carnallite

minerals and concluded from this investigation together with general geological considerations that the branching ratio, λ_K/λ_β, was 0.05 ± 0.02.

An investigation of the gas evolved from four potassium minerals has been made with a high sensitivity mass spectrometer. In all cases small amounts of argon were discovered and in each case the A^{40}/A^{36} ratio was appreciably greater than that observed for atmospheric argon. Figure 1 shows a comparison between the spectra observed for atmospheric argon and for the argon found in one of the minerals.

The procedure employed in extracting the argon from the minerals was as follows: A weighed amount of the mineral was introduced into a high temperature vacuum furnace which was heated to a temperature above 1000°C. The condensable vapors were removed by a liquid oxygen trap. If the amount of gas remaining after this treatment was less than 20 std. cc, an investigation of the A^{40}/A^{36} ratio was made directly without further purification. The extremely high sensitivity of the spectrometer permitted one to do this. For samples larger than 20 std. cc the argon concentration was too low to permit accurate isotope analysis and the gas was further purified with hot lithium metal. The principal impurity which diluted the argon was nitrogen. That the lithium treatment had no effect on the A^{40}/A^{36} ratio was shown in a test in which this ratio was determined in a sample of air before and after purification.

Table I contains pertinent data for the minerals investigated. From the data in column 6 together with the ages of the minerals and the decay constant for $K^{40} \rightarrow Ca^{40}$ one may compute the branching ratio. The results of such calcula-

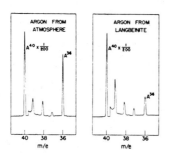

FIG. 1. Mass spectra for atmospheric argon and for argon from mineral Langbeinite. Note that this mineral has an A^{40}/A^{36} ratio greater than three times that for atmospheric argon. Peaks at 37 and 39 and part of 38 are due to residual impurities in spectrometer. There were no detectable residual impurity peaks at either mass 36 or 40.

[1] C. v. Weizsäcker, Physik. Zeits. **38**, 623 (1937).
[2] F. C. Thompson and S. Rowlands, Nature **152**, 103 (1943).
[3] L. H. Gray and G. T. P. Tarrant, Proc. Roy. Soc. **A143**, 681 (1934).
[4] O. Hirzel and H. Wäffler, Helv. Phys. Acta **19**, 216 (1946).
[5] E. Gleditsch and T. Graf, Phys. Rev. **72**, 640 (1947).
[6] H. A. Meyer, G. Schwachheim, and M. D. deSousa Santos, Phys. Rev. **71**, 908 (1947).
[7] E. Bleuler and M. Gabriel, Helv. Phys. Acta **20**, 67 (1947).
[8] H. E. Suess, Phys. Rev. **73**, 1209 (1948).

TABLE I. Data for several potassium minerals.

Mineral	Total wt. grams	Potassium* grams	Argon std. cc	Radiogenic A^{40} std. cc†	$\dfrac{\text{g radio. } A^{40}\S}{\text{g K-40}}$	Age of mineral $\times 10^{-8}$ yr.	Branching ratios for three half-lives		
							4×10^8 yr.	7×10^8	14×10^8
Orthoclase	450	62	0.31	0.25	7.0×10^{-2}	14	0.007	0.02	0.07
Microcline	450	72	0.10	0.035	0.8×10^{-2}	3.5	0.010	0.02	0.04
Sylvite	250	135	0.08	0.011	0.13×10^{-2}	2.0	0.003	0.006	0.013
Langbeinite	400	80	0.065	0.045	0.9×10^{-2}	2.0	0.02	0.04	0.09

* This column was estimated from the accepted composition of the mineral. Slight impurities would lower this number.
† This column computed from the measured A^{40}/A^{36} ratio for the mineral and that for air. It was assumed that the non-radiogenic argon had the composition of present-day atmospheric argon.
§ There is good reason to feel that the errors in analysis together with the assumptions made in computation should not affect this ratio by more than 25 percent.

tions are shown for three different assumed decay constants.

Since an exact determination of the potassium content of the mineral was not made, the amount of potassium present is at most that given. Moreover, since it is assumed that the non-radiogenic argon had the composition of present-day atmospheric argon and that no leakage of radiogenic argon occurred during the life of the mineral, the ratio given in column 6, if in error, is too small and would therefore give a minimum value for the branching ratio. Since the branching ratio computed here depends upon the assumed value for the half-life of $K^{40} \rightarrow Ca^{40}$, no definite value can be decided upon until this constant is better established. All three values used here have appeared in the literature.

It is seen from columns 8–10 of the table that there is reasonable agreement among the values of the branching ratio obtained except in the case of sylvite. The low value for the branching ratio for sylvite could be attributed to the leakage of argon from the mineral. The fair agreement obtained for the branching ratio from the other three minerals indicates that with improved techniques in carrying out experimental work of this type, together with a careful redetermination of the half-life of K^{40}, the $K^{40} \rightarrow A^{40}$ decay might become extremely useful in the measurement of geological time.

ACKNOWLEDGMENTS

The mineral samples used in this study were very generously supplied through the courtesy of Professor E. L. Tullis of the South Dakota School of Mines, Professor Adolph Knopf of Yale University, Dr. W. F. Foshag of the U. S. National Museum, and Dr. J. P. Marble, Chairman of the National Research Council Committee on Measurement of Geological Time. The apparatus employed was constructed through a grant from the Graduate School and the Minnesota Technical Research Fund subscribed to by General Mills, Inc., Minneapolis Star-Journal and Tribune, Minnesota Mining & Manufacturing Company, Northern States Power Company, and Minneapolis Honeywell Regulator Company.

The research was made possible by Navy Contract N5ori-147, T. O. III, between the Office of Naval Research and University of Minnesota.

32

Argon Determinations in Potassium Minerals I. Determination in Tertiary Potassium Salts†

F. SMITS and W. GENTNER

According to measurements by Nier,[1,13] in addition to the isotopes ^{39}K and ^{41}K, potassium has a rare isotope having a mass number of 40, whose abundance is 0.0119 per cent. Smythe and Hemmerdinger[2] examined the isotope mass-spectrographically for β rays. It was found that the well known β activity is due to the isotope ^{40}K. Following the discovery by Alvarez[3] of radioactivity associated with the capture of an electron from the K shell, von Weizsäcker[4] suggested that in addition to the β activity, ^{40}K also has an activity associated with K capture. The β decay of ^{40}K would thus lead to ^{40}Ca and the K capture to ^{40}A. This would explain the high argon content of the air. Attempts to find a positron activity, which would lead to the same final nucleus ^{40}A, were unsuccessful.[5] On the other hand, Kohlhörster[6] was the first to observe a γ radiation. However, considerable difficulties hindered the experimental detection of the K capture of ^{40}K, since the corresponding K radiation is very soft and is also of extremely low intensity. Though Thompson and Rowlands[7] reported the detection of the K radiation in the Wilson chamber, they were unable to claim very high accuracy for their intensity data. In 1947, Bleuler and Gabriel[8] published measurements of the K radiation with counter tubes. They found the branching ratio $\beta : K = 1 : 1.9$. However, subsequent work by Suess[9] and by Harteck and Suess[10] showed that this high rate of K capture certainly does not correspond to the true situation. In argon determinations on sylvinite and carnallite, these authors were unable to find corresponding quantities of argon, and they therefore gave only an upper limit of $\beta : K = 1 : 0.1$. The first definite detection of argon in potassium minerals of various ages was reported by Aldrich and Nier,[11] whose results are discussed below.

†Translated from Argonbestimmungen an Kalium-Mineralien I. Bestimmungen an tertiären Kalisalzen, *Geochim. Cosmochim. Acta*, **1**, 22–27 (1950).

Age determinations based on the potassium : argon ratio are possible in principle if the decay constant is known with sufficient accuracy. In addition to direct measurements of the decay constant for the K decay of potassium, this value can also be found by determination of argon in minerals whose age is accurately known from the uranium–lead method. To establish whether argon determinations on potassium minerals can be used for age determinations on the one hand and for the determination of the decay constant for the K capture on the other, measurements were first carried out on potassium salts.

Experimental Arrangement

The determination of argon can be carried out by principles similar to those normally used in the determination of helium (Paneth[12]). However, since contamination by argon from the air cannot be reliably avoided, the argon obtained must be checked mass-spectroscopically. In addition to the isotope ^{40}A, the argon in the air contains the isotopes ^{36}A and ^{38}A. The ^{40}A/^{36}A ratio in air is 296, and ^{36}A/^{38}A is 5.4.[13] The contamination by air can therefore be determined from the change in the isotope abundance. Since about 10 mm^3 of argon was required for a mass-spectroscopic determination, fairly large quantities of substance had to be used.

The salt is first introduced into a round-bottomed flask (Fig. 1) and kept under high vacuum for 30 hours with a diffusion pump. If the apparatus is then left closed

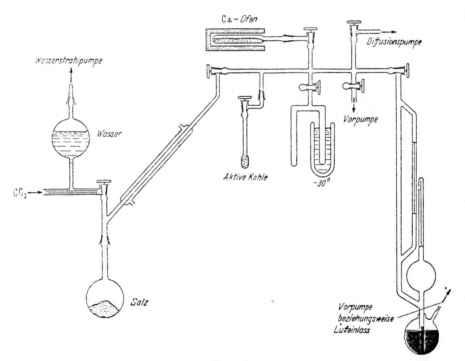

Figure 1.

overnight, no more than 0.1 mm^3 of inert gas can be detected. The salt is then dissolved in water that has been freed from air by passage of CO_2 under vacuum. The trapped gases are liberated by boiling under reflux. The gases are quantitatively transferred into the measuring section by admission of steam in bursts and subsequent freezing with liquid air. Argon is very readily absorbed in the freezing pocket. The water frozen out is, therefore, subsequently distilled over into a second limb of the freezing pocket, which is only cooled to about $-30°C$. The noninert gases are then bound with calcium, and the total quantity of inert gas is measured in a 500-cm^3 McLeod gauge.

To determine the ratio of the volume of the McLeod bulb to the rest of the apparatus, a known quantity of argon was introduced into the McLeod capillary and measured. Wth the mercury still high, the rest of the apparatus was evacuated, while the flask contained degassed salt solution. The mercury was then released and a complete determination was carried out. The quantity of argon then measured in the McLeod gauge gives the fraction of the total quantity of argon present. Five calibration measurements of this type gave the following ratios:

$$\frac{4.41 \text{ mm}^3}{1.92 \text{ mm}^3} = 2.30 \qquad \frac{3.86 \text{ mm}^3}{1.69 \text{ mm}^3} = 2.28$$

$$\frac{1.92 \text{ mm}^3}{0.83 \text{ mm}^3} = 2.32 \qquad \frac{1.69 \text{ mm}^3}{0.74 \text{ mm}^3} = 2.29$$

$$\frac{8.95 \text{ mm}^3}{3.86 \text{ mm}^3} = 2.32 \qquad \text{Average: } 2.30 \pm 0.02$$

Thus the quantity of argon measured in the McLeod gage must be multiplied by a factor of 2.30 to find the total quantity of argon.

After this determination of the total inert gas content, argon was adsorbed on activated charcoal, and helium plus neon was then determined separately. However, less than 0.1 mm^3 of gas cannot be measured accurately. The gas adsorbed in the charcoal was then investigated mass-spectrometrically by M. Pahl and J. Hilby.*

The potassium content of the solution was subsequently measured in the laboratory of the Gewerkschaft Baden (Buggingen).

Results

Salts from the Kaliwerk Buggingen (Oligocene from the Upper Rhine[14]) were investigated. Pink to red sylvinite layers continuously alternate with white halite layers at Buggingen. The total thickness of the potassium-bearing salt stratum is 3 to 5 m with about 50 sequences of sylvinite layers of various thicknesses. The thickest sylvinite

---

*The mass-spectroscopic method will be described shortly elsewhere (Z. *Naturforsch.*).

layer is about 10 cm. A specimen from the Kaliwerk Friedrichshall I, Sehnde, Hanover, was also examined.

The results obtained so far are listed in Table 1. The quantity of salt used for each measurement was 200 g. The quantity of argon obtained from this was insufficient for a mass-spectroscopic investigation. It was therefore necessary to use the argon from three to four samples for one mass-spectrometric examination.

According to earlier measurements by Lord Rayleigh[15] and Paneth[16] an abnormally high helium content was found in some potassium salts. We have unfortunately not yet been able to obtain corresponding samples. In the potassium salts investigated by us, the quantity of gas after adsorption on active charcoal was always < 0.1 mm^3, and no helium was detected mass-spectroscopically in the argon adsorbed on the active charcoal. The helium content was thus always less than 0.05 mm^3 per 100 g of salt.

Discussion

As can be seen from Table 1, the quantities of argon found for the Buggingen salt varied from 3.4 to 7.1 mm^3 of A per 100 g of K. These variations certainly lie outside the limit of accuracy of the entire measuring process. The layer with the highest argon content lies between layers with much lower argon contents. However, it is the thickest layer, and it is also coarsely crystalline. According to the work of Sturmfels,[14] the rock pressure over the potassium deposit is higher than the yield point of sylvinite. The loss of argon, which has evidently occurred in individual layers, may therefore be due either to diffusion or to geological processes.

The material from the same layer in two cases was examined both in coarse grains and in fine grains (Nos. 7–9 and 11–13; Nos. 14 and 18), since it was suspected that considerable loss of argon might have occurred during the crushing of the salts; this would provide a simple explanation for the fluctuations. A distinct difference is in fact found between the coarse and fine materials. It thus appears that under certain conditions the argon formed collects in very fine cracks and faults in the course of time. Moreover, no atmospheric contamination was found in the sylvinite of the Bi layer with the highest argon content (Nos. 7–9 and 11–13). There thus appears to be no exchange between atmospheric argon and argon of radioactive origin in this case. On the other hand, the Bh layer exhibits a much lower argon content even in coarse grains (No. 17). The fluctuations cannot therefore be entirely due to the grain size.

It is still striking and very important that the Bi layer with the highest argon content also has the lowest atmospheric contamination. The absence of atmospheric argon thus appears to be a direct measure of undisturbed conditions. It is our intention to investigate larger quantities of halite rock for argon, so that mass-spectroscopic examination of this argon will also be possible. Measurements of this nature may

Table 1

No.	Nature of sample 200 g of salt	g of potassium		mm³ of argon 0°C/760 mm Hg		$^{40}A/^{36}A$	Atmospheric argon content %	Atmospheric argon content mm³	Argon of radioactive origin mm³	mm³ of radiogenic A 100g of K
	Argon from air			ca. 10		293 ± 10				
1	Buggingen Sylvinite (only 150 g)	62	153	3.10	7.4	3400 ± 1000	9 ± 3	0.65	6.75	4.4 ± 0.5
2	Buggingen Sylvinite	91		4.30						
6	Buggingen BmNa	1.4		0.50						
7	Buggingen BiK coarse < 5 mm	87		5.70						
8	Buggingen BiK coarse < 5 mm	83	243	6.30	17.3	>5000	<6	<1.0	17.3	7.1 ± 0.9
9	Buggingen BiK coarse < 5 mm	73		5.30						
10	Buggingen BKNa	19		1.15						
11	Buggingen BiK fine < 0.3 mm	90		3.75						
12	Buggingen BiK fine < 0.3 mm	93.5	276	4.10	12.25	>3000	<10	<1.2	12.25	4.45 ± 0.8
13	Buggingen BiK fine < 0.3 mm	92.5		4.40						
14	Buggingen BmK coarse	100		5.20						
15	Buggingen BkK	49		2.82						
17	Buggingen BhK coarse	59	304	2.42	14.66	1000 ± 200	30 ± 10	4.4	10.3	3.4 ± 0.4
18	Buggingen BmK fine	95.5		4.22						
16	Buggingen BmNa fine	8.5		0.44						
3	Friedrichshall I Sylvinite	18		3.40						
4	Friedrichshall I Sylvinite	17	54	3.35	10.1	1720 + 320	17 + 3.5	1.7	8.4	15.5 ± 1.8
5	Friedrichshall I Sylvinite	19		3.35						

The designation of the Layers is based on the nomenclature of E. Sturmfels [14]. The appended letter K denotes mainly sylvinite, and Na denotes mainly halite. If no indication is given, the material was unsorted.

in certain circumstances provide information on the diffusion of the argon. The measurements carried out so far on halite samples 6, 10, and 16 cannot allow any further conclusions until their content of atmospheric argon is known.

The sylvinite Friedrichshall I (Nos. 3–5) gave a higher argon/potassium ratio, but this may not correspond to the age of the Zechstein (200×10^6 yr). Aldrich and Nier obtained similar results. Since the salt of the Zechstein was subjected to multiple recrystallizations (horst formation, etc.), very pronounced fluctuations in the argon content must also occur; these will be investigated more thoroughly.

To use the present measurements on the Buggingen salt for age determinations, one must use the highest argon/potassium ratio. Even this could represent the value after some loss of argon, so that only a lower limit for the geologic age can be given at present.

As was mentioned earlier, the decay constant for the K capture has not yet been determined directly. There are powerful reasons for assuming that the observed γ radiation is coupled with the K capture, i.e., that the number of γ quanta should be equated to the number of K disintegrations. The values given so far in the literature lie between 1.1 and 3.6 quanta per g of potassium per sec.[17-21] According to careful measurements by Spiers,[22] the value is 3.0 quanta per g of K per sec. If this value is used as a basis of the decay constant of the K capture, an age of 20×10^6 years is found for the Buggingen deposit. However, the assignment of the γ quanta to the K decay has not yet been established with sufficient certainty. Aldrich and Nier investigated orthoclase and microcline as well as salts and gave ages for these. The probability of argon escaping from these minerals should be very low, since argon has 10 times the atomic weight of helium and the diffusion should therefore be much less than in the case of helium. Moreover, the radioactive argon is uniformly distributed in [potassium] minerals, which are not damaged by γ rays in contrast to uranium minerals. Calculation of the number of K captures per g of potassium per sec on the basis of the data reported by Aldrich and Nier gives the values given in Table 2. The error limits represent the measuring accuracy of 25 per cent reported by Aldrich and Nier, without taking into account any errors in the age. Even if the sylvinite value is not taken into account for the reasons given earlier, the other three values are also lower than the value of 3.0 K captures per g of potassium per sec given above. If it should be confirmed that the number of K captures per g of potassium per sec is in fact in the vicinity of 2.0, the [age] value for the Buggingen deposit would increase from the above value to 30×10^6 years.

Table 2

Mineral	Age	mm³ A/100 g K	K captures/g K/sec
Orthoclase	14×10^8	400	1.6 ± 0.5
Microcline	3.5×10^8	49	1.1 ± 0.3
Sylvinite	2.0×10^8	8	(0.3 ± 0.1)
Langbeinite	2.0×10^8	56	2.3 ± 0.4

We are very grateful to Prof. F. Kirchheimer, Prof. M. Pfannenstiel, Prof. H. Tobien, and Dr. R. Mehnert for geological advice. We thank Dr. H. Simon, Director of the Gewerkschaft Baden, Buggingen, for his constant help, as well as the Gewerkschaft Wintershall, Heringen, and Kalichemie A.G., Sehnde, for supplying samples.

References

1. A. O. Nier, *Phys. Rev.*, **48**, 283 (1945); **50**, 1041 (1936).
2. W. R. Smythe and A. Hemmerdinger, *Phys. Rev.*, **51**, 178 (1937).
3. L. W. Alvarez, *Phys. Rev.*, **52**, 134 (1937).
4. C. F. von Weizsäcker, *Phys. Z.*, **38**, 623 (1937).
5. W. Bothe, *Naturwissenschaft*, **29**, 194 (1941), footnote 3.
6. W. Kohlhörster, *Naturwissenschaft*, **16**, 28 (1928).
7. F. C. Thompson and S. Rowlands, *Nature*, **152**, 103 (1943).
8. E. Bleuler and M. Gabriel, *Helv. Phys. Acta*, **20**, 67 (1947).
9. H. E. Suess, *Phys. Rev.*, **73**, 1209 (1948).
10. P. Harteck and H. Suess, *Naturwissenschaft*, **34**, 214 (1947).
11. L. T. Aldrich and A. O. Nier, *Phys. Rev.*, **74**, 876 (1948).
12. F. Paneth and K. Peters, *Z. Phys. Chem.*, **134**, 353 (1928); F. Paneth, H. Gehlen, and K. Peters, *Z. Anorg. Chem.*, **175**, 383 (1928).
13. A. O. Nier, *Phys. Rev.*, **77**, 789 (1950).
14. E. Sturmfels, *Neue Jahrb. Mineral., Abhandl. Abtl. A Bd.*, **78**, 134 (1943).
15. R. J. Strutt, *Proc. Roy. Soc., Ser. A*, **81**, 278 (1908).
16. F. Paneth and K. Peters, *Z. Phys. Chem.*, **B1**, 170 (1928).
17. O. Hirze and H. Wäffler, *Phys. Rev.*, **74**, 1553 (1948).
18. T. Graf, *Phys. Rev.*, **74**, 831 (1948); E. Gleditsch and T. Graf, *Phys. Rev.*, **74**, 989 (1948).
19. J. J. Floyd and L. B. Borst, *Phys. Rev.*, **75**, 1106 (1949); **74**, 989 (1948).
20. L. H. Ahrens and R. D. Evans, *Phys. Rev.*, **74**, 279 (1948).
21. H. A. Mayer, G. Schwachheim, and M. D. de Souza Santos, *Phys. Rev.*, **71**, 908 (1947).
22. F. W. Spiers, *Nature*, **165**, 356 (1950).

Geochimica et Cosmochimica Acta, 1955, Vol. 7, pp. 51 to 60.   Pergamon Press Ltd., London

A⁴⁰-K⁴⁰ dating*

$$33$$

G. J. Wasserburg

Institute for Nuclear Studies, Department of Physics, and Department of Geology,
University of Chicago, Chicago, Illinois

and

R. J. Hayden

Argonne National Laboratory, Lemont, Illinois

(*Received October* 1954)

ABSTRACT

The A^{40}/K^{40} ratios and the Pb-U ages of various co-existing potassium feldspars and uraninites were determined. It was found that with a branching ratio $\lambda_e/\lambda_\beta = 0.085 \pm 0.005$ and a decay constant $\lambda = 0.55 \times 10^{-9}/\mathrm{yr}$, the A^{40}/K^{40} ages could be brought into agreement with the Pb-U ages for samples ranging from 260 to 1860 m.y. No evidence was found for the loss of argon by diffusion from potassium feldspars.

Introduction

The purpose of this investigation was to determine the feasibility of using the radioactive decay of K^{40} into A^{40} as a method of geologic dating. It is generally agreed, at the present time, that the most reliable age determinations are those obtained by the Pb-U method when there is agreement between any two of the Pb^{206}/U^{238}, Pb^{207}/U^{235}, or Pb^{206}/Pb^{207} ages. In the present work, agreement to within experimental error of the Pb^{206}/U^{238} and Pb^{207}/U^{235} ages has been accepted as the criterion for the true age of the samples studied. In order to test the A^{40}/K^{40} method, it was therefore necessary to compare A^{40}/K^{40} ages with good Pb-U ages.

The age (τ) of a potassium mineral is given by the following expression

$$\tau = \frac{1}{\lambda} \ln \left\{ 1 + \frac{A^{40}}{K^{40}} \frac{(1 + R)}{R} \right\}$$

where λ is the total decay constant of K^{40} and is equal to the sum of the β^- decay constant (λ_β) and the electron-capture decay constant (λ_e). R is the branching ratio (λ_e/λ_β), and A^{40}/K^{40} is the ratio of the number of radiogenic A^{40} atoms to the number of K^{40} atoms now present.

The total decay constant of K^{40} is essentially determined by the rate of β^--emission because of the relatively small number of electron captures. The β^- decay constant has been determined by a number of workers with satisfactory agreement (F. Birch, 1951). There is however considerable uncertainty in λ_e and it is therefore not possible to make a direct comparison of the Pb-U and A^{40}/K^{40} dating methods. If, however, it is possible to find a unique value for the branching ratio such that the calculated A^{40}/K^{40} ages agree with the Pb-U ages for samples covering a wide span of geologic time it is then clear that the argon method is applicable. Since it was assumed that this method is absolute, the value

* This work was supported in part by AEC contract number AT(11-1)-101 and in part by National Science Foundation grant number G-208.

This work was done in partial fulfilment of the requirements for the degree of Doctor of Philosophy in the Department of Geology, University of Chicago.

of the branching ratio so obtained should agree with the value of λ_e/λ_β determined by a precise counting experiment.

The present work proposes to show that with a branching ratio of $0\cdot085$, A^{40}/K^{40} ages can be obtained which agree quite well with Pb-U ages for samples ranging in age from 260 m.y. to 1860 m.y.

It was expected that the argon method would yield better results than the helium method because of the negligible radiation damage incurred in potassium minerals due to K^{40} decay and because the much larger atomic radius of argon compared with that of helium should result in less diffusion loss.

METHOD

The feldspar to be analyzed was ground, and a convenient sieve fraction taken. Samples for the argon and potassium analyses were taken from this fraction by using a small sample splitter. The potassium analyses were done by K. JENSEN of the Argonne National Laboratory. The

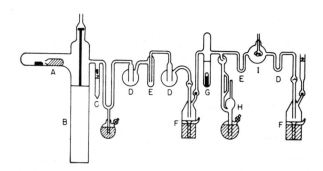

Fig. 1. Argon extraction apparatus.

A. Sample and iron slug.	F. Toepler pump.
B. Nickel furnace.	G. Calcium furnace.
C. A³⁸ tracer.	H. McLeod.
D. Cold trap.	I. Tungsten filament.
E. CuO furnace.	

samples were dissolved in aqueous HF and perchloric acids. The potassium was separated by an ethyl acetate extraction and weighed as the perchlorate. The $KClO_4$ was analyzed for rubidium, and the final results were corrected.* A sample of potassium feldspar, obtained from the Bureau of Standards, was analyzed along with each batch of samples for control. The agreement obtained on this standard was better than 1 per cent.

The radiogenic argon content was determined by an isotopic dilution technique (INGHRAM, 1953, 1954). An aliquot of the feldspar enclosed in a thin-walled Pyrex tube was inserted in the glass side-arm of the nickel reaction vessel (see Fig. 1). The system was evacuated and out-gassed until a satisfactory vacuum (less than 10^{-6} mm Hg) was achieved. The reaction vessel was then isolated by use of mercury valves, which were used throughout the vacuum system. The sample was then pushed into the furnace, and the A^{38} tracer introduced. The nickel furnace containing NaOH was heated for twenty-four hours at 600°C. At the end of this time the evolved gases were passed through a CuO furnace at 500°C to convert H_2 to H_2O. The H_2O was condensed in a cold trap, and the remaining gases were transferred to a calcium furnace. The furnace consisted of a stainless-steel crucible containing calcium metal, enclosed in a quartz tube. The

---

* The possibility of Rb contamination in the perchlorate method was pointed out to the authors by O. JOENSUU.

52

calcium furnace was heated by an external resistance furnace until a mirror was formed. The pressure was then measured with a McLeod gauge. The gas was then allowed to pass through a hot tungsten-filament lamp, CuO at 500°C, and a cold trap. The calcium-furnace and tungsten-filament treatments were repeated until optimum cleanup was obtained. The sample of rare gas remaining was then transferred by means of a Toepler pump to a sample tube and placed on the sample system of the mass spectrometer.

The mass spectrometer used has a 12-inch radius of curvature and a 60° deflection. A low background was obtained in the mass region of interest by frequent baking without opening the instrument to atmospheric pressure. With good sample cleanup, no hydrocarbons were observed at mass 36, 38, or 40 when the sample tube was kept at liquid-nitrogen temperature. On the few occasions when hydrocarbons were observed the resolution of the instrument was sufficient to completely separate them from the argon. It was not possible to resolve the A[36] and A[38] peaks from HCl, and hence, care was taken to avoid any contamination with chlorine.

In order to correct for discrimination, normal argon was run before and after each radiogenic sample, and the discrimination was calculated from the observed A^{36}/A^{40} ratio in the normal. NIER's values (NIER, 1950) for the isotopic composition of normal argon were assumed. The amount of radiogenic A[40] extracted from the feldspar was computed from the observed A^{36}/A^{38} and A^{38}/A^{40} ratios and from the known volume of A[38] tracer which had been added. It was assumed that the peak at mass 36 was due to normal argon contamination. In most cases 95 per cent or more of the argon analyzed was radiogenic.

The lead, uranium, and thorium analyses of the uraninites were done by K. JENSEN of the Argonne National Laboratory. The lead was separated by electro-deposition as PbO_2 from a nitric-acid solution. The thorium was separated by solvent extraction from the uranium and rare earths and weighed as ThO_2. The uranium was separated from the rare earths by an ether extraction and weighed as U_3O_8.

The lead isotopic composition of the uraninites was determined by using a surface ionization technique with a 12-inch mass spectrometer which was similar to the gas instrument. Pb[+] ions were formed by surface ionization from a tungsten filament which had been loaded with $Pb(NO_3)_2$ in a borax binder. This procedure is described by TILTON et al. (in press).

TRACER PREPARATION

A[38] was obtained by neutron irradiation of NaCl.* The argon was extracted from the salt by melting in an evacuated vessel. The gases were then purified, and the A[38] was transferred to the tracer line (see Fig. 2). The tracer line consisted of a series of interconnecting capillary Y tubes which were connected to a mercury reservoir. Break-off tubes of known volumes were attached

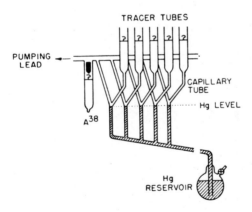

Fig. 2. Tracer preparation line.

---

* A[38] tracer later became available from the Oak Ridge National Laboratory. This was used in the latter part of this work.

4A

Table 1. Microcline, Bessner Mine, Lot 5, Con. B, Henvey Twp., Ontario

Experiment	Sample Wt. (gm)	Sieve fraction	%K	A^{38} tracer (cc STP)	A^{38}/A^{40}	A^{36}/A^{38}	Radiogenic A^{40} (cc STP) per g	A^{40}/K^{40}
1	14·522	40–80	11·00	$1·210 \pm 0·02 \times 10^{-5}$	$1·92 \pm 0·02 \times 10^{-3}$	$0·0506 \pm 0·001$	$4·25 \pm 0·12 \times 10^{-4}$	$0·0565 \pm 0·0014$
2	27·283	18·40	10·92	$1·055 \pm 0·02 \times 10^{-5}$	$9·03 \pm 0·1 \times 10^{-4}$	$0·044 \pm 0·001$	$4·27 \pm 0·12 \times 10^{-4}$	$0·0573 \pm 0·0014$
3	6·902	18·40	10·92	$1·125 \pm 0·02 \times 10^{-5}$	$3·72 \pm 0·02 \times 10^{-3}$	$0·032 \pm 0·001$	$4·22 \pm 0·12 \times 10^{-4}$	$0·0567 \pm 0·0014$
4	16·444	80–180	10·43	$1·111 \pm 0·02 \times 10^{-5}$	$1·73 \pm 0·01 \times 10^{-3}$	$0·0279 \pm 0·0001$	$3·85 \pm 0·12 \times 10^{-4}$	$0·0542 \pm 0·0014$

Table 2.

Locality	No.	Sample Wt. (gm)	%K	A^{38}/A^{40}	A^{36}/A^{38}	A^{38} tracer (cc STP) $\times 10^{+5}$	A^{40}/K^{40}	Pb-U age (yrs.)	λ_e/λ_β	A^{40}/K^{40} age for $\lambda_e/\lambda_\beta = 0\cdot085$
Strickland Quarry (Collins Hill), Portland, Conn.	1	22·904	10·51 ± 0·02	0·01603 ± 0·00004	0·0074 ± 0·0020	3·483	0·01282	$2\cdot67 \times 10^8$	0·0894	$2\cdot74 \times 10^8$
Blackstone Lake Pit Coñger Twp., Parry Sound, Ontario	2	16·701	11·98 ± 0·02	0·001475 ± 0·00001	0·056 ± 0·003	1·136	0·0556	$9\cdot94 \times 10^8$	0·0827	$9\cdot75 \times 10^8$
		12·472	11·98 ± 0·02	0·00196 ± 0·00001	0·120 ± 0·002	1·141	0·0543	$9\cdot94 \times 10^8$	0·0806	$9\cdot57 \times 10^8$
	3	28·177	1·32 ± 0·03	0·282 ± 0·0004	0·0124 ± 0·0010	3·611	0·0456	$9\cdot94 \times 10^8$	0·0660	$8\cdot34 \times 10^8$
	4	37·439	10·62 ± 0·02	0·00229 ± 0·00003	0·0348 ± 0·0020	3·739	0·0592	$9\cdot94 \times 10^8$	0·0886	$1\cdot02 \times 10^9$
Concession XVI Cardiff Twp., Ontario	5	24·778	3·02 ± 0·06	0·00995 ± 0·00003	0·0436 ± 0·0020	3·624	0·0629	$1\cdot01 \times 10^9$	0·0926	$1\cdot07 \times 10^9$
	6	34·026	4·79 ± 0·05	0·00553 ± 0·00004	0·0314 ± 0·0020	3·779	0·0587	$1\cdot01 \times 10^9$	0·0858	$1\cdot02 \times 10^9$
Tory Hill, Ontario	7	12·065	6·15 ± 0·02	0·00368 ± 0·00004	0·0303 ± 0·0010	1·035	0·0540	$1\cdot03 \times 10^9$	0·0763	$9\cdot53 \times 10^8$
	8	32·729	6·86 ± 0·04	0·00428 ± 0·00002	0·0146 ± 0·0008	3·605	0·0542	$1\cdot03 \times 10^9$	0·0765	$9\cdot55 \times 10^8$
Viking Lake, Beaver-lodge Dist., Saskatchewan	9	24·516	11·68 ± 0·02	0·00298 ± 0·00004	0·0196 ± 0·0004	8·859	0·1502	$1\cdot87 \times 10^9$	0·0912	$1\cdot95 \times 10^9$

55

Table 3. Ages of uraninites (yrs.)

Material	Locality	% Pb	% Th	$\frac{207}{206}$	$\frac{208}{206}$	$\frac{204}{206}$	$\frac{Pb^{206}}{U^{238}}$	$\frac{Pb^{207}}{U^{235}}$	$\frac{Pb^{208}}{Th^{232}}$
		% U							
Uraninite	Viking Lake, Beaverlodge Dist., Sask.	17·14 / 52·9	5·24	0·1155 ± 0·0020	0·0290 ± 0·0004	0·000057	$1·85 \times 10^{9}$	$1·88 \times 10^{9}$	$1·67 \times 10^{9}$
Uraninite	Blackstone Lake Pit, Conger Twp., Parry Sound, Ont.	10·72 / 69·3	2·99	0·0724 ± 0·0008	0·0150 ± 0·0002	0·000075	$9·94 \times 10^{8}$	$9·93 \times 10^{8}$	$8·97 \times 10^{8}$
Uraninite	Pit in Con. XVI Cardiff Twp. Ont.	9·92 / 62·3	6·61	0·0749 ± 0·0004	0·0346 ± 0·0003	0·00015	$1·00 \times 10^{9}$	$1·02 \times 10^{9}$	$0·87 \times 10^{9}$
Uraninite	Strickland Quarry Portland, Conn.	3·07 / 79·1	2·98	0·0529 ± 0·00035	0·0160 ± 0·00015	0·000139	$2·68 \times 10^{8}$	$2·66 \times 10^{8}$	$2·39 \times 10^{8}$

to the apparatus, and the system was evacuated. The system was then sealed off, and the A^{38} introduced. The mercury level was adjusted to a point immediately below the junction of the Y's. The gas was allowed to equilibrate for twelve hours, after which the mercury level was raised to isolate the tubes. The tracer tubes were then sealed off. Corrections were made for the volume of capillary between the tracer tube and the Y and for the gas trapped between the mercury and the point of sealing off.

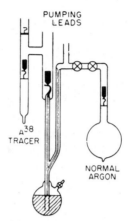

Fig. 3. Capillary pipette and tracer equilibration apparatus.

The tracers were calibrated by an isotopic dilution with a known volume of normal argon of spectroscopic purity. The volume of normal argon was measured at a known temperature in a capillary pipette with a break-off top (see Fig. 3). The volume of normal argon was then mixed with the argon from one of the previously prepared tracers by breaking the break-off end of the pipette. After sufficient time for equilibration was allowed, a mass spectrometric analysis was made of the argon mixture, and the volume of A^{38} in the tracer was computed from the A^{36}/A^{38} ratio. The discrimination of the spectrometer was determined from the A^{36}/A^{40} ratio. The amount of A^{38} in the remaining tracers was then computed by assuming that the pressure in the tracer line was everywhere equal.

EXPERIMENTAL RESULTS

In order to test the method, four samples of microcline from Bessner, Ontario, were analyzed. The samples were taken from various sieve fractions. The results agree with each other within experimental error (see Table 1). The A^{40}/K^{40} ratio obtained in run 4 is slightly lower than that obtained in the other experiments. This is possibly due to argon loss because of grinding.

In runs 1 and 2, the reaction and tracer equilibration time was twenty-four hours. In run 3, the reaction time was twenty-four hours and the equilibration time twelve hours. In run 4, the reaction and tracer equilibration time was thirty-four hours. The results indicate that the time allowed for tracer equilibration was sufficient. The agreement for different reaction times and sieve fractions suggests that the reaction was complete.

The values given in Table 1 for the volume of A^{38} tracer and the A^{40}/K^{40} ratio differ from those reported previously by WASSERBURG and HAYDEN (1954) by 6 per cent. The earlier values are incorrect owing to an error in tracer calibration.

The errors listed in the tracer-volume determination are the probable error of the isotopic ratios and the estimated errors.

57

A sample of Bessner microcline was also run with Na_2CO_3 as a flux in order to determine if more argon would be released by other fluxing methods. The material was reacted for eight hours at slightly over 900°C. The radiogenic argon yield was 12 per cent less than that obtained with NaOH. Because there was a large amount of normal argon contamination in the Na_2CO_3 experiment, this discrepancy may not be real.

Table 2 gives the results obtained for samples from several different localities. All samples were extremely fresh. large pegmatitic feldspars with the exception of No. 8. which was a sample of granite.

Samples No. 2 and No. 3 were from the same pegmatite. Sample No. 4 was from part of the same pegmatite complex as No. 2 and No. 3. Sample No. 4 was collected from an outcrop one mile north of Blackstone Lake Pit. Sample No. 3 has a factor 10 less potassium than No. 2 and No. 4. The reason for the low A^{40}/K^{40} ratio in No. 3 is unknown. Samples No. 5 and No. 6 were from the same pegmatite dike. Although they differ in potassium content by 60 per cent. their A^{40}/K^{40} ratios are in good agreement.

Sample No. 7 is from a feldspar crystal in a pegmatite dike cutting a paragneiss complex one mile north of Tory Hill. Ontario. This pegmatite is one of many pegmatitic bodies surrounding the granite body at Essonville. Ontario. and it is presumably. although not certainly. of the same age as the Essonville granite body immediately to the north of Tory Hill. Sample No. 8 is from the eastern flank of this granite body. It is a highly deformed granite with a distinct planar structure due to the presence of tabular quartz grains. There was no mica present in the specimen which was analyzed.

Lead-uranium-thorium ages were determined for uraninites associated with the various feldspars. The samples were fresh material with no alteration products present. The results are given in Table 3. The last three columns are the calculated Pb^{206}/U^{238}, Pb^{207}/U^{235} and Pb^{208}/Th^{232} ages. respectively. The values of the decay constants used in the calculations were: $\lambda Th^{232} = 4 \cdot 99 \times 10^{-11} \text{ yr}^{-1}$; $\lambda U^{238} = 1 \cdot 54 \times 10^{-10} \text{ yr}^{-1}$; and $\lambda U^{235} = 9 \cdot 72 \times 10^{-10} \text{ yr}^{-1}$. The U^{238}/U^{235} ratio was assumed to be $137 \cdot 7$.

In all cases excellent agreement between the Pb^{206}/U^{238} and Pb^{207}/U^{235} ages was obtained. The Pb^{208}/Th^{232} ages were consistently low by about 10 per cent.

The Viking Lake area. which has not previously been dated. gives ages a few per cent higher than the Pb^{207}/Pb^{206} ages reported by RUSSELL et al. (1953) for Charlebois Lake. Sickle Lake. and Lee Lake. Saskatchewan.

The results obtained for the Conger Township material agree with the data of NIER (1941). The Cardiff Township dates are also in accord with the results of NIER (1939) and COLLINS et al. (1953).

The ages which we have determined for Strickland Quarry are slightly lower than the chemical ages reported by several other workers for this locality (RODGERS, 1952). They agree within 5 per cent with the Spinelli Quarry dates obtained by NIER (1941) and show that the Spinelli and Strickland Quarries are of essentially the same age, as suggested by RODGERS (1952).

The age given for samples No. 7 and No. 8 of Table 2 is taken from the data of TILTON et al. (in press). They were determined on zircons taken from the same

58

granite body as No. 8. The Pb^{206}/U^{238} and Pb^{207}/U^{235} ages which they obtained are 1.06×10^9 and 1.03×10^9 years, respectively.

The A^{40}/K^{40} ages calculated for these samples by using a branching ratio of 0.126 (FAUST, 1950; SAWYER and WIEDENBECK, 1949, 1950; P. R. J. BURCH, 1953; INGHRAM et al., 1950; HOUTERMANS et al., 1950) and a decay constant $\lambda = 0.55 \times 10^{-9}$ yr^{-1} are consistently low as compared to the Pb-U ages. Column 10 of Table 2 gives the calculated branching ratio for each sample assuming the Pb-U age. It is seen that the spread in the calculated value is between 0.076 and 0.093 with the exception of sample No. 3. Sample No. 3 was not used in the final branching ratio calculation because it was felt that this was anamolously low due to argon loss. An X-ray diffraction picture of sample No. 3 shows it to be an albite. All other samples were microclines.* The low argon content may be in some way related to this. The sample from Bessner, Ontario, was not used in calculating the branching ratio because of the great uncertainty in its age (WASSERBURG and HAYDEN, 1954; SHILLIBEER et al., 1954).

Column 6 of Table 2 gives the A^{40}/K^{40} age calculated with a branching ratio of 0.085. This value brings the A^{40}/K^{40} ages into good agreement with the Pb-U ages over a considerable range of geologic time.

The A^{40}/K^{40} age calculated for the Bessner sample by using the above branching ratio is 9.94×10^8 years. This value is in agreement with the Pb^{207}/Pb^{206} age of $9.4 \pm 0.5 \times 10^8$ years determined by RUSSELL et al. (1953).

The value of 0.085 is considerably lower than the values of 0.11 to 0.13 reported by other workers (FAUST, 1950; SAWYER and WIEDENBECK, 1949, 1950; P. R. J. BURCH, 1953; INGHRAM et al., 1950; HOUTERMANS et al., 1950). More recent data obtained by SUTTLE and LIBBY (in press) give a branching ratio of 0.10 ± 0.01 and a decay constant $\lambda = 0.56 \times 10^{-9}$ yr^{-1}. SAWYER and WIEDENBECK (1949, 1950) using K^{42} to calibrate their scintillation counter, obtained a branching ratio of 0.127. The value for the percent gamma yield of K^{42} which they used was 25 per cent. As pointed out by HOUTERMANS et al. (1950), a more recent determination by SIEGBAHN gave a value of 16 ± 4 per cent. KAHN and LYON (1953) have remeasured the gamma yield for K^{42} and obtained a value of 20 ± 1 per cent. By recalculating the data of SAWYER and WIEDENBECK and using KAHN's value of 20 per cent, a branching ratio of 0.102 ± 0.01 is obtained.†

It is evident from the above data that a more precise determination of the decay constants of K^{40} by counting methods is needed.

It is not at present possible to distinguish between small systematic diffusion loss of radiogenic argon and a small change in the branching ratio. However, as there is no apparent trend in the calculated branching ratio with time, and because of the good agreement obtained for significantly different samples of the same age, we feel that there is no evidence for loss of argon due to diffusion.

GENTNER et al. (1953) have reported a value of $D = 10^{-19}$ cm^2/sec for argon in sylvite. Recent data by CHEMLA (1954) give a calculated value of $D = 10^{-31}$ cm^2/sec for the diffusion coefficient of Na$^+$ in NaCl at 20°C. The data for sodium were determined in the temperature range from 600°C to 700°C. Although

---

* J. R. GOLDSMITH, private communication.
† The authors are indebted to R. D. RUSSELL for calling this to their attention.

structural defects due to radiation damage may account for the much larger diffusion coefficient for argon, it is none the less somewhat difficult to reconcile these two results. If the diffusion constant for argon in feldspar were of the same order of magnitude as that reported for argon in sylvite, the effects of diffusion should have been detected in our experiments.

Conclusions

With a branching ratio of 0.085 and a decay constant $\lambda = 0.55 \times 10^{-9}$ yr^{-1}, A^{40}/K^{40} ages are obtained which agree with Pb-U ages within a few per cent for a considerable span of geologic time. These results indicate that a precise determination of the geologic age of potassium feldspars may be obtained by the A^{40}/K^{40} method. It should be possible to determine the age of feldspars as young as 2×10^7 years by the technique described, without any loss in precision.

The agreement obtained by the two methods, one of which depends on the rate of alpha-emission and the other of which depends on the rate of beta-emission and electron capture, places a rather strict limitation on the possible time dependence of the universal constants involved (Houtermans et al., 1946).

The authors wish to thank Professor M. G. Ingram and Professor H. C. Urey of the University of Chicago for advice and discussion during the progress of the work. Dr. S. C. Robinson of the Canadian Geologic Survey, and Dr. J. Rodgers of Yale University, supplied most of the samples analyzed. We are especially indebted to Dr. Kenneth Jensen of the Argonne National Laboratory, who performed the chemical analyses necessary for this work.

References

Birch, F. (1951) *J. Geophys. Res.* **56**, 107.

Burch, P. R. J. (1953) *Nature* **172**, 361.

Chemla, M. (1954) *Compt. rend. Acad. Sci. (Paris)* **238**, 82.

Collins, C. B., Russell, R. D., and Farquhar, R. M. (1953) *Can. J. Phys.* **31**, 402.

Faust, W. R. (1950) *Phys. Rev.* **78**, 150.

Gentner, W., Prag, R., and Smits, F. (1953) *Geochim. et Cosmochim. Acta* **4**, 11.

Houtermans, F. G., Haxel, O., and Heintze, J. (1950) *Z. Physik* **128**, 657.

Houtermans, F. G., and Jordan, P. (1946) *Z. Naturforsch.* **1**, 125.

Ingram, M. G. (1953) *J. Phys. Chem.* **57**, 809; (1954) *Annual Rev. Nuclear Science* **4**.

Ingram, M. G., Brown, H., Patterson, C., and Hess, D. C. (1950) *Phys. Rev.* **80**, 916.

Kahn, B., and Lyon, W. S. (1953) *Phys. Rev.* **91**, 1212.

Nier, A. O. (1939) *Phys. Rev.* **55**, 153; (1950) *Phys. Rev.* **77**, 789.

Nier, A. O., Thompson, R. W., Murphey, B. F. (1941) *Phys. Rev.* **60**, 112.

Rodgers, J. (1952) *Amer. J. Sci.* **250**, 411.

Russell, R. D., Shillibeer, H. A., Farquhar, R. M., and Mousuf, A. K. (1953) *Phys. Rev.* **91**, 1223.

Sawyer, G. A., and Wiedenbeck, M. L. (1949) *Phys. Rev.* **76**, 1535; (1950) *Phys. Rev.* **79**, 490.

Shillibeer, H. A., Russell, R. D., Farquhar, R. M., and Jones, E. A. W. (1954) *Phys. Rev.* **94**, 1793.

Suttle, A. D., and Libby, W. F. (in press) *Anal. Chem.*

Tilton, G. R., Patterson, C., Brown, H., Ingram, M. G., Hayden, R. J., Hess, D. C., and Larsen, E. (in press) *Bull. Geo. Soc. of Amer.*

Wasserburg, G. J., and Hayden, R. J. (1954) *Phys. Rev.* **93**, 645.

60

Geochimica et Cosmochimica Acta, 1955, Vol. 8, pp. 171 to 172. Pergamon Press Ltd., London

34

A^{40}/K^{40} ratios of feldspars and micas from the same rock

G. W. WETHERILL, L. T. ALDRICH, and G. L. DAVIS

Carnegie Institution of Washington, Washington, D.C.

ABSTRACT

A^{40}/K^{40} ratios have been determined for feldspars and micas from the same rock from seven geologic settings. The A^{40}/K^{40} ratio for micas is higher than that for the feldspar in every case.

Because of the widespread occurrence of potassium minerals, the possibility of dating rocks by the K^{40}-A^{40} disintegration has received considerable attention in the past several years. If this is to be used as an absolute dating method, it is necessary to assume that the radiogenic argon has been quantitatively retained in the mineral since its formation. In order to test the validity of this assumption we have measured the ratio of radiogenic argon to potassium from feldspars and micas separated from the same rock. If the two minerals are cogenetic and have retained argon quantitatively, the ratios should be the same for the two minerals. The results of these measurements are shown in Table 1.

Table 1. *Comparison of* A^{40}/K^{40} *ratios of feldspars and micas from the same rock*

Location	Rock type	A^{40}/K^{40}	
		feldspar	mica
Quartz Creek, Colo.	Granite	0·0725	0·107
Uncompahgre, Colo.	Granite	0·0709	0·109
Capetown, S. Africa	Granite	0·0105	0·0337
Ohio City, Colo.	Pegmatite	0·0590	0·0974
Black Hills, S. Dak.	Pegmatite	0·0862	0·118
Jakkalswater, S. Africa	Pegmatite	0·0359	0·0695
Wilberforce, Ontario	Pegmatite	0·0614	0·0695

The radiogenic argon content of the minerals was measured by the conventional techniques of isotope dilution for gaseous samples as developed by INGHRAM and HAYDEN, and described by WASSERBURG (1954). The potassium analyses were made by isotope dilution as well as with a flame photometer.

The feldspars are seen to have a lower A^{40}/K^{40} ratio than the micas in every case. With the exception of the feldspar from the Cape Granite, which was partially kaolinized, all the samples were fresh. In the case of the Ohio City, Colorado sample the flakes of mica were intergrown with the feldspar used in the determination.

At present we are trying to find the cause of this discrepancy. A comparison of these results with Rb-Sr determinations on the same mineral samples and with U-Pb and Th-Pb determinations on minerals from the same rock, indicates

171

that the argon age calculated for the mica probably represents a better measure of the true age of the rock than that calculated for the feldspar. These comparisons will be reported in a lengthier paper.

In view of these measurements, calculations of the branching ratio of K^{40} made by comparing A^{40}/K^{40} ratios of feldspars with ages of uraninites, as has been done by Wasserburg and Hayden (1955), must be reconsidered. In the event that the effect reported in this note is caused by the inability of the feldspar to retain argon as completely as the mica, their value of 0·085 for the branching ratio is probably too low.

References

Wasserburg, G. J. (1954) *Nuclear Geology*, edited by Henry Faul; pp. 346–349, New York: John Wiley and Sons.

Wasserburg, G. J., and Hayden, R. J. (1955) *Geochim. et Cosmochim. Acta* **7**, 51–60

Dating Young Volcanic Rocks and Geomagnetic Reversals

XII

Editor's Comments on Papers 35–39

During the early 1960s attention was directed toward determining the argon retentivities of a wide variety of geologic materials. Experimental investigations of diffusion coefficients,* and the determination of radiometric ages of coexisting minerals, both helped to establish the suitability of a number of silicate minerals for K–Ar dating. The studies of S. R. Hart initiated at MIT and completed at the Department of Terrestrial Magnetism in Washington, D.C., were particularly important in this regard.

Hart (1961, 1964) made a careful study of the discordant mineral patterns produced in Precambrian (> 1000 million years old) rocks adjacent to a 54 million year old quartz–monzonite intrusion known as the Eldora Stock, in Colorado. He was able to show that discordant radiometric ages in gneisses adjacent to the stock were due entirely to diffusive loss of radiogenic daughter products caused by thermal effects occurring at the time of intrusion of the quartz–monzonite. He found that radiogenic strontium and argon were both lost from biotite at temperatures which produced no other observable effects in the host rocks. In contrast to the biotites, Hart found the K–Ar ages of hornblendes and Rb–Sr ages of feldspars were affected only where mineralogical changes due to heating in the vicinity of the quartz–monzonite contact could be observed.

The discovery that hornblende was a suitable mineral for K–Ar dating and studies of the diffusion characteristics of argon in pyroxene, led Hart to compare the argon retentivities of hornblendes, pyroxenes and micas from a number of different localities (Paper 35). In this study he found no evidence of excess argon in hornblendes as had been postulated earlier by Damon and Kulp (1958), and he concluded that both

---

*See Mussett (1969) for a review of this subject.

hornblende and pyroxene, despite their low potash contents, would be suitable minerals for K–Ar dating.*

Hart's work led the way to K–Ar dating of mafic igneous rocks, which contain little or no mica but abundant hornblende and pyroxene. Being low in rubidium and uranium, such rocks could not be easily dated radiometrically by any other method. Of particular interest were the basalts and associated volcanic rocks which were too fine-grained for individual minerals to be separated but which could possibly be dated using whole-rock samples. Several studies were undertaken to investigate these possibilities and all met with surprising success (Erickson and Kulp, 1961; McDougall, 1961; Miller and Mussett, 1963). The K–Ar method was shown to yield reliable dates on quite young, fine-grained, basaltic material. Herein lay perhaps its greatest potential, for the widespread distribution of basaltic volcanic rocks in space and time, and their frequent occurrence in stratigraphically well defined sequences, made them ideal material for studies related to the geologic time scale.

In 1947, Arthur Holmes had published a time scale which, for the Phanerozoic, was based on only five U–Pb and Th–Pb age determinations. Helium ages had at that time been discredited (see p. 70). As more radiometric age data accumulated during the 1950s, it became obvious that a revision of the time scale was necessary (Faul, 1960). Holmes (1959) published a revision of his earlier scale using about twenty radiometric ages for the Phanerozoic, and two years later Kulp (1961) published a Phanerozoic time scale using even more data. Many of the radiometric ages on which these revised time scales were based were K–Ar determinations, and this was particularly true for the Cenozoic portions, where K–Ar ages proved to be more reliable than either Rb–Sr or U–Pb ages for the youngest rocks. With careful sample selection and sensitive analytical techniques, it was found that K–Ar ages could be determined on successively younger and younger materials until an overlap with the carbon-14 method seemed in sight.

At the Department of Geology in the University of California at Berkeley, the team of J. F. Evernden and G. H. Curtis perfected the techniques necessary for the measurement of K–Ar ages on low-potassium minerals having very young ages.† With high-potassium minerals such as sanidine, they could date rocks as young as 5000–30,000 years. Evernden and Curtis were in large measure responsible for the establishment of an accurate time scale for the Cenozoic (see Paper 36), and they achieved world-wide attention for their successful determination of the age of volcanic minerals associated with the fossil hominids (Zinjanthropus) discovered by Leakey at Olduvai Gorge, Tanzania (Evernden and Curtis, 1965). Both were skeptical about the reliability of K–Ar ages determined on whole-rock samples of basalt, pointing to the susceptibility of the glassy matrix and small crystallites to devitrification and alternation with subsequent radiogenic argon loss.

---

*Subsequent studies by Hart and Dodd (1962) revealed that some pyroxenes from deep-seated metamorphic rocks incorporate Ar-40 during crystallization and yield anomalously high ages.
†A review of the methods used for dating young rocks by K–Ar has been published by Ian McDougall (1966) of the Australian National University, who studied under Curtis and Evernden at Berkeley.

In spite of the possible uncertainties in whole-rock basalt ages, K–Ar determinations on all manner of young volcanic materials accumulated rapidly, many studies being directed toward elucidating the geochronology of oceanic islands (e.g., McDougall, 1964). Most important of all, changes in the polarity of the earth's magnetic field were recorded in basaltic volcanic rocks, and by dating these, the times of magnetic reversals could be determined and used to establish a geomagnetic-polarity time scale. (See Cox et al., 1963, 1968; McDougall and Chamalaun, 1966; Watkins, 1972; Dalrymple, 1972). Here again there were problems, and in their report on the Cenozoic time scale (Paper 36) Evernden and Evernden expressed the opinion that accurate determination of the times of geomagnetic polarity reversals could only be obtained by dating pure crystal concentrates, particularly where the resolution of short-term polarity events was concerned. They emphasized that proof of reproducibility is not proof of accuracy (Paper 36, p. 88).

With regard to the precision of K–Ar dating, Cox and Dalrymple at the US Geological Survey in Menlo Park developed a statistical method for analyzing the magnetic polarity of rocks as a function of their K–Ar ages for the purpose of dating the boundaries between geomagnetic epochs* (Paper 37). They concluded that ages of boundaries between polarity epochs in the range 0–3.6 million years would be clearly resolvable by the K–Ar method. For 2.5 million year old rocks, a dating precision of 3.6 per cent was found, but with this precision it would be extremely difficult to identify polarity events in rocks 10 million years old or older.

Addressing the problem from a different point of view, Baksi, York, and Watkins (Paper 38) demonstrated that a geomagnetic polarity event could be successfully dated by determining the K–Ar ages of a sequence of basaltic lavas that recorded the transition between opposite polarities at either the beginning or end of any polarity event. Baksi and York found that devitrified glass and slightly altered feldspar did not effect the argon retention properties of the whole-rock samples, as might have been expected. Commenting on the reliability of K–Ar ages, they stated that this could best be evaluated by dating samples from various parts of a single rock unit and examining the consistency of the data.

Application of the K–Ar method to young volcanic rocks and minerals has been studied extensively by Paul Damon and his co-workers at the University of Arizona's Department of Geochronology (e.g., Damon et al., 1967). K–Ar dating of very young rocks and minerals is limited ultimately by the presence of what Damon has called "extraneous argon" (Paper 39). Extraneous argon is radiogenic argon "inherited" by incorporation of older xenolithic material or by incomplete outgassing, and/or "excess" argon-40 occluded within a sample by processes other than in situ radioactive decay. The presence of a tiny amount of extraneous argon in any reasonably old sample would not significantly effect the measured K–Ar age, but with decreasing age the relative proportion of inherited to radiogenic argon would increase, and Damon expressed doubts that any meaningful comparison could ever be made between K–Ar

---

*By convention "epochs" are relatively long periods of a constant geomagnetic polarity, while "events" are relatively short periods.

dating and carbon-14 dating. Such a comparison was recently attempted by McDougall and others (1969) at the Australian National University in Camberra and K–Ar ages were found to be anomalously high when compared to carbon-14 ages, just as Damon had predicted.

Selected Bibliography

Cox, A., Doell, R. R., and Dalrymple, G. B. (1963). Geomagnetic polarity epochs and Pleistocene geochronometry. *Nature,* **198**, 1049–1051.

Cox, A., Doell, R. R., and Dalrymple, G. B. (1968). Radiometric time scale for geomagnetic reversals. *Quart. J. Geol. Soc. London,* **124**, 53–66.

Dalrymple, G. B. (1972). Potassium–argon dating of geomagnetic reversals and North American glaciations. *In* "Calibration of Hominid Evolution" (W. W. Bishop and J. A. Miller, eds.), pp. 107–134. Scottish Academic Press, Edinburgh.

Damon, P. E. (1970). A theory of "real" K–Ar clocks. *Eclogae Geol. Helv.,* **63**, 69–76.

Damon, P. E., and Kulp, J. L. (1958). Excess helium and argon in beryl and other minerals. *Amer. Mineral.,* **45**, 433–459.

Damon, P. E., Laughlin, A. W., and Percions, J. K. (1967). Problem of excess argon-40 in volcanic rocks. *In* "Radioactive Dating and Methods of Low-level Counting," pp. 463–481. International Atomic Energy Agency, Vienna.

Erickson, G. P., and Kulp, J. L. (1961). Potassium–argon measurements on the Palisades Sill, New Jersey. *Geol. Soc. Amer. Bull.,* **72**, 649–652.

Evernden, J. F., and Curtis, G. H. (1965). Potassium–argon dating of late Cenozoic rocks in East Africa and Italy. *Current Anthropol.,* **6**, 343–385.

Faul, H. (1960). Geologic time scale. *Geol. Soc. Amer. Bull.,* **71**, 637–644.

Hart, S. R. (1961). Mineral ages and metamorphism. *Ann. N.Y. Acad. Sci.,* **91**, 192–197.

Hart, S. R. (1964). The petrology and isotopic-mineral age relations of a contact zone in the Front Range, Colorado. *J. Geol.,* **72**, 493–525.

Hart, S. R., and Dodd, R. T. (1962). Excess radiogenic argon in pyroxenes. *J. Geophys. Res.,* **57**, 2998–2999.

Holmes, A. (1947). The construction of a geological time-scale. *Trans. Geol. Soc., Glasgow,* **21**, 117–152.

Holmes, A. (1959). A revised geological time scale. *Trans. Edinburgh Geol. Soc.,* **17**, 183–216.

Kulp, J. L. (1961). Geologic time scale. *Science,* **133**, 1105–1114.

McDougall, I. (1961). Determination of the age of a basic igneous intrusion by the potassium–argon method. *Nature,* **190**, 1184–1186.

McDougall, I. (1964). Potassium–argon ages from lavas of the Hawaiian Islands. *Geol. Soc. Amer. Bull.,* **75**, 107–128.

McDougall, I. (1966). Precision methods of potassium–argon isotopic age determination on young rocks. *In* "Methods and Techniques in Geophysics," pp. 279–304. Interscience, New York.

McDougall, I., and Chamalaun, F. H. (1966). Geomagnetic polarity scale of time. *Nature,* **212**, 1415–1418.

McDougall, I., Polach, H. A., and Stipp, J. J. (1969). Excess radiogenic argon in young subaerial basalts from the Auckland volcanic field, New Zealand. *Geochim. Cosmochim. Acta.,* **33**, 1485–1520.

Miller, J. A., and Mussett, A. E. (1963). Dating basic rocks by the potassium–argon method: the Whin sill. *Geophys. J.,* **7**, 547–553.

Mussett, A. E. (1969). Diffusion measurements and the potassium–argon method of dating. *Geophys. J. Roy. Astron. Soc.,* **18**, 257–303.

Watkins, N. D. (1972). Review of the development of the geomagnetic polarity time scale and discussion of prospects for its finer definition. *Geol. Soc. Amer. Bull.,* **83**, 551–574.

35

The Use of Hornblendes and Pyroxenes for K–Ar Dating [1]

Stanley R. Hart

*Department of Terrestrial Magnetism, Carnegie Institution, Washington, D. C., and
Department of Geology and Geophysics, Massachusetts Institute of Technology,
Cambridge, Massachusetts*

Abstract. The K–Ar ages of 12 hornblendes, one actinolite, and two pyroxenes were determined. When these ages are compared with ages of associated biotite, feldspar, or zircon, good agreement is found in most cases. No evidence is found for the existence of 'excess' radiogenic argon in these hornblendes. A maximum limit of 5×10^{-7} cc STP/g can be placed on possible 'excess' radiogenic argon in one sample. The potassium content of the amphiboles and pyroxenes is high enough so that Paleozoic or older samples can be easily dated, using present techniques. The rubidium content of the hornblendes is too low to be generally utilized for Rb–Sr dating.

Introduction

Micas have provided the basis for most K–Ar isotopic mineral age determinations done since 1956. Their use has become accepted almost to the exclusion of other possible minerals. With the available techniques, it is feasible to measure K–Ar ages on minerals in which potassium occurs as a minor or even a trace element. It is possible that some of these low potassium minerals could prove more resistant to argon loss under metamorphic conditions than do micas.

A study of mineral ages across a contact metamorphic zone [*Hart, 1961*] showed that the hornblende there retained argon much better than the biotite. *Amirkhanov, Bartnitskii, Brandt, and Voitkevich* [1959] measured the diffusion coefficient of argon in a pyroxene and found it to be very small at geologic temperatures. This evidence seemed to warrant further study of amphiboles and pyroxenes.

The alkali cation position in hornblende is only partly filled, and *Damon and Kulp* [1958] suggested that this might contain original 'excess' radiogenic argon, as for the beryl, tourmaline, and cordierite that they studied. To investigate this possibility, and to compare the argon retentivity of hornblendes and pyroxenes with micas, a number of samples from reasonably well-dated localities were analyzed for potassium and argon.

[1] MIT Age Studies No. 24

Sample Preparation

In most instances mafic mineral concentrates were available as starting materials. Various size fractions from −60 to +200 were used. These were purified on a magnetic separator and in adjusted methylene iodide. Biotite was the dominant impurity. Grain counts were made on all samples. In addition, X-ray diffraction patterns were run on all samples and compared with patterns made from known mixtures of hornblende and biotite. The detection limits for biotite using the X-ray method was about 1 per cent. In several cases, the X-ray examination revealed intergrown biotite impurity where little free biotite was detected in the grain counts. This impurity was partly removed in one instance by dealing with a −270 +325 mesh-size fraction. More extensive purification procedures were not attempted, and in all cases the biotite impurity contributed less than 40 per cent of the total potassium in the sample. Portions for K and Ar analysis were taken by mixing and sampling at random with a spatula.

The location, rock type and estimated purity of the samples are given in Table 1.

Experimental Techniques

Argon analysis. An alundum crucible containing 3 to 5 grams of sample was placed in a Kanthal-wound resistance furnace. This was enclosed with a water-jacketed bell jar. The samples were heated in vacuum at 300° to 350°C

2995

TABLE 1. Location and Description of Samples

Sample Number	Locality	Rock Type	Estimated Biotite Impurity, %	Contribution of Potassium from Biotite Impurity, %
HB47*	Wheaton, Md.	Tonalite	2	35
HB21*	Ellicott City, Md.	Granite	8	30
HP20*	Devault, Pa.	Baltimore gneiss	0.5	3
HSK*	Bear Mtn., N. Y.	Storm King granite	<0.2	< 1
HG15*	Crossnore, N. C.	Granite gneiss	2	25
HB15†	Shenandoah, Virginia	Gneiss	<0.2	< 1
H3089	Dill Sta., Ontario	Gneiss	<0.2	< 1
P3073	Kenogami, Quebec	Gabbro	<0.1	< 5
H3070	Chicoutimi, Quebec	Syenite	6	30
A3426	Neelon Twp., Ontario	Sudbury gabbro	<0.1	< 8
H3006	Rockport, Mass.	Cape Ann granite	5	20
H3136	Oak Bay, N. B.	Granite	<0.2	< 3
P3069	Mont Royal, Quebec	Tinguaite dike	<0.1	< 7
H3451	Dryden, Ontario	Amphibolite	5	40
H4068	Eldora, Colorado	Idaho Springs schist	<0.2	< 1

* Hornblende concentrates furnished by G. R. Tilton and G. L. Davis, Geophysical Laboratory, Washington, D. C.

† Hornblende concentrate furnished by B. R. Doe, Geophysical Laboratory, Washington, D. C.

All other samples obtained from the MIT Age Studies project.

H, hornblende; A, actinolite; P, pyroxene (3069-augite, 3073-hypersthene).

for about an hour. Tests of this procedure showed negligible radiogenic argon loss from a hornblende when heated for 30 minutes at 425°C, and less than 2 per cent loss when heated for 4 hours at 625°C. After this outgassing, the samples were fused directly by heating at 1100° to 1200° for 30 minutes. Enriched argon 3S, calibrated against the argon content of air, was added, and the released gases were purified using hot copper oxide and titanium sponge.

The isotope ratios were measured using a 4½-inch 60° Reynolds-type mass spectrometer. The standard errors of 6 to 8 scans for the 40/3S and 36/3S ratios were 0.1 and 0.3 per cent, respectively. A correction for the residual at mass 36 was always less than 10 per cent. Argon 36 from the spike accounted for about half the total argon 36 peak. Tests of the 36 residual with hydrogen indicated that no enhancement took place under normal conditions. The discrimination of the ion source was frequently checked, using purified air argon. Air argon blanks in the extraction system were normally in the range 1 to 2 × 10⁻⁵ cc STP and accounted for most of the air argon observed in the purified argon samples.

One of the samples (P3069) was also run in a system that allowed dropping 1 to 2 gram samples onto outgassed NaOH flux in a nickel furnace. The air-argon level of samples run in this system ranges from 1 to 5 × 10⁻⁶ cc STP.

The reproducibility of argon analyses was in the range 1 to 2 per cent. The analyzed argon content of the MIT standard biotite (3.85 × 10⁻⁴ cc STP/g) agrees within 1 per cent with that reported by other investigators [*Hurley, Fairbairn, Pinson, and Faure*, 1960].

Potassium-rubidium analysis. The potassium concentrations were determined in each case by both flame photometry and isotope dilution. The flame photometer solutions were prepared by decomposing 0.5 to 1.0 gram samples in sulfuric and hydrofluoric acids and diluting to 1 liter. As an internal standard 800 ppm lithium was added, and the solutions were compared with standards on a Perkin Elmer model-146 flame photometer, using propane flame. The unknowns and standards were both made up in 0.045 N H_2SO_4. Blanks for the procedure yielded about 25 micrograms of potassium. Corrections made for this were always less than 2 per cent. Six duplicate solutions indicated a precision of about 2½ per cent for the flame photometer potassium analyses.

TABLE 2. Analytical Data for Amphiboles and Pyroxenes

Sample Number	Potassium, wt. %		Rubidium, ppm	Sodium, wt. %	Ar^{40}* (in 10^{-5} cc STP/g)	$\dfrac{Ar^{40}*}{Ar^{40}\ Total}$
	Flame Phot.	Isotope Dil.				
HB47	0.394, 0.403	0.397	83.5	0.92	0.543	0.29
HB21	2.12	2.05	68.8	...	2.65	0.71
HP20	1.35	1.34, 1.29	16.3, 15.6	0.92	6.65	0.83
HSK	1.66, 1.69	1.68	36.1	1.25	7.82	0.82
HG15	0.532, 0.542	0.508, 0.531	12.1, 13.4	0.54	2.20	0.62
HB15	...	1.66	31.0	1.15	7.66	0.91
H3089	1.31	1.29	15.1	...	6.31	0.87
P3073	0.155	0.148, 0.148	0.56, 0.53	0.24	1.04	0.50
H3070	1.05, 1.15	1.33, 1.33	56.9, 61.0	...	4.58	0.80
A3426	0.0914	0.0872	2.05	...	0.925	0.26
H3006	1.49	1.68, 1.68	82.2, 79.8	1.14	2.26	0.56
H3136	0.482	0.506	10.3	...	0.740	0.28
P3069	0.112	0.102	4.40	...	0.0693, 0.0694	0.36, 0.31
H3451	0.663	0.835, 0.824	54.5, 51.7	0.65	15.7	0.92
H4068	1.01, 1.06	1.00	9.84	0.96	6.56	0.85

* Indicates radiogenic component.

The isotope dilution procedure involved decomposing 0.5 gram samples in perchloric and hydrofluoric acids and spiking suitable aliquots with a mixed K^{41} and Rb^{87} spike. A portion of this aliquot was applied to the single filament of a 6-inch solid-source mass spectrometer and the K^{39}/K^{41} and Rb^{85}/Rb^{87} ratios were determined in a single run. Each ratio is the mean of 8 to 10 scans. The standard error of the potassium ratios ranged from 0.05 to 0.25 per cent, and of the rubidium ratios, from 0.1 to 0.4 per cent.

The K^{39}/K^{41} ratio was measured at low current, before the rubidium ratios, and later at high current after the rubidium ratios. The low current ratios averaged about 1 per cent higher than the high current ratios. The reason for this is unknown but it could arise either from initial surface contamination or sample fractionation. The high current ratios were used in the calculations. In some cases, the rubidium ratios were obtained only by using an electron multiplier. Duplicate runs from the same aliquot generally reproduced the isotope ratios to within 1 per cent.

As a further test of accuracy, the alkalis from a number of these same aliquots were separated using a sodium tetraphenyl boron precipitation. The isotope ratios of these separated alkalis also agreed within 1 per cent with the ratios obtained from the unseparated portions. Six complete duplicate isotope dilution analyses indicate a standard deviation of 1.7 per cent for the potassium analyses and 4.5 per cent for the rubidium analyses.

Contamination blanks for the isotope dilution procedure showed from 3 to 6 micrograms of potassium and from 5×10^{-3} to 5×10^{-4} microgram of rubidium. The sodium tetraphenyl boron step added less than 0.1 microgram of potassium and 0.01 microgram of rubidium.

Sodium analysis. After taking the aliquot for the K + Rb analysis, the remaining solution was diluted so that the sodium concentrations were in the range 1 to 10 micrograms/ml. These were compared with standards on a Beckman model DU spectrophotometer with flame attachment. Duplicate samples indicate the standard deviation to be about 5 per cent. No attempt was made to evaluate the accuracy of the sodium determinations.

Constants used.

$$\lambda K^{40} = 5.30 \times 10^{-10}\ yr^{-1}$$

$$\lambda e K^{40} = 0.585 \times 10^{-10}\ yr^{-1}$$

$$K^{40} = 1.22 \times 10^{-4}\ g/gK$$

Other mineral ages recomputed to these constants, and

$$\lambda Rb^{87} = 1.39 \times 10^{-11}\ yr^{-1}$$

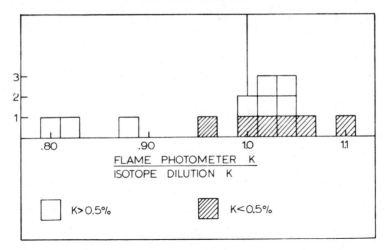

Fig. 1. Agreement of flame photometer and isotope dilution analysis of potassium in amphiboles and pyroxenes.

RESULTS AND DISCUSSION

The analytical data are presented in Table 2. Figure 1 shows a comparison of the flame photometer and isotope dilution potassium results. It is apparent that variations exist which are well-outside the precision of the individual measurements. Flame photometer analyses for potassium in micas have been shown to be quite reliable [Hurley, and others, 1958]. However, interference from other elements may be more serious in hornblendes where, for example, the Ca/K ratio is commonly 200 times larger than in micas. Without further studies of interference and absorption phenomena the flame photometer analyses for potassium in these hornblendes and pyroxenes are considered less reliable than the isotope dilution analyses. The isotope dilution results are believed to be accurate within 1 or 2 per cent and are used in the age calculations.

It is probable that much of the rubidium in these hornblendes is contributed by the biotite impurity. Because of the general low level of rubidium in the pure hornblendes, they will generally be unsuitable for Rb–Sr dating. Pinson, Fairbairn, and Cormier [1958] did find hornblendes that contained as little as 10 ppm normal strontium. These might yield meaningful Rb–Sr ages if sufficiently old.

The results of sodium analyses (Table 2) on nine of the samples show that the atom ratio, Na/K, is greater than unity in each case. The

sum of sodium plus potassium represents the extent to which the alkali cation position is filled. In terms of an ideal tremolite structure, the alkali cation positions for these nine hornblendes range from 30 to 80 per cent filled. This is the range usually observed for natural hornblendes [Boyd, 1959] and indicates that these samples do have structural holes where 'excess' radiogenic argon could be held.

The calculated K–Ar ages are presented in Tables 3 and 4, along with published ages of coexisting minerals. In comparing the hornblende ages with ages on these coexisting minerals, two points should be considered. First, are any of the hornblende ages anomalously high, suggesting 'excess' radiogenic argon? And second, what resistance to argon loss during metamorphic events do the hornblende ages imply, relative to the other minerals?

The first two samples in Table 3 are thought to be Paleozoic intrusives from the Maryland Piedmont, where the last metamorphic event affecting biotite ages seems to have occurred about 300 m.y. ago [Tilton, Wetherill, Davis, and Hopson, 1958; Tilton, Davis, Wetherill, Aldrich, and Jäger, 1959]. The hornblendes were apparently also affected by this metamorphism, giving K–Ar ages equal to K–Ar and Rb–Sr ages on the biotite. Because of the relatively low potassium and low age of the Wheaton, Maryland, hornblende, its age would be notice-

TABLE 3.   Hornblende K-Ar Ages Compared with Ages of Associated Minerals

Sample Number	Locality	Hornblende K-Ar	Biotite K-Ar	Biotite Rb-Sr	Zircon 238 206	235 207	207 206	232 208
HB21	Granite Ellicott City, Md. (a)	300 ± 10	315	290	355	370	450	310
HB47	Tonalite Wheaton, Md. (b)	315 ± 15	...	...	495	510	570 + 50	485
HP20	Baltimore gneiss Devault, Pa. (c)	960 ± 30	1010	485	...	...	...	...
HSK	Storm King granite Bear Mtn., N. Y. (c)	910 ± 25	840	930	960	990	1060	850
HG15	Granite gneiss Crossnore, N. C. (a)	840 ± 25	...	...	690	720	800 ± 50	680
HB15	Gneiss Shenandoah, Va. (c)	900 ± 25	800	880	1070	1100	1150	1110

References for ages other than hornblende K-Ar: (a) Tilton, Davis, Wetherill, Aldrich, and Jäger, 1959; (b) Davis, Tilton, Aldrich, Wetherill, and Bass, 1960; (c) Tilton, Wetherill, Davis, and Bass, 1960.

ably affected by 5×10^{-7} cc STP/g of 'excess' radiogenic argon. This may be considered a maximum limit for possible excess argon in this particular sample.

The last four samples of Table 3 are from the Appalachian orogenic belt. A number of Precambrian rocks from this belt have been shown to consistently give mineral ages in the 900–1100 m.y. range [Tilton, Wetherill, Davis, and Bass, 1960]. Allowing for some decrease in apparent age because of later, possibly Paleozoic, metamorphism, these four hornblendes are in general agreement with this range.

The biotite from Devault, Pennsylvania, is one of several known instances where the K–Ar age is significantly higher than the Rb–Sr age; see also Sudbury gabbro, Table 4. The amphibole K–Ar age in both cases is the same as the biotite K–Ar age. An explanation of the anomalous biotite age pattern based on incorporation of radiogenic argon at the time of the later metamorphism therefore does not seem likely. It is possible that a high external activity of radiogenic argon during metamorphism kept the argon diffusion rate low enough so that appreciable loss was prevented from both the hornblende and the biotite.

The Bear Mountain, New York, and Shenandoah, Virginia, samples show mica and hornblende ages that are slightly lower than the nearly concordant zircon ages. Both hornblende ages are higher than the corresponding biotite K–Ar ages and both are the same as the biotite Rb–Sr ages.

There are no biotite ages for the Crossnore, North Carolina sample. However, a number of biotites from western North Carolina have shown Rb–Sr and K–Ar ages in the range 300–400 m.y. [Long, Kulp, and Eckemann, 1959]. The Crossnore zircon ages are discordant but consistent with an original age in the range 900–1100 m.y. The hornblende age is somewhat lower than this presumed original age, but it is significantly higher than biotite ages in the near vicinity.

The geologic ages of the following localities are less well-documented, but still provide useful comparisons. The first three samples in Table 4 are from the Grenville Province of eastern Canada. Mineral ages from this province invariably fall in the range 900–1200 m.y. The hornblende age of the Dill Station, Ontario, sample is in reasonable agreement with the biotite ages and within the usual range of Grenville ages. The sample from Kenogami, Quebec, has no coexisting mineral ages. The pyroxene K–Ar age from this sample, however, is within the usual range of Grenville ages. The biotite and feldspar Rb–Sr ages on the Chicoutimi, Quebec, sample are typically Grenville, but the hornblende age of 700 m.y. is clearly low. No unique explanation for this can be given on the basis of the single sample.

The sample of Sudbury gabbro from Neelon

TABLE 4. Hornblende, Actinolite, and Pyroxene K–Ar Ages Compared with Ages of Associated Minerals

Sample Number	Locality	Hornblende K–Ar	Biotite K–Ar	Biotite Rb–Sr	Feldspar Rb–Sr
H3089	Gneiss Dill Sta., Ontario (a)	940 ± 30	960	915	...
P3073	Gabbro (pyroxene) Kenogami, Quebec	1240 ± 75	...	...	...
H3070	Syenite Chicoutimi, Quebec (b)	710 ± 30	...	950*	840
A3426	Sudbury Gabbro (actinolite) Neelon Twp., Ontario (a)	1660 ± 100	1780†	1330†	...
H3006	Cape Ann granite Rockport, Mass.	310 ± 10	...	...	...
H3136	Granite Oak Bay, New Brunswick (c)	335 ± 15	...	...	300 ± 60
P3069	Tinguaite Dike (pyroxene) Mont Royal, Quebec	163 ± 10	...	...	...
H3451	Amphibolite Dryden, Ontario	2370 ± 80	...	...	...
H4068	Idaho Springs schist Eldora, Colorado (d)	1180 ± 30	80	410	...

* Reported previously as hornblende Rb/Sr Age.
† Biotite is from different sample of Sudbury gabbro. McKim Twp., Ont. (3196).
References for ages other than hornblende K–Ar: (a) Fairbairn, Hurley, and Pinson, in press; (b) Pinson, Fairbairn, and Cormier, 1958; (c) Hurley, 1958; (d) Hart, 1961.

Township, mentioned before, is another instance where the biotite K–Ar age is markedly older than the biotite Rb–Sr age. The actinolite K–Ar age is also high and, within experimental error, the same as the biotite K–Ar age.

The Cape Ann granite has been correlated with the Quincy granite, which is overlain unconformably by Pennsylvanian sediments. In the absence of other mineral ages, the hornblende K–Ar age of 300 m.y. can only be said to be reasonable.

The hornblende age from Oak Bay, New Brunswick, is in essential agreement with the feldspar Rb–Sr age. Other mineral ages are not available.

The Mont Royal, Quebec, sample is from a tinguaite dike on St. Joseph's Boulevard, Montreal East. The tinguaite is associated with the Monteregian intrusive there. No other mineral ages are available from Mont Royal, but several ages have been published for other of the Monteregian intrusives. A biotite K–Ar age of 122 m.y. was reported by the Canadian Geological Survey [Lowdon, 1960] for the Brome Mountain nordmarkite. A biotite from the essexite at Mt. Johnston gave a K–Ar age of 110

m.y. [Hurley, 1960]. The original age reported for the Mont Royal pyroxene of 100 m.y. [Hart, Fairbairn, Pinson, and Hurley, 1960] was in error because of the large atmospheric argon correction. The sample was rerun in duplicate using the sample dropping system described above. The results were in close agreement at 163 m.y. The difference between this age and the biotite ages from the other intrusives is in the direction of radiogenic argon contamination, either original 'excess' argon, or from impurity introduced during sample processing and purification. The excess is about 2 × 10⁻⁷ cc STP/g. The fact that 0.05 per cent of a 1000 m.y. biotite impurity could contribute this excess points up one of the major difficulties inherent in working with samples low in potassium. It is also possible, of course, that the Mont Royal intrusive is actually older than the other Monteregian intrusives.

The Dryden, Ontario, hornblende again has no ages of coexisting minerals for comparison. However, Aldrich and Wetherill [1960] have reported biotite ages from Sioux Lookout and Kenora, Ontario. Both localities are within 60 miles of Dryden. The ages (m.y.) reported for biotites from the two localities are, respectively, 2420 K-Ar, 2200 Rb-Sr, and 2490 K-Ar, 2550 Rb-Sr. There is evidence in the biotite ages for a post-2600 m.y. effect that has caused some loss of argon and strontium. The hornblende K-Ar age 2370 m.y. may also show evidence of this effect.

The Eldora, Colorado, sample was collected in the Precambrian Idaho Springs formation 250 feet from the contact of a 54 m.y. Laramide stock. A series of ages in this contact zone has been reported previously [Hart, 1960]. The exceptional argon retentivity shown by this hornblende was the first indication that hornblendes might prove useful for K-Ar dating.

Conclusions

The determinations of (Na + K) demonstrate that these hornblendes have alkali-cation positions which are only partly filled. These structural holes provide sites where occluded or inherited radiogenic argon could be held. However, in no event does the K-Ar age of the hornblendes or the actinolite exceed the probable original age of the host rock as deduced from ages on associated mica, feldspar, and zircon. Thus there is no evidence from this study for the existence of 'excess' radiogenic argon in amphiboles as postulated by *Damon and Kulp* [1958].

It is not clear from this study that hornblende will always retain argon better than micas. In some cases, this is notably true; in others, the reverse seems to be true. The relationship of pyroxenes to micas is even less clear. In any event, the use of these minerals, and possibly others low in potassium, will permit dating of many intermediate and mafic rocks which could not previously be dated.

Acknowledgments. I wish to thank those who furnished samples for this study. The interest and help of W. H. Pinson, H. W. Fairbairn, L. T. Aldrich, and G. L. Davis is gratefully acknowledged. The research was supported in part by AEC contract AT(30-1)—1381. Finally, I am indebted to P. M. Hurley, whose encouragement and support made this work possible.

References

Aldrich, L. T., and G. W. Wetherill, Rb-Sr and K-Ar ages of rocks in Ontario and northern Minnesota, *J. Geophys. Research, 65*, 337–340, 1960.

Amirkhanov, Kh. I., E. N. Bartnitskii, S. B. Brandt, and G. V. Voitkevich, On the migration of argon and helium in several rocks and minerals, *Doklady Akad. Nauk SSSR, 126*, 160–162, geochemistry series, 1959.

Boyd, F. R., Hydrothermal investigations of amphiboles, *Researches in Geochemistry*, Wiley and Sons, New York, 377–396, 1959.

Damon, P. E., and J. L. Kulp, Excess helium and argon in beryl and other minerals, *Am. Mineralogist, 43*, 433–459, 1958.

Davis, G. L., G. R. Tilton, L. T. Aldrich, G. W. Wetherill, and M. N. Bass, The ages of rocks and minerals, *Carnegie Inst. Wash. Year Book, 59*, 147–158, Washington, D. C., 1960.

Fairbairn, H. W., P. M. Hurley, and W. H. Pinson, Mineral and rock ages at Sudbury-Blind River, Ontario, *Proc. Geol. Assoc. Can.*, 1961.

Hart, S. R., Mineral ages and metamorphism, Conference on geochronology of rock systems, *Ann. N. Y. Acad. Sci., 91*, 192–197, 1961.

Hart, S. R., H. W. Fairbairn, W. H. Pinson, and P. M. Hurley, Use of amphiboles and pyroxenes for K-Ar dating, Abstract, *Geol. Soc. Am. Bull., 71*, 1882, 1960.

Hurley, P. M., and others, Variations in isotopic abundances of strontium, calcium, and argon and related topics, *MIT, 6th Ann. Prog. Rept. U. S. A. E. C. NYO-3939*, 1958.

Hurley, P. M., H. W. Fairbairn, W. H. Pinson, and G. Faure, K-Ar and Rb-Sr Minimum Ages for the Pennsylvania Section in the Narraganset Basin, *Geochim. et Cosmochim. Acta, 18*, 247–258, 1960.

Long, L. E., J. L. Kulp, and F. D. Eckelmann, Chronology of major metamorphic events in the southeastern United States, *Am. J. Sci., 257*, 585–603, 1959.

Lowdon, J. A., Age determinations by the Geological Survey of Canada, Rept. 1, Isotopic ages, Paper 60–17, *Dept. Mines and Tech. Surveys, Canada*, 1–51, 1960.

Pinson, W. H., H. W. Fairbairn, and R. F. Cormier, Sr/Rb age measurements on hornblende and feldspar and the age of the syenite at Chicoutimi, Quebec, Canada, *Bull. Geol. Soc. Am., 69*, 599–601, 1958.

Tilton, G. R., G. W. Wetherill, G. L. Davis, and C. A. Hopson, Ages of minerals from the Baltimore gneiss near Baltimore, Maryland, *Bull. Geol. Am., 69*, 1469–1474, 1958.

Tilton, G. R., G. L. Davis, G. W. Wetherill, L. T. Aldrich, and Emilie Jäger, The ages of rocks and minerals, *Carnegie Inst. Wash. Year Book, 58*, 170–178, Washington, D. C., 1959.

Tilton, G. R., G. W. Wetherill, G. L. Davis, and M. N. Bass, 1000-million-year-old minerals from the eastern United States and Canada, *J. Geophys. Research, 65*, 4173–4179, 1960.

(Manuscript received May 9, 1961; revised June 13, 1961.)

GEOLOGICAL SOCIETY OF AMERICA, INC.
SPECIAL PAPER 124
1970
Printed in the United States of America

36

The Cenozoic Time Scale

JACK F. EVERNDEN

Advanced Research Projects Agency, Washington, D.C.

ROBERTA K. SMITH EVERNDEN

Smithsonian Institution and Howard University, Washington, D.C.

ABSTRACT

The Cenozoic time-scale, down to Stage-Age level for North American mammal-bearing stratigraphic sequences and for some marine sequences, and to major cultural levels in the Pleistocene, has been established by use of the potassium-argon method of radiometric dating. Materials useful for dating Cenozoic rocks must have a fixed potassium content since time of genesis or cooling of the mineral or rock and must have suffered no loss of radiogenic argon. Authigenic or primary minerals must be used because dating of detrital grains dates the source rock, not the sedimentary deposit. Most major minerals of volcanic or intrusive rocks of the Cenozoic Era are suitable for dating purposes. Materials to be avoided for various reasons and to various degrees include intrusive feldspars, volcanic feldspars with intermediate potassium values, glauconite, illite, whole-rock basalt samples, and glass shards. Care should be exercised to insure that only unaltered mineral grains are used. When samples are properly prepared and analytical care is exercised, argon and potassium measurements of sample splits are reproducible to a fraction of 1 percent.

71

CONTENTS

INTRODUCTION

This paper offers an opportunity to summarize data on the Cenozoic time-scale. Much of the information on the Cenozoic time-scale has been obtained by the senior author and others working at the University of California, Berkeley, California (*see* Funnell, 1964). Previously, these data mainly have been detailed in a series of papers beginning slightly over a decade ago with the inception of the radiometric dating laboratory at Berkeley (*see* Evernden and others, 1957; Evernden and others, 1960; Evernden and others, 1961; Leakey and others, 1961; Evernden and others, 1964; Evernden and James, 1964; Evernden and Curtis, 1965; Turner, 1968; Evernden and Kistler, in press). The major results and discussion of some of the more important problems are brought together here for the first time. Topics discussed will be materials useful for dating of Cenozoic rocks, accuracy of measurement of potassium and argon, Cenozoic time-scale based on mammalian chronology of North America, time-scale of Pleistocene glaciations, time-scale of evolution of man, Cenozoic time-scale and marine invertebrates of western North America and Europe, and potassium-argon dates and magnetic reversals of the Earth's field. So many persons, both at Berkeley and elsewhere, have contributed toward making possible the researches touched upon here that it is impossible to thank them at this time, except to say that without their assistance many of the radiometric dates now available would not have come to light. L. B. Isham drew the figures in this paper.

No attempt is made to give a complete reference list of those who have worked upon the matters discussed here, particularly relative to evaluation of argon retentivities of various minerals. This paper is intended only as a "state-of-the-art" presentation and is based largely on the results obtained at Berkeley. All useful Cenozoic dates known to the authors are included in discussions of the geologic time-scale.

MATERIALS USEFUL FOR DATING OF CENOZOIC ROCKS

The major conditions to be fulfilled by material used for potassium-argon dating are: (a) fixed potassium content since time of genesis or cooling of the mineral or rock; (b) no loss of radiogenic argon; (c) argon content of the mineral or rock, a measure of the age of the stratigraphic horizon being dated. The last condition can be fulfilled conceptually either by the growth of authigenic minerals within a detrital deposit (that is, glauconite or illite), or by cooling and solidification of a volcanic deposit interlayered with fossil-bearing strata (Table 1).

At the initiation of research into potassium-argon dating, it was hoped that authigenic minerals, such as glauconite and illite, would yield many reliable potassium-argon dates, and, therefore, that most sedimentary fossil-bearing sequences would be directly datable. However, such was not to be. Investigations into the argon-retention characteristics of micaceous minerals indicated that the conditions of low temperature required for high retention of generated radiogenic argon in glauconite were too restrictive to be fulfilled often in a geological environment for time scales of more

TABLE 1. INDICATED AGES OF BASAL KREYENHAGEN CORES

Sample No.	Depth of Core	Indicated Age
KA 189	Surface	45×10^6 years
153-1	4600 feet	44
153-2	4600	44
153-3	4600	44
285	6300	36
154	8050	38
265	9200	29
160	10790	26
267	12175	20

than 30 m.y. (*see* Evernden and others, 1960, 1961). The burial history of glauconite samples is one of the major items to be evaluated prior to their dating. The heating which occurs concomitant with burial may be sufficient to cause excessive argon loss. Thus, data presented by Evernden and others (1961) on dating of basal Kreyenhagen glauconite samples from California show clearly the relation between heating history of a sample and the argon content of that sample. If the geological environment has been satisfactory for argon retention, an accurate potassium-argon age can be obtained for rocks of any Cenozoic age. However, since it is so difficult to establish unequivocally the temperature regimes which a glauconite sample cf 50-m.y. age has experienced, the dating of such materials should not be done. (Recently, Obradovich, 1968, has re-evaluated the potential usefulness of glauconite for late Cenozoic geochronology. He concludes that it can be used if certain restrictive criteria are satisfied.) The authors' experience with illite has been restricted to the dating of Paleozoic materials (Evernden and others, 1961), which indicate that the argon retention characteristics of illite are similar to those of glauconite.

The useful materials for establishing stratigraphic age of fossil-bearing sequences are the minerals of volcanic rocks. Under extremely restrictive conditions the entire volcanic rock or volcanic glass shards can be used to obtain reliable potassium-argon ages. The use of whole-rock ages will be discussed below. Investigations of virtually all major minerals of volcanic rocks (*see* Evernden and others, 1960, 1964; Evernden and James, 1964; Evernden and Curtis, 1965) indicate that, on a Cenozoic time-scale, all of them are sources of reliable ages unless excessive reheating or alteration has occurred. The heating associated with normal burial to

10,000 to 15,000 feet is insufficient to affect argon content of these minerals. The major potassium-bearing minerals such as biotite and sanidine were, of course, the first considered. However, these minerals occur only in the more acidic volcanic rocks; therefore, it was important to evaluate the usefulness of the minerals of the more basic rocks as these rocks are much more prevalent and widely distributed. The primary minerals of this type are the plagioclases and hornblendes. Early investigations of feldspars of plutonic igneous rocks had suggested that feldspars were undesirable for potassium-argon dating because of poor argon-retention characteristics. Whether such considerations applied to the feldspars of volcanic rocks required clarification. It was shown by Evernden and James (1964) that high-argon retentivities are characteristic of volcanic feldspars of low- and high-potassium content, whereas feldspars with intermediate potassium values (that is, high-temperature feldspars showing pronounced unmixing tendencies at low temperatures) display lower retention factors. All of the feldspars of basalts and andesites, therefore, are highly satisfactory minerals for potassium-argon dating. It was concluded that low-potassium (less than 0.9 percent) feldspars should yield excellent potassium-argon ages on a Cenozoic time scale and even on a Paleozoic time scale (presuming, of course, no excessive reheating of the samples after time of extrusion). If argon retentivity is desired to be at least 94 percent at 30 m.y. age, volcanic feldspars with intermediate potassium percentages between 2.5 and 6.5 are of questionable use.

The use of low-potassium (less than 0.9 percent) materials for ages of a few million years requires perfection of argon extraction and measurement techniques. The perfection of the necessary techniques was one of the major research efforts carried out at Berkeley over a period of years, and the procedures developed are thoroughly described by Evernden and Curtis (1965). By following these procedures, including the washing of feldspars with hydrofluoric acid, ages of high accuracy can be obtained on 1-m.y.-old feldspars containing 0.1 percent potassium.

An absolutely vital consideration in the selection of material for potassium-argon dating is the state of alteration of the material. Significant detectable alteration normally should be immediate cause for rejection of the material for dating considerations. Unless techniques for removal of the altered portions (that is, hydrofluoric-acid washing of feldspars on occasion) can be followed, the

347

investigator's desire for even an "approximate" age ultimately may cause chaos in the literature. These ages which are "approximate" at the time of the initial investigation generally become quite positive at first printing and become gospel at first referencing. For all samples, the only safe procedure, particularly for whole-rock samples, is for the initial investigator to accept the responsibility of describing in the literature the characteristics of the samples dated to the level necessary to demonstrate the reliability of the date obtained (*see*, for example, Evernden and James, 1964; Evernden and Curtis, 1965). Without this substantiation, dates on possibly questionable materials should not be accepted as reliable and should not be published. It is further strongly recommended that a committee of geologists and geochronologists should be appointed with the responsibility of evaluating the reliability and meaningfulness of all radiometric dates purporting to establish the age of stratigraphic horizons. No date should be considered in stratigraphic discussions if it has not survived such scrutiny.

The possible use of whole-rock basalt, or andesite samples, or volcanic glass shards for determination of reliable potassium-argon ages is very limited. There are, of course, numerous fine-grained basalts and andesites whose ages cannot be determined by concentrating discrete mineral grains, but whose ages are greatly desired. Unfortunately, virtually all of these rocks have much of their potassium content concentrated either in the glassy matrix of the rocks or in extremely fine-grained crystallites (*see* Evernden and others, 1964, p. 154). The susceptibility of the glassy matrix to devitrification and the susceptibility of the crystallites to alteration, as well as their small size, make all such rocks virtually useless for the determination of accurate potassium-argon ages. Glass shards share this extreme susceptibility to devitrification and argon loss. This is highly unfortunate. One may assume that any laboratory reporting numerous whole-rock-basalt or andesite or volcanic-glass-shard ages must be using altered material in many cases. The results of investigations at Berkeley were that very few basalts are free of extensive alteration of potassium-bearing components of the rock, and, likewise, very few glasses are not devitrified Criteria that should be fulfilled by such materials have been described by Evernden and others (1964). Great care in selection of material for dating cannot be stressed too strongly. At the beginning of potassium-argon dating, the warning was issued by workers in radiocarbon dating that, unless extreme caution was practiced, the same unfortunate events would overtake potassium-argon dating that overtook radiocarbon dating. The proliferation of dates on whole-rock basalts, glass shards, and andesites can only mean that the literature on geologic dating soon may be nearly incomprehensible to most geologists, and the entire field of radiometric dating may come into disrepute.

ACCURACY OF MEASUREMENT OF POTASSIUM AND ARGON CONTENTS

By careful observance of analytical detail, potassium and argon measurements in the time scale under consideration here are both reproducible to a small fraction of 1 percent. Data supporting this statement have been presented by Evernden and others (1960, 1961, 1964), Evernden and James (1964), and Evernden and Curtis (1965). Evernden and Kistler (in press) demonstrate clearly the necessity for high purity in mineral concentrates if the maximum possible reproducibility is desired. Thus, the high purity required for potassium and argon measurements on biotite to be reproducible to 1 percent is difficult to achieve. Normal procedures of concentration such as crushing, gravity and magnetic separation may not achieve concentrates of the desired uniformity. For example, only by the handpicking of samples was it possible to achieve reproducibility of potassium values to within 1 percent with biotite samples from igneous rocks of the Sierra Nevada. Even carefully picked samples which happened to contain appreciable chlorite within the individual grains still showed variations in value of repeat runs of approximately 1 percent. Interlaboratory comparisons of argon-potassium determinations have, in most cases, been made by use of materials which showed a variation of potassium and argon content of 2 percent or more in individual batches of the concentrate. The use of such materials in interlaboratory comparisons is essentially pointless. It is possible to purify biotite concentrates to the point that potassium measurements are reproducible to 1 or 2 parts in 1000 (Evernden and Kistler, in press). The purity conditions adequate to achieve high reproducibility of potassium measurements are also adequate to achieve high reproducibility of argon measurements. Techniques of measurements of these two quantities are now so refined that the level of uncertainty remaining is insufficient to influence in any significant manner the ages determined for Cenozoic rocks.

CENOZOIC MAMMALIAN CHRONOLOGY OF NORTH AMERICA

The existence in North America of numerous volcanic tuffs of Tertiary age whose stratigraphic position is well controlled by fossil mammals affords the means for establishing most details of the mammalian chronology of that region. The available samples range in age from Early Puercan (Danian equivalent) through the Blancan (Villafranchian equivalent) to late interglacial. The potassium-argon ages and the Mammal Age designations are essentially in agreement, supporting the conclusion that the defined Mammal Ages have true evolutionary significance. Reference to Evernden and others (1964) should be made for a complete discussion of the paleontologic and radiometric data. The resultant time-scale of the North American Stage-Ages is indicated in the left-hand column of Figure 1. All radiometric ages controlling the

10^6 YEARS	NORTH AMERICAN LAND-MAMMAL AGES [1,2,3]	EUROPEAN AGES [1,4,5]	PACIFIC COAST BENTHONIC FORAMINIFERAL AGES [9,10,11,12]	OTHER DATES	10^6 YEARS
0		VILLAFRANCHIAN [5]		= LOMITA MARL [13]	0
	BLANCAN *[2]	ASTIAN / PLAISANCIAN	REPETTIAN	= [8] "PONTIAN" OF + NEW GUINEA	
	HEMPHILLIAN *	PONTIAN [6]			
10	CLARENDONIAN *	SARMATIAN [6] +[7]	DELMONTIAN		10
	BARSTOVIAN *	VINDOBONIAN +[7]	MOHNIAN		
			LUISIAN		
20	HEMINGFORDIAN *	+[7]	RELIZIAN		20
		BURDIGALIAN = [8]	SAUCESIAN	= [8] + SANTACRUCIAN PATAGONIA	
	ARIKAREEAN *	AQUITANIAN	ZEMORRIAN		
	WHITNEYAN	CHATTIAN			
30	ORELLAN	RUPELIAN	REFUGIAN	+ HSANDA GOL MONGOLIA	30
	CHADRONIAN *	SANNOISIAN			
	DUCHESNIAN *	LUDIAN			
40			NARIZIAN		40
	UINTAN *	BARTONIAN	= [8,12]		
			ULATISIAN		
	BRIDGERIAN *	LUTETIAN = [8]			
50		YPRESIAN	PENUTIAN	= [8] U. WILCOX	50
	WASATCHIAN *	SPARNACIAN			
	CLARKFORKIAN	THANETIAN = [8]	BULITIAN	+ [8,12]	
	TIFFANIAN				
60	TORREJONIAN	MONTIAN	YNEZIAN		60
	PUERCAN? DRAGONIAN		"DANIAN"		
	"MAESTRICHTIAN" *		"MAESTRICHTIAN"	= [8] RIPLEY	
70					70

mammalian Ages are based upon crystal concentrates from either volcanic flows or tuffs. Most Stage-Age boundaries of the mammalian sequence are probably accurate to within one or two m.y.

CHRONOLOGY OF EUROPEAN STAGE-AGES AND CORRELATION WITH NORTH AMERICAN STAGE-AGES

The dearth of radiometric ages directly related to stratigraphically critical European sequences makes firm establishment of the time-scale of the European standard sequence difficult and subject to disagreement at present. In estimating this time-scale we must relate the large number of North American radiometric dates to the European stratigraphy through paleontological correlations of one type or another. Two distinct correlation routes are

---

← Figure 1. Correlation and time-placement diagram of North American Land-Mammal Ages, European Ages, and Pacific Coast benthonic foraminiferal Ages.

* Physical dates within these Ages are given *in* Evernden and others, 1964, and Folinsbee and others, 1961.

\+ Physical date obtained on volcanic material.

= Physical date obtained on glauconite.

1 Modified *from* Wood and others, 1941.

2 Savage, 1951. Savage defines two Mammal Ages above Blancan (Irvingtonian and Rancholabrean).

3 Time-placement of North American Land-Mammal Ages is based on data of Evernden and others, 1964, and "Maestrichtian" date *from* Folinsbee and others, 1961.

4 Modified *from* Simpson, 1947a, b.

5 Scale of figure too small to indicate post-Villafranchian intervals. Base of Villafranchian 3 or 3.5 m.y., as determined by Savage and Curtis, *see* text.

6 The temporal relationship of Sarmatian and late Vindobonian follows Crusafont Pairó (1951). This correlation is supported by the similarity of some early Clarendonian genera in North American faunas (such as Mathews Ranch fauna, Cuyama Valley, California, James, 1963) to those of late Vindobonian faunas of Europe (La Grive-St. Alban, Sansan, and Vieux-Collonges). The base of the Pontian with respect to the Clarendonian is evinced by the presence of primitive *Hipparion* horses in the early Pontian (Vallesian *of* Crusafont Pairó, 1958) of Spain, which appear more primitive than late Clarendonian *Hipparion* species of North America.

7 Lippolt, ms, 1961.

8 Evernden and others, 1961. U. Wilcox and Ripley are Gulf Coast formations.

9 Kleinpell, 1938.

10 Durham, Kleinpell, and Savage, 1954. Three foraminiferal Ages are given above Repettian (Venturian, Wheelerian, and Hallian.)

11 Durham, Jahns, and Savage, 1954.

12 Mallory, 1959.

13 Obradovich, 1968, dates the base of the Lomita Marl, considered by many base of Pleistocene in southern California, at about 3 m. y.

possible. The first route depends upon present understanding of the fossil mammalian ties between North America and Europe and correlations between European mammal-bearing strata and the standard Stage-Age sequence based upon benthonic marine fossils. The existence of land connections between North America and Eurasia during several intervals of the Cenozoic, the rapid morphologic evolution of the mammals during the same period, and the demonstrated meaningfulness in a stratigraphic sense of the established fine-scale divisions of the North American mammal sequence were the basis for placing reliance upon the mammalian correlations by Evernden and others (1964). The weakest point in this correlation chain may be the tie from European mammals to European marine fossils. The time-scale of the European sequence suggested by this correlation route is as shown in the second column of Figure 1.

The second possible correlation route is from the North American mammalian Stage-Ages through the Pacific Coast foraminiferal Ages (based upon benthonic foraminifera), through tropical planktonic foraminiferal sequences, and finally into the European benthonic sequences. The first step in this series, as presented in 1964, is shown in column 3 of Figure 1. At that time, few radiometric dates were available on rocks closely related to marine biostratigraphic sequences, but the over-all picture, based on radiometric dates and biostratigraphic sequences, still appears essentially correct. Kleinpell's (1938) Miocene Ages were tentatively given nearly equal lengths between control points. The short Miocene Ages apparently represent a time of rapid evolution and speciation among the benthonic foraminifera of the Pacific Coast. In the 1964 paper, the directly controlled points of the Pacific Coast foraminiferal sequence were top of the Repettian, the Zemorrian-Saucesian boundary, the Narizian-Ulatisian boundary, and the Bulitian-Ynezian boundary.

More recently, Turner (1968) has determined additional radiometric dates related to the foraminiferal Stages of Kleinpell. Figure 2, column 4, presents a revised time scale of the Pacific Coast foraminiferal sequence. Turner has found what appear to be complications in the distribution of time represented by Kleinpell's Stages, especially the Relizian and Luisian. In spite of possible paleontologic objections, placement by Turner of the Saucesian between 15.3 and 22.5 m.y. ago seems valid. The strata used by him to establish the Zemorrian-Saucesian boundary are those first

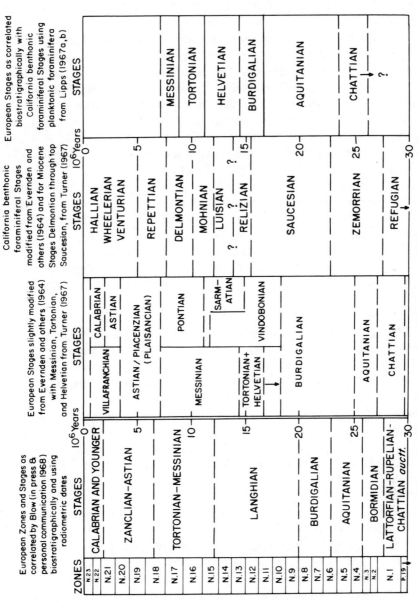

Figure 2. Correlations of some European and North American Stages.

used by Evernden and others in 1964. The stratigraphic control of his dated samples appears adequate.

Two possible correlations of the Pacific Coast foraminiferal Stages with the European Stages are shown in Figure 2. Lipps (column 5) based his correlations solely upon fossil evidence, using tropical planktonic foraminiferal "Zones" to tie together the Pacific Coast and Europe. On the other hand, Blow (column 2) has attempted an integration of available radiometric data with his paleontologic correlations between Europe and the western hemisphere. The resultant paleontologic correlations from the Pacific Coast to Europe are distinctly different.

Figure 2 also contains (column 3) the previously suggested time scale of the European sequence based upon mammalian correlations. The figure suggests quite clearly that paleontologic correlations are not at the desired state of refinement. The time scale of the European sequence must remain somewhat uncertain until more radiometric dates are available in that sequence or until paleontologic correlations are more definitely established or accepted.

Some brief comment on an unfortunate discussion that has arisen in the literature should be made. In 1961 (Evernden and others, 1961), a "Burdigalian" date was published based upon a glauconite horizon from the Vienna Basin. Lipps' (1967a) reiteration of the questionable value of that date of 25 m.y. for a Burdigalian horizon has been necessary only because of the importance placed on the date by some workers. Evernden and others (1961) originally questioned its significance on the basis of the mineral used and on the uncertain stratigraphic relationship of the "Burdigalian" of the Vienna Basin to that of the type area. The date (though questioned when first published) and its subsequent entry into extensive dispute and discussion is an excellent example of the dangers inherent in publishing in any form or under any qualification imperfectly controlled data.

Another quite different point relative to this same general subject needs to be stressed. The European Miocene Stages recognized by the Comité du Néogène Mediterranéen and used by Lipps (Fig. 2) in comparing Pacific Coast benthonic foraminiferal Stages, tropical planktonic foraminiferal zones, and European Stages, include the Aquitanian, Burdigalian, Helvetian, Tortonian, and Messinian Stages. Evernden and others (1964), however, used Aquitanian, Burdigalian, Vindobonian, and Sarmatian to cover approximately the same interval. Anyone familiar with biostratigraphic correlations will realize the problems of correlating "Stages" of different names, or of the same names but defined in different areas, that are thought to represent the "same" interval of geologic time.

PROBLEMS OF ESTABLISHMENT OF "TIME SCALE" BY BIOSTRATIGRAPHIC CRITERIA

A general tendency exists among biostratigraphers to use the criteria of presumed rates of evolution, or appearance of new species or assemblages, or the ranges of given species to demark biostratigraphic or Time-Stratigraphic units (Stages and Zones) of approximately the same time length. The units themselves, their recognizability, and correlation certainly are most important. They are not, however, intrinsically of approximately the same time length, whether based on one group of organisms (such as evolving series of foraminifera or land mammals) or on more than one group.

Rates of evolution do not measure time. Time, on the contrary, measures rates of evolution. Thus, in 1964 (p. 168), Evernden and others were able to clarify with radiometric dates the rates of evolution of certain Paleocene and Eocene land mammals, which Simpson (1947b) had used as a basis for describing Paleocene time as 17 m.y. long. Rates of evolution vary within one group and among groups of organisms. For example, planktonic foraminifera or particular groups of land mammals may evolve relatively rapidly for a time and then more slowly for a time, and, at any given time, it is unlikely that diverse groups will have been evolving at approximately the same rate.

Initially, it may not seem necessary to emphasize this point of lengths of time and contemporaneity "determined" by biostratigraphic criteria, but it is. Today, almost all workers state that when depicting biostratigraphic or time-stratigraphic units as equal in vertical extent on charts, it is solely as a convenience. Nevertheless, this useful "convenience" often conveys old biostratigraphic dogma wherein units were considered of approximately the same time length or the lengths of the units were "determined" by biostratigraphic criteria. Granted that not enough radiometric dates exist today to satisfactorily control all biostratigraphic units, but available dates should be used when possible. Failure to do so results in unnecessarily erroneous correlations and silently perpetuates the outdated "equal space-equal time" concept.

TIME SCALE OF GLACIATIONS AND THE
PLEISTOCENE SERIES

Evernden and Curtis (1965) extended the data of Figure 1 to late Quaternary. Presently available radiometric-age data suggest that the classic four glaciations of the Alps are distributed according to the time scale of Figure 3. Such a scheme, in conjunction with all available data, says: (1) Main Terraces of the Rhine are in some sense correlative with Mindel Glaciation; (2) evidence of glaciations of Mt. Kilimanjaro goes back only as far as Gunz to Mindel time; (3) the time of transition from late Blancan fauna to Irvingtonian fauna was concurrent with the Nebraskan Glaciation; (4) the time-scale of the four classic Alpine glaciations is essentially that proposed by Zeuner (1945) on the basis of the astronomical theory; (5) the Villafranchian faunas of Olduvai Gorge, Tanzania and Valros, France, are two to three times older than the Gunz Glaciation.

The radiometric age of the Villafranchian Age may be of particular interest. Dates on mid-Villafranchian mammals at Olduvai Gorge are 1.75 m.y. (*see* Evernden and Curtis, 1965). This date has been extensively confirmed. The late Blancan mammals of North America which are considered to have lived contemporaneously with the Villafranchian or early Villafranchian mammals of Europe are 2.4 m.y. old. Savage and Curtis have reported dates on upper, middle, and lower Villafranchian horizons in France of 1.3, 1.8, 1.9, 2.5, and 3.4 m.y., respectively (presented at the Society of Economic Paleontologists and Mineralogists Convention, Los Angeles, California, 1967). There seems little doubt that the base of the Villafranchian Stage is 3 to 3.5 m.y. old and that the

NORTH AMERICA	ALPS	ITALY	AGE
Wisconsin	Riss and Wurm		10,000 to approximately 250,000 years
Illinoisan	Mindel	Flaminian	approximately 400,000 years
Kansan	Gunz	Cassian	approximately 600,000 years
Nebraskan	Late Donau	Aquatraversan	1,000,000 - 1,200,000 years

Figure 3. Suggested time scale and correlations of late Pleistocene glaciations.

top of the Villafranchian is something less than 1 m.y. old. The age of the base of the Villafranchian is of immediate relevance in establishing the age of late Cenozoic glaciations, and the reason is that deposits thought by Curtis to be unequivocally indicative of glacial conditions are associated with the lower Villafranchian deposits of southern France. Previously, it has been presumed on the basis of such evidence that glaciations evidenced by these lower Villafranchian deposits were correlative with Gunz glaciations. It now appears indisputable that this is not the case and that the sequence of so-called Donau glaciations of the Alps may be the residual montane evidence of glaciations extending back at least 3 m.y. If the base of the Pleistocene is to be considered as equivalent to the time of initiation of late Cenozoic glaciation, the age of base of the Pleistocene must be 3 to 3.5 m.y. On the other hand, if the base of the Pleistocene is to be equated with the base of the Calabrian (*see* Fig. 2) or with marine events occurring concurrently with deposition of the base of the Calabrian, the base of the Pleistocene will be approximately 1.8 m.y.

Obradovich (1968) has dated the base of the Lomita Marl at approximately 3 m.y. In spite of biostratigraphic uncertainties, this base is considered by many to mark the Pliocene-Pleistocene boundary in southern California (*see* Woodring and others, 1946; Woodring, 1952, 1957). Interestingly, a 3 m.y. age would be in keeping with recent findings in Europe on the early Villafranchian. Obradovich's date is based on nine potassium-argon dates backed by rubidium-strontium analyses on glauconites of the (underlying) Fernando Formation and Lomita Marl and foraminifera of Repettian age. On the basis of the rubidium-strontium analyses, the age of the glauconite from the Fernando is approximately 6 m.y. (6.1 ± 1.2 m.y.), which Obradovich thinks represents a valid minimum age for the late early Pliocene of California.

TIME SCALE OF EVOLUTION OF MAN

The perfection of argon extraction and measurement techniques described by Evernden and Curtis (1965) allows the determination of radiometric ages of high potassium minerals, such as sanidine, on rocks as young as 5000 to 30,000 years. This satisfactorily overlaps the range of radiocarbon dating. Therefore, of itself and in conjunction with radiocarbon dates, potassium-argon allows dating the entire time-scale of evolution of man. The primary dates are those on the volcanic sequence at Olduvai Gorge, Tanzania. These results have been substantiated by an extensive series of dates from Bed I, Olduvai Gorge (*see* Leakey and others, 1961; Evernden and Curtis, 1965). Bed I and Bed II, Olduvai, including Olduwan and Chellean cultural remains, cover a time interval of at least 1.35 m.y., extending from approximately 1.85 m.y. ago to something less than 500,000 years ago. Each of these cultural epochs encompass nearly the same length of time. Ages on several minerals and numerous horizons confirm the correctness of these ages. In addition, to quote from F. Clark Howell (*in* Evernden and Curtis, 1965, p. 369):

> Ages of this magnitude should not really be surprising in the light of either stratigraphic or paleontologic evidence for early ranges of the early Quaternary, the extensive continental and littoral marine sedimentation, orogeny, tectonics, and volcanism of that range of time have been noted for many years, not only in the better-studied area of Europe, but also in the Maghreb of Africa and in the southerly foothills of the Himalayas. Those investigators acquainted with this range of the Quaternary have long noted its very substantial duration by comparison with the so-called Glacial Pleistocene.

Also, it is relevant to quote Adolph Knopf (*in* Evernden and Curtis, 1965, p. 373):

> It is of interest to recall that the eminent paleontologist Schuchert wrote as long ago as 1936 that Stille's estimate of 600,000 years is too short, and the length of the Pleistocene is at least 1,000,000 years and will eventually be shown to have been two or three times as long. This prediction is now amply verified by use of the high-precision potassium-argon technique. . . .

Figure 4 gives a suggested time-scale of evolution of man and his cultures. An extensive discussion of the data leading to that figure and of the figure itself was given in comments appended to the primary paper presented by Evernden and Curtis in 1965.

POTASSIUM-ARGON DATES AND MAGNETIC REVERSALS OF THE EARTH'S FIELD

A few comments on this subject are probably appropriate because of current efforts to use magnetic reversals as criteria for establishing stratigraphic age in deep-sea sedimentary cores. Evernden and others (1964) presented data that established un-

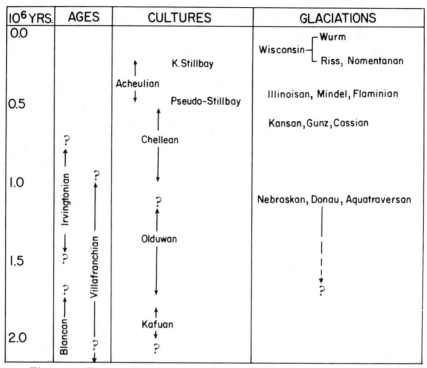

10⁶ YRS.	AGES	CULTURES	GLACIATIONS

Figure 4. Time scale of latest Pliocene-Pleistocene cultural events.

equivocally the existence of three distinct periods of contrasting magnetism during the last 3 m.y. of Earth history. If the present period is termed one of normal magnetic direction, data presented indicate two periods of normal and one of reversed magnetic polarization during the last 3.3 m.y. and suggest that the field was of reversed polarization approximately 3.4 m.y. ago. Much of the data was based on mineral concentrates, the remainder being on basalt whole-rock samples which were felt satisfactory for dating purposes. Cox and others (1967) and Cox (1969) have done much more work in this area and have extensively confirmed the general time-scale presented by Evernden and Curtis in 1965, and also the reversed epoch ending 3.4 m.y. ago. However, Cox and others, seem to be obtaining evidence for the existence of numerous very short intervals of time when the Earth's magnetic polarization was opposite to that dominant during that general interval of time, particularly for periods older than the first broad period of reversal. Thus, they have defined a Mammoth reversed event with a duration of approximately 50,000 years inserted in the middle of their Gauss normal epoch extending for approximately 850,000 years. Two short-term normal events are thought to exist in the Matuyama reversed epoch extending from 0.7 m.y. ago to 2.5 m.y. ago. The proof of the existence of such short intervals requires data of the highest reliability. Any assertions as to the reality of these second-order phenomena require presentation of substantiating data on the adequacy of the material used for dating. Many of the samples that must be used in such investigations as this are whole-rock samples. There is no way out of this basic problem. The use of whole-rock samples is necessary, and the unreliability of such samples is so general that the extension of our knowledge of the detailed pattern and magnetic reversals backward in time into the Cenozoic is very difficult. With a pattern of major reversals occurring every 0.7 to 2.0 m.y., the uncertainty in whole-rock basalt ages will result in a melange of normal and reversed horizons. Accurate knowledge of the pattern of these reversals will be obtained only by the laborious accumulation of ages determined from pure crystal concentrates. Knowing the grave problems associated with dating of whole-rock basalts, we consider that dating laboratories should not report such dates without full documentation on the materials used. Proof of reproducibility is not proof of accuracy. Short magnetically reversed epochs may exist. However, if the proliferation and meaningfulness of such epochs are to be believed, it is necessary to have the basic analytical data available for evaluation. Such data are not published by most investigators.

REFERENCES CITED

Blow, W. H., in press, Late middle Eocene to recent planktonic foraminiferal biostratigraphy: 1st Planktonic Conference Proc. (Geneva), 1967.

Cox, A., 1969. Geomagnetic reversals: Science, v. 163, no. 3864, p. 237-245.

Cox, A., Dalrymple, G. B., and Doell, R. R., 1967, Reversal of the Earth's magnetic field: Sci. American, v. 216, p. 44-54.

Crusafont Pairó, M., 1951, El sistema Miocenico en la Depression Española del Valles-Penedes: 18th Internat. Geol. Congr. Rept., Great Britain, pt. 11, p. 33-42.

——1958, Endemism and paneuropeism in Spanish fossil mammalian faunas with special regard to the Miocene: Soc. Sci. Fennica, Commentationes Biologicae, v. 18, no. 1, 31 p.

Durham, J. W., Jahns, R. H., and Savage, D. E., 1954, Marine-nonmarine relationships in the Cenozoic section of California: California Div. Mines and Geology Bull. 170, p. 59-71.

Durham, J. W., Kleinpell, R. M., and Savage, D. E., 1954, Mammalian Provincial ages, megafaunal "stages" and microfaunal stages on the Pacific Coast: California Div. Mines and Geology Bull. 170, p. 25.

Evernden, J. F., and Curtis, G. H., 1965, The Potassium-Argon dating of late Cenozoic rocks in East Africa and Italy: Current Anthropology, v. 6, p. 343-385.

Evernden, J. F., Curtis, G. H., and Kistler, R. W., 1957, Potassium-Argon dating of Pleistocene volcanics: Quaternaria, v. 4, p. 1-5.

Evernden, J. F., Curtis, G. H., Kistler, R. W., and Obradovich, J., 1960, Argon diffusion in glauconite, microcline, sanidine, leucite, and phlogopite: Am. Jour. Sci., v. 258, p. 583-604.

Evernden, J. F., Curtis, G. H., Obradovich, J., and Kistler, R. W., 1961, On the evaluation of glauconite and illite for dating sedimentary rocks by the potassium-argon method: Geochim. et Cosmochim. Acta, v. 23, p. 78-99.

Evernden, J. F., and James, G. T., 1964, Potassium-argon dates and the Tertiary floras of North America: Am. Jour. Sci., v. 262, p. 945-974.

Evernden, J. F., and Kistler, R. W., in press, Chronology of emplacement of Mesozoic batholithic complexes in California and western Nevada: U.S. Geol. Survey Prof. Paper.

Evernden, J. R., Savage, D. E., Curtis, G. H., and James, G. T., 1964, Potassium-argon dates and the Cenozoic mammalian chronology of North America: Am. Jour. Sci., v. 262, p. 145-198.

Folinsbee, R. E., Baadsgaard, H., and Lipson, J., 1961, Potassium-argon dates of Upper Cretaceous ash falls, Alberta, Canada: New York Acad. Sci., v. 91, art. 2, p. 352-359.

Funnell, B. M., 1964, The Tertiary Period, in Harland, W. B., Smith, A. G., and Wilcox, B., Editors, The Phanerozoic time-scale: Geol. Soc. London Quart. Jour., v. 120 s, 458 p.

James, G. T., 1963, The paleontology and nonmarine stratigraphy of the Cuyama Valley Badlands, California, (Pt. I): California Univ Pubs. Geol. Sci., v. 45, p. 1-154.

Kleinpell, R. M., 1938, Miocene stratigraphy of California: Tulsa, Oklahoma, Am. Assoc. Petroleum Geologists, 450 p.

Leakey, L.S.B., Evernden, J. F., and Curtis, G. H., 1961, Age of Bed I. Olduvai Gorge, Tanganyika: Nature, v. 191, n. 4787, p. 478-479.

Lippolt, H. J., ms, 1961, Altersbestimmungen nach der K-Ar-Methode bei Kleinen Argon und Kaliumkonzentrationen: Ph.D. dissert., Heidelberg, Germany, Univ. Heidelberg, 82 p.*

Lipps, J. H., 1967a, Planktonic foraminifera, intercontinental correlation and age of California mid-Cenozoic microfaunal stages: Jour. Paleontology, v. 41, n. 4, p. 994-999.

——1967b, Miocene calcareous plankton, Reliz Canyon, California, in Gabilan Range and adjacent San Andreas fault: Pacific Sections

Am. Assoc. Petroleum Geologists and Soc. Econ. Paleontologists and Mineralogists Guidebook, 110 p.

Mallory, V. S., 1959, Lower Tertiary biostratigraphy of the California Coast Ranges: Tulsa, Oklahoma, Am. Assoc. Petroleum Geologists, 416 p.

Obradovich, J. D., 1968, The potential use of glauconite for late-Cenozoic geochronology, *in* Means of correlation of Quaternary successions: Congress International Assoc. Quat. Res., v. 8, proc. 7, p. 267-279.

Savage, D. E., 1951, Late Cenozoic vertebrates of the San Francisco Bay region: California Univ. Pubs. Geol. Sci., v. 28, n. 10, p. 215-314.

Simpson, G. G., 1947a, Holarctic mammalian faunas and continental relationships during the Cenozoic: Geol. Soc. America Bull., v. 58, p. 613-687.

——1947b, A continental Tertiary time chart: Jour. Paleontology, v. 21, no. 5, p. 480-483.

Turner, D. L., ms, 1968, Potassium argon dates concerning the Tertiary foraminiferal time scale and San Andreas fault displacement: Ph.D. dissert., Berkeley, California, Univ. California, 99 p. (*see* Turner paper, this volume).

Wood, H. E., and others, 1941, Nomenclature and correlation of the North American continental Tertiary: Geol. Soc. America Bull., v. 52, p. 1-48.

Woodring, W. P., 1952, Pliocene-Pleistocene boundary in California Coast Ranges: Am. Jour. Sci., v. 250, p. 401-410.

——1957, Marine Pleistocene of California: Geol. Soc. America Mem. 67, p. 589-598.

Woodring, W. P., Bramlette, M. N., and Kew, W.S.W., 1946, Geology and paleontology of Palos Verdes Hills, California: U.S. Geol. Survey Prof. Paper 207, 145 p.

Zeuner, F. E., 1945, The Pleistocene period; its climate, chronology, and faunal succession: Ray Soc., (London), 322 p.

————

*Excerpts from Lippolt's thesis are found in the following publications:

Bout, P., Frechen, J., and Lippolt, H. J., 1966, Stratigraphic and radio chronologic dating of several basaltic flows of Limagne: Revue d'Auvergne, v. 80, no. 4, p. 207-230.

Frechen, J., and Lippolt, H. J., 1965, Kalium-Argon-Daten zum Alter des Laacher Vulkanismus, der Rheinterrassen und der Eiszeiten: Eiszerfalder and Gegenwart, v. 16, p. 5-30 (öbringen/Württ).

Gentner, W., Lippolt, H. J., Schaeffer, O. A., 1963, Argonbestimmungen an Kaliummineralien. H. Die Kalium-Argon-alter der gläser des Nördlinger Rieses und der böhmusch-mährischen tektite: Geochim. et Cosmochim. Acta, v. 27, p. 191-200.

Lippolt, H. J., and Genfner, W., 1963, Alters hestimmungen nach der kalium-Argon-Methode an tertiären Eruptivgestinen Sudwesdeutschlands: Geol, Jährb. Landesamt, Baden-Württemberg, v. 6, p. 507-538.

Journal of Geophysical Research   Vol. 72, No. 10   May 15, 1967

Statistical Analysis of Geomagnetic Reversal Data and the Precision of Potassium-Argon Dating

Allan Cox and G. Brent Dalrymple

U. S. Geological Survey, Menlo Park, California 94025

37

A new statistical method has been developed for analyzing the magnetic polarity of rocks as a function of their potassium-argon ages for the purpose of determining the ages of the boundaries between geomagnetic polarity epochs. The analysis also yields an estimate of the precision of the potassium-argon dating. A value of 3.6% is found by this analysis for the dating precision of rocks about 2.5 m.y. old, which is in agreement with an independent estimate of the precision of the dating obtained from an analysis of analytical errors. The following are the best statistical estimates of the ages of the boundaries between geomagnetic polarity epochs: Gilbert-Gauss boundary, 3.36 m.y.; Gauss-Matuyama boundary, 2.5 m.y.; Matuyama-Brunhes boundary, 0.70 m.y. The duration of polarity events is estimated to vary from 0.07 to 0.16 m.y., and the best estimate of the time required for the earth's field to undergo a complete change in polarity is 4600 years.

Introduction

During the past 4 m.y. the geomagnetic field has changed polarity at least nine times, producing a polarity structure consisting of polarity epochs, polarity events, and polarity transition intervals, all with different characteristic times. The longest are the geomagnetic polarity epochs, defined as intervals during which the predominant direction of the field was either normal, as it is today, or reversed by 180° from the present direction. The durations of epochs during the past 3.6 m.y. range from 0.7 to 1.8 m.y., although in the more distant past they have been much longer than this. Polarity events are much briefer polarity fluctuations lasting only about 10^5 years, but otherwise they are exactly like epochs. Finally, transition intervals are even briefer time intervals that bound epochs and events. The behavior of the geomagnetic field during polarity transitions is poorly known. At the few localities where transitions have been paleomagnetically recorded the field has assumed different directions intermediate between normal and reversed and has decreased in intensity to about one-fifth its usual value during transitions. Whether the transitional field represents the nondipole field or a reduced geomagnetic dipole is uncertain.

The chronology of reversals shown in Figure 1 was determined from combined paleomagnetic measurements and potassium-argon age determinations on volcanic rocks. Until recently,

subjective methods have proven adequate for interpreting the rather small amount of experimental data in terms of polarity epochs and events. As the number of paleomagnetic and radiometric measurements has increased, however, a need has arisen for an objective statistical method for selecting the best-fit boundaries between polarity epochs. The main problem in developing a method for doing this is closely related to the existence of errors in the radiometric dating. If such errors were zero so that experimental results corresponded exactly to points on the true reversal time scale, the ages of polarity epoch boundaries could be found by simply listing the magnetic polarity data in sequence of age. Because of dating errors, the sequential order of a set of ages calculated from radiometric data will in general be somewhat different from that of the corresponding set of true ages. As a result, inconsistencies are generated in the pattern of magnetic polarity as a function of radiometric age, the extent of the inconsistencies being determined by the precision of the dating. The plan of this paper is to evaluate the precision of the radiometric dating and then to develop a model that relates this precision to the amount of inconsistency expected in polarity data. This model provides a method for calculating the best-fit boundary between polarity epochs and also yields an independent estimate of the dating precision. The model is extended to obtain indirect estimates

2603

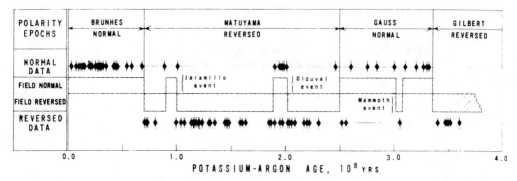

Fig. 1. Time scale for reversals of the earth's field. The observed magnetic polarities and potassium-argon ages of volcanic rocks are shown as diamond-shaped symbols. Only the more reliable primary group of data is included. The inferred time scale is shown by the bar at the center.

of the duration of events and of the time required for polarity transitions, which because of their brevity are difficult to date directly.

PRECISION OF POTASSIUM-ARGON DATING

Errors in the over-all accuracy of the dating, such as errors due to uncertainties in the radioactive decay constants, are not relevant to the present analysis because, although such errors would produce a systematic shift in the polarity time scale, they would not change the degree of internal consistency of the polarity data. This is controlled by the precision of the dating, which can be evaluated by analyzing the contributions of the individual sources of experimental error in the following way.

If the molar content of the potassium 40 and radiogenic argon 40 in a rock or mineral is equal to K^{40} and Ar_{rad}^{40}, the age is given by

$$ t = \frac{1}{\lambda_\epsilon + \lambda_\beta} \ln \left[\frac{Ar_{rad}^{40}}{K^{40}} \left(\frac{\lambda_\epsilon + \lambda_\beta}{\lambda_\epsilon} + 1 \right) \right] $$

where λ_ϵ and λ_β are the decay constants for K^{40} to Ar^{40} and Ca^{40}, respectively. Substituting $\lambda_\epsilon = 0.585 \times 10^{-10}$ yr^{-1} and $\lambda_\beta = 4.72 \times 10^{-10}$ yr^{-1},

$$ t = 1.885 \times 10^9 \ln \left[9.071 \frac{Ar_{rad}^{40}}{K^{40}} + 1 \right] $$

$$ \cong 1.710 \times 10^{10} \frac{Ar_{rad}^{40}}{K^{40}} $$

where the approximation is valid for $t < 10^7$ years.

The analytical precision of the argon determination depends strongly on the amount of

contamination by atmospheric Ar^{40}. In the isotope dilution method for determining argon, the gas sample measured in the mass spectrometer consists of (1) an unknown amount of radiogenic Ar^{40}, (2) X moles of an argon tracer which, for simplicity, we will assume is pure Ar^{38}, and (3) an unknown amount of atmospheric (or normal) argon consisting of 99.6% Ar^{40}, 0.063% Ar^{38}, and 0.337% Ar^{36}. The measurements made with the spectrometer are of three ion beam currents corresponding to the three argon isotopes, and after instrumental corrections the ratios of the beam currents are equal to the ratios of argon isotopes in the gas mixture, M_{38}^{40}, M_{38}^{36}, etc., for which the corresponding percentage errors in the ratios are σ_{38}^{40}, σ_{38}^{36}, etc. In contrast with Ar^{40} and Ar^{38} the Ar^{36} beam current is usually small and has a low ratio of signal to noise. The percentage uncertainty in measuring the Ar^{36} beam current is thus much greater, so that the percentage errors of M_{38}^{36} and M_{40}^{36} are much greater than the error of M_{38}^{40}.

The total amount of Ar^{40} in the experiment is given by

$$ Ar_E^{40} = X \, M_{38}^{40}/(1 - M_{38}^{36} N_{36}^{38}) $$

where $N_{36}^{38} = 0.187$ is the ratio Ar^{38}/Ar^{36} in normal air. For small percentage errors σ_{38}^{40} and σ_{38}^{36}, an estimate of the percentage error in Ar_E^{40} is given by

$$ \sigma_E^{40} \cong [(\sigma_X)^2 + (\sigma_{38}^{40})^2 $$

$$ + (\sigma_{38}^{36})^2 F^2/(1 - F)^2]^{\prime} $$

where σ_X is the percentage error in X and $F = M_{38}{}^{36}N_{36}{}^{38}$. The quantities σ_X, $\sigma_{38}{}^{40}$, and $\sigma_{38}{}^{36}$ are analytical errors that can be determined independently from replication experiments. The quantity F depends on the outcome of the particular dating run for which we wish to assess the error, so that the above expression for $\sigma_E{}^{40}$ is not exact. However, the percentage variation of F is equal to that of $\sigma_{38}{}^{36}$, which is of the order of 2%, so that the uncertainty in F as determined from one run is small. Moreover, in almost all experiments the value of $M_{38}{}^{36}$ is 0.01 or less, so that F is less than 0.002 and the contribution of the last term is negligible.

Under assumptions similar to the ones made above, an estimate of the total percentage error σ in the age determination after correcting for the amount of Ar^{40} due to atmospheric contamination is given by

$$\sigma \cong \left[(\sigma_K)^2 + (\sigma_X)^2 + (\sigma_{38}{}^{40})^2 \left(\frac{1}{r}\right)^2 \right.$$
$$\left. + (\sigma_{38}{}^{36})^2 \left(\frac{1-r}{r} - \frac{F}{1-F}\right)^2 \right]^{1/2}$$
$$\cong \left[(\sigma_K)^2 + (\sigma_X)^2 + (\sigma_{38}{}^{40})^2 \left(\frac{1}{r}\right)^2 \right.$$
$$\left. + (\sigma_{38}{}^{36})^2 \left(\frac{1-r}{r}\right)^2 \right]^{1/2} \qquad (1)$$

where $r = 1 - M_{38}{}^{36}N_{36}{}^{40}/M_{38}{}^{40}$ is the fraction of the $Ar_E{}^{40}$ that is radiogenic. The value of r, like that of F, depends on the particular dating run for which we wish to assess the error, so that this expression is not exact. It breaks down entirely for values of r approaching zero.

For our laboratory, σ_K is known from replication experiments to be about 0.5% standard deviation [Lanphere and Dalrymple, 1965; Dalrymple and Hirooka, 1965]. The error $\sigma_{38}{}^{40}$ may be evaluated from instrumental discrimination checks which are made regularly in our laboratory using normal argon extracted from air. Nineteen such measurements of the ratio $M_{40}{}^{36}$ made over the past two years have a total standard deviation of 0.77%, part of which is due to instrumental discrimination and part to errors in measuring $M_{40}{}^{36}$. Eighteen of these measurements were made in pairs, so that $\sigma_{40}{}^{36}$ can be found from a pooled estimate of the standard deviation of the paired data. The value

found for $\sigma_{40}{}^{36}$ is 0.17%. In these discrimination experiments to determine $M_{40}{}^{36}$, the high signal-to-noise ratio is comparable to that observed in determining $M_{38}{}^{40}$ for dating experiments, so that the value of 0.17% provides a reasonable estimate of $\sigma_{38}{}^{40}$ in the dating error (equation 1). The signal to noise ratio for $M_{38}{}^{36}$ observed in dating experiments is much smaller than that of $M_{40}{}^{36}$ in the discrimination measurements and varies according to the quality of the experiment, so that $\sigma_{38}{}^{36}$ varies and is difficult to evaluate experimentally. On the basis of repeat runs made on gas from young rock samples containing a moderate amount of atmospheric argon, we estimate that $\sigma_{38}{}^{36}$ is usually about 1.5–2%. For small values of r, however, it may be as low as 0.5% because of the increase in signal to noise for $M_{38}{}^{36}$ as r decreases. The precision σ_X of the calibration of the Ar^{38} tracer depends on the method of tracer preparation and may be minimized by using a pipette system [Wasserburg et al., 1962]. In our analyses, for which we used a pipette system that is not connected directly to the extraction line [Lanphere and Dalrymple, 1966], the precision σ_X is estimated to be 0.3%. This estimate is based on an analysis of the errors in a set of 14 argon determinations for a standard biotite for which the over-all analytical precision due to errors of all types, including sample inhomogeneity, was 0.6% [Lanphere and Dalrymple, 1965]. In a typical dating experiment for which $r = 0.5$, the following values are found for the four terms in equation 1:

$$\sigma = [(0.5)^2 + (0.3)^2 + (0.17)^2(2)^2 + (2)^2]^{1/2}$$
$$= 2.11\%$$

The last term in the brackets, which is the dominant term, corresponds to the estimate made by Lipson [1958] of the error due to atmospheric contamination.

A useful graphical representation of equation 1 is given in Figure 2. The curves were calculated using $\sigma_K = 0.5\%$, $\sigma_X = 0.3\%$, $\sigma_{38}{}^{40} = 0.2\%$, and various values of $\sigma_{38}{}^{36}$. Because all the quantities may vary depending on the material analyzed and the quality of the experiment, curves such as these can never furnish a truly rigorous estimate of precision. For this, replicate age determinations on the same sample are needed. Where we have performed such replica-

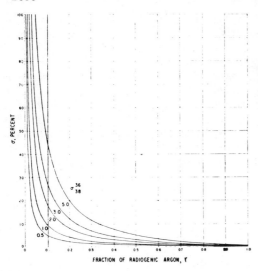

FRACTION OF RADIOGENIC ARGON, γ

Fig. 2. Percentage error σ in potassium-argon dating assuming that $\sigma_K = 0.5\%$, $\sigma_x = 0.3\%$, and $\sigma_{38}{}^{40} = 0.2\%$ in equation 1.

tion experiments, however, the estimates obtained from the curves in Figure 2 are in general agreement with the replication results (Table 1).

In reporting data used for determining the reversal time scale, we have used the above method to estimate the precision of each potassium-argon date. In Table 2 these estimates are summarized, excluding the data that are insufficiently precise to be useful for deter-

mining the reversal time scale. For ages in the range 0 to 2.0 m.y. a cutoff of 0.1 m.y. was used, and for ages greater than 2.0 m.y. a cutoff of 5%. From the values in Table 2 it is apparent that the precision of potassium-argon dating is adequate to resolve some of the phenomena associated with reversals, but not all of them. The ages of the boundaries between polarity epochs are clearly resolvable for the range of ages considered, 0 to 3.6 m.y. The duration of polarity events of about 10^5 years is the same as the standard deviation of the dating within a factor of 2 in this age range, so that events can be resolved only with difficulty. The transition time of about 5000 years required to complete a change in polarity is clearly too short to be dated directly.

THEORETICAL MODEL FOR POLARITY
INCONSISTENCIES

In this section we consider the consistency to be expected between the magnetic polarities of volcanic rocks and their radiometric ages. Data from transitional intervals are not considered in this analysis, and it is assumed that reversed magnetism produced mineralogically rather than by a reversal of the earth's field has been identified. An important preliminary point is that the consistency between polarities and ages does not depend on whether the volcanic rocks are from a local area or whether they are worldwide in their distribution, because

TABLE 1. Comparison of Precision of Potassium-Argon Ages Determined from
Replication Experiments with Estimates of Precision from Figure 2
Values from Figure 2 were read to nearest 0.5%.

Description	K-Ar Age, 10^6 years	Average, r	σ from Repeated Age Determinations	σ Estimated Using Figure 2
12 Whole-rock basalt specimens from same hand sample [Dalrymple and Hirooka, 1965]	3.4	0.72	1.90	1.5
7 Whole-rock basalt samples from same flow [Dalrymple and Hirooka, 1965]	3.3	0.66	2.11	2.0
12 Ar and 10 K_2O analyses on muscovite P-207 [Lamphere and Dalrymple, 1965]	81.0	0.90	1.12	1.0
14 Ar and 3 K_2O analyses on biotite 62ALe1 [Lanphere and Dalrymple, 1966]	159	0.92	0.77	1.0
Multiple Ar and K_2O analyses on 13 sanidine and obsidian samples (Doell, Dalrymple, Smith, and Bailey, unpublished data)	0.4–1.0	0.56	3.98	4.0

the earth's field is never a normal dipole field over parts of the earth and simultaneously a reversed field over the remainder. This follows from the fundamental properties of a dipole field and from the abundant paleomagnetic evidence that during the time interval of interest here, 0 to 3.6 m.y. ago, the geomagnetic field had been predominantly dipolar. Therefore, the inconsistencies to be expected between magnetic polarities and radiometric ages are due to dating errors and should occur in the vicinity of polarity transitions. For the data presently available it is, in fact, near polarity transitions that inconsistencies do occur, indicating that dating errors rather than undetected self-reversals are responsible.

Assume a population of true ages T distributed with uniform density. Radiometric ages t are generated from this population by a radiometric dating technique with a variance σ^2. Assume that $T_{N/R}$ is the true age of a polarity transition such that the field was reversed for $\infty > T > T_{N/R}$ and normal for $T < T_{N/R}$. If repeated age determinations are made on a normally magnetized sample with a true age $T < T_{N/R}$, the radiometric ages will be distributed as shown in Figure 3, and polarity inconsistencies will result from all radiometric ages $t > T_{N/R}$. The probability P_R that a datum with radiometric age t will have reversed polarity is given by

$$P_R = \int_{T_{N/R}}^{\infty} (1/\sqrt{2\pi}\,\sigma)$$

$$\cdot \exp\left(-(t - T)^2/2\sigma^2\right)\,dT$$

$$= \tfrac{1}{2}\,\mathrm{erfc}\left[\frac{T_{N/R} - t}{\sqrt{2}\,\sigma}\right] \qquad (2)$$

TABLE 2.   Average Estimated Dating Errors for Primary Time Scale Data

Range of Ages, m.y.	Average Estimated Standard Deviation of K-Ar Ages	
	Millions of Years	Per Cent
0–1.0	0.041	18
1–3.0	0.076	4.3
3.0–3.6	0.097	2.9

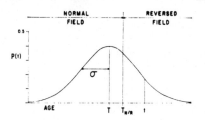

Fig. 3.   Probability model for polarity inconsistencies. A lava flow is assumed to have formed at a time T when the field was normal, and the geomagnetic field is assumed to have changed polarity at time $T_{N/R}$. Radiometric dates t on the flow are assumed to have normal distribution with variance σ^2. Dates with a radiometric age $t > T_{N/R}$ correspond to a time when the field was reversed and hence constitute inconsistent data points.

Similarly the probability of a normal polarity at t is

$$P_N = \int_{-\infty}^{T_{N/R}} (1/\sqrt{2\pi}\,\sigma)$$

$$\cdot \exp\left(-(t - T)^2/2\sigma^2\right)\,dT$$

$$= \tfrac{1}{2}\,\mathrm{erfc}\left[\frac{t - T_{N/R}}{\sqrt{2}\,\sigma}\right] \qquad (3)$$

These expressions are exact for a single polarity transition occurring on an infinite time scale. For multiple polarity changes, such as those observed in nature, the numerical results are the same to within 1%, provided that the time between polarity epoch boundaries is 3 times larger than the dating precision, which is true in the time range considered here.

The probability distribution P_N is shown in Figure 4 for different values of the dating precision σ. The radiometric age t for part of this population lies in the range $t > T_{N/R}$, producing apparent inconsistencies between magnetic polarity and K-Ar age. A symmetrical situation exists for the samples of reversed polarity so that the probability distribution P_I for the subset of the entire population with inconsistent polarities is given by

$$P_I = \tfrac{1}{2}\,\mathrm{erfc}\left|(t - T_{N/R})/\sqrt{2}\,\sigma\right| \qquad (4)$$

An important objective in analyzing the observational data is to obtain an accurate estimate of the time $T_{N/R}$ when the polarity changed. This may be done by assuming a trial

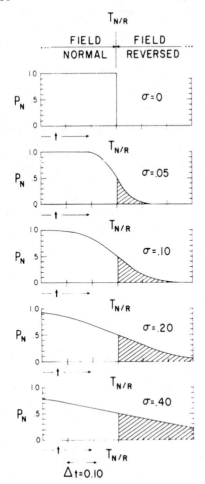

Fig. 4. Probability P_N that the polarity of a rock will be normal, given that its radiometric age is t, that the dating precision is σ, and that a polarity transition occurred at time $T_{N/R}$. For all values of dating precision σ, P_N approaches unity and zero at times far removed from $T_{N/R}$. Data for which the radiometric ages and magnetic polarities are inconsistent are indicated by the shaded pattern.

boundary τ and then calculating the standard deviation E_D of the radiometric ages for those data points that are inconsistent with this assumed boundary. On varying τ, both the number of inconsistencies and their distance from the boundary changes. The value of τ that minimizes E_D provides an accurate estimate of $T_{N/R}$. To include only the polarity inconsistencies that are related to the boundary of one particular change in polarity, only data with

radiometric ages in an interval $\tau \pm \Delta\tau$ are used to find the boundary. The value of $\Delta\tau$ is not critical provided that it is small enough to include only one polarity transition yet large enough, relative to the dating precision, to include most of the polarity inconsistencies. For the present data, a value of 0.3 m.y. was used. To provide theoretical models to compare with the observational data, values of E_τ analagous to E_D were found for the theoretical probability distributions. Setting $T_{N/R} = 0$ so that $P_N = \frac{1}{2}$ erfc $(t/\sqrt{2}\,\sigma)$,

$$E_\tau{}^2 = \int_\tau^{\tau+\Delta\tau} (t - \tau)^2 \tfrac{1}{2} \text{ erfc } (t/\sqrt{2}\,\sigma)\; dt$$

$$+ \int_{\tau-\Delta\tau}^{\tau} (t - \tau)^2 (1 - \tfrac{1}{2} \text{ erfc } t/\sqrt{2}\,\sigma)\; dt \quad (5)$$

For $\tau = T_{N/R}$, $E_\tau{}^2$ closely approximates the variance

$$\int_{-\infty}^{\infty} (t - T_{N/R})^2 \tfrac{1}{2} \text{ erfc } |t/\sqrt{2}\,\sigma|\; dt$$

of the population distribution of inconsistent polarities, provided that the dating precision σ is several times smaller than $\Delta\tau$. In Figure 5 the boundary between polarity epochs is sharply defined by a minimum in E_τ which occurs when

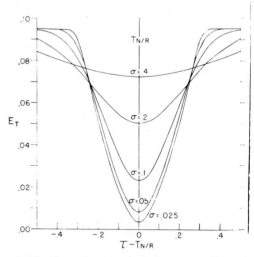

Fig. 5. Theoretical standard deviations E_τ for inconsistent magnetic polarities and radiometric ages as a function of assumed age of the trial boundary τ between polarity epochs (equation 5). A minimum in E_τ occurs when the assumed boundary τ is set equal to the true boundary $T_{N/R}$.

$\tau = T_{N/R}$. This minimum provides an objective basis for determining the age of the polarity epoch boundary.

POLARITY EPOCH BOUNDARIES

Turning to observational data, a total of 133 paleomagnetic polarity measurements and associated potassium-argon ages in the range 0 to 3.6 m.y. have been reported for a wide variety of volcanic rock types from many parts of the world [*Rutten*, 1959; *Cox et al.*, 1963a, b, 1964a, b; *McDougall and Tarling*, 1963; *Grommé and Hay*, 1963; *Evernden et al.*, 1964; *Doell et al.*, 1966; *Cox and Dalrymple*, 1966; *Cox et al.*, 1966; *Chamalaun and McDougall*, 1966; *McDougall and Wensink*, 1966]. In these investigations, the completeness of the magnetic measurements has varied widely, as has the accuracy and precision of the dating, so that the entire data set is very heterogeneous. For the present analysis, a more homogeneous primary subset of 88 data was selected on the basis of the following criteria. (1) Each magnetic polarity determination is based on laboratory measurements of the magnetic polarity and its stability. (2) The estimated precision of the K-Ar age determination is less than 0.1 m.y. for ages of 0 to 2.0 m.y and less than 5% of the age for ages greater than 2.0 m.y. (3) The magnetic and age measurements were made on the same volcanic cooling unit. (4) The rock or mineral used for the polarity and age determination is of a type known to yield reliable results. These criteria are essentially those used by *Doell et al.* [1966] for selecting a primary data set useful for refining the time scale, except that here a cutoff of 5% is used rather than one of 0.1 m.y. for ages older than 2.0 m.y. This permits extension of the criteria to older rocks.

Following the method previously used for theoretical distributions of polarity inconsistencies, for this data set the standard deviation of the inconsistent data, E_D, was calculated in the following way. Assume that N radiometric ages t_i occur in the interval $\tau \pm \Delta\tau$. For each age we define two parameters B_i and C_i as follows. For $t_i \leq \tau$, $B_i = 0$, and for $t_i > \tau$, $B_i = 1$. For data with normal polarity $C_i = 0$, and for data with reversed polarity $C_i = 1$. Then for a reversed-to-normal polarity change

$$E_D{}^2 = \frac{2\,\Delta\tau}{N} \sum_{i=1}^{N} (t_i - \tau)^2$$

$$\cdot [(1 - B_i)C_i + B_i(1 - C_i)] \qquad (6a)$$

and for a normal-to-reversed change

$$E_D{}^2 = \frac{2\,\Delta\tau}{N} \sum_{i=1}^{N} (t_i - \tau)^2$$

$$\cdot [(1 - B_i)(1 - C_i) + B_i C_i] \qquad (6b)$$

where the first term, which is the inverse of the average number of dates per unit time, normalizes E_D to permit direct comparison with the values of E_T as a function of τ for the theoretical distributions.

Curves of E_D as function of τ for the three polarity epoch boundaries are given in Figure 6. All the data from the primary group were included in this analysis except the data from polarity events, which are discussed separately in the next section. The boundaries between polarity epochs are defined by the E_D minima at the following times:

Matuyama-Brunhes boundary	0.7 m.y.
Gauss-Matuyama boundary	2.5 m.y.
Gilbert-Gauss boundary	3.36 m.y.

The apparent polarity inconsistencies observed near polarity transitions provide an independent measure of the dating precision. At the Gauss-Matuyama boundary, the minimum value of E_D is 0.020, corresponding to a dating error of 0.09 m.y. or a standard deviation of 3.6%. This value does not differ significantly from the independent estimates obtained by the method described previously, which are summarized in Table 2. This result is especially important because the previous estimates of dating precision are essentially measures of the reproducibility of the analyses, whereas the value obtained from the minimum value of E_D includes, in addition, the effect of variations in the retention of argon by rocks and minerals from different parts of the world. The good agreement between the two estimates of precision indicates that variations in the retention of argon are very small for the data included in the primary group.

At the Gilbert-Gauss and Matuyama-Brunhes boundaries, the minimum value of E_D is zero because the density of data points near the boundaries is too small to produce polarity inconsistencies. Although it can be shown that the

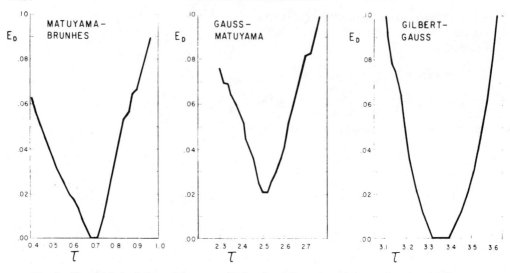

Fig. 6. Standard deviation of apparent dating inconsistencies, E_D, as a function of the age τ, the assumed polarity epoch boundary.

perfect consistency of the present data set would be very improbable if the errors in the radiometric dating were much larger than the estimates given in Table 2, additional data will be needed before a precise estimate can be made of the dating precision at these boundaries.

LENGTHS OF POLARITY EVENTS

Consider a normal polarity event lasting from T_1 to T_2, with the field normal during this interval and reversed outside of it. Then the probability P_E that a sample with a radiometric age t will have true age in the interval T_1 to T_2 and, hence, that it will be of normal polarity is given by

$$P_E = \int_{T_1}^{T_2} (1/\sigma \sqrt{2}\pi) \exp\left[-(t-T)^2/2\sigma^2\right] dT$$

$$= \tfrac{1}{2} \operatorname{erfc}\left[(t-T_2)\sqrt{2}\,\sigma\right]$$

$$- \tfrac{1}{2} \operatorname{erfc}\left[(t-T_1)/\sqrt{2}\,\tau\right] \qquad (7)$$

This distribution is shown in Figure 7 for a normal event lasting 0.1 m.y. as it would be observed experimentally with different dating errors. If the dating error is 0.4 m.y., at most 10% of the data in any radiometric dating interval will be of normal polarity, even if the radiometric dates are close to the true age of the event. With large dating errors, a polarity event will thus appear experimentally as a broad band of polarity inconsistencies.

The only polarity event with sufficient data to permit statistical analysis is the Olduvai event, for which there are paleomagnetic and radiometric data from Africa [*Grommé and Hay*, 1963], Alaska [*Cox et al.*, 1966], and from the island of Réunion in the Indian Ocean [*Chamalaun and McDougall*, 1966]. In the part of the Matuyama polarity epoch between the older boundary at 2.5 m.y. and the Jaramillo event at 0.91 m.y., there are 42 reliable paleomagnetic-radiometric data pairs, of which six are normal. The radiometric ages of the normal samples are all in the range 1.79 to 2.01 m.y. (Figure 8). This remarkably tight grouping is consistent with a polarity event 0.1 m.y. long and with a dating error of about 0.05 m.y. or 2.5%. The dating errors for the data in the vicinity of the Olduvai event thus appear to be no higher than the independent estimates previously discussed.

Even with dating errors this small, the duration of the events cannot be determined by direct dating. The theoretical probability distributions shown in Figure 7, suggest, however, an indirect method for doing this. For an interval of length $2I$ about the central point T_E of a normal event lasting from time T_1 to time

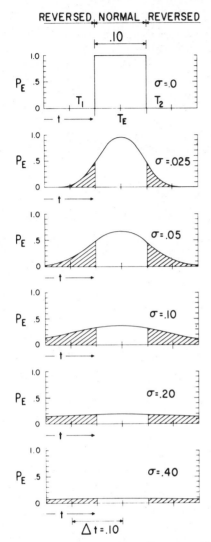

Fig. 7. Probability P_E that a lava flow with a true age in the interval T_1 and T_2 corresponding to an event lasting 0.1 m.y. will have a radiometric age t. Data for which the radiometric ages and magnetic polarities are inconsistent with the true polarity of the event are shaded.

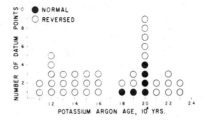

Fig. 8. Histogram showing the normally magnetized Olduvai lava flows and reversed flows of Matuyama epoch. Data are from primary group and include results from Africa [*Grommé and Hay*, 1963], from Alaska [*Cox et al.*, 1966], and from Réunion Island [*Chamalaun and McDougall*, 1966].

$I = 0.25$ m.y. is large enough to include all the radiometric dates belonging to each event while at the same time small enough to exclude irrelevant data from adjacent polarity epoch boundaries. Evaluating R as the proportion of all the reliable radiometric dates in this interval for which the polarity is the same as the polarity of the event, equation 8 yields the estimates given in Table 3 for the duration of the polarity events.

An independent estimate of the duration of the Olduvai normal event and the Mammoth reversed event may be made from the paleomagnetic study of Icelandic lavas by *Wensink* [1964, 1965]. Although the Icelandic sequences provide one of the most nearly continuous geomagnetic records yet made in a volcanic area, the polarity events are so short that they appear only in some of the sections sampled. From Wensink's data for the complete stratigraphic sections as Jökuldalur, Hofsárdalur, Fljótsdalur, and Tjörnes we have found the proportion of

T_2, the proportion R of normal polarities within the interval $2I$ is given by

$$R \cong (T_2 - T_1)/2I \qquad (8)$$

where the approximation is valid to within 1%, provided that I is several times larger than the dating precision σ. For evaluating observational data in the interval 0 to 3.6 m.y., a value of

TABLE 3. Lengths of Polarity Events

Name of Event	Jaramillo Normal	Olduvai Normal	Mammoth Reversed
Mean age of event	0.91	1.96	3.06
Duration by equation 8	0.12	0.16	0.07
Duration from results of Wensink	...	0.14	0.05
Duration from deep-sea sediments	0.05	0.14	...

all flows in the Gauss polarity epoch that acquired reversed polarity during the Mammoth reversed event and similarly the proportion of flows in the Matuyama reversed epoch that recorded the Olduvai normal event. The duration of each event was then estimated by multiplying these proportions by the durations of the respective epochs (Table 3).

The possibility that the Olduvai event may be a double event has been suggested by *McDougall and Wensink* [1966], partly on the basis of the stratigraphy of the Icelandic basalts and partly on the basis of a potassium-argon date of 1.6 m.y. obtained from one of the Icelandic flows that formed during the Olduvai event. The dating evidence for a double event is rather weak because glass is present in the dated basalt flow, and basaltic glass does not always retain argon well [*Cox et al.*, 1966; *Dalrymple*, 1964]. On the other hand, in view of the small amount of data for this part of the time scale, a double event as suggested by McDougall and Wensink is a distinct possibility. If the Olduvai is a double event, the estimate of its duration obtained from the Icelandic data applies to both parts of the event.

Geomagnetic polarity epochs and events are also recorded paleomagnetically in deep-sea cores [*Harrison and Funnel*, 1964; *Opdyke et al.*, 1966]. *Ninkovich et al.* [1966] have obtained the estimates given in Table 3 for the duration of the Jaramillo and Olduvai events, using the known age of the Matuyama-Brunhes boundary as a point of calibration to determine the rate of sedimentation. The advantage of this approach is that the continuous nature of deep-sea deposition provides an opportunity to obtain a much more complete paleomagnetic record than is available from volcanic rocks. The data from the volcanic rocks, on the other hand, provide the absolute time scale needed to calibrate the detailed records obtained from the deep-sea cores. The estimates for the duration of the Olduvai event are in agreement. For the Jaramillo event, the available radiometric dates are few in number and irregular in distribution, so that the estimate obtained from the deep-sea cores is probably superior. Although all these estimates are subject to obvious uncertainties, they appear to be sufficiently consistent to indicate that the duration of polarity events is about 10^5 years.

TIME REQUIRED FOR POLARITY CHANGES

An estimate of the time required for polarity transitions may be made from the proportion of observed intermediate directions of magnetization on the basis of the following probability model. Assume that over some time interval θ the field has changed polarity M times and that the time required for the ith change is $\Delta\theta_i$ (Figure 9), so that the total time $\Delta\theta_T$ during which the field was transitional is given by

$$\Delta\theta_T = \sum_{i=1}^{M} \Delta\theta_i$$

If N age determinations have been made in the interval θ and if they are randomly distributed with respect to the transition, the probability that exactly R of the age determinations will occur in transitional intervals is given by the binomial distribution

$$P_B = \binom{N}{R}(\Delta\theta_T/\theta)^R(1 - \Delta\theta_T/\theta)^{N-R}$$

The best estimate of $\Delta\theta_T$ is given by the quantity $\theta R/N$, and confidence limits for this distribution may be calculated by standard methods [*Clopper and Pearson*, 1934]. Note that it has not been necessary to assume that all values of θ_i are equal.

In applying this model to observational data, we have used the following two criteria for identifying those data points that are from transitional intervals. (1) The average direction of magnetization of the volcanic unit differs from the direction of the dipole field at the sampling locality by at least 2δ, where δ is the angular standard deviation of geomagnetic field directions produced by long-period geomagnetic secular variation. (2) The radiometric age of the volcanic unit differs by no more than 2σ from the age of the nearest epoch or event

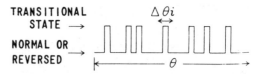

Fig. 9. Model for estimating time required for polarity transitions. During time interval θ the geomagnetic field is either normal or reversed (base line) except for brief time intervals θ_i during which the direction of the field is intermediate between normal and reversed.

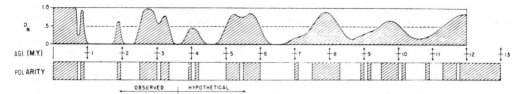

Fig. 10. Proportion P_N of samples that would have normal polarity as a function of their radiometric age, assuming a dating precision of 3%. The bottom bar shows the assumed polarity of the geomagnetic field, normal polarity being indicated by shading. The assumed polarities prior to 3.6 m.y. are entirely hypothetical and are intended only to show the loss of resolving power of the radiometric dating for the earlier part of the reversal time scale.

boundary, where σ is the standard error of the age determination. The second criterion is necessary because large angular deviations from the dipole field direction may occur at times well removed from polarity transitions. For example, a lava flow on St. Paul Island, Alaska, has an average direction of magnetization deviating from the direction of the dipole field by 65°, which exceeds the value of 2δ and hence satisfies criterion 1, yet the age of the flow is less than 10^5 years and hence not close to a polarity transition. Data of this type are rejected by criterion 2. It would be desirable to employ a third criterion based on a reduction of geomagnetic field intensity during polarity transitions, but as yet there are too few paleomagnetic intensity determinations available from radiometrically dated rocks.

Only one datum from the primary data set satisfies these criteria. This is unit 4D057, a rhyolite dome in New Mexico with an age of 0.88 m.y., which occurs at the boundary between the Jaramillo normal event and the Matuyama reversed polarity epoch [*Doell and Dalrymple*, 1966]. Setting $N = 88$, $R = 1$, and $\theta = 3.6$ m.y., the best estimate of $\Delta\theta_T$ is 41,000 years. Upper and lower confidence limits at the 95% confidence level are 14,000 and 190,000 years. The best estimate of the average time required for each of the nine transitions is 4600 years, with upper and lower confidence limits at the 95% level of 1600 and 21,000 years.

An independent estimate of the time required for a polarity transition has been obtained by *Ninkovich et al.* [1966] from their paleomagnetic investigation of deep-sea sediment cores. They estimate that during reversals the field was reduced in intensity for about 20,000 years, whereas the reversal in direction occurred in the much shorter interval of 1000 years. Although the estimate obtained from the potassium-argon dating results and from the sediment cores are both subject to numerous uncertainties, they are in order-of-magnitude agreement.

DISCUSSION

The present analysis has implications for future attempts to extend the geomagnetic polarity time scale beyond its present limit of 3.6 m.y. by means of potassium-argon dating. For rocks 10 m.y. old, for example, a dating precision of 3% corresponds to a dating error of 0.3 m.y., which would make it extremely difficult to identify polarity events.

The loss of resolving power that would be produced by dating errors of 3% may be seen in Figure 10, which shows the proportion P_N of samples that would have normal polarity as a function of their radiometric age, assuming that the actual polarity has changed as shown in the bar at the bottom of the figure. (Changes for ages >3.6 m.y. are purely hypothetical.) Values for P_N were calculated using an equation similar to equation 7, but integrating over entire sequence of polarity changes shown in the lower part of Figure 10. These results indicate that polarity events will probably not be identifiable on the basis of the potassium-argon dating of rocks older than about 6 m.y. A useful result of the present calculation is that at the boundaries of the polarity epochs, the proportion of normal polarities remains close to 0.50, even where events are close to the boundaries. Thus given sufficient data, the boundaries of the longer polarity epochs can probably be identified, although the amount of data required would be formidable. Knowledge of the detailed polarity structure of the geomagnetic field prior to 6 m.y. ago will probably come from approaches other than potassium-argon dating.

Acknowledgment. Publication is authorized by the Director, U. S. Geological Survey.

REFERENCES

Chamalaun, F. H., and Ian McDougall, Dating geomagnetic polarity epochs in Réunion, *Nature, 210,* 1212–1214, 1966.

Clopper, C. J., and E. S. Pearson, The use of confidence or fiducial limits illustrated in the case of the binomial, *Biometrika, 28* (3–4), 404–413, 1934.

Cox, Allan, and G. B. Dalrymple, Paleomagnetism and potassium-argon ages of some volcanic rocks from the Galapagos Islands, *Nature, 209,* 776–777, 1966.

Cox, Allan, R. R. Doell, and G. B. Dalrymple, Geomagnetic polarity epochs and Pleistocene geochronometry, *Nature, 198,* 1049–1051, 1963a.

Cox, Allan, R. R. Doell, and G. B. Dalrymple, Geomagnetic polarity epochs—Sierra Nevada, 2, *Science, 142,* 382–385, 1963b.

Cox, Allan, R. R. Doell, and G. B. Dalrymple, Geomagnetic polarity epochs, *Science, 143,* 351–352, 1964a.

Cox, Allan, R. R. Doell, and G. B. Dalrymple, Reversals of the earth's magnetic field, *Science, 144,* 1537–1543, 1964b.

Cox, Allan, D. M. Hopkins, and G. B. Dalrymple, Geomagnetic polarity epochs—Pribilof Islands, Alaska, *Bull. Geol. Soc. Am., 77,* 883–910, 1966.

Dalrymple, G. B., Cenozoic chronology of the Sierra Nevada, California, *California Univ. Publ. Geol. Sci., 47,* 41 pp., 1964.

Dalrymple, G. B., and Kimio Hirooka, Variation of potassium, argon, and calculated age in a late Cenozoic basalt, *J. Geophys. Res., 70,* 5291–5296, 1965.

Doell, R. R., and G. B. Dalrymple, Geomagnetic polarity epochs—a new polarity event and the age of the Brunhes-Matuyama boundary, *Science, 152,* 1060–1061, 1966.

Doell, R. R., G. B. Dalrymple, and Allan Cox, Geomagnetic polarity epochs—Sierra Nevada data, 3, *J. Geophys. Res., 71,* 531–541, 1966.

Evernden, J. F., D. E. Savage, G. H. Curtis, and G. T. James, Potassium-argon dates and the Cenozoic mammalian chronology of North America, *Am. J. Sci., 262,* 145–198, 1964.

Grommé, C. S., and R. L. Hay, Magnetization of basalt of Bed I, Olduvai Gorge, Tanganyika, *Nature, 200,* 560–561, 1963.

Harrison, C. G. A., and B. M. Funnell, Relationship of paleomagnetic reversals and micropaleontology in two late Cenozoic cores from the Pacific Ocean, *Nature, 204,* 556, 1964.

Lanphere, M. A., and G. B. Dalrymple, P-207—an interlaboratory standard muscovite for argon and potassium analyses, *J. Geophys. Res., 70,* 3497–3503, 1965.

Lanphere, M. A., and G. B. Dalrymple, Simplified bulb-tracer system for argon analyses, *Nature, 209,* 902–903, 1966.

Lipson, Joseph, Potassium-argon dating of sedimentary rocks, *Bull. Geol. Soc. Am., 69,* 137–149, 1958.

McDougall, Ian, and D. H. Tarling, Dating of polarity zones in the Hawaiian Islands, *Nature, 200,* 54–56, 1963.

McDougall, Ian, and H. Wensink, Paleomagnetism and geochronology of the Pliocene-Pleistocene lavas in Iceland, *Earth Planetary Sci. Letters, 1,* 232–236, 1966.

Ninkovich, D., N. Opdyke, B. C. Heezen, and J. H. Foster, Paleomagnetic stratigraphy, rates of deposition, and tephrachronology in north Pacific deep-sea sediments, *Earth Planetary Science Letters, 1,* 476–492, 1966.

Opdyke, N. D., B. Glass, J. D. Hays, and J. Foster, Paleomagnetic study of antarctic deep-sea cores, *Science, 154,* 349–357, 1966.

Rutten, M. G., Paleomagnetic reconnaissance of mid-Italian volcanoes, *Geol. Mijnbouw, 21,* 373–374, 1959.

Wasserburg, G. J., G. W. Wetherill, L .T. Silver, and P. T. Flawn, A study of the ages of the Precambrian of Texas, *J. Geophys. Res., 67,* 4021–4047, 1962.

Wensink, H., Secular variation of earth magnetism in Plio-Pleistocene basalts in eastern Iceland, *Geol. Mijnbouw, 43,* 403–413, 1964.

Wensink, H., Paleomagnetic stratigraphy of younger basalts and intercalated Plio-Pleistocene tillites in Iceland, *Geol. Rundschau, 54,* 364–384, 1965.

(Received December 5, 1966.)

JOURNAL OF GEOPHYSICAL RESEARCH   VOL. 72, No. 24   DECEMBER 15, 1967

Age of the Steens Mountain Geomagnetic Polarity Transition

A. K. BAKSI AND D. YORK

Geophysics Division, Department of Physics, University of Toronto
Toronto, Ontario, Canada

N. D. WATKINS

Department of Geology, Florida State University, Tallahassee, Florida 32306

38

The paleomagnetism of a series of lavas outcropping on Steens Mountain, southeastern Oregon, is consistent with a change in the geomagnetic field from reversed to normal polarity. Potassium-argon analyses of nineteen specimens from six lavas indicate that this polarity change occurred at $t = 15.1 \pm 0.3$ m.y. Devitrified glass and slightly altered feldspars in one of the lavas have not adversely affected argon retention properties.

INTRODUCTION

The major features of the geomagnetic polarity changes during the past 3.5 million years are now known, as the result of combined paleomagnetic measurements and potassium-argon dating of terrestial igneous rocks (summary by *McDougall and Chamalaun* [1966]). These polarity changes have been verified by measurements on submarine sedimentary cores [*Opdyke et al.*, 1966; *Watkins and Goodell*, 1967].

Pitman and Heirtzler [1966] and *Vine* [1966] have analyzed the linear magnetic anomalies associated with some mid-oceanic ridges, in terms of the crustal spreading model of *Hess* [1962] and the known geomagnetic polarity changes [*Cox et al.*, 1966], following closely the idea of *Vine and Matthews* [1963]. It appears that, since the magnetic profiles normal to the surveyed oceanic ridges extend in some cases beyond the limits of hypothesized spreading of oceanic crust of 3.5-m.y. age, the magnetic profiles can be used to predict the major geomagnetic polarity changes to at least the base of the Pliocene and possibly as far back as the lower Mesozoic era [*Vine*, 1966], if a constant spreading rate is assumed. It follows that testing the crustal spreading model can be facilitated by delineation of the geomagnetic field polarity history. There is little question that the experimental error involved in isotope dating methods, when compared with the suggested polarity history, renders the possibility of precise delinea-tion of the durations of short periods of constant geomagnetic polarity of ages in excess of 6 m.y. extremely difficult at this stage [*Cox and Dalrymple*, 1967]. This is especially true if the method used is to date randomly occurring igneous rocks of known magnetic polarity and to infer polarity boundaries by interpolation between assembled data.

We present here an alternative method of delineating a limit of a geomagnetic polarity 'epoch' or 'event' (which by convention [*Cox et al.*, 1964] refer to relatively long and short periods, respectively, of constant geomagnetic polarity). This comprises the dating of lavas associated with the finite period of transition between opposite geomagnetic polarities, which is suspected to occupy a period of the order of 5,000 to 10,000 years. Known epochs are of the order of 1,000,000 years duration, whereas the events so far described persist for a period of the order of 50,000 years.

The paleomagnetic methods are now well established [*Collinson et al.*, 1967]. Since the K-Ar method of dating basalts is still at the development stage, however, we shall describe our dating method in detail.

GEOLOGICAL DETAILS

Steens Mountain (latitude, 118°33′W; longitude, 42°40′N) is one of the world's largest single exposures of successive Tertiary lavas. Part of the extension of the Basin and Range province into southeastern Oregon, the mountain is an easterly facing normal fault with over

6299

6,000 feet of nearly vertical exposure. Although, as *Fuller* [1931] shows, slumping is a feature of the steep fault scarp, the outcrop is so perfect that little difficulty is encountered in obtaining samples from successive lavas, except where inaccessibility occurs. The lavas are thin, averaging less than 20 feet in thickness, show almost no interbasaltic sediment, and are dominantly a diktytaxitic porphyritic basalt [*Fuller*, 1931]. Some petrological details resulting from the present study will be discussed later.

Evernden and James [1964] have obtained isotope dates of 14.5 and 14.7 m.y. for whole rock basalt and plagioclase separates for a basalt immediately overlying the section. These dates assist in confirming the Upper Miocene age of vertebrates found in volcanically derived sediments southwest of Steens Mountain [*Walker and Repenning*, 1965].

PALEOMAGNETIC RESULTS

Watkins [1963, 1965a] has integrated paleomagnetic data from lavas at the summit of Steens Mountain into a regional paleomagnetic survey. In a later survey, at least six cores were taken from each of seventy-one successive lavas on Steens Mountain. The samples were taken with a portable gasoline-powered drill and were oriented in geographic coordinates to an accuracy of 2½°. The paleomagnetic properties were obtained with a spinner magnetometer and a triaxial demagnetizing unit using alternating magnetic fields up to 800 oe [*Doell and Cox*, 1967]. Preliminary results revealed that the lower nineteen lavas were extruded during a geomagnetic polarity transition [*Watkins*, 1965b]. The inclinations of natural remanent magnetism (NRM) for these lavas are illustrated in Figure 1. *Strangway and Larson* [1965] have also, by implication, reported the existence of this geomagnetic polarity transition, since they have detected an upper section of normal polarity and a lower section of reversed polarity on Steens Mountain. *Watkins* [1967] has presented the paleomagnetic results for the lavas above the polarity transition. None of these lavas, from which more than 300 oriented cores were collected, are of reversed polarity.

Polished sections have been manufactured from four cores in each magnetically transitional lava for examination of the iron-titanium

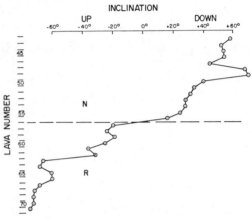

Fig. 1. Mean inclination of natural remanent magnetism for each lava in the lower part of the sampled Steens Mountain section, showing the transition from reversed to normal ambient geomagnetic polarity. The natural remanent magnetism for each lava results from at least six separate cores, following treatment in an alternating magnetic field of 200 oe. The ordinate is the number of the lava from the top of the section. In the northern hemisphere, positive (downward) inclination is normal (N) and negative (upward) inclination is reversed (R). After *Watkins* [1965b].

oxides (using a Reichert Zetopan-pol reflection microscope, at ×1200 under oil), specifically for integration into a study of magnetic and associated petrological properties in a large suite of lavas in southeastern Oregon [*Wilson and Watkins*, 1967]. Curie points were also obtained from each of the cores examined in polished section by means of a modified Chevallier balance [*Ade-Hall et al.*, 1965]. A correlation between magnetic polarity and oxidation state of the iron-titanium oxides in the transitional lavas was observed by *Wilson and Watkins* [1967], but there is little question that the NRM is due to the lavas' having been extruded during a geomagnetic polarity transition. *Coe* [1967] has strengthened this conclusion in obtaining experimental evidence [*Thellier and Thellier*, 1959] on lava 57 (Figure 1) consistent with a value of 0.11 ± 0.02 gauss for the ambient geomagnetic field intensity during extrusion. This relatively low value (about 25% of the present dipole intensity) may be expected during the polarity transition [*Rikitake*, 1966, p. 81]. The magnetic data are presented in Table 1.

TABLE 1A. Paleomagnetic Data for Steens Mountain Basalts

Specimen numbering convention is as follows: 11-3(i) means lava number 11 (numbers increase with increasing age; see Figure 1), 3 means core number 3, (i) means specimen number 1. D and I are declination and inclination of NRM following treatment in an alternating magnetic field of 200 oe. D increases from geographic north ($D = 0$) eastward, and I is positive downward and negative upward. J_0 is the original intensity of magnetization. J_{200} is the intensity of magnetization (in emu cm^{-3} × 10^{-3}) after treatment in an alternating magnetic field of 200 oe. Oxidation index is an optically defined relative oxidation state on a scale of 1 (minimum oxidation) to 5 (maximum oxidation). For discussion of this index see *Wilson and Watkins* [1967]. Curie points are given to the nearest 25°C.

Specimen Number	D, deg	I, deg	J_0, emu cm^{-3} × 10^{-3}	J_{200}/J_0	Oxidation Index	Curie Point, °C	Position with Respect to Polarity Transition
11-3(i)	11.0	+65.0	2.32	0.19	3	575	
11-4(i)	13.5	+73.5	1.73	0.21	3	525	Above
11-7(i)	7.0	+58.5	2.19	0.34	3	575	
17-4(i)	339.0	+58.5	3.86	0.67	3	575	
17-4(ii)	343.0	+59.0	3.71	0.64	2	525	Above
17-6(i)	352.5	+62.8	4.85	0.83	4	575	
51-1(i)	117.5	+42.0	2.87	0.04	2	225, 525	
51-2(i)	118.5	+43.5	3.16	0.05	1	225, 525	In transition
51-5(i)	121.5	+36.5	1.60	0.34	3	575	
51-6(i)	131.5	+35.5	1.45	0.57	3	575	
61-1(i)	219.0	−41.0	1.05	0.87	2	575	In transition
61-6(i)	234.0	−51.0	2.13	0.91	2	525	
61-7(i)	238.0	−21.5	1.25	0.27	2	525	
68-1(i)	219.0	−76.5	8.03	0.74	4	575	
68-4(i)	234.0	−69.0	4.62	0.88	3	575	Below
68-5(i)	232.5	−74.0	4.50	0.84	4	575	
68-6(i)	246.5	−66.5	7.35	0.80	4	575	
70-3(i)	215.5	−69.5	5.59	0.84	4	575	
70-5(i)	200.0	−70.0	5.58	0.84	5	575	Below

POTASSIUM-ARGON EXPERIMENTAL METHODS

Potassium-argon whole rock analyses have been made of specimens from six lavas, representing two lavas from above the polarity transition, two lavas within the polarity transition, and two lavas below the polarity transition, as indicated in Table 1. This sample spacing was chosen in an attempt to detect a measurable age difference in the sampled section, prior to focusing efforts on delineation of the age of the polarity transition. As we shall demonstrate, however, the entire sampled sec-

TABLE 1B. Paleomagnetic Data for Steens Mountain Basalts

N is number of cores per lava. D and I are the declination and inclination as defined in Table 1A. K_0 and K_{200} are *Fisher's* [1953] precision parameters ($K = (N − 1)/(N − R)$, where R is the resultant vector assigning unit vector per specimen) for original NRM and for NRM following treatment of each specimen in an alternating magnetic field of 200 oe.

Lava Number	N	D, deg	I, deg	K_{200}	K_{200}/K_0
11	7	2.5	+64.0	151.7	2.32
17	6	343.0	+61.6	522.1	8.17
51	6	126.2	+34.0	11.7	2.15
61	7	213.2	−36.1	15.8	4.19
68	6	228.5	−72.5	149.0	0.39
70	6	212.9	−73.2	375.0	2.40

tion appears to have been extruded during a very short period, and the age of the polarity transition was therefore obtained by the initial dating.

Sample preparation. Basalt specimens were the original paleomagnetic specimens, in cylinders 2.5 cm long and 2.3 cm in diameter. Nineteen cores from six lavas (as indicated in Tables 1 and 2) were used in the following way:

1. Eleven cores were ground to a 10–15 mesh homogeneity. Potassium and argon analyses were made on aliquots of this mixture.

2. Thin (1 to 2mm) slices were taken from opposite ends of five cores and ground to 100 mesh before potassium analysis. Pieces, approximately 3 cm³ in volume, of the rest of each core were then fused for argon analysis.

3. Thin slices were taken from opposite ends

of three cores and ground to 100 mesh for potassium analysis. The remainder was ground to 10–15 mesh for argon analysis.

Thin slices of the specimens or the adjacent specimen in the same core were retained for thin and polished section manufacture.

Argon analysis methods. The basalt samples were fused by radio-frequency induction heating in a vacuum. Two fusion systems were used, one of which could be baked. The gases evolved from the basalts were purified by Cu-CuO and titanium furnaces, and a liquid air cold trap. Ar^{40} volumes were determined by isotope dilution with Ar^{38}. Spikes approximately 7×10^{-6} cm³ at NTP were manufactured at uniform pressure in groups of sixty, using a manifold system [*Inghram*, 1954]. Spikes from three different groups were used, and each group was

TABLE 2. K-Ar Data and Ages for Steens Mountain Basalts

Methods of analysis are described under section on sample preparation in text. Ar^{40*} indicates radiogenic Ar^{40}. Errors in Ar^{40*} are estimated from Figure 2. Errors in K = $\pm2\%$. $\lambda_e = 0.584 \times 10^{-10}$ yr^{-1}; $\lambda_\beta = 4.72 \times 10^{-10}$ yr^{-1}. $K^{40} = 0.0119$ atom % of total K.

Specimen Number	Method of Analysis	K, %	Atmospheric Ar^{40}, %	Ar^{40*}, ppm $\times 10^{-3}$	Calculated Age, m.y.	Estimated Error, m.y.
11-3	1	1.53	18.7	1.674	15.3	0.3
			24.3	1.657	15.1	0.3
11-4	1	1.51	40.1	1.641	15.2	0.3
			31.6	1.612	14.9	0.3
11-7	2	1.50	15.6	1.622	15.1	0.4
17-4(i)	1	0.84	47.6	0.903	15.0	0.4
			43.7	0.909	15.1	0.3
17-4(ii)	1	0.83	41.5	0.887	14.9	0.3
17-6	1	0.835	47.1	0.885	14.8	0.4
51-1	2	1.308	60.6	1.385	14.8	0.5
51-2	2	1.335	63.7	1.471	15.4	0.4
51-5	2	1.271	57.1	1.404	15.4	0.4
			34.9	1.380	15.2	0.4
51-6	2	1.270	93.0	1.348	14.9	2.0
			17.6	1.342	14.8	0.3
61-1	1	0.985	95.2	1.054	15.0	1.8
			94.2	1.047	14.9	1.6
61-6	1	0.870	68.3	0.962	15.5	0.4
			54.0	0.925	14.9	0.4
61-7	1	0.885	49.2	0.949	15.0	0.4
			66.6	0.950	15.0	0.4
68-1	3	0.890	75.8	0.957	15.0	0.5
			93.1	0.960	15.0	1.1
68-4	1	1.00	90.7	1.090	15.2	0.9
68-5	1	0.924	93.6	1.014	15.3	1.4
68-6	1	1.00	70.6	1.056	15.5	0.4
70-3	3	0.953	74.5	1.003	14.7	0.5
70-5	3	0.833	65.5	0.892	15.0	0.4

calibrated in identical fashion. From each group, three spikes were randomly chosen and mixed with spec-pure argon of atmospheric composition, before mass-spectrometric analysis of the Ar^{40}/Ar^{38} ratios, in order to determine the spike volumes. Division of the spike volume by the appropriate break-seal volumes provided three estimates of the volume of Ar^{38} per unit volume of a spike break seal. The mean of these three estimates was assumed to be the calibration factor for all the spikes. The calibrations in the three groups of spikes for this study fluctuated about the mean values by ±1%. As a check on any systematic error, one spike from each group was used in argon fusion runs with a rock standard (a homogeneous slate, Oxford 20576A) provided by M. H. Dodson, of the Department of Earth Science, University of Leeds, England. The Oxford geochronology laboratory value for the rock standard results from eight argon analyses, which agree within ±1.5% with our results. The interlaboratory muscovite sample U.S.G.S. P207 [*Lanphere and Dalrymple*, 1965] provided values of 5.12 × 10^{-2} and 5.08 × 10^{-2} ppm radiogenic argon with our spikes [*Farrar*, 1966]. These results differ

from the mean value for ten reported values by 0.4 and 1.2%, respectively. These results, together with our numerous previous comparative analyses during the past three years, provide evidence that any systematic error in argon analyses is less than ±1%. The A. E. I. MS10 mass spectrometer [*Farrar et al.*, 1964] was used throughout. In this device ions are deflected through 180° in a semicircle of 2 inches radius of curvature. The magnetic field is produced by a small permanent magnet and electrostatic scanning is used. An ultrahigh vacuum was maintained in the mass spectrometer tube after baking, and the pressure before an analysis was usually in the range 10^{-9} to 10^{-10} mm of mercury. All analyses were made in the static mode [*Reynolds*, 1956], no correction being required for memory effects. Ar^{40} and Ar^{38} peak heights were determined to ±0.2%. The much smaller Ar^{36} peak was less accurately known, and the error depended on the amount of atmospheric argon present. We estimate our Ar^{36} measurements were accurate to within 0.4 to 2.0%. Ar^{36} error estimates were made for each mass spectrometric run. Figure 2 illustrates the importance of Ar^{36} measurement uncertainties, par-

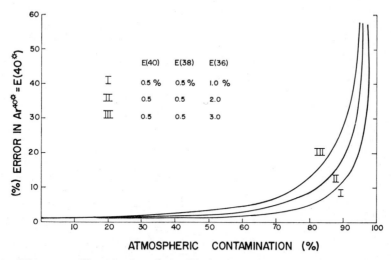

Fig. 2. Diagram to illustrate the relationship between the error in the Ar^{40*} determination and the atmospheric argon contamination as a function of the mass spectrometric errors. The curves were drawn from the equation

$$E(40^*) = \left\{ \left(1 + \frac{A}{R}\right)^2 E^2(40) + E^2(38) + \left(\frac{A}{R}\right)^2 E^2(36) \right\}^{1/2}$$

where $E(40^*)$ = % error in Ar^{40*}; $E(40)$, $E(38)$, and $E(36)$ are the % errors in Ar^{40}, Ar^{38}, and Ar^{36} peak heights, respectively; A = % atmospheric Ar^{40} contamination; R = % Ar^{40*}.

ticularly when atmospheric contamination exceeds 90%. A careful check was maintained on the discrimination of the mass spectrometer, by analyzing atmospheric argon samples for Ar^{40}/Ar^{36}. Ten such analyses during a year produced a standard deviation for a single measurement of the Ar^{40}/Ar^{36} ratio of approximately 0.3%. Within this degree of uncertainty, the discrimination of the MS10 may be taken as constant.

Potassium analysis methods. Potassium analyses were made with two different flame photometers: a Perkin-Elmer 52c and an Instrumentation Laboratories model 143, following the procedure described by *Cooper* [1963], with a lithium internal standard and a sodium buffer. According to our results, precisions for single determinations for the instruments are ±2 and ±1%, respectively, although dual instrument determinations on single samples yield agreements to better than ±1%. Possible systematic errors were monitored by means of the Oxford slate standard sample with both photometers. This provided a potassium concentration of 3.95%, in comparison with Dodson's value of 4.00% obtained using an E.E.L. flame photometer without internal standard or buffering and A. Stelmach's (University of Alberta) value of 3.99% obtained by the tetraphenyl boron precipitation method. From these results and several other analyses, we conclude that any

systematic errors in potassium values will be less than ±1%. All potassium analyses were made in triplicate.

POTASSIUM-ARGON RESULTS

Table 2 shows the results of the potassium-argon analyses for each specimen and other relevant data. A histogram of the ages from all cores is shown in Figure 3. No measurable age difference exists between the lavas, the within-lava and between-lava variation being similar. The data exhibit a normal distribution. The weighted mean for all the dates is 15.08 ± 0.04 m.y., but possible systematic errors in potassium and argon analyses suggest that at this stage an age of 15.1 ± 0.3 m.y. is more realistic.

The age of the lavas is constant despite within-specimen and within-lava potassium concentration differences of up to 5.5 and 12.0%, respectively (Table 3). A similar observation has been made by *Dalrymple and Hirooka* [1965]. These results illustrate the precision of the data and the fact that no significant amount of argon was inherited during initial cooling of the lavas.

The consistency in ages obtained using the different sample preparation methods (stated above) supports the conclusion of *Amaral et al.* [1966] that no measurable argon loss results from the grinding of specimens.

Although the atmospheric argon contamina-

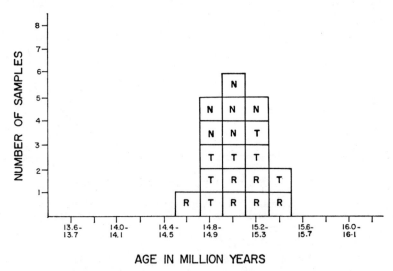

AGE IN MILLION YEARS

Fig. 3. Histogram of the potassium-argon ages of the nineteen specimens from the sampled Steens Mountain section. The average age of each specimen is plotted. R, T, N refer to lavas that exhibit reversed, transitional, and normal polarity, respectively.

TABLE 3. Variation of Potassium Concentration in Individual Cores of Steens Basalt
Specimen numbering convention same as in Table 1A. For explanation of sample preparation see text. Potassium results are on separate lines for each slice.

Specimen Number	Method of Sample Preparation	Flame Photometer*	Individual Results K, %			Average at Each End K, %	Average for Specimen K, %
11-7(i)	2	I.L.	1.467	1.484	1.464	1.472	1.495
			1.531	1.531	1.537	1.533	
51-1(i)	2	I.L.	1.298	1.284	1.290	1.291	1.308
			1.339	1.318	1.318	1.325	
51-2(i)	2	I.L.	1.316	1.316	1.319	1.317	1.335
			1.344	1.359	1.352	1.352	
51-5(i)	2	I.L.	1.239	1.256	1.252	1.249	1.271
			1.305	1.285	1.296	1.292	
51-6(i)	2	I.L.	1.242	1.285	1.255	1.252	1.270
			1.282	1.301	1.280	1.288	
68-1(i)	3	P.E.	0.882	0.870	0.874	0.875	0.890
			0.898	0.915	0.902	0.905	
70-3(i)	3	P.E.	0.930	0.920	0.927	0.926	0.953
			0.986	0.974	0.981	0.980	
70-5(i)	3	P.E.	0.818	0.821	0.831	0.823	0.833
			0.839	0.846	0.843	0.843	

* I. L. indicates Instrumentation Laboratories flame photometer, Model 143; P. E. indicates Perkin-Elmer photometer, Model 52-C.

tion ranges from 15.6 to 95.2% (Table 2), no correlation exists between age and atmospheric contamination (Figure 4), therefore providing additional evidence of the precision of the data.

PETROLOGICAL SIGNIFICANCE

The consistency in ages, despite the considerable variation in grain size within and between lavas, precludes any possibility that long-term diffusional losses of argon have occurred, and also indicates that weathering has not significantly affected the results.

Thin section examination shows that the lavas are on the average 60% feldspar, 20% pyroxene, 8–10% olivine, and 8% iron-titanium oxides, which vary considerably between lavas in their oxidation index as determined by polished section examination under oil at $\times$ 1200. (For a discussion of the oxidation index classification, magnetic significance, and photomicrographs of identification and classification examples see *Wilson and Watkins* [1967] or *Watkins and Haggerty* [1967].) Feldspars occur mainly as laths, but phenocrysts sometimes constitute up to 4% of the thin sections. Despite some alteration in the olivines, the feldspars are, except in one lava, unaltered, and the specimens can be considered to be essentially fresh.

The exception to the generally fresh aspect of the samples is lava 11. Specimens 11–4 and 11–7 show about 15% brown devitrified glass, and specimen 11–3 shows slightly cloudy feldspars. The mean age of lava 11 is 15.1 ± 0.1 m.y., however, and this value is in exact agreement with the average age for all the measured lavas. *Dalrymple* [1963] and others have pointed out that some degree of alteration in the olivines can be tolerated in the potassium-argon dating of basalts. *Evernden and James* [1964] and *Cox et al.* [1966] consider, however, that samples exhibiting alteration of feldspars or devitrification of glassy material should not be used. Our results for lava 11, on the other hand, indicate that thin section analysis is not an infallible criterion for selection of suitable samples for whole-rock dating of basalts, as has been previously pointed out by *Amaral et al.* [1966]. It appears that the most reasonable method of assessing the reliability of the age of a lava is to date samples from various parts of the body and examine the consistency of the data. Such perturbing effects as inherent variation in cooling rates throughout a lava [*Watkins and Haggerty*, 1967] and localized weathering effects should be readily detected if sampling is widely spaced, vertically and hori-

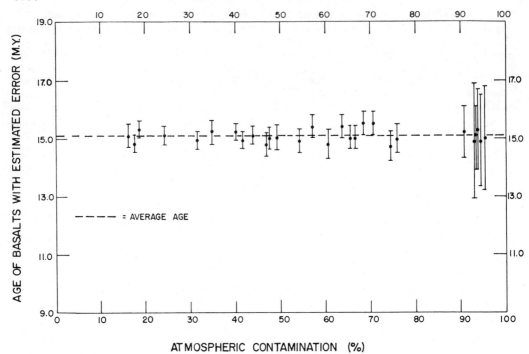

Fig. 4. Potassium-argon age with errors as a function of amount of atmospheric contamination for lavas of the sampled Steens Mountain section.

zontally. It appears that pronounced oxidation variation as indicated by the oxidation index (Table 1) cannot be correlated with argon retention characteristics, but, because of the limited number of specimens used, this observation is not conclusive.

COMPARISON OF RESULT WITH INDEPENDENT DATA

The result of 15.1 ± 0.3 m.y. for the Steens Mountain geomagnetic polarity transition compares closely with *Evernden and James's* [1964] potassium-argon ages for two lavas close to the lavas discussed in this communication. They obtain 14.5 m.y. (whole rock) and 14.7 m.y. (plagioclase separate) for a lava lapping onto the slightly dipping Steens section, and 15.0 m.y. (sanidine separate) for a lava that is toward the top of the Steens scarp. This consistency with our results is additional evidence of the validity of the age obtained.

The only comparable dated paleomagnetic polarity that has been published is from the Nordlinger Ries crater in southern Bavaria. *Genter et al.* [1963] obtained an average potas-

sium-argon age of 14.8 ± 0.5 m.y. from eight samples of glass from the crater, but, if the one determination that gave a very low age is neglected, their average age is 15.0 ± 0.2 m.y. *Fleischer et al.* [1965], using fission track methods on impact glass from the crater, obtained an age of 15.0 ± 1.6 m.y. *Angenheister and Pohl* [1964] later demonstrated the reversed magnetic polarity of the material. It is not impossible that the reversed paleomagnetic polarities of the Steens section and the Nordlinger Ries crater represent the same reversed Upper Miocene geomagnetic epoch or event. Until further data are available, however, and the possibility of systematic laboratory errors are further investigated, this conclusion is, at the very best, only tentative.

CONCLUSIONS

The geomagnetic field changed from reversed to normal polarity 15.1 ± 0.3 m.y. ago.

Sixty successive lavas on Steens Mountain, Oregon, were extruded during a period of not more than 300,000 years.

No significant argon loss was associated with

the occurrence of devitrified glass and slightly altered feldspars in a single lava.

The dating precision obtained suggests that, in the absence of stratigraphic control, polarity events of less than 300,000 years duration occurring approximately 15 m.y. ago cannot at present be readily distinguished from each other when using basalts. This emphasizes the desirability of dating actual polarity transitions, since a reversed to normal polarity is separated from the next reversed to normal polarity boundary by two complete separate polarity epochs or events, and this approach may therefore assist considerably in refining the polarity scale.

Based on analyses of marine magnetic data from mid-oceanic areas in terms of the crustal spreading hypothesis, J. R. Heirtzler (personal communication) predicts a change from reversed to normal geomagnetic polarity 14.8 m.y. ago. Our results do not conflict with this tenuous prediction.

Future work on the geomagnetic polarity time scale will by necessity require replicate analyses of lavas of different polarity but known stratigraphic relationships, rather than of randomly chosen lavas. It may be preferable for the laboratories concerned with this research, if at all possible, to individually compile data for polarity time scale definition covering periods of about 5 m.y. or so, rather than to produce scattered results. In this way possibly significant between-laboratory systematic errors may be detected. Most desirable, of course, is the establishment of interlaboratory basalt standards for the age range 0–20 m.y.

Appendix

The following thin-section commentary was provided by C. T. Harper.

The slides were similar in many respects. They are listed below with the most fine-grained textured slides first (17–4), then the intermediate ones, and finally the coarsest textured slides.

17–4, 17–6. Fine-grained basalt, characterized by equigranular pyroxene, olivene, and opaques. Plagioclase feldspars display poikilitic texture with the mafics. No feldspar phenocrysts present. Olivine slightly altered, but other mafics and feldspars fresh.

11–4, 11–7. Plagioclase feldspars and pyroxenes fresh. Little olivine present. 11–7 slightly more coarse-grained than 11–4. *11–4* has *10–15%* and *11–7 15–20%* of brown devitrified glass.

11–3. A few quite large laths of pyroxenes present. Granular opaques. Equigranular feldspars appear cloudy.

61–7. Medium-grained rock containing zoned feldspar phenocrysts. Medium-sized olivine granules slightly altered to iddingsite. Pyroxenes and feldspars fresh.

51–1, 51–2, 51–5, and 51–6. Apart from increase in grain-size from 1 through 2 and 5 to 6, these slides are very similar in appearance. Olivine slightly altered; pyroxenes and feldspars are fresh.

61–1 and 61–4. Mineralogically similar to 61–7. Olivine slightly altered to iddingsite. Fine interstitial iron ore. Plagioclase mostly fresh.

68–4. Large crystals of olivine, some altered to iddingsite. Plagioclase laths are quite fresh.

70–3 and 70–5. Coarse-grained rock with no plagioclase phenocrysts. Plagioclase laths present. Most of the olivine altered to iddingsite. Large fresh pyroxene grains.

Acknowledgments. Dr. C. T. Harper kindly provided comments on the thin sections, and Mr. W. J. Kenyon gave considerable assistance in the K-Ar laboratory.

The paleomagnetic survey was supported by National Science Foundation grant GF76. The Leverhulme foundation is acknowledged for support during opaque mineralogical examination and Curie point measurements. Support for the K-Ar laboratory was received from The National Research Council of Canada and the Ontario Research Foundation. During the period of this research, A. K. Baksi held a fellowship of the Ontario Provincial Government.

References

Ade-Hall, J., R. L. Wilson, and P. J. Smith, The petrology, Curie points, and natural magnetizations of basic lavas, *Geophys. J., 9,* 323–336, 1965.

Amaral, G., U. G. Cordani, K. Kawashita, and J. H. Reynolds, Potassium-argon dates of basaltic rocks from southern Brazil, *Geochim. Cosmochim. Acta, 30*(2), 159–190, 1966.

Angenheister, G., and J. Pohl, The remanent magnetization of the suevite from the Ries area (southern Germany), *Z. Geophys., 30*(5), 258–259, 1964.

Coe, R. S., Paleo-intensities of the earth's magnetic field determined from Tertiary and Quaternary rocks, *J. Geophys. Res., 72*(12), 3247–3262, 1967.

Collinson, D. W., K. M. Creer, and S. K. Runcorn (Editors), *Methods in Palaeomagnetism*, 609 pp., Elsevier, Amsterdam, 1967.

Cooper, J. A., The flame photometric determination of potassium in geological materials used for potassium-argon dating, *Geochim. Cosmochim. Acta*, 27(5), 525–546, 1963.

Cox, A., and G. B. Dalrymple, Statistical analysis of geomagnetic reversal data and the precision of potassium-argon dating, *J. Geophys. Res.*, 72, 2603–2614, 1967.

Cox, A., D. M. Hopkins, and G. B. Dalrymple, Geomagnetic polarity epochs: Pribilof Islands, Alaska, *Bull. Geol. Soc. Am.*, 77(9), 883–910, 1966.

Dalrymple, G. B., Potassium-argon dates on some Cenozoic volcanic rocks of the Sierra Nevada, California, *Bull. Geol. Soc. Am.*, 74(4), 379–390, 1963.

Dalrymple, G. B., and K. Hirooka, Variation of potassium, argon, and calculated age in a late Cenozoic basalt, *J. Geophys. Res.*, 70(20), 5291–5296, 1965.

Doell, R. R., and A. Cox, Measurement of the remanent magnetization of igneous rocks, *U. S. Geol. Surv. Bull.*, 1203A, 1–32, 1965.

Doell, R. R., and A. Cox, Analysis of alternating field demagnetization equipment, in *Methods in Paleomagnetism*, edited by D. W. Collinson, K. M. Creer, and S. K. Runcorn, pp. 241–253, Elsevier, Amsterdam, 1967.

Evernden, J. F., and G. T. James, Potassium-argon dates and Tertiary floras of North America, *Am. J. Sci.*, 262(8), 945–974, 1964.

Farrar, E., The extraction and ultra-high vacuum mass spectrometry of argon from rocks, Ph.D. thesis, University of Toronto, Toronto, Canada, 1966.

Farrar, E., R. M. Macintyre, D. York, and W. J. Kenyon, A simple mass spectrometer for the analysis of argon at ultra-high vacuum, *Nature*, 204(4958), 531–533, 1964.

Fisher, R., Dispersion on a sphere, *Proc. Roy. Soc. London, A*, 217, 295–305, 1953.

Fleischer, R. L., P. B. Price, and R. M. Walker, On the simultaneous origin of tektites and other natural glasses, *Geochim. Cosmochim, Acta*, 29(3), 161–166, 1965.

Fuller, R. E., The geomorphology and volcanic sequence of Steens Mountain in southeastern Oregon, *Univ. Washington (Seattle) Publ. Geol.*, 3(1), 1–130, 1931.

Gentner, W., H. J. Lippolt, and O. A. Schaeffer, Die Kalium-Argon-Alter der Glässer des Nördlinger Rieses und der böhmischmahrischen Tektite, *Geochim. Cosmochim. Acta*, 27(2), 191–200, 1963.

Hess, H. H., History of ocean basins, *Geol. Soc. Am., Buddington Volume*, 599–620, 1962.

Inghram, M. G., Stable isotope dilution as an analytical tool, *Ann. Rev. Nucl. Sci.*, 4, 81, 1954.

Lanphere, M. A., and G. B. Dalrymple, An interlaboratory standard muscovite for potassium and argon analyses, *J. Geophys. Res.*, 70(14), 3497–3503, 1965.

McDougall, I., and F. H. Chamalaun, Geomagnetic polarity scale of time, *Nature*, 212(5069), 1415–1418, 1966.

Opdyke, N. D., B. Glass, J. D. Hays, and J. Foster, Paleomagnetic study of Antarctic deep-sea cores, *Science*, 154(3748), 349–357, 1966.

Pitman, W. C., and J. R. Heirtzler, Magnetic anomalies over the Pacific-Antarctic ridge, *Science*, 154(3753), 1164–1171, 1966.

Reynolds, J. H., High sensitivity mass spectrometer for noble gas analysis, *Rev. Sci. Inst.*, 27, 928–934, 1956.

Rikitake, T., *Electromagnetism and the Earth's Interior*, 308 pp., Elsevier, Amsterdam, 1966.

Strangway, D. W., and E. E. Larson, A paleomagnetic study of some late Cenozoic basalts from Oregon (abstract), *Trans. Am. Geophys. Union*, 46(1), 66–67, 1965.

Thellier, E., and O. Thellier, Sur l'intensite du champ magnetique terrestre dans le passe historique et geologique, *Ann. Geophys.*, 15, 285–376, 1959.

Vine, F. J., Spreading of the ocean floor: New evidence, *Science*, 154, 1405–1415, 1966.

Vine, F. J., and D. H. Matthews, Magnetic anomalies over oceanic ridges, *Nature*, 199(4897), 947–949, 163.

Walker, G. W., and C. A. Repenning, Reconnaissance geologic map of the Adel quadrangle, Lake, Harney, and Malheur counties, Oregon, *U. S. Geol. Surv. Misc. Geol. Invest.*, *Map 1–446*, 1965.

Watkins, N. D., Behaviour of the geomagnetic field during the Miocene period in south-eastern Oregon, *Nature*, 197(4863), 126–128, 1963.

Watkins, N. D., Paleomagnetism of the Columbia plateaus, *J. Geophys. Res.*, 70(6), 1379–1406, 1965a.

Watkins, N. D., Frequency of extrusions of some Miocene lavas during an apparent transition of the polarity of the geomagnetic field, *Nature*, 206(4986), 801–803, 1965b.

Watkins, N. D., Paleomagnetic evidence from Oregon for a return to shallow 'transitional' geomagnetic inclination following an Upper Miocene geomagnetic polarity change (abstract), *Trans. Am. Geophys. Union*, 48(1), 83, 1967.

Watkins, N. D., and H. G. Goodell, Geomagnetic polarity change and faunal extinction in the Southern Ocean, *Science*, 156, 1083–1087, 1967.

Watkins, N. D., and S. E. Haggerty, Primary oxidation variation and petrogenesis in a single lava, *Contr. Mineral. Petrol.*, 15, 251–271, 1967.

Wilson, R. L., and N. D. Watkins, Correlation of petrology and natural remanent magnetic polarity in the Columbia Plateau basalts, *Geophys. J.*, 12(4), 405–424, 1967.

(Received May 22, 1967.)

Potassium–Argon Dating of Igneous and Metamorphic Rocks

39

12

Paul E. Damon

E *Extraneous argon*

It appears that analytical techniques *per se*, including the air ^{40}Ar correction, are not the ultimate limitation on K–Ar dating of Pleistocene minerals. This ultimate limitation will be set by three factors: (1) retention of ^{40}Ar in xenolithic minerals, (2) contamination of samples by small amounts of older minerals, (3) excess ^{40}Ar in minerals.

Dalrymple (1964) has demonstrated that the feldspars within a xenolith of granite in basalt retained 2 to 5% of their original radiogenic ^{40}Ar. Assuming 0.1% by weight contamination of feldspars from a Precambrian granite originally containing 4×10^{-8} mole/g radiogenic ^{40}Ar (a typical amount for Precambrian feldspars) and 2% retention of this argon, the 30,000 year old date for the hypothetical feldspar sample (table 4) would be increased to 77,000 years. Even 0.01% contamination of such xenolithic feldspars would make the quoted precision meaningless as an estimate of the accuracy of the K–Ar age. Furthermore, only 70 micrograms of a Lower Cambrian feldspar contaminant containing 1.4×10^{-8} mole/g ^{40}Ar would increase the apparent age by 4600 years. This is equivalent to one grain in a 28 to 35 mesh feldspar sample weighing about the same as in our example (12.6 g). Six to seven grains would double the apparent age. These problems, plus the excess ^{40}Ar problem, which will be discussed next, leave little room for optimism concerning the possibility of a meaningful comparison of K–Ar dating with carbon-14 dating. However, it does seem possible to obtain meaningful dates for most of Pleistocene time. That this is by no means an easy task is exemplified by studies of the Bishop tuff (Evernden, 1959; Evernden and others, 1957, 1964; Dalrymple and others, 1965).

When different minerals from the same deposit are dated by the potassium–argon method and by other methods or by stratigraphic dating with reference to the geologic time scale, it occasionally

Table 4 Analysis of hypothetical 30,000 year old feldspar

Weight (gram)	K (%)	^{40}Ar (radiogenic $\times 10^{-12}$ mole)	^{40}Ar (atmospheric $\times 10^{-12}$ mole)	^{40}Ar (atmospheric %)	K–Ar date (year)
12.6	9.48	6.36	20.6	76.3	$30,000 \pm 2400$

happens that one or more minerals may yield discordantly high apparent K–Ar ages. Beryl and cordierite, for example, invariably yield an absurdly high apparent age (Damon and Kulp, 1958b). This can best be explained by reference to the argon radioactive production equation (5) from which a term has been tacitly omitted, i.e. the term expressing the initial amount of argon-40 ($^{40}Ar_0$ present in the mineral at zero time measured from some geologic event for which independent evidence exists). Adding this term, we obtain the following modified equation:

$$^{40}Ar = {}^{40}Ar_0 + \frac{\lambda_K}{\lambda}\,^{40}K[\exp(\lambda t) - 1] \tag{12}$$

When the event to be dated is, for example, a postcrystallization thermal metamorphic event, ^{40}Ar produced in the mineral by radioactive decay and retained from the premetamorphic history of the mineral is said to be 'inherited' ^{40}Ar. Argon-40 retained within xenoliths in igneous rocks which was accumulated by radioactive decay in the xenolith prior to solidification of the igneous rock is also referred to as 'inherited' ^{40}Ar with respect to the time of solidification. On the other hand, ^{40}Ar occluded within a mineral by processes other than radioactive decay in the mineral is referred to as 'excess' ^{40}Ar. The sum of these two components of $^{40}Ar_0$, excess and inherited ^{40}Ar, is commonly referred to as 'extraneous' ^{40}Ar. Obviously, inherited ^{40}Ar cannot always be operationally distinguished from excess ^{40}Ar.

Beryl and cordierite are clathrate-like minerals which contain tubular chains formed by six-membered silica tetrahedron rings. Argon is occluded within these tubular channels (Damon and Kulp, 1958b; Smith and Schreyer, 1962). Argon occluded in beryl and cordierite within pegmatitic or metamorphic environments is very pure radiogenic ^{40}Ar produced in the geologic environment exterior to the minerals and trapped in the minerals at the time of crystallization (Damon and Kulp, 1958b). There is also the possibility that, under confining pressure, environmental ^{40}Ar may diffuse into the minerals subsequent to crystallization.

Although the excess ^{40}Ar phenomenon is most extreme for the cyclosilicates, beryl, cordierite and also tourmaline, it has also been observed in other minerals (table 5). The error in potassium–argon dating will be most evident for minerals in which potassium is not a major component, i.e. pyroxene, plagioclase, fluorite and quartz. Excess ^{40}Ar has been looked for but not observed in the calcium mica, margarite. Fortunately, potassium micas, which are in other

Table 5 Excess [40]Ar in minerals

Mineral	Excess [40]Ar ($\times 10^{-10}$ mole/g)	Reference[a]
Chlorite	<0.009	1
Fluorite with fluid inclusions	0.0 to 0.17	2, 3
Margarite (calcium mica)	<0.45	4
	<0.031	1
Phlogopite (in xenolith)	4.0 and 11.3	5
Albite	0.19 to 0.57	6
Plagioclase	0.15 to 0.35	7
Plagioclase–quartz	0.0 to 1.21	7
Orthoclase	0.0 to 2.71	7
Sodalite	1.74 to 6.68	8
Pyroxene	5 to 50	9, 10
Pyroxene (xenolithic)	0.0 to 0.85	5
Quartz with fluid inclusions	0.016 to 5.4	3
Tourmaline	8 to 52	11
Cordierite	61 to 360	11
Beryl	20 to 14,360	11

[a] 1. Hart (1966); 2. Lippolt and Gentner (1963); 3. Rama and others (1965); 4. Damon and Kulp (1957); 5. Lovering and Richards (1964); 6. Laughlin (1966); 7. Livingston and others (1967); 8. York and MacIntyre (1965); 9. Hart and Dodd (1962); 10. McDougall and Green (1964); 11. Damon and Kulp (1958b).

respects good potassium–argon geochronometers, also usually contain negligible amounts of excess argon. Also, the degassing process upon extrusion of lava appears to be an efficient mechanism for removing most extraneous [40]Ar (Evernden and Curtis, 1965).

It is necessary to view the excess [40]Ar problem in proper perspective in order to evaluate its effect on the accuracy of K–Ar dating (figure 3). In figure 3 a range of concentrations of excess [40]Ar are given for different fractions of 10% K content in minerals, 10%/n K, where $n = 1, 2, 3, 4, \ldots$, i.e. any integral number from one to infinity. It can immediately be seen from table 5 and figure 3 that the cyclosilicates, beryl, cordierite and tourmaline, which in addition to occluding large amounts of excess [40]Ar also have low potassium contents, cannot be used for K–Ar dating. The uncertainty is also very large for low potassium minerals which contain fluid inclusions, e.g. quartz and fluorite (table 5), even for Precambrian dating work. Pyroxenes from intrusive environments also yield high ages even for Precambrian rocks, although in some intrusive igneous rocks, formed under

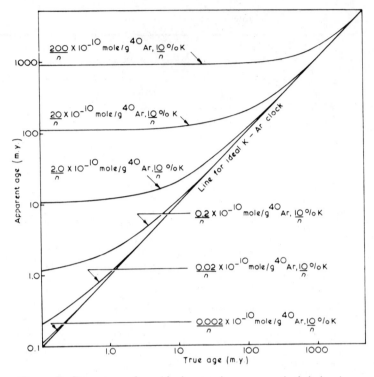

Figure 3 Departures from ideal potassium–argon clock behavior resulting from different amounts of excess ^{40}Ar in minerals of different potassium content (see also table 5). n can be any integer from 1 to ∞ (1, 2, 3, 4, 5, ..., ∞).

conditions of low partial argon pressure or in lavas, K–Ar dates for Precambrian or even younger pyroxenes may quite closely approximate the true cooling age.

It should be pointed out again that in general the degassing process upon extrusion of a magma as lava or pyroclastic ejecta is a relatively efficient process. The extraneous argon retained after extrusion is probably one to two orders of magnitude lower than the preextrusion content (Dalrymple, 1964; Evernden and Curtis, 1965; Damon, 1965). The excess ^{40}Ar data in table 5 are for intrusive igneous or hydrothermal vein environments except when otherwise indicated.

Plagioclase feldspars from intrusive igneous environments also yield maximum potassium–argon ages although for some Precambrian rocks the age may closely approximate those obtained for other

radiometric geochronometers (Livingston and others, 1967). Plagioclase, with a relatively high potassium content, from volcanic rocks
appears to yield reasonable dates for all but Plio–Pleistocene time
(Evernden and others, 1964; Evernden and James, 1964; unpublished
data from this laboratory).

Potash feldspars would be good geochronometers for Paleozoic
and Precambrian dating problems involving intrusive igneous rocks,
but unfortunately pre-Mesozoic feldspars rarely, if ever, quantitatively retain argon (Goldich and others, 1961). The occasionally
high excess ^{40}Ar found in potash feldspars from post-Paleozoic
plutonic rocks by Livingston and others (1967) together with their
relative sensitivity to argon loss does not inspire confidence in K–Ar
ages for these rocks when based upon potash feldspars alone. On the
other hand, sanidine and other potash feldspars associated with volcanic rocks can yield reliable K–Ar dates from late Mesozoic time to
late Pleistocene time (Evernden and others, 1964; Evernden and
James, 1964; Evernden and Curtis, 1965; Doell and others, 1966),
although, as noted above, the uncertainty increases greatly for vol-
canic rocks less than 100,000 years old. The lowest curve $(0.02/n \times 10^{-10}$ mole/g ^{40}Ar, $10\%/n$ K) in figure 3 appears to correspond
approximately to the potash feldspars from volcanic environments.

The phyllosilicates (mica, chlorite, etc.) rarely appear to contain
significant quantities of excess ^{40}Ar even when from the intrusive
igneous environment. This makes the micas ideal potassium–argon
geochronometers for most igneous environments for all but Pleistocene time. The tendency of the phyllosilicates to adsorb relatively
large amounts of air argon frequently results in very large air corrections for very young samples.

Hornblende, an inosilicate, has very high argon retentivity and,
as yet, there is no reliable evidence for excess ^{40}Ar in this mineral
(Hart, 1961). However, quite certainly it will eventually be demonstrated. Excess ^{40}Ar appears to be a ubiquitous phenomenon of
greater or lesser magnitude particularly in intrusive igneous environments. For example, Lovering and Richards (1964, see table 5) have
observed relatively large amounts of excess ^{40}Ar in phlogopite mica
from inclusions in abyssal intrusives.

In order to further increase the accuracy and precision of K–Ar
dating, it will be necessary to study the excess ^{40}Ar problem in
different environments. Igneous rocks from certain environments,
for example the extrusive and to a lesser extent the hypabyssal environments, seem to be suitable for K–Ar dating work. On the other

hand, the problem is magnified for abyssal igneous rocks. In general, the problem seems to be most severe under the following conditions: (1) the mineral has a low content of potassium and a large number of 'holes' such as tubular channels, fluid inclusions, vacancy defects or dislocations; (2) the mineral is crystallized and cooled under high pressure conditions in a dry environment, i.e. under a high partial pressure of ^{40}Ar; and (3) the time between crystallization and dating is so short that excess ^{40}Ar is not negligible in comparison to the argon produced within the mineral by radioactive decay. Or, positively stated, the accuracy of K–Ar dating is greatest for potash minerals with few defects which crystallized or cooled under conditions of low partial ^{40}Ar pressure. Added confidence in the results of K–Ar dating can be attained by the following: (1) dating different minerals with varying potassium contents from the same rock; (2) comparison of the results of K–Ar dating with results by other methods, e.g. the rubidium–strontium and uranium–isotopic lead methods; and (3) using helium-4 as an indicator of excess ^{40}Ar (Damon and Kulp, 1958b; Damon and Green, 1963; see also discussion by Damon of paper by Kulp and Engels, 1963).

There is an interesting age effect which has been observed for both helium and argon in beryl and cordierite. The magnitude of the excess ^{4}He and ^{40}Ar is greater by several orders of magnitude in the very old minerals (>2300 m.y.) than in minerals of Cenozoic age (Damon and Kulp, 1958b; Gerling, 1961). Damon and Kulp (1958b) have concluded that this is not solely the result of deeper burial or the increased rate of production of helium and argon in the past. They have suggested that this effect may have been the result of increased heat production in both the mantle and crust and consequent greater mobilization of the inert gases. Recycling of water, nitrogen and other chemically active volatiles in the crust by sedimentation and metamorphism would maintain a high proportion of these volatiles relative to the inert gases in that environment. On the other hand, the ratio of inert gases to chemically active volatiles in the mantle should constantly increase as a result of degassing because ^{4}He and ^{40}Ar will be continuously replenished by radioactive decay. Thus, mantle gas and crustal gas may have markedly different composition. This effect may possibly help to explain the high contents of excess helium and argon observed in minerals derived from deep-seated environments (Cherdyntsev, 1956, p. 86; Lovering and Richards, 1964; Naughton and others, 1966). This excess inert gas age effect may also result in occasional anomalous K–Ar ages for minerals from Archean rocks although the effect is masked in potash minerals by the large amounts of radiogenic argon produced within the minerals during such a long expanse of time.

*   *   *   *   *   *   *

The Argon-39 Method

XIII

Editor's Comments on Paper 40

By the end of the 1960s the full potential of the potassium–argon method had been realized. Most of the problems encountered involved regional metamorphic rocks, where loss of radiogenic argon from minerals held at elevated temperatures for long periods of time meant that the measured K–Ar ages of metamorphic minerals recorded cooling histories rather than crystallization histories. This problem could be circumvented by dating whole-rock samples of low-grade metasedimentary rocks such as slate and phyllite (Harper, 1967; Dodson and Rex, 1971) which crystallize at moderately low temperatures and are usually not involved in an attenuated cooling history. Unlike radiogenic strontium (p. 245), argon-40 diffusing from potassium minerals at high temperatures is not retained within the whole-rock system, and only in cases where equal concentrations of radiogenic argon have been lost from a cogenetic suite of minerals with different potassium contents can an isochron plot be used to obtain the true age of crystallization (Harper, 1970).

Another problem related to the migration of argon was the discovery of excess argon-40 in certain metamorphic biotites and hornblendes, which invalidated the measured K–Ar ages and indicated that pressure as well as temperature controlled the apparent "retentivity" of any mineral with respect to radiogenic argon-40 (Brewer, 1969; Giletti, 1971; Wilson, 1972).

Imperfect retention of radiogenic argon and/or the presence of excess or inherited Ar-40 may be detected by a variation of the K–Ar method, known as the Ar-39 method, which has recently been applied with great success to meteoritic and lunar materials, and seems likely to supersede the conventional K–Ar for terrestrial materials in the near future. The Ar-39 method, suggested independently by Sigurgeirsson (1962) and Merrihue (1965), was established by Merrihue and Turner in 1966 (Paper 40). It involves the conversion, by irradiation with fast neutrons, of a proportion of the existing K-39 atoms to Ar-39. After irradiation, the Ar-39 remains fixed in the sample until released along with the radiogenic Ar-40 by heating. Provided the proportion of K-39 atoms converted to Ar-39 is known, the age of the sample can be calculated directly from the measured Ar-40/Ar-39 ratio.

As Merrihue and Turner pointed out, the Ar-39 method has two important advantages over the conventional K–Ar method. Both potassium and argon are measured simultaneously by mass spectrometry on the same aliquot of sample, and a knowledge of their absolute abundances is not required. Isotope dilution procedures and the need for a separate potassium analysis can be dispensed with. Furthermore, partial loss of radiogenic argon from nonretentive sites and/or the presence of excess or highly retentive inherited argon in any sample may be revealed by step-heating the irradiated samples and measuring the argon ratios for each step. With increasing temperatures radiogenic argon-40 held in the more retentive sites will be released along with its parent, allowing an argon retention age to be calculated for each aliquot of gas released.

It is too early yet to make an accurate forecast, but the Ar-39 method holds great promise for the future. Further discussion pertaining to this new method may be found in Mitchell (1968), Fitch et al. (1969), Turner (1970), Berger and York (1970), Dalrymple and Lanphere (1971), and Lanphere and Dalrymple (1971).

Selected Bibliography

Berger, G. W., and York, D. (1970). Precision of the Ar-40/Ar-39 dating technique. *Earth Planet. Sci. Lett.,* **9**, 39–44.
Brewer, M. S. (1969). Excess radiogenic argon in metamorphic micas from the Eastern Alps, Austria. *Earth Planet. Sci. Lett.,* **6**, 321–331.
Dalrymple, G. B., and Lanphere, M. A. (1971). Ar-40/Ar-39 technique of K–Ar dating: a comparison with the conventional technique. *Earth Planet. Sci. Lett.,* **12**, 300–308.
Dodson, M. H., and Rex, D. C. (1971). Potassium–argon ages of slates and phyllites from south-west England. *Quart. J. Geol. Soc. London,* **126**, 405–499.
Fitch, F. J., Miller, J. A., and Mitchell, J. G. (1969). A new approach to radioisotope dating in orogenic belts. *In* "Time and Place in Orogeny" P. E. Kent, G. E. Satterthwaite, and A. M. Spencer, eds.), pp. 157–195. Geol. Soc. London, Special Pub. No. 3.
Giletti, B. J. (1971). Discordant isotopic ages and excess argon in biotites. *Earth Planet. Sci. Lett.,* **10**, 157.
Harper, C. T. (1967). The geological interpretation of potassium–argon ages of metamorphic rocks from the Scottish Caledonides. *Scottish J. Geol.,* **3**, 46–66.
Harper, C. T. (1970). Graphical solutions to the problem of radiogenic argon-40 loss from metamorphic minerals. *Eclogae Geol. Helv.,* **63**, 119–140.
Lanphere, M. A., and Dalrymple, G. B. (1971). A test of the ^{40}Ar age spectrum technique on some terrestrial materials. *Earth Planet. Sci. Lett.,* **12**, 359–372.
Merrihue, C. (1965). Trace-element determinations and potassium–argon dating by mass spectroscopy of neutron-irradiated samples. *Trans. Amer. Geophys. Union,* **46**, 125 (abstract).
Mitchell, J. G. (1968). The argon-40/argon-39 method for potassium–argon age determination. *Geochim. Cosmochim. Acta,* **32**, 781–790.
Sigurgeirsson, T. (1962). Dating recent basalt by the potassium–argon method (in Icelandic). Rept. Physical Laboratory of the Univ. Iceland, 9 pp.
Turner, G. (1970). Argon-40/argon-39 dating of lunar rock samples. *Science,* **117**, 466–468.
Wilson, M. R. (1972). Excess radiogenic argon in metamorphic amphiboles and biotites from the Sulitjelms region, Central Norwegian Caledonides. *Earth Planet. Sci. Lett.,* **14**, 403–412.

Potassium-Argon Dating by Activation with Fast Neutrons

4O

CRAIG MERRIHUE[1] AND GRENVILLE TURNER[2]

Department of Physics, University of California, Berkeley

When a potassium-bearing mineral is irradiated by a neutron flux containing a significant fraction of fast neutrons, 270-year Ar^{39} is produced by the K^{39} (n, p) reaction, and this may be used as a basis for measuring the potassium-argon age of the mineral. *Wänke and König* [1959] described such a method in which counting techniques were used to detect the Ar^{39}, as well as Ar^{41} produced by the Ar^{40} (n, γ) reaction. A calculation of the potassium content for a single sample would require a knowledge of the flux-energy distribution in the reactor and the excitation function of the K^{39} (n, p) reaction. The uncertainties of this calculation can be avoided, however, by comparing the (Ar^{40}/Ar^{39}) ratio in the unknown sample with that quantity in a sample of known age, given the same irradiation. In this case we may write down the following relationships.

$$\frac{(Ar^{40}/K^{40})}{(Ar^{40}/K^{40})_s} = \frac{(Ar^{40}/Ar^{39})}{(Ar^{40}/Ar^{39})_s} = \frac{(Ar^{41}/Ar^{39})}{(Ar^{41}/Ar^{39})_s}$$

$$= \frac{(\exp(T/\tau) - 1)}{(\exp(T_s/\tau) - 1)} \quad (1)$$

T is the unknown potassium-argon age, τ is the mean life of K^{40}, and the subscript s refers to the sample of known age.

A correction for atmospheric contamination is not possible with this technique, and more recently *Merrihue* [1965] has suggested using mass spectrometric detection of the Ar^{39} and radiogenic Ar^{40}. The method has the advantage of allowing Ar^{36} to be measured, and, conse-

[1] Dr. Craig Merrihue was killed in a climbing accident on Mount Washington, New Hampshire, on March 14, 1965.
[2] Present address: Department of Physics, Sheffield University, England.

TABLE 1. Potassium-Argon Ages from Irradiation 1[a]

Meteorite	Sample Weight, g	A^{40}/A^{39}	A^{40}/A^{36} This Paper	A^{40}/A^{36} Literature[a]	K-A Age, b. y. This Paper	K-A Age, b. y. Literature[b]
Richardton	0.910	4860 ± 200	1810 ± 60	1850 1910 2050	≡4.32	4.35 4.15[c] 4.47[d]
Abee	0.157	4660 ± 200	178 ± 5	185	4.25	4.71
Holbrook	0.942	4320 ± 150	3340 ± 80	4200 5280	4.12	4.44 4.4[c]
St. Marks	0.354	4110 ± 170	127 ± 3	102	4.04	3.78
Indarch	0.379	4030 ± 150	967 ± 30	892	4.01	4.29
Murray	0.077	3190 ± 180	8.92 ± .25	4.63 3.65 8.54	3.62	1.58 1.90[e] 2.77[d]
Bruderheim (1) (2) See Figure 5 and text.	0.223	742 ± 30	605 ± 20	740	1.61	1.85

[a] Integrated thermal neutron flux, $1.7 \times 10^{19} n/cm^2$.
[b] *Kirsten et al.* [1963].
[c] *Geiss and Hess* [1958].
[d] *Reynolds* [1960].
[e] *Stauffer* [1961].

2852

TABLE 2. Potassium-Argon Ages from Irradiation 2[a]

Meteorite	Sample Weight, g	A^{40}/A^{39}	A^{40}/A^{36}		K-A Age, b. y.	
			This Paper	Literature[a]	This Paper	Literature[b]
Bjurböle	0.590	9850 ± 350	4230 ± 60	1830	≡4.33	4.34
				9150		
				1310		
Pantar light[e]	0.600	9700 ± 900	5600 ± 100		4.30	
Peseyanoe						
(900°C)	0.322	9110 ± 450	27.4 ± .5		4.20	
(1500°C)		9140 ± 450	16.6 ± .4		4.21	
Total		9130 ± 450	19.1 ± .5	18.2	4.21	4.50
				28.0		4.2
				26.8		
Bruderheim[b] (1)	0.855	1850 ± 100	630 ± 50	740	1.86	1.85
(2)	0.88	2860 ± 150	407 ± 15		2.41	

[a] Integrated thermal neutron flux, $5.6 \times 10^{18} n/cm^2$.
[b] *Kirsten et al.* [1963].
[c] Temperature run—see graph.
[d] *Geiss and Hess* [1958].
[e] *Gerling and Levskii* [1956].
[f] *Stauffer* [1961].

quently, a correction for atmospheric contamination can be made in some circumstances. The correction is particularly simple for terrestrial samples under the assumption that all the Ar^{36} is of atmospheric origin, so that the Ar^{40} contamination is just 296 times the Ar^{36} value. In meteorites the situation is more complex because of two additional sources of argon: primordial, present in the material since its solidification, and cosmogenic, produced by cosmic-ray-induced nuclear reactions within the past 100 m.y. or so. For the sake of convenience we shall refer throughout the paper to all argon other than Ar^{39} and radiogenic Ar^{40} as contamination.

This technique offers two important advantages over conventional potassium-argon dating methods, in which argon abundance is measured by mass spectrometry and isotope dilution, and the potassium, by the flame photometry of a 'split' of the sample. First, both potassium and argon are determined simultaneously by a single measurement on the same sample. Secondly, neither potassium nor argon absolute abundances are required. The age is derived from the (Ar^{40}/Ar^{39}) isotope ratio, a quantity that can be measured more accurately than the absolute abundances of either potas-

sium or argon. In principle, therefore, the technique is ideal for dating very small samples, such as single meteorite chondrules, or for particularly rare samples.

The purpose of this letter is to present some preliminary data and to show how the method may be usefully extended by observing variations in the isotopic ratio with temperature during gas release. This extension is similar to the technique of correlated release employed in I-Xe dating. [*Jeffery and Reynolds*, 1961; *Reynolds*, 1963].

The results of some argon measurements carried out routinely in the course of I-Xe dating experiments are shown in Tables 1 and 2, the two tables representing two distinct irradiations. In Table 1, the measurements refer to the total gas released from the meteorite in a single melting of one-hour duration. The K-Ar ages have been calculated relative to the meteorite Richardton, the age of which has been assumed to be 4.32 b.y. on the basis of three earlier determinations [*Kirsten et al.*, 1963]. In future experiments of this kind it might be preferable to include a terrestrial sample of more accurately known age. No correction for air contamination has been made in Table 1 because of the uncertainties already

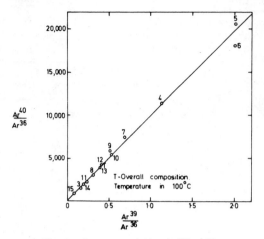

Fig. 1. Argon correlation in Bjurböle.

ent, Ar^{40} and Ar^{39} derived from potassium (subscript k) and Ar^{40} and Ar^{36} present as contamination (subscript c), one may write

$$(Ar^{40}/Ar^{36}) = (Ar^{40}/Ar^{36})_c$$
$$+ (Ar^{40}/Ar^{39})_k(Ar^{39}/Ar^{36}) \quad (2)$$

Thus a plot of (Ar^{40}/Ar^{36}) against (Ar^{39}/Ar^{36}) would be a straight line. For meteorites $(Ar^{40}/Ar^{36})_c$ will not be constant, unfortunately, unless one of the three forms of contamination predominates. However, since this ratio in meteorites will in general lie somewhere between the atmospheric value of 296 and zero, the plot will be, for all practical purposes, a straight line if the observed values of (Ar^{40}/Ar^{36}) are much larger than 296 or than the variations in $(Ar^{40}/Ar^{36})_c$.

The first condition is fulfilled in Figures 1 and 2, and from the two slopes a K-Ar age of (4.30 ± 0.15) b.y. is calculated for Pantar (light) relative to an assumed 4.33 b.y. for Bjurböle [*Kirsten et al.*, 1963]. We should point out that in both cases the highest (Ar^{40}/Ar^{36}) ratio observed was a factor of 5 to 6 times the average or total meteorite value. Thus, even in the case of a meteorite for which the average (Ar^{40}/Ar^{36}) value is low and the contamination correction normally difficult to apply, we would expect that the separation of components brought about by a temperature run would reduce by almost an order of magnitude the uncertainty in this correction.

mentioned in attempting to do this for meteorites. However, (Ar^{40}/Ar^{36}) ratios were measured, and these are presented for comparison with other experimental determinations on unirradiated samples to see whether any systematic change in air contamination is brought about by the irradiation. Richardton, Holbrook, St. Marks, Murray, and Bruderheim show (Ar^{40}/Ar^{36}) ratios closer to the atmospheric value in the irradiated samples than in the unirradiated ones, whereas for Abee, Indarch, Bjurböle, and Peseyanoe the reverse is true. Consequently, no definite conclusions regarding contamination can be drawn. No information can be obtained from Ar^{38}, which is affected by Cl^{37} (n, γ) reactions. Nevertheless, the calculated K-Ar ages seem to be in reasonable agreement with earlier determinations except for the case of Murray. Air contamination is almost certainly the reason for this discrepancy.

Table 2 summarizes the results of more detailed experiments in which the irradiated sample was outgassed at a series of temperatures up to the melting point. For Bjurböle and Pantar (light) this was done in 100°C steps, heating for one hour at each temperature, and Figures 1 and 2 are graphs of (Ar^{40}/Ar^{36}) against (Ar^{39}/Ar^{36}) obtained. The measurement errors on the individual points are typically $\pm 1\text{-}2\%$ in the (Ar^{40}/Ar^{36}) ratio and $\pm 5\%$ in the (Ar^{39}/Ar^{36}) ratio.

If one makes the assumption that two types of gas of distinct composition are pres-

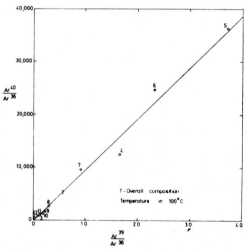

Fig. 2. Argon correlation in Pantar (light).

The possible presence of Ar^{39} from Ca^{42} (n, α) reactions has not been overlooked in these experiments. However, by comparing (Figure 3) the quite different release patterns of Ar^{39} and Ar^{37} (from the Ca^{40} (n, α) reaction), we are able to conclude that a negligible proportion of Ar^{39} has been produced in this way.

A less detailed experiment on Peseyanoe at just two temperatures yielded comparable (Ar^{40}/Ar^{39}) ratios corresponding to a K-Ar age, relative to Bjurböle, or 4.21 b.y. The agreement of the (Ar^{40}/Ar^{39}) ratios for different (Ar^{40}/Ar^{36}) ratios indicates that Ar^{40} air contamination was negligible and the Ar^{36} mainly primordial.

The results of outgassing experiments on Bruderheim are superficially less clear cut and indicate a further avenue for useful research. Diffusion loss of Ar^{40} has occurred in this meteorite whose K-Ar age based on total gas content is 1.85 b.y. according to *Kirsten et al.* [1963]. Because of this, the distribution within the meteorite of radiogenic Ar^{40} and K^{40} may not be the same and the $(Ar^{40}/Ar^{39})_k$ ratio observed in a temperature run would therefore be expected to vary. If the loss of Ar^{40} has been gradual or incomplete, one might intuitively expect the observed $(Ar^{40}/Ar^{39})_k$ ratio to increase monotonically from some value below the average to a value nearer to that corresponding to its 'true' age at higher temperatures. On the other hand, if the loss of Ar^{40} had occurred at a distinct time in the past and

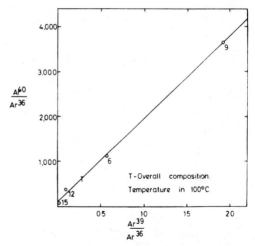

Fig. 4. Argon correlation in Bruderheim (irradiation 2, sample 1).

was complete, the observed $(Ar^{40}/Ar^{39})_k$ ratio would be expected to be constant and to indicate the time of outgassing.

In Figure 4 (Table 2, sample (1)) the results of a shortened outgassing of Bruderheim, with 300°C steps, lie on a straight line with a slope corresponding to 1.86 b.y. An attempt at a more detailed temperature run (Figure 5), however, indicates that this correlation is probably fortuitous, and, in fact, a variation of the (Ar^{40}/Ar^{39}) ratio is observed in a fashion consistent with incomplete degassing of the meteorite at some time in the past. While the overall (Ar^{40}/Ar^{39}) ratio in Figure 5 corresponds to 1.7 b.y., the 'age' of the various fractions rises from less than 1 b.y. at the low temperatures to more than 3 b.y. at higher temperatures. We can be more specific in an interpretation of this variation if we consider the possibility that partial outgassing occurred (or ended) at some definite time in the past. At this time the (Ar^{40}/K^{40}) ratio would be expected to be zero at the surface of mineral grains and practically zero for some depth into the grains. Subsequently, accumulation of Ar^{40} would lead to an (Ar^{40}/K^{40}) ratio in these surface regions corresponding to the time of outgassing, and the argon released at low temperatures in the present experiments should indicate this ratio. While stressing the preliminary nature of our results, we might point out that in Figure 5 the 200°C, 300°C, 400°C, and

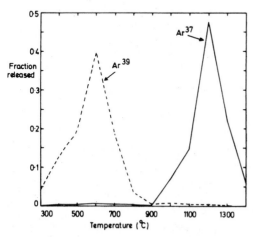

Fig. 3. Release of pile-produced argon from Bruderheim (irradiation 1, sample 2).

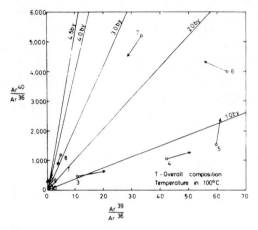

Fig. 5. Argon release from Bruderheim (irradiation 1, sample 2).

500°C points, together with the atmospheric point (A), lie on a line whose slope corresponds to 0.55 ± .05 b.y. This is particularly interesting in view of the observed tendency for (U, Th)—He ages of hypersthene chondrites to cluster close to 0.5 b.y. [*Anders*, 1964], possibly indicating a major parent body collision around this time. A similar interpretation of the present data is at this stage speculative, although one may certainly produce arguments to justify it. There is, for instance, a strong tendency for the loosely held atmospheric contamination to be released in the initial stages of an outgassing experiment, and we would therefore expect $(Ar^{40}/Ar^{36})_c$ to be 296 in the low temperature release, particularly in an ordinary chondrite with its very low primordial gas content. We may also set an upper limit of 0.7 b.y. on the age of the 400°C and 500°C gas, by drawing a line through these points and the origin, $(Ar^{40}/Ar^{36})_c = 0$. We expect $(Ar^{40}/Ar^{39})_k$ to increase monotonically with gas release so 0.7 b.y. is an upper limit for the 300°C point also. More detailed experiments on the hypersthene chondrites are in progress to check our initial findings.

In view of the above discussion, it is apparent that experiments of this kind could provide an extremely useful tool for investigating the thermal histories of both meteorites and terrestrial minerals. We are at present investigating the possibilities of a more detailed quantitative interpretation.

These preliminary experiments have indicated the potential usefulness of the correlated release method used in conjunction with neutron activation of potassium bearing minerals or groups of minerals. Very small samples may be used, and, in the case of meteorites, the uncertain effects of argon from other sources reduced by a large factor. There exists also the exciting possibility of deducing information of the past thermal history of the sample. For terrestrial materials with moderate or low air contamination simple one-shot ratio determinations on a small specimen would be sufficient to give the K-Ar age. The limit of detection of Ar^{39} in a mass spectrometer is set by the presence of the much larger adjacent Ar^{40} peak, the tail of which tends to obscure the Ar^{39}. The problem of producing sufficient Ar^{39} for detection in this way should therefore be much easier for terrestrial samples with lower K-Ar ages. For extremely young terrestrial samples with large air contamination, the separation of air and radiogenic argon in a many temperature outgassing experiment could be profitable in lowering by almost an order of magnitude the detectable (Ar^{40}/K^{40}) ratio.

Acknowledgments. We wish to express our thanks to Dr. J. H. Reynolds in whose laboratory at the University of California, Berkeley, this work was carried out. The neutron irradiations were received through the courtesy of Dr. Oliver Schaeffer and the Brookhaven National Laboratory.

The work was supported in part by the U. S. Atomic Energy Commission.

REFERENCES

Anders, E., Origin, age, and composition of meteorites, *Space Sci. Rev., 3*, 583–714, 1964.
Geiss, J., and D. C. Hess, Argon-potassium ages and the isotopic composition of argon from meteorites, *Astrophys. J., 127*, 224–236, 1958.
Gerling, E. K., and L. K. Levskii, Origin of inert gases in stone meteorites, *Dokl. Akad. Nauk SSSR, 110*, 750–754, 1956.
Jeffery, P. M., and J. H. Reynolds, Origin of excess Xe^{129} in stone meteorites, *J. Geophys. Res., 66*, 3582–3583, 1961.
Kirsten, T., D. Krankowsky, and J. Zähringer, Edelgas- und Kalium- Bestimmungen an einer grösseren Zahl von Steinmeteorites, *Geochim. Cosmochim. Acta, 27*, 13–42, 1963.
Merrihue, C. M., Trace-element determinations and potassium-argon dating by mass spectroscopy of neutron-irradiated samples (abstract), *Trans. Am. Geophys. Union, 46*, 125, 1965.

Reynolds, J. H., Isotopic composition of primordial xenon, *Phys Rev. Letters, 4,* 351–354, 1960.
Reynolds, J. H., Xenology, *J. Geophys. Res., 68,*
2939–2956, 1963.
Stauffer, H., Primordial argon and neon in carbonaceous chondrites and ureilites, *Geochim.
Cosmochim. Acta, 24,* 70–82, 1961.

Wänke, H., and H. Konig, Eine neue Methode
zur Kalium-Argon-Alterbestimmung und ihre
Anwendung auf Steinmeteorite, *Z. Naturforsch.,
14a,* 860–866, 1959.

(Received December 30, 1965;
presentation revised February 18, 1966.)

Current Applications

XIV

Editor's Comments on Papers 41, 42, and 43

The last three papers selected for this volume need little introduction. They represent the culmination of over sixty years of research with radiometric age methods, their application and interpretation.

In the first of three (Paper 41), Stephen Moorbath of Oxford University discusses the geologic interpretation of radiometric ages obtained by each of the three principal methods in use today, the K–Ar, the Rb–Sr, and the U–Pb methods. He points out that the analytical precision associated with a radiometric age is no guarantee of its validity in terms of geological processes. Geochemically, radiometric ages are related to the time of closure of the analyzed system with respect to migration of the radiogenic daughter and its parent. Although experimental evidence is scarce, it seems probable that radiogenic accumulations of geochemically different daughter elements are initiated under quite different environmental conditions. Radiometric ages determined by different methods on the same sample are therefore likely to be discordant.*

Assuming that any daughter element incorporated in the analyzed sample at the time of closure can be recognized, the main difficulty associated with the interpretation of radiometric data, as Moorbath points out, is the problem of correlating the measured time of closure with a geological event. Diffusion characteristics, in particular the exponential variation of the diffusion coefficient D with temperature, indicate that closure will take place over a very narrow temperature range,† so that with high-level intrusive rocks and volcanic materials which have cooled rapidly after crystallization the measured time of closure may be equated without error to the time of crystallization.

---

*Detailed studies of the discordant age patterns associated with the thermal effects of contact metamorphism have been reported by Hart et al., 1968; and Hanson and Gast, 1967.
†Sometimes referred to as the "blocking temperature."

On the other hand, with deep-seated plutonic igneous rocks and regional metamorphic rocks which may have remained at elevated temperatures at depth for significantly long periods of time, the measured time of closure to radiogenic daughter migration may be more closely related to post-crystallization events, such as epeirogenic uplift and cooling.

Another problem in interpretation involves the question of partial loss of the accumulating radiogenic daughter elements at some time subsequent to closure. In the case of the uranium–lead method this will result in internally discordant Pb-207/U-235 and Pb-206/U-238 age data, which, using the concordia plot (Papers 19, 20), can often be extrapolated back to yield the time of initial closure, provided cogenetic systems with sufficient variation in uranium content can be found. Partial loss of radiogenic strontium from rubidium-rich minerals has been found to be associated in many cases with isotopic homogenization on a larger scale (Paper 26), so that the time of initial closure can be found by analyzing larger (whole-rock) aliquots of the unit involved. Disturbances of the K–Ar system is more difficult to correct for, unless the same concentration of radiogenic argon is lost from a cogenetic suite of samples with different potassium contents (see p. 394). However, radiogenic argon held in the most retentive sites in a sample can be analyzed separately, as with the Ar-39 spectrum technique (Paper 40).

In his report Moorbath concludes that with careful selection of sample and dating technique, measured closure times can in many cases be successfully correlated with recognizable geological events, and that discordant age patterns, so often encountered in orogenic belts, do have real meaning in terms of the thermal histories of the regions involved (for example, see Dewey and Pankhurst, 1970).

The last two papers included here summarize radiometric age patterns on a global scale using data obtained from the only two planetary bodies sampled by man, the earth and the moon. In their report on continental radiometric ages (Paper 42) dated July, 1968, Hurley and Rand provide a comprehensive review of all the available age data covering about 68 per cent of the continental crust of the earth, together with a selected bibliography. They do not attempt to reconstruct the original positions of the continents,* but point to what appears to be evidence for an increase in the rate of formation of stable continental crust with time, acknowledging the fact that measured radiometric ages from the continents may not be exactly equated with the time of formation of the continental crust (due to geologic recycling of crustal material). In the first 450 million years of recorded time, a total area of a little more than one million square kilometers of continental crust was stabilized, whereas in the last 450 million years more than 38 million square kilometers formed. To a first approximation a linear increase in the growth rate is indicated, with little evidence for episodic activity.†

---

*This was done in two subsequent publications; see Hurley and Rand, 1969; Hurley, 1970.
†Compare Gastil, 1960; Sutton, 1963; Dearnley, 1966; Fitch and Miller, 1965.

In the final paper (Paper 43), George Wetherill, now at the Department of Planetary and Space Science, University of California in Los Angeles, summarizes the geochronologic data obtained on the first two successful manned flights to the moon, Apollo 11 and 12. The age of formation of the earth and moon are found to be similar. Genesis occurred between 4,500 and 4,600 million years ago. Wetherill emphasizes that information regarding the origin of the earth and the solar system and a record of the earliest events can only be obtained from the moon, where much of the evidence for the earliest events is still preserved. In contrast to the moon, much of the earth's crust is made up of reworked material, and evidence for the early history has consequently been lost. Interpretation of the radiometric and geochemical data from the moon is still controversial, but one thing is certain: following the formation of the mare basalts, between 3.3 and 3.7 billion years ago, there has been little crustal activity on the moon, with the obvious exceptions of formation of some lunar craters and solar and cosmic-ray bombardment.

Preliminary reports on samples collected on the successful Apollo 14 and 15 missions have since appeared (see *Science,* **173**, No. 3998, and **175**, No. 4020), and more detailed information regarding the Apollo 11 and 12 samples is contained in the proceedings of the Lunar Science Conferences. Many reports on the geochronology of lunar samples have appeared in recent issues of *Earth and Planetary Science Letters,* and we may expect the geochronology of the moon and other planets (Mars?) to occupy the attention of many geochronologists in the future. A recent conference held in Philadelphia sponsored by the US National Aeronautics and Space Administration (NASA) considered the possibility of undertaking radiometric age determinations on other planets by remote control. Geochronologists participating at the conference concluded that, although more information could be extracted from samples returned to earth, remote-controlled radiometric age determinations by unmanned landing vehicles was quite feasible. The future for geochronology looks bright ahead.

Selected Bibliography

Dewey, J. F., and Pankhurst, R. J. (1970). The evolution of the Scottish Caledonides in relation to their isotopic age pattern. *Trans. Roy. Soc. Edinburgh,* **68**, 361–389.

Dearnley, R. (1966). Orogenic fold-belts and a hypothesis of earth evolution. *In* "Physics and Chemistry of the Earth" (L. H. Ahrens et al., eds.), Vol. 7, pp. 1–114. Pergamon Press, New York.

Fitch, F. J., and Miller, J. A. (1965). Major cycles in the history of the earth. *Nature,* **206**, 1023–1027.

Gastil, G. (1960). The distribution of mineral dates in time and space. *Amer. J. Sci.,* **258**, 1.

Hanson, G. N., and Gast, P. W. (1967). Kinetic studies in contact metamorphic zones. *Geochim. Cosmochim. Acta.,* **31**, 1119–1154.

Hart, S. R., Davis, G. L., Steiger, R. H., and Tilton, G. R. (1968). A comparison of the isotopic mineral age variations and petrologic changes induced by contact metamorphism. *In* "Radiometric Dating for Geologists" (E. I. Hamilton, and R. M. Farquhar, eds.), pp. 73–110. Interscience, London.

Hurley, P. M. (1970). Distribution of age provinces in Laurasia. *Earth Planet. Sci. Lett.,* **8**, 189–196.

Hurley, P. M., and Rand, J. R. (1969). Pre-drift continental nuclei. *Science,* **164**, 1229–1242.

Sutton, J. (1963). Long-term cycles in the evolution of continents, *Nature,* **198**, 731–735.

Earth-Science Reviews – Elsevier Publishing Company, Amsterdam – Printed in The Netherlands

RECENT ADVANCES IN THE APPLICATION AND INTERPRETATION OF RADIOMETRIC AGE DATA

41

S. MOORBATH

Department of Geology and Mineralogy, University of Oxford, Oxford (Great Britain)

SUMMARY

Strictly speaking, radiometric dates on minerals and rocks relate to a time and temperature, during cooling, when diffusion of radiogenic nuclide out of the system ceased, and not to the time of crystallization. In deep-seated igneous and metamorphic terrains there may be a significant time-lag between crystallization and termination of a diffusion episode. This review is mainly concerned with the different types of concordant and discordant mineral and whole rock age patterns produced in terrains with a simple cooling history and in those which have been subjected to more than one thermal episode. This leads to a discussion of the true geological significance, if any, of radiometric dates obtained on different types of minerals and whole rocks by the various age methods, as well as the validity of histograms in geochronology. Several published geochronological case histories are described, including the British Caledonides, the western Alps, the Appalachian Belt of North America and the eastern Great Basin of Utah and Nevada in the western United States, which serve to illustrate these principles.

INTRODUCTION

It is evident from the ever increasing number of published radiometric age determinations that most geologists and geochronologists are still paying insufficient attention to the exact significance of geochronometric data. In the great majority of cases it is simply assumed that the frequently very precise *analytical dates* relate to major geological events such as intrusion, metamorphism, orogeny, etc. This assumption may be justifiable for high-level intrusive rocks, as well as for extrusives; whether it holds for analytical dates from plutonic igneous rocks or from regionally metamorphosed terrains is much more questionable. A good deal of evidence, to be reviewed in this article, suggests that analytical dates from such rock units must be treated with great caution, since it can be demonstrated in some cases that they do not relate to the time of crystallization of the analysed mineral or rock; rather do they date some other event in its thermal history. In the case of discordant mineral ages from polymetamorphic terrains, analytical dates may have no meaning at all in terms of time.

The basic precept of this article is that for igneous and metamorphic rocks, which have had a simple history since the time of crystallization, analytical dates represent the time when the temperature of the system passed a threshold value, above which atoms of radiogenic daughter nuclides were able to diffuse quantitatively out of their host mineral or rock, but below which they were quantitatively retained. This is a rather naive oversimplification of the laws governing the diffusion of foreign atoms and ions through crystal lattices, but it probably holds good to a first approximation for the purposes of geochronometry. For most minerals this threshold temperature is certainly several hundred degrees centigrade below that of magmatic or metamorphic crystallization. Furthermore, no reasonable doubt can remain that frequently the attainment of the threshold temperature may post-date the time of magmatic or metamorphic crystallization by many millions of years. The existence of this important time-lag is now beginning to be more widely appreciated. Unless geologists and geochronologists squarely face up to the importance of the correct interpretation of analytical dates, the whole subject is in danger of deteriorating into an elaborate number-game. However, there are no grounds for undue pessimism, because certain methods have been introduced in recent years where the analytical dates often do approximate rather closely to the true age of crystallization and, hence, to major geological events. These methods include Rb/Sr on whole rocks, U/Pb on zircons and, in some cases, K/Ar and Rb/Sr on certain types of minerals *other* than fine-grained micas.

If one is to claim with any confidence that the analytical date from a mineral or rock indicates the time of crystallization, the following conditions are necessary: (*1*) that the decay constant of the radioactive nuclide is accurately known; (*2*) that proper correction has been made for the amount of radiogenic nuclide (if any) incorporated into the mineral or rock at the time of crystallization; (*3*) that there have been no gains or losses of parent or daughter nuclide in the mineral or rock since the time of crystallization by processes other than radioactive decay.

Conditions *1* and *2* have figured prominently in the geochronological literature of recent years and it may be stated that, for all intents and purposes, they can be satisfactorily met despite an annoying little uncertainty of 6% in the decay constant of ^{87}Rb. However, condition *3* is usually taken for granted without further discussion. It is with certain important aspects of the failure of condition *3* that this article is principally concerned—in particular, the diffusive loss of radiogenic nuclide after crystallization.

Detailed descriptions of the U/Pb, Rb/Sr and K/Ar methods of age determination including analytical techniques, methodology, applicability, etc., have been dealt with in numerous research papers of the past 10 years. No attempt is made in this article at a comprehensive survey of all this information. Instead,

Earth-Sci. Rev., 3 (1967) 111–133

reference is made at this stage to a number of recent review papers and books which, between them, cover much of the field (HAMILTON et al., 1962; TILTON and HART, 1963; HARLAND et al., 1964; HAMILTON, 1965; MOORBATH, 1965; CAHEN and SNELLING, 1966; MCDOUGALL, 1966; SCHAEFFER and ZÄHRINGER, 1966). In addition, several further text-books on geochronometry and geochronology are in the process of publication.

DIFFUSION OF RADIOGENIC NUCLIDES IN MINERALS

From the time that it became evident that different minerals from a given rock may give discordant radiometric ages, many workers have studied the diffusion of radiogenic argon in some of the commoner rock-forming minerals. Such studies, as will be seen later, can give a close insight into some of the more commonly observed radiometric age patterns. A review of argon diffusion studies has recently been published by FECHTIG and KALBITZER (1966). Although there are still fairly large differences in the measurement of diffusion constants and activation energies on any particular mineral amongst different research groups, certain generalizations of geochronological importance are possible. A summary of the most relevant experimental diffusion measurements is given in Table I. The data offer a rough guide to argon retentivities of different minerals in a geological environment. For example, the higher the value of the activation energy, the more firmly is the argon retained.

In some minerals, such as the potassium feldspars (other than sanidine), several reservoirs of radiogenic argon exist—all energetically different from each other—some with exceptionally low activation energies. Orthoclase and microcline can lose argon quite readily at atmospheric temperatures due to the existence of gross structural defects in the form of dislocations and grain boundaries, resulting from exsolution and alteration phenomena. SARDAROV (1957), for example, demonstrated that the apparent K/Ar age decreases with increasing degree of perthitisation. The unsuitability of orthoclase and microcline for K/Ar age work has been known for many years, since they almost always yield anomalously low K/Ar ages when compared with coexisting micas (WETHERILL et al., 1955; FOLINSBEE et al., 1956; GENTNER and KLEY, 1956; GOLDICH et al., 1957; REYNOLDS, 1957). However, in the majority of minerals in which potassium is homogeneously distributed as a major constituent, practically all the argon can escape by volume (lattice) diffusion only, and all workers are agreed that diffusion at atmospheric temperatures is so small that no significant argon loss can occur. Since the activation energies usually exceed 40 kcal./mole, it follows that the diffusion parameter D/a^2 (diffusion constant divided by the square of the crystal radius, or other significant diffusion dimension) is very sensitive to changes of temperature, especially in the lower temperature ranges. This means that, for all practical purposes,

Earth-Sci. Rev., 3 (1967) 111–133

TABLE I

EXPERIMENTALLY DETERMINED ARGON DIFFUSION CONSTANTS D OR DIFFUSION PARAMETERS D/a^2 AT ROOM TEMPERATURE AND ACTIVATION ENERGIES E FOR MINERALS OF GEOCHRONOLOGICAL IMPORTANCE

Minerals	Reference	$D, D/a^2$	E (Kcal./mole)
Micas			
Biotite	GERLING and MOROZOVA (1957)		57
Biotite	GERLING et al. (1963)		69, 48, 33
Biotite	FRECHEN and LIPPOLT (1965)	$(D/a^2)_{20°C} = 10^{-29}/sec$	69
Biotite	FRECHEN and LIPPOLT (1965)	$(D/a^2)_{20°C} = 10^{-34}/sec$	86
Muscovite	GERLING et al. (1961)		72, 37
Muscovite	GERLING and MOROZOVA (1957)		85
Muscovite	BRANDT and VORONOVSKY (1964)		56
Phlogopite	EVERNDEN et al. (1960)	$D_{20°C} = 10^{-29} cm^2/sec$	28
Phlogopite	GERLING and MOROZOVA (1957)		67
Feldspars			
Microcline	GERLING and MOROZOVA (1962)		130, 99, 42, 26, 15
Microcline	GERLING et al. (1963)		100, 46, 32
Microcline	EVERNDEN et al. (1960)	$D_{20°C} = 10^{-26} cm^2/sec$	24
Microcline	BAADSGAARD et al. (1961)	$D_{25°C} = 10^{-20} cm^2/sec$	18
Anorthite	FECHTIG et al. (1960)	$D_{20°C} = 10^{-30} cm^2/sec$	56
Sanidine	FECHTIG et al. (1961)	$(D/a^2)_{20C°} = 2·10^{-28}/sec$	40
Sanidine	BAADSGAARD et al. (1961)	$D_{25°C} = 10^{-38} cm^2/sec$	52
Sanidine	FRECHEN and LIPPOLT (1965)	$(D/a^2)_{20°C} = 4·10^{-28}/sec$	41
Sanidine	FRECHEN and LIPPOLT (1965)	$(D/a^2)_{20°C} = 3·10^{-33}/sec$	48
Pyroxenes			
"Pyroxene"	AMIRKHANOV et al. (1959)	$D_{25°C} = 4·10^{-55} cm^2/sec$	74
Augite	FECHTIG et al. (1960)	$(D/a^2)_{25°C} < 10^{-30}/sec$	83
Amphiboles			
Pargasite	GERLING et al. (1965)		> 200
"Amphibole"	GERLING et al. (1965)		172 ± 40
"Amphibole"	GERLING et al. (1965)		140 ± 20
"Amphibole"	GERLING et al. (1965)		136 ± 30 / 135 ± 45
Riebeckite	GERLING et al. (1965)		120 ± 30 / 110 ± 15

the temperature interval above which argon can diffuse nearly quantitatively out of a mineral and below which it is nearly quantitatively retained is only about 50–100°C. Thus, from their experimental results EVERNDEN et al. (1960) find that for a phlogopite mica grain 1 mm thick, the relation between the fraction of argon lost (F) and temperature (t) maintained for 100,000 years is as follows: $t = 20°C, F = 4 · 10^{-6}; t = 50°C, F = 4 · 10^{-5}; t = 95°C, F = 4 · 10^{-4}; t = 150°C, F = 4 · 10^{-3}; t = 225°C, F = 4 · 10^{-2}; t = 320°C, F = 0.4$. Evernden et al. conclude that, with phlogopite mica, regional metamorphism of low grade (350°C) should destroy the possibility of dating events prior to metamorphism. A similar calculation by LAMBERT (1964) for a highly retentive Ordovician sanidine gave

Earth-Sci. Rev., 3 (1967) 111–133

the following results: $t = 100\,^{\circ}\mathrm{C}$, $F = 3 \cdot 10^{-3}$; $t = 150\,^{\circ}\mathrm{C}$, $F = 0.1$; $t = 200\,^{\circ}\mathrm{C}$, $F = 0.24$; $t = 250\,^{\circ}\mathrm{C}$, $F = 0.85$.

The actual temperature at which different minerals become effectively closed systems to argon in nature is not known with any accuracy, but probably does not exceed about 200–300 °C for muscovite and sanidine (see above). For biotite, the diffusion threshold temperature for geochronometrically significant argon loss is generally considered to be as low as 150–200 °C.

The use of amphiboles for K/Ar dating has become fashionable in recent years as a result of greatly improved analytical techniques for measuring small amounts of argon and potassium. Activation energies for argon diffusion in amphiboles have recently been published by GERLING et al. (1965), and are summarised in Table I. The values are considerably higher than for micas and it is concluded from the laboratory data that argon will be quantitatively retained in amphiboles at temperatures below about 500 °C for a heating period of 100 million years. There also appears to be a relationship between the activation energy in amphiboles and their chemical composition. The magnesium-rich amphiboles have higher activation energies than the iron-rich ones.

The high-temperature potassium feldspar sanidine is suitable for K/Ar age work and has been used extensively in recent years, particularly for the dating of young, volcanic rocks (BAADSGAARD et al., 1961; BAADSGAARD and DODSON, 1964; BAADSGAARD et al., 1964; MCDOUGALL, 1966).

Pyroxenes, too, have reasonable activation energies (Table I), but have not been widely used for K/Ar dating. This is partly because of their extremely low contents of K and Ar. Furthermore, pyroxenes from high-grade metamorphic complexes occasionally yield anomalously high ages—sometimes greater than the age of the earth—since they appear to be able to incorporate argon during crystallization (HART and DODD, 1962; MCDOUGALL and GREEN, 1964; EVANS, 1965). As the structural sites are not large enough to admit argon, Hart and Dodd postulate that the excess argon is held in crystal imperfections such as dislocations and grain-boundaries. There is, as yet, no absolutely conclusive evidence for the presence of excess argon in other minerals of geochronological interest. If it exists it is presumably swamped by the much greater amount of radiogenic $^{40}\mathrm{Ar}$ formed within the crystal lattice of most other types of minerals commonly used for dating.

Very little, if any, experimental work has been published on the diffusion of strontium in minerals. However, there is plenty of indirect geological evidence from Rb/Sr age studies (see below) that the diffusion behaviour of radiogenic strontium is very similar to that of radiogenic argon.

Another important and fruitful approach for the understanding of relevant diffusion processes is the study of discordant mineral age patterns produced in a geological environment by a given thermal impulse. HART (1964) studied discordant mineral patterns produced in 1,300 m.y. (million years)-old basement rocks in Colorado, U.S.A., in the vicinity of a large, granitic intrusive (the Eldora stock)

Earth-Sci. Rev., 3 (1967) 111–133

intruded just over 50 m.y. ago. Minerals from the basement rock were sampled up to a distance of about 20,000 ft. from the igneous contact, and the apparent ages determined by both the K/Ar and Rb/Sr methods. Some of the results are shown in Fig.1, taken from a paper by KULP (1963), which combines Hart's work with that of Kulp and his co-workers (e.g., CATANZARO and KULP, 1964). It is evident that the zircon ^{207}Pb/^{206}Pb age is least affected, whilst the hornblende K/Ar age is rather intensive to the heating effect, changing significantly only within about 100 ft. from the contact. Coarse muscovite Rb/Sr is next in order of resistance to the thermal effect. K/Ar and Rb/Sr ages from coarse, pegmatitic biotites increase steadily with distance from the contract, but the Rb/Sr ages are greater than the K/Ar ages at all points. Only beyond about 10,000 ft. from the contact were the ages unaffected by the intrusive, yielding characteristic basement dates of about 1,300 m.y. The important effect of grain size is shown by the K/Ar curve for fine-grained biotites from the country rock schists, which do not reach the true basement date even within about 20,000 ft. from the contact. HART (1964) also presented feldspar K/Ar data from the contact zone, but for various reasons (see above) these are extremely erratic and always much lower than the true age. On the other hand, Hart found that feldspar Rb/Sr ages approach the true basement age value only about 80 ft. from the igneous contact. The feldspar Rb/Sr ages, however, were analytically somewhat imprecise due to low radiogenic ^{87}Sr enrichment.

The observed discordant pattern of apparent ages within the contact zone of the Eldora stock is closely analogous to those observed in other geological environments, particularly in those that have undergone a post-crystallization metamorphism. Quite commonly, part or all of the following sequence of ap-

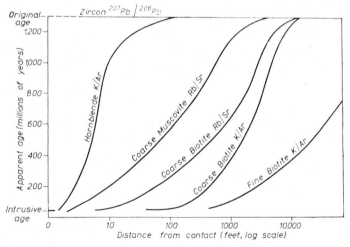

Fig.1. Effects of a 50 million-year old, 10,000 ft. diameter, intrusive on 1,300 million-year-old minerals up to a distance of about 20,000 ft. from the intrusive contact. Based mainly on the work of HART (1964) and also on work by KULP (1963) and his associates (by kind permission of Professor J. L. Kulp).

Earth-Sci. Rev., 3 (1967) 111–133

parent ages is actually observed in such an environment: U/Pb (zircon) > Rb/Sr (potash feldspar) > K/Ar (hornblende) > Rb/Sr (muscovite) > Rb/Sr (biotite) ≥ K/Ar (biotite) > K/Ar (potash feldspar). Very detailed work of this type has recently been published by ALDRICH et al. (1965).

It is of great interest to note that in the case of the Eldora stock, samples at a distance greater than only about 10 ft. from the contact exhibit no petrographical, mineralogical or textural characteristics clearly related to the contact metamorphism, except for a triclinic–monoclinic inversion of potash feldspar within about 1,000 ft. of the contact. This emphasises what is so frequently observed in geochronological studies, namely that the most sensitive and sometimes the only indicator and guide to the thermal history of a suite of rocks is provided by the observed age pattern.

HART (1964) also investigated the effect of grain size in determining the losses of radiogenic nuclide. Simple diffusion theory shows that fractional loss of radiogenic isotope from a crystal at a given temperature is directly proportional to the square root of the diffusion constant and of the time, and inversely proportional to the significant diffusion dimension. Hart studied the radial distribution of Rb and radiogenic ^{87}Sr in large flakes of biotite about 250 ft. from the contact of the Eldora stock, where the heating effect of the intrusion has strongly influenced the ages (Fig.1). Whilst Rb was essentially uniform throughout the grain, the content of radiogenic ^{87}Sr decreased sharply from the centre to the edge, confirming that the low Rb/Sr age was due to loss of radiogenic isotope and not to gain of the radioactive parent element, and that diffusion was controlled by the actual grain size of the particle and not by sub-microscopic factors. Theoretical calculations based on these observations and on a reasonable heat-flow model indicate that the activation energies for Sr and Ar diffusion in the range 80–550 °C are essentially equal (≯ 30 kcal./mole), but that the diffusion constant for Ar is five times larger than that for Sr. This explains the difference between the biotite K/Ar and Rb/Sr age curves in the contact zone (Fig.1).

From studies of diffusion of argon from the biotite of a mica schist due to intrusion of a dyke, WESTCOTT (1966) obtained an activation energy of 40 kcal./mole. HURLEY et al. (1962) give a geologically determined value of 27 kcal./mole. From published values of the activation energy of argon, it appears that laboratory values are usually higher than geologically determined values. Discussion of possible reasons for this is outside the scope of this article.

GENERAL DISCUSSION OF OBSERVED MINERAL AGE PATTERNS AND THEIR GEOLOGICAL SIGNIFICANCE

HURLEY et al. (1963) suggested that the very late stages in the history of a mountain belt, such as uplift and cooling, may be of great importance for obtaining a truer understanding of the significance of radiometric dates. After an orogenic uplift is completed, with base-level peneplanation and restoration of gravitational equilibrium, rocks exposed at the surface may well represent original depths of up to 20 km or more. Clearly, the geological relationships mapped in such a belt occurred at depth in regions which may have been at elevated temperatures for a long period of time. From calculations of probable rates of uplift in an orogenic belt combined with plausible thermal gradients, Hurley et al. concluded that most shield areas that are base-levelled could have remained at a depth greater than 10 km for 100 million years, and above 150 °C for 200 million years. Taking into account the kind of diffusion parameters discussed earlier, Hurley et al. furthermore suggest that an error of about 100 million years on the low side would not be surprising in age measurements on minerals from basement complexes because of diffusion of radiogenic isotopes during cooling, and that this error may even amount to several hundred million years if some of the more recent indications of low-temperature loss are found to be more general. Of particular relevance is HURLEY et al.'s (1962) discovery of very low K/Ar and Rb/Sr ages (< 5 m.y.) in biotite-gneisses of probable Jurassic age in the centre of the Recent uplift zone of the New Zealand Alps.

As a result of the foregoing discussions, we may now consider some geo-chronological situations which could (and do) actually arise in a geological environment, with special reference to extensive, deep-seated basement complexes. Some geological case histories will be described later.

Basement complex with a one-stage thermal history

Only one major thermal event—such as regional metamorphism—is envisaged as having occurred. As regards mineral age patterns, there are two possibilities, the first of which is merely the limiting condition of the second:

(*1*) All mineral ages from the basement complex may be concordant within the analytical error, as a result of sufficiently rapid cooling for all minerals to become closed systems for the radiogenic nuclide(s) at the same time. This situation may correspond to comparatively rapid uplift of an orogenic belt.

(*2*) Different minerals dated by one or more radiometric age methods may yield a discordant age pattern, as a result of the minerals becoming radiogenically closed systems at different temperatures and, therefore, at different times. This corresponds to slow cooling of the orogenic belt as a result of very gradual uplift and/or reduction of the thermal gradient at depth. In this situation the sequence of apparent mineral ages from a given rock unit will be in the order of the individual retentivities and activation energies for the radiogenic nuclide(s). In particular, potash feldspar Rb/Sr, hornblende K/Ar, muscovite Rb/Sr may yield older analytical dates, or apparent ages, than biotite K/Ar, Rb/Sr. The effect of grain size is also extremely important, and coarse-grained, pegmatitic minerals will become closed systems before fine-grained minerals of the same type. It is obvious that the oldest apparent ages from a given rock unit will be the ones that approximate most closely to the age of crystal, although they must still be regarded as minimum ages for the latter. Age discordances produced in this way are usually quite small. In very ancient rock units an intrinsically discordant mineral age pattern may be partially or completely masked by the inherently greater absolute analytical error of individual age measurements.

If mineral dates from basement complexes really represent cooling and uplift rather than peak of deep-seated metamorphic or igneous activity, it follows that ideally they must also reflect the time of erosion and sedimentation in the vicinity of a newly-forming mountain belt. Indeed, in the case of very prolonged uplift and erosion, the measured ages may significantly post-date the onset of primary sedimentation (HADLEY, 1964; LAMBERT, 1964; MOORBATH, 1964, 1965; ARMSTRONG, 1966). Clearly, radiometric mineral dates from deep-seated basement complexes cannot be used as unambiguous marker points for the definition of the Phanerozoic time-scale. Several cases are known where minerals from such an environment yield apparent ages which are definitely less than the true geological age of stratigraphically younger sediments. Of much greater importance for construction of the Phanerozoic time-scale are those radiometric dates obtained from rocks that cooled very rapidly, such as small, high-level igneous intrusives, lavas, ash-flows, or authigenic minerals in sediments. This whole topic has been discussed in great detail in the proceedings of a symposium on the geological time-scale published by the Geological Society of London (HARLAND et al., 1964).

Basement complex with a multi-stage thermal history

Two or more major geological events, such as metamorphism and/or igneous intrusion, are envisaged as having affected the basement complex. Grossly discordant mineral age patterns are extremely common in such terrains, and are well understood in an empirical way. Interpreted sensibly and statistically they can yield much more information than could possibly have been obtained from a completely concordant set of ages.

It frequently happens that the effects of a further thermal event on a plutonic basement complex are very variable. In areas less affected by the later event, minerals may survive from the earlier episode. In such cases, radiogenic nuclides may not be completely expelled from the original minerals by the later thermal event. A whole range of apparent ages may then be observed falling between the time at which the minerals became closed systems after the earlier event and after the later event. More than two major events would complicate the situation farther. In geochronological terminology, this is often referred to as "overprinting" of ages. The most retentive and coarsest-grained minerals will yield the oldest apparent ages. The least retentive and finest-grained minerals will give the youngest

Earth-Sci. Rev., 3 (1967) 111–133

apparent ages. If sufficient radiometric data are available from a variety of minerals, limiting age values can be obtained for two or more major thermal events, such as regional metamorphism and/or igneous activity.

A commonly observed phenomenon in such a situation, which appears to puzzle many geologists unnecessarily, is where coarse-grained minerals such as potash feldspar, muscovite and hornblende from cross-cutting, post-tectonic pegmatites yield older Rb/Sr and/or K/Ar dates than the typically finer-grained micas from the adjacent metamorphic country rocks. The pegmatites were clearly emplaced prior to the latest metamorphism which was, however, thermally inadequate to expel all, or any, pre-existing radiogenic nuclide from the coarse, retentive pegmatitic minerals. On the other hand, the fine-grained minerals from the country rock schists or gneisses give the time of radiogenic closure after the latest metamorphism irrespective of when the minerals actually crystallized.

To conclude this section, a possibly somewhat overpessimistic, yet generally valid, quotation from a recent geochronological study by GABRIELSE and REESOR (1964, p.127) is appropriate: "... any individual age may be a compound of many elements, a plutonic event, a structural event, or simply loss of daughter isotope during quiescent cooling, or any combination or superimposition of any number of such individual events. No specific method exists for identifying the precise degree to which any one or any number of the above elements has contributed to a particular isotopic ratio that is measured as an age. Thus, specific interpretation of a few age determinations must be made with caution and only if much associated, contributory evidence is available".

CAN ONE DATE MAJOR GEOLOGICAL EVENTS IN AN OROGENIC BELT?

It is clear from the above discussion that even the most retentive minerals such as zircon (U/Pb), potash feldspar (Rb/Sr), hornblende (K/Ar), coarse muscovite (Rb/Sr), can still only be expected to yield minimum ages for major geological events such as metamorphism and intrusion which occurred at depth within an orogenic belt. In favourable cases the highest minimum ages may closely approach the true date of crystallization within the limits of analytical error, particularly in older Precambrian terrains, where the absolute analytical errors on a measured age are much greater than, say, for Phanerozoic terrains. On the whole, however, the present author tends to agree with ARMSTRONG (1966) that "we do not yet know of any way of unequivocally and precisely dating any instant in time related to the time span of metamorphism". Nevertheless, a number of dating methods has been developed in recent years which *can*, in principle, approach more closely to the true age of a major geological event than mineral dates. These methods utilise the dating of certain types of whole rocks by the K/Ar or Rb/Sr methods and they are now briefly described.

Earth-Sci. Rev., 3 (1967) 111–133

K/Ar dating of low-grade metamorphic rocks, such as slates

K/Ar ages approaching the true age of metamorphism should be obtainable from materials which recrystallized completely during metamorphism under conditions leading to complete outgassing of inherited argon, but at temperatures close to that at which subsequently formed radiogenic argon can accumulate. HARPER (1964) showed that recrystallization resulting in the formation of slaty cleavage occurs in this range of temperatures and that slates from a high stratigraphic level which have cooled rapidly after metamorphism may yield K/Ar whole rock dates approximating to the true age of metamorphism within the orogenic belt. Minerals from deeper zones of the orogenic belt, which actually underwent metamorphism at the same time as the slates, may give younger, cooling ages. Harper's work was carried out on slates from the Upper Dalradian part of the Caledonian mountain belt of central Scotland (see later). He provided convincing evidence that all pre-existing argon was expelled from these slates during recrystallization of the pelitic sediments, which produced new mica minerals with a high degree of crystallinity. Furthermore, the slates do not appear to have lost significant amounts of argon since the time of cooling following recrystallization.

Rb/Sr dating of suitable whole rocks, such as granite and granite gneiss

The basic principles and methodology underlying Rb/Sr whole rock dating were described by COMPSTON and JEFFERY (1959, 1961), COMPSTON et al. (1960), FAIRBAIRN et al. (1961), HART (1961), NICOLAYSEN (1961). Since then, the method has gained wide acceptance. Recent reviews of the Rb/Sr whole rock method have been given by MOORBATH (1964, 1965), HAMILTON (1965), CAHEN and SNELLING (1966). For full details of methodology and methods of graphical representation (isochron and intersection diagrams) the reader is referred to the quoted works, since space does not permit further repetition here. However, the basic ideas behind the method in as far as they concern the subject matter of this review are briefly described below.

Consider firstly the case of a post-tectonic plutonic granite. Following upon the emplacement of a large, deep-seated intrusion, radiogenic Sr may continue to diffuse out of those minerals in which it is not easily accommodated, such as biotite and potash feldspar, until the whole system has cooled sufficiently for the individual minerals to attain radiogenic closure, which may significantly post-date the time of emplacement and crystallization. The radiogenic Sr which diffuses out of mica and potash-feldspar crystals enters Rb-poor, Sr-rich crystals in the immediate vicinity, such as plagioclase feldspars and apatite, and mixes with the reservoir of common strontium there. Provided that isotopic diffusion and mixing occurs essentially between neighbouring crystal grains, i.e., over distances of the order of millimetres, any reasonably sized hand-specimen of rock can remain a

Earth-Sci. Rev., 3 (1967) 111–133

closed system with regard to Rb and Sr from an early stage in the consolidation history of the rock. A whole rock Rb/Sr date should, therefore, in principle approximate closely to the time of emplacement of the intrusive. There are numerous examples in the literature where plutonic granites have been dated either by means of separated minerals, or by the Rb/Sr whole rock isochron method. However, very few detailed studies have been reported in which such intrusives have been dated by both approaches. There is scope here for a good deal of fundamental work bearing on the post-consolidation and cooling history of large intrusive bodies.

Secondly, we consider a pre-tectonic granitic intrusive, which later became converted to a granite-gneiss. It is now known that a granitic intrusive may remain virtually a closed system with regard to Rb and Sr even during a subsequent metamorphism of sufficient intensity to convert it into a gneiss. Individual minerals, such as micas and potash feldspars lose much of their radiogenic Sr during the metamorphic heating, or more strictly between the times of attainment of the diffusion threshold temperature during heating and cooling. During this period, radiogenic Sr mixes with the reservoir of common Sr in minerals such as plagioclase, apatite, etc. As in the previously discussed case, individual hand-specimens of whole rock may remain closed systems with regard to Rb and Sr. Whole rock analytical dates will then yield a close approximation to the time of original crystallization and emplacement, whilst minerals record the times of cooling and radiogenic closure after the metamorphism, in the order of their respective retentivities for radiogenic Sr.

Many such examples have been reported in the literature, the results usually being presented graphically in the form of isochrons, i.e., a plot of $^{87}Rb/^{86}Sr$ vs. $^{87}Sr/^{86}Sr$ for different whole rock samples from the same intrusion. Close alignment of the points to a straight line indicates that each individual hand-specimen system remained closed with respect to Rb and Sr during metamorphism (LONG, 1964). Anomalous whole rock Rb/Sr ages could result, particularly by the introduction of Rb by metasomatic processes, thus destroying the regularity of the isochrons (MOORBATH, 1965). Up to now most geochronologists have avoided dating obviously metasomatised rocks by whole rock methods.

It is probably an oversimplification to regard large granitic bodies as remaining *completely* closed systems with regard to Rb and Sr during regional metamorphism. Some radiogenic Sr probably *does* diffuse into the country rock. However, the fractional loss of radiogenic Sr is likely to be very small indeed, since the whole intrusive may be regarded to a first approximation as a single grain whose significant diffusion dimension is essentially infinite. Probably a great deal also depends upon the type of country rock and the geochemical nature of the diffusion gradients which are set up. Cases are certainly known in which radiogenic Sr has, indeed, migrated over rather greater distances within a metamorphic terrain. Thus, Precambrian hornblende-diorite dykes and sills in the Panamint Mountains of California were enriched in radiogenic Sr during a Late

Earth-Sci. Rev., 3 (1967) 111–133

Mesozoic metamorphism of the adjacent Precambrian granitic basement (LAN-PHERE et al., 1964; WASSERBURG et al., 1964).

USE OF HISTOGRAMS IN GEOCHRONOLOGY

It is evident from the foregoing discussions that the popular exercise of plotting analytical dates on histograms is quite unjustifiable unless (*a*) the analytical dates refer to the true age of a geological event, and (*b*) the nature and significance of that event is fully understood. Most published date histograms certainly contain a high proportion of cooling dates or overprinted dates. Cooling dates may be geologically significant, of course, in that they can refer to the time of uplift, with accompanying erosion and sedimentation. The most significant intrusive and metamorphic dates from a basement complex are obtained from particularly retentive minerals, from low-grade metamorphic zones, and from whole rocks (Rb/Sr). Such types of measurement are, on the whole, poorly represented in most published date histograms. The great majority of the points has usually been obtained by K/Ar and Rb/Sr measurements on micas. There is no reason why the mean or mode of a mineral (and, especially, mica) date histogram should give the time of intrusion or metamorphism in an orogenic belt.

For the purposes of defining episodes of crystallization in a basement complex, a single U/Pb, Rb/Sr, or K/Ar date on a really resistant (i.e., in terms of diffusion of radiogenic isotope), coarse-grained mineral, or a single Rb/Sr whole rock isochron date, may be far more significant than any number of K/Ar and Rb/Sr dates on fine-grained micas. Though the latter may form a convincing peak on a date histogram, this many signify no more than the termination of a radiogenic Ar or Sr diffusion episode. In contrast, a considerable spread of dates on a histogram may often be due, not to continuous plutonic activity, but to the inclusion of a large number of apparent, "overprinted" dates—particularly in a polymetamorphic terrain.

With further progress and more general understanding of the whole subject, a time will undoubtedly come when only truly meaningful ages are plotted on histograms. This will be of great importance for the truer understanding of the sequence and correlation of orogenic events than has been possible hitherto.

SOME GEOCHRONOLOGICAL CASE HISTORIES

A few case histories from recent geochronological literature are described briefly below, which serve to illustrate the principles outlined in the previous pages, with special reference to the hypothesis that mineral dates do *not* neces-

Earth-Sci. Rev., 3 (1967) 111–133

sarily give the time of a major geological event such as plutonic intrusion or regional metamorphism. This survey of case histories is by no means exhaustive.

The Caledonian orogenic belt of Scotland and Ireland

Recent summaries of the wider aspects of the geology of the British Caledonides have been published by READ and MacGREGOR (1948), PHEMISTER (1960), READ (1961), McKERROW (1962), JOHNSON and STEWART (1963), WATSON (1963), JOHNSON (1965), MERCY (1965). Numerous K/Ar age measurements have been reported by the Cambridge workers (J.A. MILLER, 1961; BROWN et al., 1965b, c; J. A. MILLER and BROWN, 1965) and Rb/Sr and K/Ar measurements by the Oxford workers (GILETTI et al., 1961; LONG and LAMBERT, 1963; BELL, 1964; HARPER, 1964). Most of these measurements were done on separated minerals, particularly fine-grained micas. In all but the most recent work it was argued that the analytical dates reflect the time of one or other of the main metamorphic or tectonic maxima which have been recognized in the Caledonides by petrologists and structural geologists. K/Ar mineral date histograms have been published by BROWN et al. (1965b) which show a strong maximum at about 420 m.y. for the regionally metamorphosed Moine metasediments (deposited in Late Precambrian times) of the northern Scottish Highlands and about 440 m.y. for the regionally metamorphosed Dalradian metasediments (Late Precambrian–partly Cambrian) of the central Scottish Highlands. The Rb/Sr mineral dates fit this pattern very closely. In many cases K/Ar and Rb/Sr dates on micas are concordant within analytical error. Some of the original Rb/Sr mica dates reported by GILETTI et al. (1961) were a little too high because inadequate correction had been made for the isotopic composition of initial common strontium in the rock (MOORBATH, 1965).

The general picture envisaged by FITCH et al. (1964) is that the main Caledonian metamorphism occurred about 420 m.y. ago, and that any older ages (including those from the Dalradian with their mineral date peak at about 440 m.y. as well as a very few isolated higher ages reaching up to ca. 540 m.y. from both Moine and Dalradian) represent incomplete overprinting of some hypothetical Late Precambrian metamorphism (> 700 m.y.) by the main 420 m.y. Caledonian "event". The present author finds it difficult to accept this view on geological and geochronological grounds. Some of the geological evidence against it has been discussed by T. G. MILLER (1964).

HARPER (1964) and WATSON (1964) suggested that the observed age pattern reflects post-metamorphic uplift and cooling, and that the Dalradian metasediments, on the whole, cooled somewhat earlier than the Moine metasediments. Particularly convincing evidence for the uplift-cooling has been provided by HARPER (1964). He found that in coexisting Dalradian muscovites and biotites, the muscovites consistently gave higher ages (+ 12 m.y.) than the biotite. He noted

Earth-Sci. Rev., 3 (1967) 111–133

that this difference occurred irrespective of the relative order of crystallization and considered it to indicate that argon began to accumulate in muscovite some 12 m.y. before it began to accumulate in biotite. Harper further found that completely recrystallized slates from a high stratigraphical and structural position within the Caledonian belt gave significantly higher whole rock K/Ar ages (in the range ca. 440–490 m.y.) than either slates from a lower structural level, or minerals from the regionally metamorphosed schists. It was concluded that the highest ages from the high-level slates, i.e., ca. 490 m.y., might be a close approximation to the true age of the main metamorphism within this part of the Caledonian belt, whilst all the numerous younger ages simply reflected the delay in cooling of the deeper levels.

Recently, BROWN et al. (1965a) have reported biotite K/Ar age in the range 460–500 m.y. from large, essentially post-metamorphic basic intrusives in the Dalradian Schists of northeastern Scotland. The schists in this area give the normal regional mineral date of about 440 m.y. (BELL, 1964). Several cases have been reported in the literature where minerals from large, basic intrusions yield older dates than minerals from the country rock. ARMSTRONG (1966) refers to these as "date asylums".

The above evidence suggests that the acme of regional metamorphism in the Caledonides of Scotland—certainly in the Dalradian sector—occurred earlier than about 490–500 m.y. ago. The pronounced maxima on the mineral date histograms of about 420 m.y. for the Moines and about 440 m.y. for the Dalradian almost certainly represent cooling dates.

An upper limit for the age of metamorphism of the Dalradian Series and for at least a part of the Moine Series is given by Rb/Sr whole rock ages on pre-metamorphic granites, some of which are highly foliated or gneissic. The most accurately dated is the regionally metamorphosed Carn Chuinneag granitic complex intrusive into the Moine Series of the northern Highlands, which yields a Cambrian Rb/Sr whole rock isochron age of 530 ± 10 m.y. (LONG, 1964). Comparable intrusives in the Dalradian, such as the Ben Vuroch, Portsoy and Windyhills Granites, also yield Rb/Sr whole rock isochron ages of about 530 m.y. (BELL, 1964). Individual K/Ar and Rb/Sr mineral ages in these granite-gneisses are concordant with those from the adjacent country rock, i.e., about 100 million years younger than the Rb/Sr whole rock ages.

It is concluded that the main metamorphism in much of the Scottish Caledonides occurred between about 530 and 500 m.y. ago and is of Upper Cambrian age, *not* Late Silurian as believed by many geologists and geochronologists.

In the Dalradian metasedimentary Series of Connemara, western Ireland, the situation is particularly intriguing. In contrast with Scotland, there is overwhelming geological evidence (DEWEY, 1961) that the main metamorphisms and foldings in the Dalradian "Connemara Schists" occurred before the beginning of Ordovician times, i.e., before 500 ± 15 m.y. ago (HARLAND et al. 1964). Further-

Earth-Sci. Rev., 3 (1967) 111–133

more, the post-metamorphic Oughterard Granite, which is intrusive into the Connemara Schists, yields a Rb/Sr whole rock isochron age of 510 $\pm$ 35 m.y. (LEGGO et al., 1966). On the other hand, K/Ar and Rb/Sr mineral dates from the Connemara Schists, and from within the Oughterard Granite itself, cluster closely around 440–450 m.y. (MOORBATH et al., 1967). Hornblendes and muscovites yield very slightly older ($+$ 10 m.y.) apparent ages than biotites, regardless of the order of crystallization. By no stretch of the imagination can these mineral dates be regarded as dating the time of metamorphism. It is, in fact, postulated that the mineral date peak reflects uplift, cooling and unroofing of the whole Connemara Massif in Mid or Upper Ordovician times, which is also the time of maximum sedimentation immediately to the north, in the great Mayo sedimentary basin. This interpretation is again in sharp contrast to that of FITCH et al. (1964) who, on the basis of a few mineral dates, postulated overprinting of a pre-Arenig metamorphism by a major metamorphism about 420 m.y. ago.

Unfortunately, no pre-metamorphic granites like those of Scotland are known from western Ireland, so that it is not yet possible to give a reliable upper limit for the age of metamorphism. However, *if* the major metamorphic and tectonic events in the Dalradian of Scotland and Ireland are contemporaneous, then they could all have occurred between about 530 and 500 m.y. ago.

Finally, an example of the extraordinary argon retentivity of hornblende, from the western part of the Caledonian mountain belt in northwestern Scotland, where the Moine metasediments are tectonically associated with inliers of ancient Precambrian "Lewisian" gneisses. Lewisian rocks form the foreland to the Caledonian orogen to the west of the great Moine thrust system, along the northwestern seaboard of Scotland (RAMSAY, 1958; SUTTON, 1962; JOHNSON, 1965; WATSON, 1965). In the well-known Glenelg area, east of the Moine thrust, it can be shown that the rocks of the Lewisian inliers underwent, firstly, high-grade regional metamorphism; secondly, basic dyke intrusion; thirdly, erosion prior to the deposition of the Late Precambrian Moine sediments. Furthermore, also to the east of the Moine thrust, the Moine and Lewisian together underwent Caledonian regional metamorphism, although, in this particular area, not of a high grade. Whilst biotite from a Lewisian gneiss inlier at Loch Duich, just north of Glenelg, gave the normal Moine mineral cooling date of about 420 m.y., already referred to earlier, hornblendes from a cross-cutting amphibolite dyke, only a few feet from the biotite, yielded locally very variable K/Ar dates in the range 1,500–2,200 m.y. (S. Moorbath, 1967, unpublished data). This indicates that the dykes were intruded at least 2,200 m.y. ago, but that the hornblende dates were also subject to some later overprinting. This value actually agrees well with the age of basic dykes in parts of the Lewisian foreland to the west. The elucidation of the complex sequence of metamorphic and igneous events in the Lewisian Complex of northwestern Scotland by means of discordant mineral dates and whole rock measurements

Earth-Sci. Rev., 3 (1967) 111–133

has been described by GILETTI et al. (1961) and EVANS (1965), and is not sum-marised here.

Western Alps

Much fundamental geochronological work on the Western Alps has been carried out in recent years by Dr. E. Jäger and colleagues at Bern University. The Alps form a particularly favourable area for study because: (*a*) the rocks and minerals have remained undisturbed since the last phase of the Alpine (Tertiary) metamorphism; (*b*) the geology of the western Alps and the degree of Alpine metamorphism have been extensively studied; (*c*) these relatively young ages can be determined with small absolute errors, making it possible to measure age differences of only a few million years.

Detailed mineral date studies have been reported by JÄGER (1966) and ARMSTRONG et al. (1966) from a particularly interesting area in the central part of the western Alps. In this area, the rocks were formed in Palaeozoic times, but were influenced in varying degrees by the Tertiary Alpine metamorphism. The principal results may be summarised as follows: micas from high tectonic units give pre-Alpine dates up to about 340 m.y., approximating to the termination of the last orogenic phase before formation of the Alps. In the stilpnomelane zone there is a gradual transition from pre-Alpine to Tertiary Alpine ages, due to partial overprinting. In the chloritoid zone, Tertiary mineral dates are most com-mon, but rocks which preserve their pre-Alpine structure can still yield pre-Alpine biotite dates. In this zone, the biotite ages depend on the *p*, *t* conditions, on the type of rock and its degree of resistance to metamorphism. In the northern Gotthard Massif, ARNOLD and JÄGER (1965) found apparent biotite ages of 266, 164, 54 and 23 m.y. all within a few feet of each other. The 266 m.y. value was from a biotite in an ultrabasic inclusion, which preserved its pre-Alpine character better than the immediately surrounding rocks.

In the highest grade zones of Alpine metamorphism, where the minerals staurolite, kyanite and sillimanite occur, there are no pre-Alpine or partially over-printed mineral dates. K/Ar and Rb/Sr biotite dates from any given specimen are essentially concordant, but vary from 11 to 25 m.y. for different specimens. This range is definitely not due to analytical error, the young ages being not randomly distributed, but changing gradually with geographic position. There is, however, no correlation between the apparent ages and the type of rock or the tectonic position in the high grade metamorphic zones. There is strong evidence that these ages must be regarded as uplift and cooling ages, and not as metamorphic ages. Thus, Rb/Sr dates on muscovites formed during the Alpine metamorphism yield about 30 m.y., providing a lower age limit for the Alpine metamorphism. STEIGER (1964) has also reported hornblende K/Ar ages of 23–30 m.y. from the central Alps, which he does not consider to be relict ages. Furthermore, JÄGER (1966) finds

Earth-Sci. Rev., 3 (1967) 111–133

423

a rather constant age difference of 8 m.y. between the Rb/Sr ages of biotite and muscovites, the latter always having the higher ages. The muscovites became a closed system to radiogenic Sr at a higher temperature than did the biotites, during the cooling process. The mineral dates essentially represent a time at which the rocks had reached a certain level below the surface.

Not unexpectedly, muscovite dates are less sensitive to metamorphism than biotite dates. Pre-Alpine muscovite dates are found in the chloritoid zone and in the exterior part of the kyanite zone. However, in regions of highest grade Alpine metamorphism, muscovite usually gives Alpine dates. In contrast, Rb/Sr whole rock analyses can yield pre-Alpine dates even in the zones of highest grade metamorphism. JÄGER (1966) quotes a Rb/Sr whole rock age of 300 m.y. for highly recrystallized Pennine gneisses. Other relevant Rb/Sr whole rock work has been published by WÜTHRICH (1965) on the Aar Massif and by GRAUERT (1966) on the orthogneisses of the Silvretta of western Switzerland. The latter yield whole rock isochrons of 428 ± 17 m.y., and 351 ± 14 m.y., whereas the minerals yield a metamorphic cooling date of 295 ± 12 m.y. Some younger mineral ages were also found in the region.

In the eastern Alps of Austria, published K/Ar dates have been interpreted in terms of uplift and cooling events, just as in the western Alps (OXBURGH et al., 1966).

Appalachians of eastern North America

It is impossible in the available space to begin to summarise the vast amounts of geochronological data published by American workers in the last 10 years in the Appalachian Mountain Belt. All the published age data have recently been collected together in one volume (ANONYMOUS, 1965). The geology and geochronology of the mountain belt are both extremely complex. One of the main problems has been the interpretation of the age data in terms of actual geological events. There may frequently exist a contradiction between the geological and geochronological evidence. The idea that many Appalachian mineral dates reflect uplift and cooling, rather than time of crystallization, is becoming more widely accepted and has been discussed particularly by HADLEY (1964), ARMSTRONG (1966) and WETHERILL et al. (1966). There is no doubt that this removes some of the earlier difficulties. Hadley has provided evidence, in the form of histograms, for the close relationship between K/Ar, Rb/Sr mineral dates and intensity of sedimentation in neighbouring depositional basins. As in the previous case histories, the dates which approximate most closely to the actual time of crystallization have been obtained from highly retentive minerals in favourably protected surroundings and by the whole rock Rb/Sr dating of granites and gneisses. In this way it has been possible to recognize the existence and measure the age of Precambrian Grenville basement (ca.1,000–1,100 m.y.) in certain areas of the

Earth-Sci. Rev., 3 (1967) 111–133

Appalachians (TILTON et al., 1958; DAVIS et al., 1962), as well as the oldest pre-metamorphic granitic intrusives of the Appalachian Mountain Belt proper, which are about 560 m.y. old (FAIRBAIRN et al., 1967).

Eastern Great Basin of Utah and Nevada, western United States

ARMSTRONG and HANSEN (1966) measured K/Ar mineral dates from a meta-morphic complex definitely produced during a mid-Mesozoic (post-Triassic, pre-Lower Cretaceous) metamorphism. All the measured K/Ar dates, however, were Tertiary, ranging mainly between 20 and 40 m.y. These results were interpreted as uplift and cooling in Tertiary times. Thus, a discrepancy of nearly 100 m.y. exists between the time of metamorphism inferred on geological grounds and the measured dates. Samples showing the greatest effects of static recrystallization have the youngest dates and may represent the material that was structurally deepest at the time of metamorphism. Evidence is presented to show that during Oligocene times, ca. 30 m.y. ago, the oldest rocks which represent lower parts of the Cordilleran geosynclinal sequence were deeply buried. The known thickness of the Palaeozoic stratigraphic sequence in the region is greater than 30,000 ft. Assuming an average geothermal gradient, burial to this depth would imply a temperature of about 300°C, certainly sufficient to cause continuous argon loss from micas. The heat flux and thermal gradient may well have been higher than normal, since the Great Basin was a volcanically active area at that time. Intrusives in the area generally give Tertiary mineral dates indistinguishable from the meta-morphic mineral dates. They are also believed to be cooling dates and not in-trusive dates.

We may conclude with ARMSTRONG and HANSEN (1966) that "the differences between expected and observed results and between different areas cannot be explained by experimental error and should give pause to those geologists working in older metamorphic terrains who are using K/Ar dates to establish 'absolute' ages of metamorphic episodes".

ACKNOWLEDGEMENT

I thank Dr. N. J. Snelling and Prof. B. J. Gilletti for commenting upon the manuscript.

REFERENCES

ALDRICH, L. T., DAVIS, G. L. and JAMES, H. L., 1965. Ages of minerals from metamorphic and igneous rocks near Iron Mountain, Michigan. *J. Petrol.*, 6: 445–472.
AMIRKHANOV, K. I., BARTNITSKII, E. N., BRANDT, S. B. and VOIT-KEVICH, G. V., 1959. The

diffusion of argon and helium in certain rocks and minerals. *Dokl. Akad. Nauk S.S.S.R.*, 126: 160–162.

ANONYMOUS, 1965. Geochronology of North America. *Comm. Nuclear Sci., Wash., NAS-NRC Publ.*, 1276: 315 pp.

ARMSTRONG, R. L., 1966. K–Ar dating of plutonic and volcanic rocks in orogenic belts. In: O. A. SCHAEFFER and J. ZÄHRINGER (Editors), *Potassium–Argon Dating.* Springer, Berlin–Heidelberg–Göttingen, pp.117–133.

ARMSTRONG, R. L. and HANSEN, E., 1966. Cordilleran infrastructure in the eastern Great Basin. *Am. J. Sci.*, 264: 112–127.

ARMSTRONG, R. L., JÄGER, E. and EBERHARDT, P., 1966. A comparison of K–Ar and Rb–Sr ages on Alpine biotites. *Earth Planetary Sci. Letters*, 1: 13–19.

ARNOLD, A. und JÄGER, E., 1965. Rb–Sr Altersbestimmungen an Glimmern im Grenzbereich zwischen voralpinen Alterswerten und alpiner Verjüngung der Biotite. *Eclogae Geol. Helv.*, 58: 369–390.

BAADSGAARD, H. and DODSON, M. H., 1964. Potassium–argon ages of sedimentary and pyro-clastic rocks. In: W. B. HARLAND, A. G. SMITH and B. WILCOCK (Editors), *The Phanerozoic Time-Scale—Quart. J. Geol. Soc. London*, 120: 119–127.

BAADSGAARD, H., LIPSON, J. and FOLINSBEE, R. E., 1961. The leakage of radiogenic argon from sanidine. *Geochim. Cosmochim. Acta*, 25: 147–157.

BAADSGAARD, H., CUMMING, G. L., FOLINSBEE, R. E. and GODFREY, J. D., 1964. Limitations of radiometric dating. In: F. FITZOSBORNE·(Editor), *Geochronology of Canada—Roy. Soc. Can., Spec. Publ.*, 8: 20–38.

BELL, K., 1964. *A Geochronological and Related Isotopic Study of the Rocks of the Central and Northern Highlands of Scotland.* Thesis, Oxford Univ., 175 pp. Also: *Bull. Geol. Soc. Am.*, in preparation.

BRANDT, S. B. and VORONOVSKY, S. N., 1964. Dependence of the energy of activation of radio-genic gases in minerals on temperature. *Geochem. Intern.*, 1: 302–305.

BROWN, P. E., MILLER, J. A., GRASTY, R. L. and FRASER, W. E., 1965a. Potassium–argon ages of some Aberdeenshire granites and gabbros. *Nature*, 207: 1287–1288.

BROWN, P. E., MILLER, J. A., SOPER, N. J. and YORK, D., 1965b. Potassium–argon age pattern of the British Caledonides. *Proc. Yorkshire Geol. Soc.*, 35: 103–138.

BROWN, P. E., YORK, D., SOPER, N. J., MILLER, J. A., MACINTYRE, R. M. and FARRAR, E., 1965c. Potassium–argon ages of some Dalradian, Moine and related Scottish rocks. *Scot. J. Geol.*, 1: 144–151.

CAHEN, L. and SNELLING, N. J., 1966. *The Geochronology of Equatorial Africa.* North-Holland, Amsterdam, 205 pp.

CATANZARO, E. J. and KULP, J. L., 1964. Discordant zircons from the Little Belt (Montana), Beartooth (Montana) and Santa Catalina (Arizona) Mountains. *Geochim. Cosmochim. Acta*, 28: 87–124.

COMPSTON, W. and JEFFERY, P. M., 1959. Anomalous "common strontium" in granite. *Nature*, 184: 1792.

COMPSTON, W. and JEFFERY, P. M., 1961. Metamorphic chronology by the rubidium–strontium method. *Ann. N.Y. Acad. Sci.*, 91: 185–191.

COMPSTON, W., JEFFERY, P. M. and RILEY, G. H., 1960. Age of emplacement of granites. *Nature*, 186: 702–703.

DAVIS, G. L., TILTON, G. R. and WETHERILL, G. W., 1962. Mineral ages from the Appalachian Province in North Carolina and Tennessee. *J. Geophys. Res.*, 67: 1987–1996.

DEWEY, J. F., 1961. A note concerning the age of metamorphism of the Dalradian rocks of western Ireland. *Geol. Mag.*, 98: 399–405.

EVANS, C. R., 1965. Geochronology of the Lewisian basement near Lochinver, Sutherland. *Nature*, 207: 54–56.

EVERNDEN, J. F., CURTIS, G. H., KISTLER, R. W. and OBRADOVICH, J., 1960. Argon diffusion in glauconite, microcline, sanidine, leucite and phlogopite. *Am. J. Sci.*, 258: 583–604.

FAIRBAIRN, H. W., HURLEY, P. M. and PINSON, W. H., 1961. The relation of discordant Rb–Sr mineral and whole rock ages in an igneous rock to its time of crystallization and to the time of subsequent $^{87}Sr/^{86}Sr$ metamorphism. *Geochim. Cosmochim. Acta*, 23: 135–144.

FAIRBAIRN, H. W., MOORBATH, S., RAMO, A. O., PINSON, W. H. and HURLEY, P. M., 1967. Rb–Sr age of granitic rocks of southeastern Massachusetts and the age of the Lower Cambrian at Hoppin Hill. *Earth Planetary Sci. Letters*, in preparation.

FECHTIG, H. and KALBITZER, S., 1966. The diffusion of argon in potassium-bearing solids. In: O. A. SCHAEFFER and J. ZÄHRINGER (Editors), *Potassium–Argon Dating*. Springer, Berlin–Heidelberg–Göttingen, pp.68–107.

FECHTIG, H., GENTNER, W. und ZÄHRINGER, J., 1960. Argonbestimmungen an Kaliummineralien. 7. Diffusionsverluste von Argon in Mineralien und ihre Auswirkung auf die Kalium–Argon-Alterbestimmung. *Geochim. Cosmochim. Acta*, 19: 70–79.

FECHTIG, H., GENTNER, W. und KALBITZER, S., 1961. Argonbestimmungen an Kaliummineralien. 9. Messungen zu den verschiedenen Arten der Argondiffusion. *Geochim. Cosmochim. Acta*, 25: 297–311.

FITCH, F. J., MILLER, J. A. and BROWN, P. E., 1964. Age of Caledonian orogeny and metamorphism in Britain. *Nature*, 203: 275–278.

FOLINSBEE, R. E., LIPSON, J. and REYNOLDS, J. H., 1956. Potassium–argon dating. *Geochim. Cosmochim. Acta*, 10: 60–68.

FRECHEN, J. und LIPPOLT, H. J., 1965. Kalium–Argon Daten zum Alter des Laacher Vulkanismus, der Rheinterrassen und der Eiszeiten. *Eiszeitalter Gegenwart*, 16: 5.

GABRIELSE, H. and REESOR, J. E., 1964. Geochronology of plutonic rocks in two areas of the Canadian Cordillera. In: F. FITZOSBORNE (Editor), *Geochronology of Canada—Roy. Soc. Can., Spec. Publ.*, 8: 96–138.

GENTNER, W. und KLEY, W., 1957. Die Frage der Argonverluste in Kalifeldspaten und Glimmermineralien. *Geochim. Cosmochim. Acta*, 12: 323–329.

GERLING, E. K. and MOROZOVA, I. M., 1957. Determination of the activation energy of argon liberated from micas. *Geochemistry*, (1957)4: 359–367.

GERLING, E. K. and MOROZOVA, I. M., 1962. Determination of the activation energy for the release of argon and helium from minerals. *Geochemistry*, (1962)12: 1255–1267.

GERLING, E. K., MOROZOVA, I. M. and KURBATOV, V. V., 1961. The retentivity of radiogenic argon in ground micas. *Ann. N.Y. Acad. Sci.*, 91: 227–234.

GERLING, E. K., LEVSKII, L. K. and MOROZOVA, I. M., 1963. On the diffusion of radiogenic argon from minerals. *Geochemistry*, (1963)6: 551–555.

GERLING, E. K., KOLTSOVA, T. V., PETROV, B. V. and ZULFIKAROVA, Z. K., 1965. On the suitability of amphiboles for age determination by the K–Ar method. *Geochem. Intern.*, 2: 148–154.

GILETTI, B. J., MOORBATH, S. and LAMBERT, R. ST. J., 1961. A geochronological study of the metamorphic complexes of the Scottish Highlands. *Quart. J. Geol. Soc. London*, 117: 233–272.

GOLDICH, S. S., BAADSGAARD, H. and NIER, A. O., 1957. Investigation in $^{40}Ar/^{40}K$ dating. *Trans. Am. Geophys. Union*, 38: 547–551.

GRAUERT, B., 1966. Rb–Sr age determinations on orthogneisses of the Silvretta (Switzerland). *Earth Planetary Sci. Letters*, 1: 139–147.

HADLEY, J. B., 1964. Correlation of isotopic ages, crustal heating and sedimentation in the Appalachian region. *Virginia Polytech. Inst., Dept. Geol. Sci., Mem. 1—Tectonics of the Southern Appalachians*, pp.33–45.

HAMILTON, E. I., 1965. *Applied Geochronology*. Acad. Press, London, 267 pp.

HAMILTON, E. I., DODSON, M. H. and SNELLING, N. J., 1962. The application of physical and chemical methods to geochronology. *Intern. J. Appl. Radiation Isotopes*, 13: 587–610.

HARLAND, W. B., SMITH, A. G. and WILCOCK, B. (Editors), 1964. *The Phanerozoic Time-Scale—Quart. J. Geol. Soc. London*, 120: 458 pp.

HARPER, C. T., 1964. Potassium–argon ages of slates and their geological significance. *Nature*, 203: 468–470.

HART, S. R., 1961. Mineral ages and metamorphism. *Ann. N.Y. Acad. Sci.*, 91: 192–197.

HART, S. R., 1964. The petrology and isotopic mineral age relations of a contact zone in the Front Range, Colorado. *J. Geophys. Res.*, 72: 493–525.

HART, S. R. and DODD, R. T., 1962. Excess radiogenic argon in pyroxenes. *J. Geophys. Res.*, 67: 2998–2999.

HURLEY, P. M., HUGHES, H., PINSON, W. H. and FAIRBAIRN, H. W., 1962. Radiogenic argon and strontium diffusion parameters in biotite at low temperatures from Alpine fault uplift in New Zealand. *Geochim. Cosmochim. Acta*, 26: 67–80.

HURLEY, P. M., FAIRBAIRN, H. W., FAURE, G. and PINSON, W. H., 1963. New approaches to geochronology by strontium isotope variations in whole rocks. In: *Symposium on Radioactive Dating, Athens, 1962—Proc. Intern. At. Energy Agency, Vienna*, pp.201–217.

JÄGER, E., 1966. Rb–Sr age determinations on minerals and rocks from the Alps. *Sci. Terre*, 10: 395–406.

JOHNSON, M. R. W., 1965. (a) Torridonian and Moinian, (b) Dalradian. In: G. Y. CRAIG (Editor), *The Geology of Scotland*. Oliver and Boyd, Edinburgh, pp.79–160.

JOHNSON, M. R. W. and STEWART, F. H. (Editors), 1963. *The British Caledonides*. Oliver and Boyd, Edinburgh, 280 pp.

KULP, J. L., 1963. Present status of geochronometry. *Isotopics, U.S. At. Energy Comm. (Oak Ridge)*, 1: 1–7.

LAMBERT, R. ST. J., 1964. The relationship between radiometric ages obtained from plutonic complexes and stratigraphical time. In: W. B. HARLAND, A. G. SMITH and B. WILCOCK (Editors), *The Phanerozoic Time-Scale—Quart. J. Geol. Soc. London*, 120: 43–25.

LANPHERE, M. A., WASSERBURG, G. J., ALBEE, A. L. and TILTON, G. R., 1964. Redistribution of strontium and rubidium isotopes during metamorphism, World Beater Complex, Panamint Range, California. In: H. CRAIG, S. L. MILLER, and G. J. WASSERBURG (Editors), *Isotopic and Cosmic Chemistry*. North-Holland, Amsterdam, pp.269–320.

LEGGO, P. J., COMPSTON, W. and LEAKE, B. E., 1966. The geochronology of the Connemara granites and its bearing on the antiquity of the Dalradian Series. *Quart. J. Geol. Soc. London*, 122: 91–118.

LONG, L. E., 1964. Rb–Sr chronology of the Carn Chuinneag intrusion, Ross-shire, Scotland. *J. Geophys. Res.*, 69: 1589–1597.

LONG, L. E. and LAMBERT, R. ST. J., 1963. Rb–Sr isotopic ages from the Moine Series. In: M. R. W. JOHNSON and F. H. STEWART (Editors), *The British Caledonides*. Oliver Boyd, Edinburgh, pp.217–247.

McDOUGALL, I., 1966. Precision methods of K–Ar isotopic age determination on young rocks. In: S. K. RUNCORN (Editor), *Methods and Techniques in Geophysics*. Interscience, New York, N.Y., pp.279–304.

McDOUGALL, I. and GREEN, D. H., 1964. Excess radiogenic argon in pyroxenes and isotopic ages on minerals from Norwegian eclogites. *Norsk. Geol. Tidsskr.*, 44: 183–196.

McKERROW, W. S., 1962. The chronology of Caledonian folding in the British Isles. *Proc. Natl. Acad. Sci.*, 48: 1905–1913.

MERCY, E. L. P., 1965. Caledonian igneous activity. In: G. Y. CRAIG (Editor). *The Geology of Scotland*. Oliver and Boyd, Edinburgh, pp.229–267.

MILLER, J. A., 1961. Age of metamorphism of Moine Schists. *Geol. Mag.*, 98: 85–86.

MILLER, J. A. and BROWN, P. E., 1965. Potassium–argon age studies in Scotland. *Geol. Mag.*, 102: 106–134.

MILLER, T. G., 1964. Age of Caledonian orogeny and metamorphism in Britain. *Nature*, 204: 358–361.

MOORBATH, S., 1964. The rubidium–strontium method. In: W. B. HARLAND, A. G. SMITH and B. WILCOCK (Editors), *The Phanerozoic Time-Scale—Quart. J. Geol. Soc. London*, 120: 87–98.

MOORBATH, S., 1965. Isotopic dating of metamorphic rocks. In: W. S. PITCHER and G. W. FLINN (Editors), *Controls of Metamorphism*. Oliver and Boyd, Edinburgh, pp.235–267.

MOORBATH, S., BELL, K., LEAKE, B. E. and McKERROW, W. S., 1967. Geochronological studies in Connemara and Murrisk, Western Ireland. In: E. I. HAMILTON and R. M. FARQUHAR (Editors), *Radiometric Dating for Geologists*. Interscience, London, in press.

NICOLAYSEN, L. O., 1961. Graphic interpretation of discordant age measurements on metamorphic rocks. *Ann. N.Y. Acad. Sci.*, 91: 198–206.

OXBURGH, E. R., LAMBERT, R. ST. J., BAADSGAARD, H. and SIMONS, J. G., 1966. Potassium–argon age studies across the southeast margin of the Tauern window, eastern Alps. *Verhandl. Geol. Bundesanstalt*, 1–2: 17–33.

Earth-Sci. Rev., 3 (1967) 111–133

PHEMISTER, J., 1960. *British Regional Geology. Scotland: the Northern Highlands*, 3rd ed. H. M. Stationery Office, London, 104 pp.

RAMSAY, J. G., 1958. Moine–Lewisian relations at Glenelg, Inverness-shire. *Quart. J. Geol. Soc. London,* 113: 487–523.

READ, H. H., 1961. Aspects of the Caledonian magmatism in Britain. *Liverpool Manchester Geol. J.,* 2: 653–683.

READ, H. H. and MACGREGOR, A. G., 1948. *British Regional Geology. The Grampian Highlands*, 2nd. ed. H. M. Stationery Office, London, 83 pp.

REYNOLDS, J. H., 1957. Comparative study of argon content and argon diffusion in mica and feldspar. *Geochim. Cosmochim. Acta*, 12: 177–184.

SARDAROV, S. S., 1957. The preservation state of radiogenic argon in microclines. *Geochemistry*, (1957)3: 233–237.

SCHAEFFER, O. A. and ZÄHRINGER, J. (Editors), 1966. *Potassium-Argon Dating.* Springer, Berlin–Heidelberg–Göttingen, 234 pp.

SCHAEFFER, O. A. and ZÄHRINGER, J., 1966. *Potassium-Argon Dating.* Springer, Berlin–Heidelberg–Göttingen, 234 pp.

STEIGER, R. H., 1964. Dating of orogenic phases in the central Alps by K–Ar ages of hornblende. *J. Geophys. Res.,* 69: 5407–5421.

SUTTON, J., 1962. The Moine Series of Scotland. *Sci. Progr.,* 50: 76–86.

TILTON, G. R. and HART, S. R., 1963. Geochronology. *Science*, 140: 357–366.

TILTON, G. R., WETHERILL, G. W., DAVIS, G. L. and HOPSON, C. A., 1958. Ages of minerals from the Baltimore Gneiss near Baltimore, Maryland. *Bull. Geol. Soc. Am.,* 69: 1469–1474.

TILTON, G. R., WETHERILL, G. W., DAVIS, G. L. and BASS, M. N., 1960. 1,000 million-year-old minerals from the eastern United States and Canada. *J. Geophys. Res.,* 65: 4173–4179.

WASSERBURG, G. J., ALBEE, A. L. and LANPHERE, M. A., 1964. Migration of radiogenic strontium during metamorphism. *J. Geophys. Res.,* 69: 4395–4401.

WATSON, J., 1963. Some problems concerning the evolution of the Caledonides of the Scottish Highlands. *Proc. Geol. Assoc. London,* 74: 213–258.

WATSON, J., 1964. Conditions in the metamorphic Caledonides during the period of late orogenic cooling. *Geol. Mag.,* 101: 457–465.

WATSON, J., 1965. Lewisian. In: G. Y. CRAIG (Editor), *The Geology of Scotland.* Oliver and Boyd, Edinburgh, pp.49–77.

WESTCOTT, M. R., 1966. Loss of argon from biotite in a thermal metamorphism. *Nature*, 210: 83–84.

WETHERILL, G. W., ALDRICH, L. T. and DAVIS, G. L., 1955. $^{40}Ar/^{40}K$ ratios of feldspars and micas from the same rock. *Geochim. Cosmochim. Acta*, 8: 171–172.

WETHERILL, G. W., TILTON, G. R., DAVIS, G. L., HART, S. R. and HOPSON, C. A., 1966. Age measurements in the Maryland piedmont. *J. Geophys. Res.,* 71: 2139–2155.

WÜTHRICH, H., 1965. Rb–Sr Altersbestimmungen am alpin metamorph überprägten Aarmassiv. *Schweiz. Mineral. Petrog. Mitt.,* 45: 875–971.

(Received June 13, 1967)

15. CONTINENTAL RADIOMETRIC AGES[1]
(July 1968)

PATRICK M. HURLEY AND JOHN R. RAND

42

1. Introduction

New evidences for sea-floor spreading and continental drift have opened up the possibility that the continents have been rifted and rejoined possibly many times before the last drifting episode. Reconstructions of earlier motions require a knowledge of the basement rock age provinces of the continents, and their structural trends and boundaries. Therefore as many of the published age data on predrift basement rocks as can be found in the accessible literature have been recorded. It is of no concern that most of the reported data are K–Ar measurements and represent "apparent ages" only, because the matching of age provinces can be done equally well utilizing these age values as long as they are not mixed with values given by other methods. This chapter also provides an indication of the areas of continental crust that are particularly in need of work, and the age patterns of those continental margins which appear to be truncated by existing oceans and may therefore logically be fitted to some other continental mass in a predrift reconstruction.

Previous surveys in continental areas include the work of Goldich, Muehlberger, Lidiak, Hedge, Walthall, Denison, Marvin, Thomas, and Bass (1966) and Wetherill, Bickford, Silver, and Tilton (1965) for the United States; of Wanless and associates (1967) of the Geological Survey of Canada for the Canadian Shield area; of Wilson, Compston, Jeffery, and Riley (1960) for Australia; and of Cahen and Snelling (1966) for Equatorial and South Africa. This survey updates and adds to these earlier compilations, and brings the material together in a unified form. Continental areas in the eastern USSR and China are omitted; there undoubtedly have been similar surveys reported in these countries, but the published works have not been obtained.

The literature covered includes all of the standard journals of wide circulation, privately published bulletins or reports from various laboratories, and some unpublished age data kindly supplied by investigators who have not yet finished their work in the areas they are investigating. The maps therefore include some information that will not be published for one or more years. It is expected that because of oversight there are also serious omissions of data that are available. Perhaps a future international data center for reports on new age measurements will eliminate this problem.

Because of the magnitude of the bibliography involved in this compilation, a selected list of publications which will serve to lead the reader back through the literature stemming from any group of authors, is provided. The selection is based on the need to list a single recent report from many different investigators, representing different laboratories, rather than several reports from any single author or group.

[1] M.I.T. Age Studies No. 84.

575

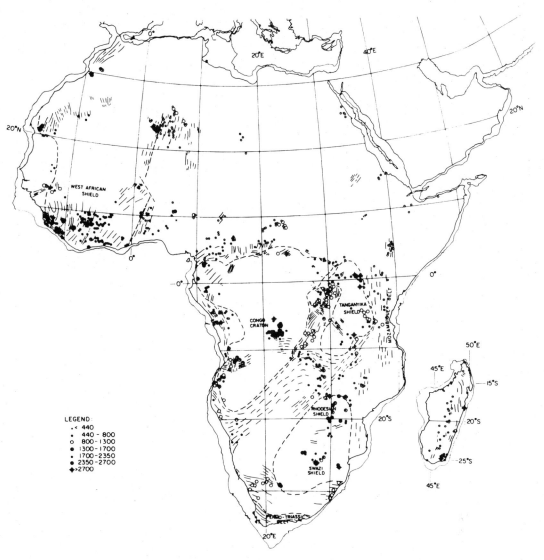

Figs. 1–7. These figures show plots of published radiometric apparent ages of basement rocks in Africa, S. America, N. America, Europe, India, Australia, and Antarctica. Groups of values are represented by the symbols given in Fig. 1. Structural trend lines and age-province boundaries are included sparcely by the lighter and darker dashed lines. Continental boundaries are indicated both by present shore lines and the 500-fathom or 1000-meter contour.

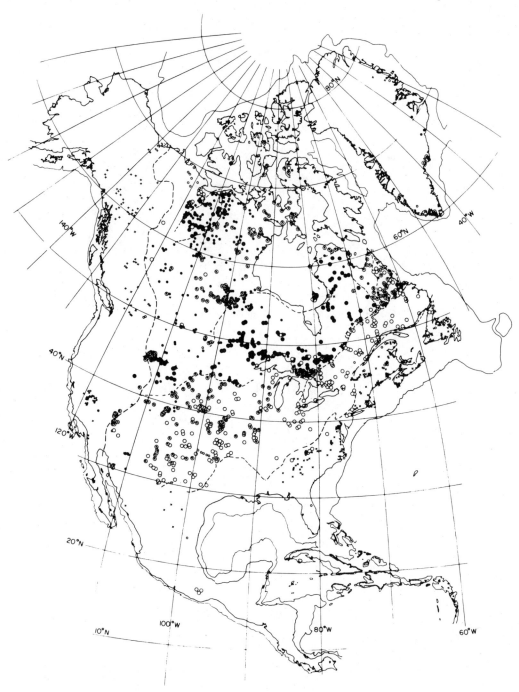

2. Mapping Procedures

Maps of age data for Africa, South America, North America, Europe, India, Australia and Antarctica are given in Figs. 1–7. The edges of the continents are indicated on the map sheets by two lines: (a) the present shore lines, and (b) the 500 fathom or 1000 meter contour lines on the continental slopes. The age symbols indicated in Fig. 1 apply to all the map sheets. These symbols are arranged

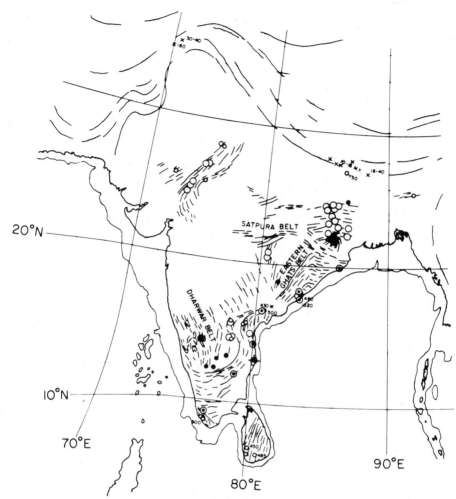

so that the older the age, the more black its appearance. It should be remembered that a thin cover of sedimentary rocks and sediments lies over much of the total area of the continents so that there are large blank areas in which no information on the age of the basement rocks can be obtained. In a few instances the ages of the basement rocks have been determined on samples obtained in core-drilling resulting from petroleum exploration. If the ages indicated are less than 440 million years these are recorded only for the case of orogenic belts within which igneous and metamorphic rocks can be taken to represent the true crustal basement. However, the criteria for this judgment are always open to debate.

No attempt is made to indicate on these maps the Phanerozoic folded and intruded mountain systems that have been adequately surveyed by ordinary geological methods unless age measurements are available. Information on the ages and locations of these Phanerozoic belts is available in the geological map

435

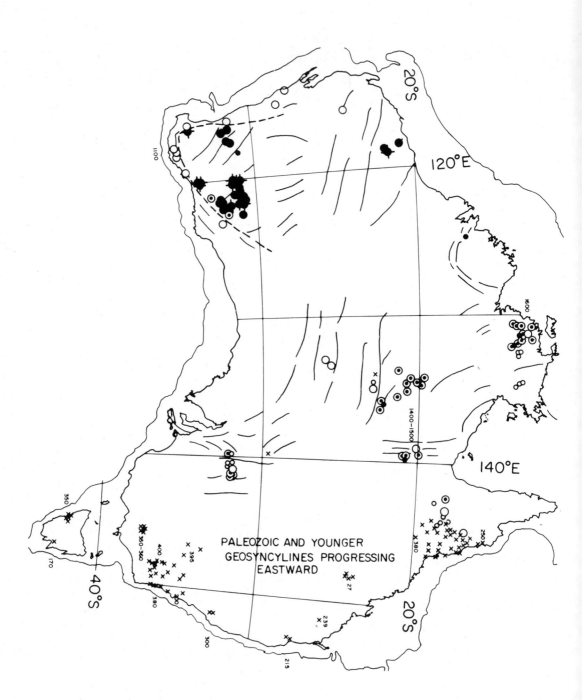

PALEOZOIC AND YOUNGER
GEOSYNCYLINES PROGRESSING
EASTWARD

436

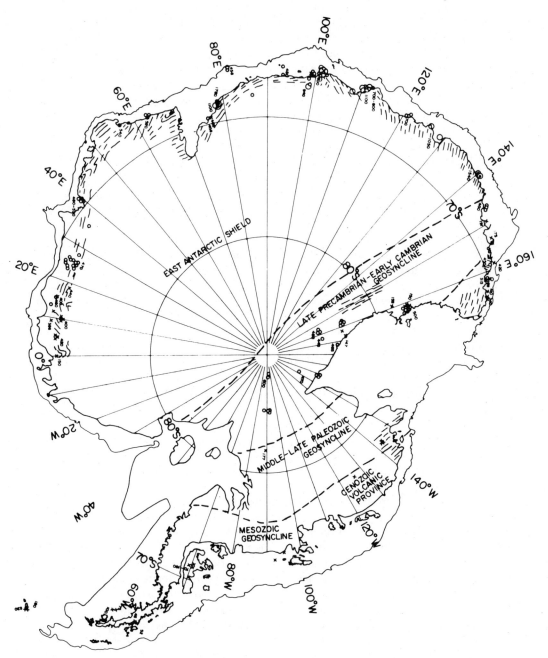

sheets on the various continents, and there would be no way to reproduce them on these maps. The continental age provinces are therefore considerably better known than indicated by the age points that have been measured radiometrically.

The age measurements are arrived at by a mixture of methods. Most of them are by K–Ar analysis, but an increasing number are being made by whole-rock Rb-Sr isochron plots. In general the K–Ar measurements indicate the time at which the temperature of the rock decreased to a value somewhere in the range of 100–200°C. This value cannot be specified because the diffusion loss of argon from minerals depends on time, temperature, and recrystallization. Generally the K-Ar age is either the time of uplift of the system into near-surface regions, or the time of the last thermal episode.

On the other hand, if it is satisfactory the whole-rock Rb-Sr isochron plot indicates the time at which the system became essentially closed to migrations of rubidium and strontium, and if the samples are large enough, the sample system could have remained essentially closed despite prolonged periods of time at elevated temperatures. Therefore the Rb-Sr method tends to show the age of metasomatism or development of a gneiss, and to record this event even if the rock mass remains at an elevated temperature over much of its lifetime or is reheated in a thermal event. Comparisons of these two age methods on a number of samples in widely different terrains involving igneous, metamorphic, and metasomatic rock types show that the Rb-Sr whole-rock age is generally older than the K-Ar age by approximately 100 million years. This means that, on the average, the basement rocks of shield areas and younger rejuvenated zones have taken roughly this amount of time to rise and cool to present elevations and temperatures. The grouping of ages in this report is broad enough so that it is relatively insensitive to the age method employed, with the exception of those particular cases in which the Rb-Sr whole-rock method may ignore a much later thermal episode which has reset the K-Ar system.

The decay constants for the recorded age values are not unified because the purpose of this report does not warrant this refinement. However, in most cases the decay constants for ^{40}K are $\lambda_e = .585 \times 10^{-10}$ yr^{-1}, $\lambda_\beta = 4.72 \times 10^{-10}$ yr^{-1}; and for ^{87}Rb they are roughly equally divided between 1.39 and 1.47×10^{-11} yr^{-1}.

Dashed-line boundaries of age provinces occasionally are included on the map sheets when there is particularly good evidence of a sharp break in the level of the age values. The use of a large number of geologically determined boundaries is purposely avoided because these frequently divide the continent into more narrow time classifications in the latter part of earth history than in the earlier part. This tends to obscure the major continental age patterns which seldom have well-mapped boundaries because of sedimentary cover or lack of pre-Cambrian stratigraphic correlations. Some boundaries marking Phanerozoic provinces in which the ages are largely determined by stratigraphy and paleontology are inserted because age measurements have been fewer in these regions. A minimum number of structure lines is presented on the map sheets, again with the purpose of indicating (in a rough way) both the general structural trends and the limited areas of exposure of the basement provinces without obscuring the primary age data.

It has been found to be almost universally true that previously accepted geological nomenclature in pre-Cambrian areas must be at least partially abandoned when the region is covered by radiometric age dating. This is particularly true of periods of orogenesis. Because the new nomenclature is not yet settled in most instances, we have omitted almost all the pre-Cambrian regional terminology and retain only the original names of crustal blocks which have, with minor exceptions, no stratigraphic or orogenic significance.

3. Discussion

Radiometric dating of basement rocks on continental areas will ultimately be sufficiently definitive so that age provinces may be clearly outlined. This will yield information on the mean age of the continental masses, their growth history if they are indeed growing, and possibly their dynamic history of splitting apart and reconstruction. Current trends in thought suggest that where the continental boundary merges with the ocean-floor without such disturbances as mountain building, volcanism, trench formation, and other features of orogenic activity, the continent and adjacent ocean-floor are coupled and move together as a unit. This appears to be the case over much of the coast lines on either side of the Atlantic and Indian Oceans. In contrast to this it is generally believed that where there is relative motion between the ocean-floor and the continental margin, there is orogenic activity such as seen along the west coast of South America; or if there is relative motion of sections of sea-floor, island arc structures develop, as in the western Pacific. If these hypotheses are true, then the presence of folded mountain belts in the interior part of a present continental mass may possibly suggest the location of previous continental margins moving against a down-sinking oceanic crust at the time represented by the age of the mountain system.

If the recent history is a key to the past, it is also evident that the rifting of continental land masses appears to follow preferentially the lines of ancient thermo-tectonic belts which leave the shore lines with basement structural trends parallel to the coastline. Therefore, although the leading edge of a moving continent may be marked by a parallel structural belt, it is also commonly true of the trailing edge. The only difference is that the orogeny at the leading edge is contemporaneous with the motion of the continent relative to the ocean-floor, whereas the belt parallel to the trailing edge is more ancient than, and not related to, the time of the continental motion. Thus the dating of thermo-tectonic or orogenic episodes is necessary in any reconstruction of the dynamic history of the earth's surface.

This report draws no conclusions regarding the reconstruction of previous land masses because enough attempts for the amount of data currently available have been made in recent literature. The purpose here is to establish a view of continental geochronology and a consolidation of the literature in order to provide a

TABLE I

Measured Areas of Continental Basement Rocks Included in Apparent Age Provinces Based on 450-million-year Intervals, in km²

Continent	Age intervals, in million years						
	0–450 m.y.	450–900 m.y.	900–1350 m.y.	1350–1800 m.y.	1800–2250 m.y.	2250–2700 m.y.	2700–3150 m.y.
Africa	2,371,300	17,349,800	1,953,100	315,000	6,559,500	1,826,700	289,800
South America	5,139,800	10,995,600	699,300	—	4,769,100	—	42,100
North America	12,831,700	3,825,400	3,988,300	3,912,500	5,091,300	3,162,100	—
Australia	2,806,600	1,607,400	3,359,400	1,192,100	490,400	322,200	694,500
India	1,761,100	1,679,100	1,194,300	506,200	606,000	356,500	39,200
Antarctica	5,751,600	5,378,600	768,300	511,300	—	—	—
Europe	7,562,500	312,200	2,677,200	2,287,300	1,914,400	516,400	—
Total	38,224,600	41,148,100	14,639,900	8,724,400	19,430,700	6,183,900	1,065,600

base for new evidence derived from studies of sea-floor spreading and paleo-magnetism.

Despite the fact that apparent ages may not be directly equated with the time of development of new sialic crust, it is of interest to determine the areal extent of apparent age provinces. These have been measured and are listed in Table I. The total area measured is 129,417,200 square kilometers, which is about 68% of the area of all the continents and their shelves. A histogram of these data is plotted in Fig. 8. Time intervals used are 450 million years, extending backward from the

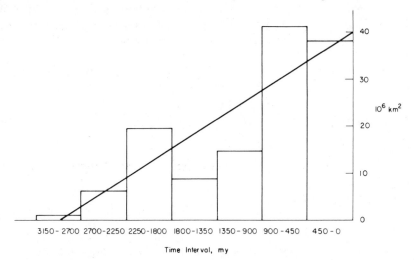

Fig. 8.   The figure gives a histogram of measured areas represented by age intervals of 450 million years. The total area of the survey is 68% of the exposed continental area of the earth. Although these areas are based mostly on apparent age values (mineral age values) the usually small difference between these and Rb-Sr isochron ages on whole rock samples makes it clear that the process of formation of the continental crust is accelerating with time. A linear increase in the growth rate is indicated by the heavy line as a first approximation. Earlier suggestions of a strong grouping of age values at certain times in the pre-Cambrian have been at least partially obliterated by this more extensive survey.

present. It is clear that the process of development of stable continents is accelerating with time. A first order fit to the histogram is indicated by the straight line on the diagram; this is equivalent to a constant acceleration of 40 $km^2 \cdot m.y.^{-2}$ in the development of new geologic age provinces throughout the 3000 million year total span of time and over the entire continental crust. In recent earth history the new age provinces were developing at a rate of about 120,000 $km^2 \cdot m.y.^{-1}$ compared to 45,000 $km^2 \cdot m.y.^{-1}$ at 2000 million years ago.

A comparison between Rb-Sr age values on whole rock samples and K-Ar values on separated minerals over a wide variety of cases indicates that the two values are generally fairly comparable except for the slight lowering of the K-Ar values by about 100 million years. The Rb-Sr age values are derived from the increase in radiogenic ^{87}Sr in the total rock which in turn can be equated with the time-integrated ratio of Rb/Sr in the rock. If the generation of sialic crust is accompanied by an increase in the ratio Rb/Sr, the process can be equated in a general way with the development of the area of successive age provinces. Therefore the latter part of this discussion on the acceleration of the development of the area of age provinces applies also to the development of the continental crust. The mean apparent age of the total continental crust, using the figures in Table I, is less than 1000 million years. Because of some recycling of crustal materials the mean true age of the continental crust is somewhat greater than this, but within the limits discussed by Hurley (1968). This conclusion could be false if radiogenic ^{87}Sr is constantly lost to the mantle by the process described by R. L. Armstrong (1968).

Acknowledgments

This work has been supported by the Division of Research, U.S. Atomic Energy Commission. We are grateful to U. G. Cordani and R. E. Denison for supplying some unpublished age data.

References

Armstrong, R. L., 1968. A model for the evolution of Sr and Pb isotopes in a dynamic earth. *Rev. of Geophysics*, **6**(2) 175–199.

Cahen, L., and N. J. Snelling, 1966. *The Geochronology of Equatorial Africa*, North-Holland, Amsterdam.

Goldich, S. S., E. G. Lidiak, C. E. Hedge, and F. G. Walthall, 1966. Geochronology of the midcontinent region, United States, 2. Northern area. *J. Geophys. Res.*, **71**(22) 5389–5408.

Goldich, S. S., W. R. Muehlberger, E. G. Lidiak, and C. E. Hedge, 1966. Geochronology of the midcontinent region, United States, 1. Scope, methods, and principles. *J. Geophys. Res.*, **71**(22) 5375–5386.

Hurley, P. M., 1968. Correction to "Absolute abundance and distribution of Rb, K and Sr in the earth." *Geochim. et Cosmochim. Acta*, **32**, 1025–1030.

Lidiak, E. G., R. F. Marvin, H. H. Thomas, and M. N. Bass, 1966. Geochronology of the midcontinent region, United States, 4. Eastern area. *J. Geophys. Res.*, **71**(22) 5427–5438.

Muehlberger, W. R., C. E. Hedge, R. E. Denison, and R. F. Marvin, 1966. Geochronology of the midcontinent region, United States, 3. Southern area. *J. Geophys. Res.*, **71**(22) 5409–5426.

Wanless, R. K., R. D. Stevens, G. R. Lachance, and C. M. Edwards, 1967. Age determinations and geological studies: K–Ar isotopic ages. Report 7, Canada. Geol. Surv. Paper 66–17, 120 pp.

Wetherill, G. W., M. E. Bickford, L. T. Silver, and G. R. Tilton, 1965. Geochronology of North America. Nat. Acad. Sci., Nat. Res. Council, Publ. 1276, Nuclear Sciences Series, Report No. 41, 315 pp.

Wilson, A. F., W. Compston, P. M. Jeffery, and G. H. Riley, 1960. Radioactive ages from the pre-Cambrian rocks in Australia. *J. Geol. Soc. Australia*, **6**, Pt. 2, 179–195.

30 July 1971, Volume 173, Number 3995

SCIENCE

43

Of Time and the Moon

Dating of lunar materials reveals the
early history of an observable planetary body.

George W. Wetherill

*The American cool was becoming a
narcotic. The horror of the Twentieth
Century was the size of each new event,
and the paucity of its reverberation.*—
NORMAN MAILER, *Of a Fire on the
Moon*

The current series of Apollo lunar
missions, together with the unmanned
Surveyor, Orbiter, and Soviet missions,
permit for the first time a rather detailed study of the formative stage of a
planetary body. Prior to the Apollo
landings, many National Aeronautics
and Space Administration and National Academy of Sciences study
groups produced documents predicting
that lunar exploration would enable us
to understand the origin of the moon,
the earth, and the solar system. To
some, these statements probably seemed
to be exaggerations calculated to per-
suade Congress and the public to provide the necessary financial support
But to a large measure these predictions have proven to be true, and it may
be expected that lunar data will both
supplement and supplant meteoritic
data as the primary source of information about the origin and early history
of the earth and solar system. For this
reason, the continuing supply of data
obtained from the returned lunar sam-

The author is professor of geophysics and
geology in the Department of Planetary and
Space Science, University of California, Los
Angeles 90024.

ples, from the geophysical instruments
emplaced on the lunar surface, and
from the experiments to be carried out
in lunar orbit deserve to be followed
closely by the scientific community,
particularly by earth scientists. This
new source of information comes at a
time when earth scientists are to some
extent overwhelmed with the necessity
of incorporating into their thinking
new concepts, particularly those arising
from the discovery of sea floor spreading and plate tectonics. This wealth of
understanding concerning the origin
and evolution of the earth, as well as
fundamental earth processes, may well
lead to a renaissance in earth science,
especially at a time when great concern
as to the future of the earth should
serve as an incentive to understanding
our planet.

Rb-Sr Evidence for Primordial
Geochemical Differentiation

Measurements on samples returned
from Apollo 11 and 12 have shown
that the composition of the source of
lunar rocks differs significantly from the
average composition of the solar system, as indicated by the composition of
the sun and of the most abundant and
least differentiated class of meteorites,
the chondrites. Furthermore, this difference was established very early in the
history of the solar system, 4.5 × 10⁹

to 4.6×10^9 years ago. It is quite likely
that this primordial differentiation,
which depleted the moon of relatively
volatile elements such as potassium,
rubidium, and lead, took place in the
solar nebula, or in a protomoon, prior
to the final accretion of the moon.

Evidence for this primordial fractionation is provided by the isotopic
composition of strontium separated
from lunar rocks and minerals. Radiogenic ^{87}Sr is generated by the beta decay of ^{87}Rb with a half-life of 5×10^{10}
years, so that the ratio of ^{87}Sr to the
nonradiogenic isotope ^{86}Sr increases
with time. The rate at which this ratio
increases is proportional to the ratio
$^{87}Rb/^{86}Sr$. The lowest $^{87}Sr/^{86}Sr$ ratio
found in any natural sample is in the
achondritic meteorite Angra dos Reis
and is equal to 0.69884 ± 0.00004,
which indicates that this ratio was at
least this low at the time of contraction
of the solar nebula. Similar measurements on the more abundant basaltic
achondrite meteorites give an only
slightly higher value of 0.69898 (*I*).
Measurements on the chondrite Guareña (4.53×10^9 years old) (*2*) show
that, at the time chemical equilibrium
was established within this meteorite
sample, the ratio $^{87}Sr/^{86}Sr$ had risen
to 0.69995. For a typical chondritic
$^{87}Rb/^{86}Sr$ ratio of 0.75, this increase
of ~ 0.001 requires only 100 million
years; and for a solar $^{87}Rb/^{86}Sr$ ratio
of 2.25, it requires only about 30 million years. Furthermore, measurements
in chondrites of ^{129}Xe, formed by the
decay of the now extinct (17-million-
year half-life) natural radioactivity ^{129}I,
show that the parent body of the chondrites crystallized within approximately
100 million years of the time of formation of the solar nebula, from which
the sun and planets subsequently formed
(*3*). Consequently, any object less than
4.5×10^9 years old which has had a
ratio of $^{87}Sr/^{86}Sr$ less than about 0.700
was derived from a source which underwent a significant depletion in Rb relative to Sr within the first 100 million
years of solar system history.

Initial $^{87}Sr/^{86}Sr$ ratios for typical
lunar rocks and minerals are shown in

Fig. 1. Knowledge of their present $^{87}Sr/^{86}Sr$ ratios, together with their age (discussed later), permits extrapolation of the $^{87}Sr/^{86}Sr$ ratio of each sample back to that at the time of its formation; this initial ratio is given by the intercepts on the ordinate of the lines representing the growth of $^{87}Sr/^{86}Sr$ with time for each of these samples. All of these initial values are less than the value for the chondrite Guareña, and but slightly higher than the ratio found in the chondrite Angra dos Reis. In fact, for some of the samples (for example, the plagioclase feldspar from basalt sample 10017) even the *present* $^{87}Sr/^{86}Sr$ ratio is less than 0.700, and the extrapolation back to the even lower initial ratio is insensitive to any possible error in the age of the sample. These results show that the source from

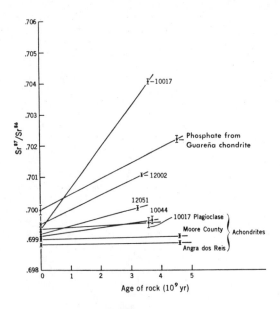

Fig. 1 (left). Extrapolation of measured $^{87}Sr/^{86}Sr$ ratios back to the value at the time of formation of the rock. The initial values for lunar rocks fall between the values found for achondritic and chondritic meteorites (Data primarily from Wasserburg and co-workers).

Fig. 2 (below). (A) Thin section of lunar rock 12013, a breccia containing light and dark regions with numerous included rock fragments. Horizontal dimension of thin section is 1.2 centimeters. Total weight of rock is 82 grams. (B) Large lunar rock (~ 50 centimeters) with light and dark portions, photographed by Surveyor I, in western Oceanus Procellarum.

Fig. 3. (A) Rubidium-strontium data of Asylum (5) for rock fragments from rock 12013. The points lying along the isochron marked "4.52 × 10⁹ years" represent material that has been enriched in Rb relative to Sr very early in lunar history. (B) Rubidium-strontium data of Asylum (5) for minerals separated from fragments of rock 12013. The mineral points lying on the two distinct isochrons are interpreted to be part of ∼ 4.5 × 10⁹ year-old rocks which were metamorphosed 4.0 × 10⁹ years ago. (C) Rubidium-strontium data for materials separated from the lunar soil indicate that these soils contain a component which was involved in differentiation very early in lunar history.

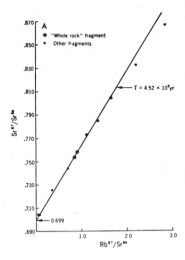

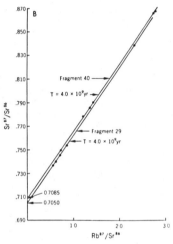

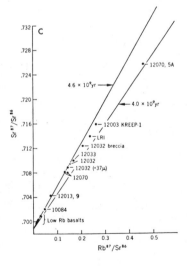

which these rocks were derived has not had a chondritic $^{87}Rb/^{86}Sr$ ratio since the formation interval of the solar system. This depletion in Rb relative to Sr could have taken place in that portion of the solar nebula which subsequently accumulated to form the moon, or alternatively, the region in the moon from which these basalts were derived was depleted in Rb immediately following formation of the moon, with the concomitant enrichment in Rb of some other portion of the moon, presumably a region nearer the surface. Because of their chemical similarity a similar depletion in K should also have occurred.

Rb-Sr Evidence for Subsequent
Early Geochemical Differentiation

Regardless of the site of lowering of this primordial Rb/Sr ratio, there is evidence that, in addition, some portions of the moon underwent a possibly subsequent, but nevertheless, early, enrichment in Rb relative to Sr. This is most clearly shown by rock 12013, collected on the Apollo 12 mission. Studies of thin sections of this rock (Fig. 2A) indicate that this is a breccia consisting of dark fragments, which have been intruded by a matrix of lighter material. Analyses of this rock (4) show that neither the light nor the dark portions of this rock are similar in composition to the more common mare basalts returned by Apollo 11 and 12. Rock 12013 is enriched in the elements K, Rb, Ba, and the rare earths by a factor of about 10.

At the time this rock was found, it seemed possible that it was an extreme differentiate occurring at the top of a layered igneous body, the average composition of which was more similar to

the basaltic rocks at this site. However, subsequent Rb-Sr measurements indicate, that this rock is considerably older than the basaltic rocks, which have an age of 3.3 × 10⁹ years. Figure 3A shows the results of Rb-Sr measurements carried out by Asylum (5) on individual fragments from this rock, plotted on a Sr evolution diagram. The open circles represent fragments of "whole rock systems," and the closed circles are fragments more likely to be similar to mineral separates. If $^{87}Rb/^{86}Sr$ ratios are plotted against $^{87}Sr/^{86}Sr$ ratios for each of an assemblage of closed chemical systems a straight line will result if all of the systems had the same $^{87}Sr/^{86}Sr$ ratio at some time in the past (6). The slope of this line is a function of the time elapsed since all of these systems had their common ratio, and the initial ratio itself is given by the intercept of this line with the ordinate. Such would be the case for a group of rocks separated from a common magma source, an assemblage of minerals crystallizing from a melt, or a metamorphic assemblage of minerals forming in chemical equilibrium as a result of a metamorphic event subsequent to the original crystallization of the rock. The "whole rock" points in Fig. 3A fall along the line marked "4.52 × 10⁹ years," so that all of these fragments were probably involved in a fractionation of Rb relative to Sr at this time in the past, and at that time they possessed a common $^{87}Sr/^{86}Sr$ ratio equivalent to that found in achondritic meteorites. This age is the same as that obtained by similar measurements on minerals separated from chondrites (1, 2, 7) or on measurements on whole meteorites (7, 8), and represents a time during the formation interval of the solar system.

Similar data obtained from individual minerals separated from two rock fragments of sample 12013 are plotted in Fig. 3B. These also fall along straight lines, but with slopes corresponding to an age of 4.0 × 10⁹ years, which may be interpreted as a time of subsequent metamorphism, possibly related to the time at which the fragments were assembled into a breccia. This age of 4.0 × 10⁹ years has also been obtained by K-Ar measurements (9).

It does not seem likely that rocks similar to 12013 are extremely rare. An 83-milligram fragment (LR-1) similar to some parts of 12013 has been found in a soil sample from Apollo 11 (10).

445

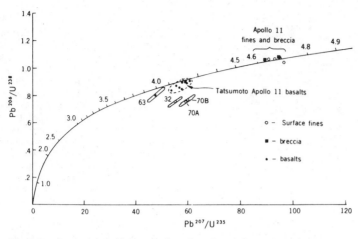

Fig. 4. Uranium-lead data for lunar rocks. All of the Apollo 11 fine soil and breccia give essentially concordant ages of 4.6×10^9 to 4.7×10^9 years. The Apollo 11 basalt data of Tatsumoto (16) gives "apparent" U-Pb ages of $\sim 4 \times 10^9$ years, which are actually not in conflict with 3.6×10^9 year Rb-Sr and K-Ar ages on these same rocks, as discussed in text. A typical Apollo 12 rock (12063) gives slightly younger apparent ages.

Measurements of various soil samples from Apollo 12 (Fig. 3C) indicate that the Rb and radiogenic Sr of soil at the Apollo 12 site are entirely dominated by a 4.5×10^9 year old component, and individual rock types [microbreccia (11, 12) and "KREEP" glass (13)] separated from these soil samples are even richer in this component. Apollo 11 soil, while containing less Rb and less radiogenic ^{87}Sr, nevertheless also falls on the 4.6×10^9 year line (isochron) in Fig. 3C (10). Rocks resembling 12013 in appearance have also been photographed by Surveyor I (Fig. 2B). It seems very likely that lunar igneous rocks are not limited to the common mare basalts, but also contain abundant differentiated rocks of varied compositions, many of which differentiated very early in the history of the moon and the solar system.

U-Pb and Th-Pb Evidence for

Early Geochemical Differentiation

Additional evidence for primordial differentiation may be obtained from measurements of radiogenic Pb from lunar materials and its associated parent isotopes of uranium and thorium. The isotopic composition of Pb from lunar rocks and soil samples is entirely different from that found in analogous terrestrial materials. Lunar rocks have much higher ratios of the radiogenic

isotopes ^{206}Pb, ^{207}Pb, and ^{208}Pb, relative to the nonradiogenic isotope ^{204}Pb (14–16). This is because the lunar abundance of ^{204}Pb is lower than that found in similar terrestrial rocks by a factor of the order of 100. These data indicate that the moon has been very much depleted in the relatively volatile Pb, and that essentially all of the Pb on the moon has formed as a consequence of the radioactive decay of U and Th.

The isotopic composition of Pb can also be used to show that this gross depletion of Pb relative to the refractory U and Th also took place very early in lunar history, quite likely during the process of lunar formation. The parent of ^{207}Pb, ^{235}U, is quite short-lived (its half-life is 0.7×10^9 years), and consequently is nearly, but not quite (unfortunately?) an "extinct" radioactivity. Highly radiogenic Pb very rich in ^{207}Pb, as is the case for lunar Pb, must therefore have formed very early in the history of the solar system, prior to the decay of most of the ^{235}U.

A quantitative extension of this qualitative argument leads to the conclusion that this early depletion in Pb relative to U and Th took place between 4.6×10^9 and 4.7×10^9 years ago. Figure 4 shows the results of U-Pb analyses of Apollo 11 soil samples plotted on a "concordia" diagram. The Pb isotopic composition has been corrected for primary Pb by subtraction

of a component with the isotopic composition of primordial Pb, as given by Pb from the troilite (FeS) of iron meteorites (17). Because the lunar ratios of $^{206}Pb/^{204}Pb$ (~ 262) and $^{207}Pb/^{204}Pb$ (~ 171) are so high, the result is insensitive to this correction. The concordia curve is the locus of chemical systems for which the ages calculated from the ratio $^{206}Pb/^{238}U$ are the same as those calculated from the ratio of $^{207}Pb/^{235}U$ (18). This common "concordant" age can be read from the scale marked along concordia. Figure 4 shows that samples of soil and breccia (essentially compacted soil) from Apollo 11 fall on concordia at an age of 4.6×10^9 to 4.7×10^9 years. Thus the Apollo 11 soil appears to represent a closed system which was enriched in U and Th relative to Pb about 4.65×10^9 years ago, and has not experienced a major U-Pb fractionation since that time. Remarkably, this same result was obtained for the $^{232}Th-^{208}Pb$ age. This age of 4.65×10^9 years obtained from these data has been interpreted as the age of the moon (15, 16). This is a plausible inference, because more detailed discussions show that this result is unlikely on a body significantly less than 4.65×10^9 years old, whereas a moon significantly older would be in conflict with the age of the solar system, as calculated from meteoritic data.

At the time this result was first reported, there was considerable discussion both in the press and within the scientific community as to how it was possible for the soil samples to be older than the rocks (3.6×10^9 years by Rb-Sr dating) from which they presumably were derived. In many of these discussions, there was at least an implicit assumption that some kind of logical paradox was thereby involved. This is not the case. For example, if lunar basalts were formed 3.6×10^9 years ago, and if, at least on the average, there were no U-Pb or Th-Pb fractionations at the time of their formation, then this is exactly the result which would be found. Therefore, the paradox, if any, is a geochemical one, and not a logical one.

Apollo 12 soils and breccia differ from those at Apollo 11 in that they do not give concordant U-Pb and Th-Pb ages (12, 19, 20). When plotted on a concordia diagram, they fall along a line approximately joining the concordant Apollo 11 points and a point on the concordia corresponding to about 1.5×10^9 years. As was the case for Rb, the

386

Apollo 12 soil is greatly enriched in U and Th relative to the Apollo 12 rocks; this enrichment is apparently associated with the presence in the soil of at least one component rich in K, Rb, Ba, U, Th, P, and rare earth elements, and this component has been shown (Fig. 3C) to have a Rb-Sr age of 4.5×10^9 years. The U-Pb data may be interpreted as indicating an age of 4.5×10^9 years for the source of this component; however this component has lost Pb relative to U and Th at some much more recent time. A simple two-stage model for this loss implies that the loss took place $\sim 1.5 \times 10^9$ years ago. This model is by no means unique. However, all plausible models imply that at least some Pb loss took place this recently. A highly speculative interpretation of these data would be that this component was ejected from a deeper layer in the moon at the time of formation of a large nearby crater—for example, Copernicus—that the loss of Pb resulted from vaporization of Pb from the hot ejecta, and that consequently the crater formation took place $\sim 1.5 \times 10^9$ years ago. This is a very interesting possibility, but it is not free from difficulties.

Time of Filling of the Lunar Maria

A key question in any discussion of lunar chronology is that of the time at which the dark mare regions were filled with the basaltic rock now present in these regions, and the related question as to whether mare filling was a single event, or took place at various times in lunar history. Data obtained from Apollo 11 and Apollo 12 samples have given some clear answers relevant to these questions.

The Apollo 11 basalts fall into two discrete groups, one characterized by Rb concentrations of about 6 micrograms per gram of sample, the other by Rb concentrations of about 1 microgram per gram of sample; corresponding differences are found for many other minor elements (21). It is possible that these two groups merely represent two different lava flows underlying the soil at this site. Measurements of Rb and Sr have been carried out on rock from both of these groups, in order to determine the time since the rock was crystallized (10, 15, 21). Although the enrichment in radiogenic Sr is very low, definitive ages have been obtained.

Both the high (Fig. 5A) and the low

Rb groups were formed 3.61×10^9 years ago. Although both of these groups were formed at nearly the same time, they cannot be interpreted as representing differentiation from a common source at the time of extrusion, because their initial $^{87}Sr/^{86}Sr$ ratios are distinctly different. In addition, the initial $^{87}Sr/^{86}Sr$ ratios of the high Rb group rocks is lower than that predicted for a 3.6×10^9 year magma source with the Rb/Sr ratio of these rocks. The initial ratio is consistent with an approximately threefold enrichment in Rb relative to Sr at the time the rocks were formed. On the other hand, the initial $^{87}Sr/^{86}Sr$ ratio of the low Rb rocks does not require this enrichment.

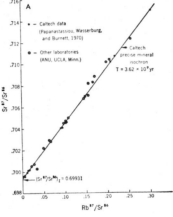

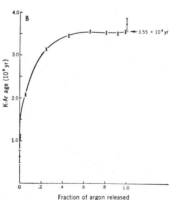

Fig. 5. (A) Rubidium-strontium data for high Rb group of Apollo 11 basalts. Both groups of Apollo 11 basalts have been dated by this method at 3.6×10^9 years. (B) Potassium-argon data of Turner (23) obtained by neutron irradiation and stepwise heating of the sample. The "plateau" at 3.55×10^9 years is interpreted as the age of the rock.

Similar Rb-Sr data for the Apollo 12 rocks indicate a distinctly younger age of $\sim 3.25 \times 10^9$ years (11, 12, 22); possibly younger rocks with an age of $\sim 2.9 \times 10^9$ years are also present (22), but this result requires further confirmation. Like the low Rb group from Apollo 11, the initial ratios found for the Apollo 12 basalts require no major Rb-Sr fractionation between the rocks and their source.

Preliminary K-Ar data indicated that the Apollo 11 and Apollo 12 basalts are considerably younger than these Rb-Sr ages (4), but more refined K-Ar measurements have been reported by Turner (23) which are in generally good agreement with the Rb-Sr data. These more accurate data have been obtained by the step-heating method of Merrihue and Turner (24), and involve irradiation of the sample with energetic neutrons converting ^{39}K to ^{39}Ar by an (n, p) reaction. The sample is then heated in vacuum to successively higher temperatures, and an age calculated from the $^{39}Ar/^{40}Ar$ ratio measured on the gas released following each heating step. In most cases (Fig. 5B), a plateau is reached; at higher temperatures a nearly uniform age is found. This high temperature data is interpreted to represent sites in which radiogenic ^{40}Ar is quantitatively retained, and hence these plateau ages represent the true age of the rock. Ages are found by this method to be from 3.5×10^9 to 3.8×10^9 years for the Apollo 11 rocks, and from 3.2×10^9 to 3.3×10^9 years for the Apollo 12 rocks. These are similar to ages obtained by the Rb-Sr method, although in some cases the K-Ar ages are apparently 200 million years older; this slight discrepancy remains to be explained.

The interpretation of the U-Pb and Th-Pb ages for these basaltic rocks is less straightforward. In principle, one should be able to separate minerals from each of these rocks, and obtain ^{238}U-^{206}Pb, ^{235}U-^{207}Pb, and ^{232}Th-^{208}Pb isochrons analogous to those found by the Rb-Sr method. This has been attempted by Tatsumoto (16), but the data have considerable scatter and are only roughly in agreement with the Rb-Sr results. The cause of the scatter has not been definitely established; however, it may be caused by terrestrial Pb contamination, in spite of the fact that Tatsumoto has achieved lower contamination levels than anyone else working on this problem. Analytical errors, possibly associated with equilibration of the isotopic tracers used in the isotope dilution analysis, may also play a role. Furthermore, in contrast to the case of Rb-Sr, for which a large number of internal mineral isochrons have been obtained there is no backlog of analogous U-Pb and Th-Pb age data for terrestrial rocks. It may be that U and Th and their Pb daughters are not sufficiently tightly bound to sites in their host minerals in basaltic rocks to fulfill the closed system requirement essential to interpretation of these data as an age. A corollary of this failure to obtain internal isochrons for lunar rocks is that we have no idea what isotopic composition should be assumed

in calculating whole rock U-Pb and Th-Pb ages for these rocks. We have seen that lunar Pb is extremely radiogenic because the ratio of $^{238}U/^{204}Pb$ in lunar rocks is ~ 1000, in contrast to the average terrestrial ratio of 9. These lunar rocks were almost certainly derived from a source which also had a very high ratio of $^{238}U/^{204}Pb$ and consequently it would be grossly incorrect to use terrestrial Pb data to estimate the initial isotopic composition of lead in a lunar basalt.

Thus, it is not possible at the present time to calculate meaningful U-Pb and Th-Pb ages for lunar rocks. It is possible, however, to discuss the extent

to which those measurements which have been made on lunar rocks are consistent with the Rb-Sr and K-Ar data which have already resolved the question of the age of these mare basalts.

The lunar basaltic magmas probably undergo U-Pb fractionation similar to that of terrestrial basalts at the time of their formation. Tatsumoto (25) has published data for modern terrestrial basalts, which are plotted on a concordia diagram in Fig. 6A. In calculating the radiogenic $^{206}Pb/^{238}U$ ratios for this figure, the Pb isotopic composition has been corrected by subtraction of the primordial Pb present at the time the earth was formed, as indicated by meteoritic data (17). It may be seen from Fig. 6A that the resulting data points from these new basalts with an age of zero do not lie at the origin of the concordia diagram, but are spread out, with some scatter, along a line extending from the origin through a point on concordia corresponding to $\sim 4.55 \times 10^9$ years. This indicates that basalts invariably contain some radiogenic Pb at the time of their formation; those points lying above the concordia curve in Fig. 6A correspond to basaltic magmas for which the $^{238}U/^{204}Pb$ ratio was fractionated so as to decrease this ratio in the magma relative to the average mantle ratio of 9; those points lying below concordia correspond to basaltic magmas which were enriched in ^{238}U relative to the value in their source. Both types of fractionation are common, and on the average the $^{238}U/^{204}Pb$ ratio of basalts may be about the same as that of their source.

Data of this kind exist only for modern terrestrial basalts, that is, those of zero age. If it is assumed that similar processes occurred during the first billion years of earth and lunar history, these results can be used to "predict" what 3.6×10^9 year lunar basalts should look like when data from them is plotted in this way (Fig. 6B). The points are spread out along a line extending from 3.6×10^9 years on the concordia curve through 4.6×10^9 years on the curve. This line passes through Tatsumoto's Apollo 11 data (Fig. 4), within their experimental errors. Tatsumoto's data appears to be less affected by terrestrial contamination than that obtained in other laboratories, and consequently it is the best data to use for this comparison. Therefore, the U-Pb data do not conflict with the Rb-Sr and K-Ar data on these rocks. A

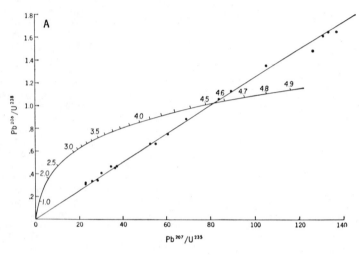

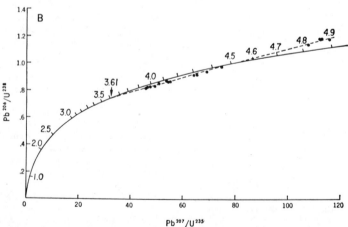

Fig. 6. (A) Data of Tatsumoto (25) for terrestrial basalts plotted on a concordia diagram after subtraction of the primordial Pb present at the time the earth was formed. The points scatter about a line passing through the origin and the point on concordia representing the age of the earth. (B) "Predicted" distribution of points representing 3.6×10^9 year old basalts, based on data of (A).

388

similar comparison for Apollo 12 basalts leads to the same conclusion.

It may be asked why all of the lunar data occupy a relatively limited portion of the line. This would imply a similar U/Pb fractionation factor relative to their source for all these rocks. It is quite likely that the Apollo 11 data represent but two lava flows. Terrestrial data do not exist for multiple samples from an identical lava flow. However, data obtained on subsequent flows from the same source give $^{238}U/^{204}Pb$ ratios similar to one another (26). Because of the limited number of flows sampled, no explanation may be required for the clustering of the lunar data nor for the fact that all the lunar data plot below the concordia curve, while the terrestrial data fall both above and below the curve. On the other hand, this latter fact could represent a systematic difference between lunar and terrestrial basalts, a systematic Pb volatilization occurring when a thin lava flow is exposed to the lunar vacuum. In considering this possibility it must be recognized that the Rb-Sr data for soil (Fig. 3C) gives no evidence that Rb was lost from the Rb-rich component of the soil, in spite of the fact that a considerable fraction of this Rb-rich component is contained in glassy material which was almost certainly exposed to the lunar vacuum in the form of liquid droplets. Significant Rb loss would increase the scatter in these points and cause at least some of them to fall well to the left of the 4.6×10^9 year isochron. Laboratory data on the volatilization of Pb relative to Rb is not definitive (12, 27), but it is quite likely that significant Pb loss must be accompanied by similar loss of Rb and K.

The above discussion suggests several possibilities about the concordant 4.6×10^9 year ages found for the Apollo 11 soil. First, if lunar basalts, like terrestrial basalts, undergo both U/Pb enrichment and depletion relative to their source, points both above and below concordia on the dashed line in Fig. 6B should occur, and these could average to give concordant ages. This would require that there be nearby basalts which would lie above concordia; the sampling at the Apollo 11 site was certainly not complete enough to eliminate this possibility. This would be particularly likely if there were a major component in the Apollo 11 soil derived from an $\sim 4.4 \times 10^9$ to 4.6×10^9 year old source, because the spread along

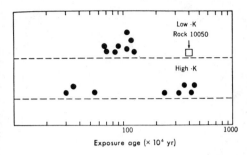

the line will become less as the age approaches 4.65×10^9 years. It is also possible, in spite of the reservation mentioned in the previous paragraph, that Pb is volatilized from the magma when it flows onto the lunar surface. If this volatilized Pb is mixed into the lunar soil together with Pb derived from the basaltic rocks, the result will be that the soil is effectively a closed U-Pb and Th-Pb system (21). This volatilized Pb may be related to the component of old radiogenic Pb which Silver (14) has separated from Apollo 11 soil. Because of these possibilities, the coexistence of a 4.6×10^9 year old soil with 3.6×10^9 year old basalts does not necessarily represent a geochemical paradox, and certainly not a logical paradox.

Solar and Cosmic-Ray Bombardment

Another aspect of lunar chronology of considerable importance concerns more recent surface processes resulting from extralunar bombardment by meteoritic bodies. Information bearing on the time scale for these processes can be obtained by techniques which measure the length of time lunar material has been near the lunar surface. These techniques make use of the fact that the lunar surface is bombarded by particles of various energies such as solar wind (1000 electron volts per nucleon), solar flare particles (10 to 100 million electron volts), and galactic cosmic rays (1 to 10 billion electron volts per nucleon). These particles implant in the lunar surface material, cause nuclear reactions in this material which result in the production of both stable and radioactive nuclei, and leave particle tracks as a result of radiation damage produced as the particles traverse lunar minerals. The depths to which these effects extend vary according to the

energy of the bombarding particle; the solar wind penetrates only to a depth of 0.1 micrometer, solar flare particles to a depth of several centimeters, and galactic cosmic rays to a depth of several meters. This makes possible the measurement of various "bombardment ages," which, depending on the type of particle studied, permit determination of such quantities as the time since a rock now lying on the surface was ejected from a nearby crater, the rate of turnover of the lunar soil to a given depth ("gardening"), and the time since a sample of lunar material, once on the surface, was buried. A complete review of the experimental data obtained by these techniques, and their theoretical basis, is beyond the scope of this article. However, there is room to outline the principal discoveries that have been made.

It has been determined that the energy spectrum and intensity of solar flare particles has been essentially constant over the last several million years, and some rocks have been within ~ 1 centimeter of the lunar surface for times of the order of 10 million years (28). Rocks and soil have been within ~ 1 meter of the lunar surface for times of approximately 450 million years (29). The rate at which rocks are eroded by micrometeorite bombardment has been found to be very low (10^{-7} to 10^{-8} centimeter per year) (30) and the solar flare bombardment record of a lunar surface rock is more likely to be terminated by burial with ejecta from nearby craters, or by total destruction by bombardment by meteoritic bodies of the order of 1/40 the dimension of the rock itself (31), rather than by micrometeorite erosion.

Both the experimental and the theoretical problems associated with obtaining and interpreting these data are complex, and it will probably be some time before many of the details are understood. However, progress to date

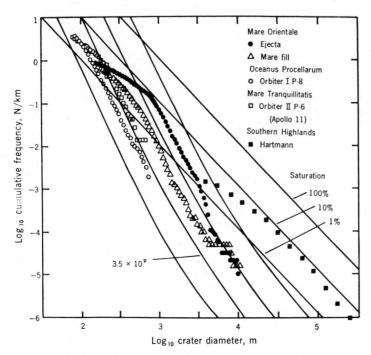

Fig. 8. Data of Gault (*32*) for the density of craters of a given size for several regions of the moon. The curve marked "3.5 × 10⁹" fits the Mare Tranquillitatis data which have been dated at this age by the Rb-Sr and K-Ar methods. This curve and the three curves parallel to it represent integral fluxes differing by a factor of 10. The curves labeled "Saturation, 100%, 10%, 1%" are theoretical curves for steady-state crater frequency distributions.

is encouraging. One very interesting example of the potential application of data obtained by these techniques is that reported by Eberhardt *et al.* (*29*) in which they were able to resolve the high and low Rb group of Apollo 11 basalts on the basis of their galactic cosmic-ray bombardment ages. These results are shown in Fig. 7 and indicate that all but one of the low Rb group rocks were exposed for 100 million years, while the high Rb group rocks were exposed for either a shorter or longer period of time. One possible interpretation of the data on the high Rb group is that some of these rocks were formed sufficiently near the surface to allow "prebombardment" prior to their more recent ejection to very near the surface.

Cratering Chronology

Prior to the availability of lunar materials, lunar chronology was entirely based on photogeology, first making use of earth-based telescopic photo-

graphs, and later, those obtained from Ranger and Orbiter spacecraft. Photogeology provided chronological data based on the principle of superposition and also by counting the density of craters as a function of diameter in various regions of the moon. On the basis of these techniques, a relative lunar stratigraphy was established. The oldest lunar terranes were in the densely cratered highland regions, which in part have been covered with ejecta blankets from the circular maria. These ejecta blankets are considerably more densely cratered than the material filling the maria, and hence are significantly older. The relatively lightly cratered mare regions are in turn older than the mare craters, such as Copernicus, and also older than some highlands craters such as Tycho. In addition, some younger volcanic regions, such as the Marius Hills, have been distinguished.

Attempts were made to place this cratering chronology on an absolute basis, using observed values for the flux of the crater-forming impacting bodies. A rather complete account of

this work has been given by Gault (*32*). The principal conclusions were that the youngest craters were about 10⁷ years old, the mare regions about 10⁸ years old, and the mare ejecta blankets about 10⁹ years old, whereas the heavily cratered highlands required an increased flux early in lunar history to account for their crater density. An important consideration in this work is the concept of saturation and equilibrium. The crater density on an initially crater-free surface will at first increase in proportion to the integrated flux of impacting bodies. However, this linear increase cannot continue indefinitely, even if one ignores the effects of material ejected from the craters; after a time the surface will become saturated for purely geometrical reasons; new craters will overlap and obscure older ones. Experimental work by Gault (*32*) has shown that before this condition of geometrical saturation is reached, an equilibrium, or steady-state condition is reached, whereby older small craters are filled by the ejecta of later craters. The experimental data shows that this effect limits the crater density to about 5 to 10 percent of the saturation value for craters of a given size. Cumulative crater count data as a function of crater diameter (Fig. 8) exhibit a characteristic break in the cumulative crater density, below which diameter the steady-state condition exists.

The curve in Fig. 8 labeled "3.5 × 10⁹" and the three other solid curves nearly parallel to it are calculated cumulative crater frequency distribution curves representing integral fluxes which differ by a factor of 10. Prior to the dating of lunar samples, the curve marked "3.5 × 10⁹" was called "10⁸ years," based on the assumed present-day flux. We now know that the time of filling of Mare Tranquillitatis was more like 3.5 × 10⁹ years ago, so this curve which approximately fits the Mare Tranquillitatis crater counts can be relabeled to represent 3.5 × 10⁹ years. A uniform flux of bombarding particles would imply that the next curve should be labeled "3.5 × 10¹⁰ years," older than the age of the moon, the solar system, and even the universe. The cratering data for the ejecta blanket for Mare Orientale fall to the right of even this curve. Clearly, it is not possible to assume a uniform rate of bombardment prior to 3.5 × 10⁹ years ago. In actuality, the southern highlands points must represent no more than 4.6 × 10⁹ years of integrated flux, and

390

SCIENCE, VOL. 173

the flux must have dropped by about three orders of magnitude during the the first billion years of lunar history.

There remains the problem of reconciling the presently observed flux at the earth with that required by lunar crater counting, the observed flux being a factor of about 35 higher than the mean lunar flux over the last 3.5×10^9 years. At the velocity of these impacting bodies, the differences in the gravitational fields and heliocentric velocities of the earth and moon will have a negligible effect. It has been suggested that the flux of meteoritic bodies on the earth and moon has gone through a minimum and has now increased to the present value. This seems unlikely, based on our present knowledge of the orbits and sources of these bodies, which appear to be primarily associated with remnants of short-period comets (33), and there is no known reason why the rate at which comets are captured by Jupiter into short-period orbits should be increasing at the present time. This discrepancy remains an unsolved problem; it appears worthwhile to critically examine the evidence for the present flux of objects over 10^6 kilograms in mass.

Future Problems

Surprisingly much has been learned about lunar history from the Apollo 11 and Apollo 12 samples. In addition, some heretofore unknown problems have come to attention, and hypotheses have been made regarding the solution to these problems.

The principal need in future work will be to provide samples from a sufficiently wide variety of lunar terranes to avoid misconceptions arising from limited sampling, and to understand the relationship of the processes occurring in these other regions of the moon to one another and to the mare regions already sampled. The number of regions which may be studied is severely constrained by the small number of remaining Apollo missions; the original number of planned Apollo landings would have permitted a far more adequate study, and any further curtailment in the Apollo program would represent a lost opportunity which would be increasingly regretted as its consequences became generally realized.

Hopefully, future missions will extend our time scale so as to include the time (or times) at which igneous rocks

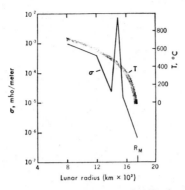

Fig. 9. Data of Sonett *et al.* (37) for electrical conductivity (σ) and temperature (T) for the interior of the moon. R_m is the lunar radius.

were formed in the lunar highlands, the relationship of this age to the age of rocks underlying the mare basalts, the time of mare formation, and the extent to which lunar volcanism has persisted since the time of mare filling dated by the samples from Apollo 11 and Apollo 12.

The answers to geochronological questions have major implications with regard to many other lunar problems. For example, the association of mascons with circular maria (34) indicates that the origin of the mascons must be in some way, perhaps indirectly, associated with the impacts which produced the maria. These impacts are generally thought to have occurred very early in lunar history ($\sim 4.5 \times 10^9$ years ago), and future geochronological work may confirm this inference. Some theories for the origin of mascons have invoked filling of the mare basins with volcanic rocks at some later time, but an objection has been that subsequent mare volcanism would not be consistent with the preservation of the departures from isostatic equilibrium indicated by the mascons. While this may be hard to understand, the geochronological data clearly indicate that relatively late mare volcanism did in fact occur, and the invocation of such volcanism cannot be considered a fault of any lunar theory.

One of the central problems is that of the thermal history of the moon. We now know that lunar igneous activity was not confined to a single epoch in lunar history, such as might be associated with a close approach to the earth at the time of capture. An adequate thermal history must explain the presence of an internal regime sufficient

to provide igneous differentiation over a period of time at least as long as 1.3×10^9 years. This heat source cannot be entirely surficial, since seismic data (35) indicates that the mare basalts extend to a depth of at least 20 kilometers. The great enrichment of these basalts in certain elements (U and Th) implies a much deeper chemically differentiated region, extending to a depth of several hundred kilometers, which would include about half of the volume of the moon. Measurements of remanent magnetism of lunar rocks indicate that they crystallized in a magnetic field of about one tenth the strength of the present field of the earth (36). Again, the geochronological data precludes an explanation in terms of a close approach to the earth, and these data may therefore imply that the moon has a small core and, therefore, extensive chemical differentiation. This evidence for deep chemical differentiation must be reconciled with measurements of the moon's electrical conductivity and present internal temperature as a function of depth reported by Sonett *et al.* (37), and shown in Fig. 9. The present internal temperature is far below the melting point of iron or basaltic rocks, and the deeper regions of the moon should not have cooled this much in the time available, if they had melted. Severe constraints are being placed on the thermal history of the moon, and further geochronological data, particularly on young igneous rocks, will further limit the range of possibilities.

It may be expected that future geochronological work will rely heavily on small rock fragments, separated from the lunar soil, which have been ejected from distant craters, as only in this way will it be possible to sample the many areas of the moon which cannot be landing sites. Measurements of these fragments should permit an understanding of the continuity or episodicity of lunar igneous activity, the extent to which it has diminished with time, and the changes in chemical and mineralogical composition of lunar rocks as a function of time. In addition to the usual quantity of rocks of about 1 kilogram mass and soil particles of less than 1 millimeter which are returned from the moon, it would be very valuable to return a similar mass of 1-millimeter to 2-centimeter fragments obtained by sieving, on the moon, the lunar soil. In this way, many thousands of samples, sufficiently

391

451

large for considerable geochronological work, as well as other studies, would be available.

Future geochronological work may also emphasize measurements of the products of extinct natural radioactivities, such as ^{129}I, and ^{244}Pu. In contrast to the importance of these radioisotopes in studies of meteorites, they have not yet played a major role in lunar work. If, as anticipated, rocks approaching the age of the moon are found in the highlands and in the mare ejecta, the high inherent resolution of these techniques may prove useful in interpreting the history of these very ancient rocks.

Summary

Considerable information concerning lunar chronology has been obtained by the study of rocks and soil returned by the Apollo 11 and Apollo 12 missions. It has been shown that at the time the moon, earth, and solar system were formed, $\sim 4.6 \times 10^9$ years ago, a severe chemical fractionation took place, resulting in depletion of relatively volatile elements such as Rb and Pb from the sources of the lunar rocks studied. It is very likely that much of this material was lost to interplanetary space, although some of the loss may be associated with internal chemical differentiation of the moon.

It has also been shown that igneous processes have enriched some regions of the moon in lithophile elements such as Rb, U, and Ba, very early in lunar history, within 100 million years of its formation. Subsequent igneous and metamorphic activity occurred over a long period of time; mare volcanism of the Apollo 11 and Apollo 12 sites occurred at distinctly different times, 3.6×10^9 and 3.3×10^9 years ago, respectively. Consequently, lunar magmatism and remanent magnetism cannot be explained in terms of a unique event, such as a close approach to the earth at a time of lunar capture. It is likely that these phenomena will require explanation in terms of internal lunar processes, operative to a considerable depth in the moon, over a long period of time. These data, together with the low present internal temperatures of the moon, inferred from measurements of lunar electrical conductivity, impose severe constraints on acceptable thermal histories of the moon.

Progress is being made toward understanding lunar surface properties by use of the effects of particle bombardment of the lunar surface (solar wind, solar flare particles, galactic cosmic rays). It has been shown that the rate of micrometeorite erosion is very low (angstroms per year) and that lunar rocks and soil have been within approximately a meter of the lunar surface for hundreds of millions of years.

Future work will require sampling distinctly different regions of the moon in order to provide data concerning other important lunar events, such as the time of formation of the highland regions and of the mare basins, and of the extent to which lunar volcanism has persisted subsequent to the first third of lunar history. This work will require a sufficient number of Apollo landings, and any further cancellation of Apollo missions will jeopardize this unique opportunity to study the development of a planetary body from its beginning. Such a study is fundamental to our understanding of the earth and other planets.

References and Notes

1. D. A. Papanastassiou, thesis, California Institute of Technology (1970); D. A. Papanastassiou and G. J. Wasserburg, *Earth Planet. Sci. Lett.* **5**, 361 (1969).
2. G. J. Wasserburg, D. A. Papanastassiou, H. G. Sanz, *Earth Planet. Sci. Lett.* **7**, 33 (1969).
3. J. H. Reynolds, *J. Geophys. Res.* **68**, 2939 (1963); C. N. Hohenberg, *Science* **166**, 212 (1969); G. J. Wasserburg, D. N. Schramm, J. Huneke, *Astrophys. J.* **157**, L91 (1969).
4. Lunar Sample Preliminary Examination Team, *Science* **167**, 1325 (1970).
5. L. Asylum, *Earth Planet. Sci. Lett.* **9**, 137 (1970).
6. L. O. Nicolaysen, *Ann. N.Y. Acad. Sci.* **91**, 198 (1961); M. A. Lanphere, G. J. Wasserburg, A. L. Albee, G. R. Tilton, in *Isotopic and Cosmic Chemistry*, H. Craig *et al.*, Eds. (North-Holland, Amsterdam, 1964), p. 269.
7. K. Gopalan and G. W. Wetherill, *J. Geophys. Res.* **75**, 3457 (1970).
8. P. W. Gast, *Geochim. Cosmochim. Acta* **26**, 927 (1962); W. H. Pinson, Jr., C. C. Schnetzler, E. Beiser, H. W. Fairbairn, P. M. Hurley, *ibid.* **29**, 455 (1965); K. Gopalan and G. W. Wetherill, *J. Geophys. Res.* **73**, 7133 (1968), *ibid.* **74**, 1439 (1969); S. K. Kaushal and G. W. Wetherill, *ibid.*, p. 2717; *ibid.* **75** 463 (1970).
9. G. Turner, *Earth Planet. Sci. Lett.* **9**, 177 (1971).
10. D. A. Papanastassiou, G. J. Wasserburg, D. S. Burnett, *ibid.* **8**, 1 (1970).
11. D. A. Papanastassiou and G. J. Wasserburg, *ibid.*, p. 269.
12. R. A. Cliff, C. Lee-Hu, G. W. Wetherill, Apollo 12 Conference, Houston (1971), unpublished data; *Geochim. Cosmochim. Acta* Suppl. 2, in press.
13. N. Hubbard, C. Meyer, P. Gast, H. Wiesman, *Earth Planet. Sci. Lett.*, in press.
14. L. T. Silver, *Geochim. Cosmochim. Acta*, Suppl. 1, **2**, 1533 (1970).
15. K. Gopalan, S. Kaushal, C. Lee-Hu, G. W. Wetherill, *ibid.*, p. 1195.
16. M. Tatsumoto, *ibid.*, p. 1595.
17. C. C. Patterson, *Geochim. Cosmochim. Acta* **7**, 151 (1955).
18. G. W. Wetherill, *Trans. Amer. Geophys. Union* **37**, 320 (1956).
19. M. Tatsumoto, Apollo 12 Conference, Houston (1971), unpublished data.
20. L. T. Silver, Apollo 12 Conference, Houston (1971), unpublished data.
21. W. Compston, B. W. Chapell, P. A. Arriens, M. J. Vernon, *Geochim. Cosmochim. Acta* Suppl. 1, **2** 1007 (1970).
22. W. Compston, H. Berry, M. J. Vernon, Apollo 12 Conference, Houston (1971), unpublished data; V. R. Murthy, N. Evenson, Bor-Ming Jahn, M. Coscio, Jr., Apollo 12 Conference, Houston (1971), unpublished data.
23. G. Turner, Apollo 12 Conference, Houston (1971), unpublished data; *Geochim. Cosmochim. Acta* Suppl. 1, **2** 1665 (1970).
24. C. M. Merrihue and G. Turner, *J. Geophys. Res.* **71**, 2852 (1966).
25. M. Tatsumoto, *ibid.*, p. 1721.
26. H. Kurosawa, *Geochem. J.* **2**, 11 (1968); M. Tatsumoto and R. J. Knight, *ibid.* **3**, 53 (1969); M. Tatsumoto and P. D. Snavely, *J. Geophys. Res.* **74**, 1087 (1969).
27. D. R. Chapman and L. C. Scheiber, *J. Geophys. Res.* **74**, 6737 (1969).
28. R. L. Fleischer, E. L. Haynes, H. R. Hart, Jr., R. T. Woods, G. M. Comstock, *Geochim. Cosmochim. Acta* Suppl. 1, **3**, 2103 (1970); P. B. Price and D. O'Sullivan, *ibid.*, p. 2351; G. Crozaz, U. Haack, M. Hair, M. Maurette, R. Walker, D. Woolum, *ibid.*, p. 2051; R. C. Finkel, J. K. Arnold, R. C. Reedy, J. S. Fruchter, H. H. Loosli, J. C. Evans, J. P. Shedlovsky, M. Imamura, A. C. Delany, Apollo 12 Conference, Houston (1971), unpublished data.
29. P. Eberhardt, J. Geiss, H. Graf, N. Grögler, U. Krähenbühl, H. Schwaller, J. Schwarzmüller, A. Stetler, *Earth Planet. Sci. Lett.* **10**, 67 (1970), and references therein.
30. G. M. Comstock, A. D. Evwaraye, R. L. Fleischer, H. R. Hart, Jr., Apollo 12 Conference, Houston (1971), unpublished data; G. Crozaz, R. Walker, D. Woolum, Apollo 12 Conference, Houston (1971), unpublished data.
31. E. M. Shoemaker, M. H. Hait, G. A. Swann, D. L. Schleicher, G. G. Schaber, R. L. Sutton, D. H. Dahlem, E. N. Goddard, A. C. Waters, *Geochim. Cosmochim. Acta* Suppl. 1, **3**, 2399 (1970); E. M. Shoemaker and M. H. Hait, Apollo 12 Conference, Houston (1971), unpublished data.
32. D. E. Gault, *Radio Sci.* **5**, 237 (1970).
33. G. W. Wetherill, in *Vinogradov 75th Anniversary Volume* (Moscow, in press); *EOS* **50**, 224 (1969).
34. P. M. Muller and W. L. Sjogren, *Science* **161**, 680 (1968).
35. G. Latham, E. Ewing, F. Press, G. Sutton, J. Dorman, V. Nakamura, R. Meissner, N. Toksoz, F. Duennebier, R. Kovach, D. Lammlein, Apollo 12 Conference, Houston (1971), unpublished data.
36. C. E. Helsley, *Geochim. Cosmochim. Acta* Suppl. 1 **3**, 2213 (1970); S. K. Runcorn, D. W. Collinson, W. O'Reilly, M. H. Battey, A. Stephenson, J. M. Jones, A. J. Manson, P. W. Readman, *ibid.*, p. 2369.
37. C. P. Sonett, B. F. Smith, D. S. Colburn, G. Schubert, K. Schwartz, P. Dyal, C. W. Parkin, Apollo 12 Conference, Houston (1971), unpublished data.
38. I thank M. Duke and J. Strand for photographs of lunar rocks.

Author Citation Index

Adams, N. I., 112
Adamson, O. J., 240
Ade-Hall, J., 383
Ahrens, L. H., 167, 170, 177, 189, 190, 213, 240, 304, 318
Albee, A. L., 247, 428, 429, 452
Aldrich, L. T., 167, 170, 189, 190, 212, 213, 218, 261, 270, 279, 318, 342, 425, 429
Allsopp, H. L., 293
Alpert, D., 304
Alpher, R. A., 120, 158
Alvarez, L. W., 318
Amaral, G., 383
Amirkanov, K. I., 189, 342, 425
Anders, E., 400
Anderson, D. H., 143
Angenheister, G., 383
Antevs, E., 82
Armstrong, R. L., 247, 426, 442
Arnold, A., 426
Arnold, J. R., 452
Arriens, P. A., 247, 452
Arrol, W. J., 131
Aston, F. W., 100, 109, 218
Asylum, L., 452

Baadsgaard, H., 143, 361, 426, 427, 428
Badash, L., 14
Bainbridge, K. T., 309
Baranovskaya, N. V., 305
Barrell, J., 9, 65
Barrer, R. M., 189
Bartnitskii, E. N., 189, 342, 425
Bass, M. N., 342, 429, 442
Bastin, E. S., 128
Bate, G. L., 167
Bateson, S., 10
Battey, M. H., 452
Bauer, H., 223, 240
Beall, G. H., 293

Beck, G., 309
Becker, G. F., 9
Beckinsale, R. D., 304
Beiser, E., 452
Bell, K., 426, 428
Bence, A. E., 247
Berg, G., 309
Berger, G. W., 395
Berggren, T., 240
Berry, H., 452
Besairie, H., 442
Bickford, M. E., 442
Birch, F., 328
Black, L. P., 120
Blake, G. S., 28
Bleuler, E., 310, 318
Blow, W. H., 360
Boadsgaard, H., 305
Bofinger, V. M., 247
Bøggild, J. K., 189
Bolinger, J., 213
Boltwood, B. B., 9, 14, 19, 26, 31, 39, 40, 45, 65, 100
Bor-Ming, J., 452
Borst, L. B., 318
Bothe, W., 318
Bout, P., 362
Boyd, F. R., 342
Bradley, W. H., 82
Bramlette, M. N., 362
Brandt, S. B., 189, 342, 425, 426
Bray, J. M., 240
Brewer, A. K., 240
Brewer, M. S., 395
Broecker, W. S., 167
Brögger, W. C., 42
Brooking, D. G., 280
Brooks, C., 247
Brown, E. W., 81, 84
Brown, H., 138, 167, 213, 328

453

Subject Index